Ion and Molecule Transport in Membrane Systems

Ion and Molecule Transport in Membrane Systems

Editors

Victor Nikonenko
Natalia Pismenskaya

MDPI • Basel • Beijing • Wuhan • Barcelona • Belgrade • Manchester • Tokyo • Cluj • Tianjin

Editors
Victor Nikonenko
Kuban State University
Russia

Natalia Pismenskaya
Kuban State University
Russia

Editorial Office
MDPI
St. Alban-Anlage 66
4052 Basel, Switzerland

This is a reprint of articles from the Special Issue published online in the open access journal *International Journal of Molecular Sciences* (ISSN 1422-0067) (available at: https://www.mdpi.com/journal/ijms/special_issues/Ion_Membrane).

For citation purposes, cite each article independently as indicated on the article page online and as indicated below:

LastName, A.A.; LastName, B.B.; LastName, C.C. Article Title. *Journal Name* **Year**, *Volume Number*, Page Range.

ISBN 978-3-0365-1359-1 (Hbk)
ISBN 978-3-0365-1360-7 (PDF)

Contents

About the Editors ix

Preface to "Ion and Molecule Transport in Membrane Systems" xi

Yury A. Trofimov, Nikolay A. Krylov and Roman G. Efremov
Confined Dynamics of Water in Transmembrane Pore of TRPV1 Ion Channel
Reprinted from: *Int. J. Mol. Sci.* **2019**, *20*, 4285, doi:10.3390/ijms20174285 1

Mark Tingey, Krishna C. Mudumbi, Eric C. Schirmer and Weidong Yang
Casting a Wider Net: Differentiating between Inner Nuclear Envelope and Outer Nuclear Envelope Transmembrane Proteins
Reprinted from: *Int. J. Mol. Sci.* **2019**, *20*, 5248, doi:10.3390/ijms20215248 15

Kai Yue, Xiaochen Sun, Jue Tang, Yiang Wei and Xinxin Zhang
A Simulation Study on the Interaction Between Pollutant Nanoparticles and the Pulmonary Surfactant Monolayer
Reprinted from: *Int. J. Mol. Sci.* **2019**, *20*, 3281, doi:10.3390/ijms20133281 31

Bao-Ying Wang, Na Zhang, Zhen-Yu Li, Qiao-Lin Lang, Bing-Hua Yan, Yang Liu and Yang Zhang
Selective Separation of Acetic and Hexanoic Acids across Polymer Inclusion Membrane with Ionic Liquids as Carrier
Reprinted from: *Int. J. Mol. Sci.* **2019**, *20*, 3915, doi:10.3390/ijms20163915 45

Wu-jhao Tien, Kun-you Chen, Fong-yin Huang, and Chi-cheng Chiu
Effects of Cholesterol on Water Permittivity of Biomimetic Ion Pair Amphiphile Bilayers: Interplay between Membrane Bending and Molecular Packing
Reprinted from: *Int. J. Mol. Sci.* **2019**, *20*, 3252, doi:10.3390/ijms20133252 65

Shujahadeen B. Aziz, Muhamad H. Hamsan, Mohd F. Z. Kadir, Wrya O. Karim and Ranjdar M. Abdullah
Development of Polymer Blend Electrolyte Membranes Based on Chitosan: Dextran with High Ion Transport Properties for EDLC Application
Reprinted from: *Int. J. Mol. Sci.* **2019**, *20*, 3369, doi:10.3390/ijms20133369 81

Shujahadeen B. Aziz, Wrya O. Karim, M. A. Brza, Rebar T. Abdulwahid, Salah Raza Saeed, Shakhawan Al-Zangana and M. F. Z. Kadir
Ion Transport Study in CS: POZ Based Polymer Membrane Electrolytes Using Trukhan Model
Reprinted from: *Int. J. Mol. Sci.* **2019**, *20*, 5265, doi:10.3390/ijms20215265 107

Babak Jaleh, Ehsan Sabzi Etivand, Bahareh Feizi Mohazzab, Mahmoud Nasrollahzadeh and Rajender S. Varma
Improving Wettability: Deposition of TiO_2 Nanoparticles on the O_2 Plasma Activated Polypropylene Membrane
Reprinted from: *Int. J. Mol. Sci.* **2019**, *20*, 3309, doi:10.3390/ijms20133309 129

Ali Abbasi, Soraya Hosseini, Anongnat Somwangthanaroj, Ahmad Azmin Mohamad and Soorathep Kheawhom
Poly(2,6-Dimethyl-1,4-Phenylene Oxide)-Based Hydroxide Exchange Separator Membranes for Zinc–Air Battery
Reprinted from: *Int. J. Mol. Sci.* **2019**, *20*, 3678, doi:10.3390/ijms20153678 139

Qianqian Ge, Xiang Zhu and Zhengjin Yang
Highly Conductive and Water-Swelling Resistant Anion Exchange Membrane for Alkaline Fuel Cells
Reprinted from: *Int. J. Mol. Sci.* **2019**, *20*, 3470, doi:10.3390/ijms20143470 . . . . **157**

Fabao Luo, Yang Wang, Maolin Sha and Yanxin Wei
Correlations of Ion Composition and Power Efficiency in a Reverse Electrodialysis Heat Engine
Reprinted from: *Int. J. Mol. Sci.* **2019**, *20*, 5860, doi:10.3390/ijms20235860 . . . . **173**

Olesya Rybalkina, Kseniya Tsygurina, Ekaterina Melnikova, Semyon Mareev, Ilya Moroz, Victor Nikonenko and Natalia Pismenskaya
Partial Fluxes of Phosphoric Acid Anions through Anion-Exchange Membranes in the Course of NaH_2PO_4 Solution Electrodialysis
Reprinted from: *Int. J. Mol. Sci.* **2019**, *20*, 3593, doi:10.3390/ijms20143593 . . . . **185**

Jianguo Li, Hongying Dong, Fan Yang, Liangcheng Sun, Zhigang Zhao, Ruixi Bai and Hao Zhang
Simple Preparation of $LaPO_4$:Ce, Tb Phosphors by an Ionic-Liquid-Driven Supported Liquid Membrane System
Reprinted from: *Int. J. Mol. Sci.* **2019**, *20*, 3424, doi:10.3390/ijms20143424 . . . . **207**

Loïc Henaux, Jacinthe Thibodeau, Geneviève Pilon, Tom Gill, André Marette and Laurent Bazinet
How Charge and Triple Size-Selective Membrane Separation of Peptides from Salmon Protein Hydrolysate Orientate Their Biological Response on Glucose Uptake
Reprinted from: *Int. J. Mol. Sci.* **2019**, *20*, 1939, doi:10.3390/ijms20081939 . . . . **221**

Guillaume Dufton, Sergey Mikhaylin, Sami Gaaloul and Laurent Bazinet
Positive Impact of Pulsed Electric Field on Lactic Acid Removal, Demineralization and Membrane Scaling during Acid Whey Electrodialysis
Reprinted from: *Int. J. Mol. Sci.* **2019**, *20*, 797, doi:10.3390/ijms20040797 . . . . **237**

Arthur Merkel and Amir M. Ashrafi
An Investigation on the Application of Pulsed Electrodialysis Reversal in Whey Desalination
Reprinted from: *Int. J. Mol. Sci.* **2019**, *20*, 1918, doi:10.3390/ijms20081918 . . . . **255**

Giuseppe Battaglia, Luigi Gurreri, Girolama Airò Farulla, Andrea Cipollina, Antonina Pirrotta, Giorgio Micale and Michele Ciofalo
Membrane Deformation and Its Effects on Flow and Mass Transfer in the Electromembrane Processes
Reprinted from: *Int. J. Mol. Sci.* **2019**, *20*, 1840, doi:10.3390/ijms20081840 . . . . **267**

Sylwin Pawlowski, João G. Crespo and Svetlozar Velizarov
Profiled Ion Exchange Membranes: A Comprehensible Review
Reprinted from: *Int. J. Mol. Sci.* **2019**, *20*, 165, doi:10.3390/ijms20010165 . . . . **289**

Pierre Magnico
Electro-Kinetic Instability in a Laminar Boundary Layer Next to an Ion Exchange Membrane
Reprinted from: *Int. J. Mol. Sci.* **2019**, *20*, 2393, doi:10.3390/ijms20102393 . . . . **305**

Lucie Vobecká, Tomáš Belloň and Zdeněk Slouka
Behavior of Embedded Cation-Exchange Particles in a DC Electric Field
Reprinted from: *Int. J. Mol. Sci.* **2019**, *20*, 3579, doi:10.3390/ijms20143579 . . . . **335**

About the Editors

Victor V. Nikonenko Professor of Physical Chemistry at the Kuban State University, Krasnodar, Russia.

Victor Nikonenko graduated from the Mathematical Faculty of the Kuban State University; defended his Doctor of science thesis at the A.N. Frumkin Institute of Electrochemistry of Russian Academy of Sciences (1996) and his HDR thesis at the Paris-Est University (2001). He is working in the field of transport phenomena in membrane systems including 'membrane structure - properties' relationships, coupling between ion and water transport as well as coupling between ion/molecule transport and chemical reactions. He is Russian co-director of the French-Russian International Associated Laboratory "Ion-exchange membranes and related processes". Member of the editorial board of "Russian Journal of Electrochemistry" and "Membranes and Membrane Technologies" journal (published in Russian and English). Member of the editorial board and guest editor of the "International Journal of Molecular Sciences" (MDPI) journal and "Membranes" (MDPI) journal. Doctor Honoris Causa at the University of Montpellier, France; associated professor at the Laval University, Québec, Canada.

Natalia D. Pismenskaya Professor of Physical Chemistry at the Kuban State University, Krasnodar, Russia.

Natalia Pismenskaya graduated from the Chemical Faculty of the Kuban State University; defended her Ph. D. thesis (1989) at the Kuban Polytechnic State University and her Doctor of Science thesis (2004) at the Kuban State University. She is working in the field of ion-exchange membranes and processes; transport phenomena in membrane systems; concentration polarization and coupled phenomena of concentration polarization (water splitting, electroconvection, gravitation convection, etc.); chemical reactions coupled with ions transfer; fouling; ion exchange membrane modification and characterization; experimental techniques for the investigation of membrane system.

Head of the Laboratory of Electromembrane Phenomena, Institute of Membrane, Kuban State University; Member of the editorial board of the "Membranes" (MDPI) journal.

Preface to "Ion and Molecule Transport in Membrane Systems"

Membranes play an enormous role in our life. A biological cell membrane is an en-closing film that acts as a selective barrier, within or around the cell. Cell membranes control the fluxes of substances in and out of cells. Artificial membranes are widely used for the treatment of water and food solutions (milk, juices, wine, ...), fractionation of organic acids, bioactive compounds and nutrients, energy production and other applications. Artificial membranes largely mimic the structure and functions of biological membranes. Like cell membranes, many kinds of artificial membranes involve macromolecules consisting of a relatively long hydrophobic polymer chain and a hydrophilic "head" at its end. Such elements allow multiple types of interactions (hydrophobic-hydrophobic, dipole-dipole, ion-dipole, ion-ion) between them and water, which provides self-assembly resulting in formation of permselective thin films. The similarity in the structure leads to the similarity in the properties and the approaches to study the laws governing the behavior of both biological and artificial membranes. It is of interest that Kedem and Katchalsky deduced their famous equations for the description of transport processes in biological membranes [1,2]. Now, these equations are largely used for all types of membranes, in particular, for modelling ion and water transport in a promising technological process named Pressure Retarded Osmosis (PRO), which is employed for energy harvesting from salinity variations [3].

The idea of this special issue is to recollect the papers describing physico-chemical and chemico-physical aspects of ion and molecule transport, which are common for both biological and artificial membrane systems. The scope of the issue involves: Experimental studies and mathematical modeling providing new knowledge on the mechanisms of ion and molecule transport in artificial and living systems; Similarities in behavior of biological and artificial membranes; Biomimetic structural features of artificial membranes and their impact on membrane properties and performance for separation processes; Generalities and case studies in the field of material structure–properties relationships; Thermodynamics and irreversible thermodynamics description; Equilibriums and kinetics of transport processes in membrane systems; Coupling of ion and molecule transport with chemical reactions and catalysis; Impact of forced and natural convection on ion and molecule transport; Mechanisms of electric current-induced convection and its impact on ion and molecule transport across membranes; Concentration polarization and coupled effects occurring in membrane systems under the action of external pressure and electric driving forces; The physico-chemical and chemico-physical aspects of all kinds of separation, purification, and fractionation in membrane systems. In all cases, analysis of phenomena at the molecular level is encouraged.

Within this issue, there are papers devoted to studying thin mechanisms of ionic and molecular transport in cell membranes [4–6] as well as in artificial ones, which mimic biological membranes [7,8]. Y. Trofimov et al. reported the results of molecular dynamics simulations of water confined in the pore of a cell membrane, taking into account that microscopic properties of water near the molecular surface are radically different from those in the bulk [4]. K. Yue et al. [6] applied molecular dynamics to simulate the interactions of inhaled pollutant nanoparticles with the pulmonary surfactant monolayer. The review by M. Tingey et al. [5] evaluates the current tools and methodologies available to study the role of transmembrane proteins in some kinds of cell membranes. The team of Y. Zhang [7] described the mechanism of selective separation of volatile fatty acids (VFAs) using polymer inclusion membranes (PIMs) containing ionic liquids as the carrier.

This process mimics the selective transport of some compounds (such as phenols, amino acids) by facilitated diffusion through cell membranes. W. Tien et al. [8] described the effects of cholesterol on water permittivity of biomimetic ion pair amphiphile bilayers, which are used to fabricate vesicles with various pharmaceutical applications.

The problems of preparation and properties of artificial membranes are considered in [9–13]. S.B. Aziz et al. [9] developed polymer blend electrolyte membranes based on chitosan, a biopolymer. It is shown that new membranes have a high performance in electrical double-layer capacitor applications, such as water desalination. The use of the Truhan model allowed detailed analysis of ion transport parameters of the chitosan-based polymer membrane [10]. B. Jaleh et al. [11] use deposition of TiO2 nanoparticles to improve the wettability of the O2 plasma activated polypropylene membrane. A new hydroxide exchange membrane was synthesized by A. Abbasi et al. [12]. Low zincate crossover and high discharge capacity of this kind of membranes make them promising to be used in zinc–air batteries, which represent an alternative to lithium–ion batteries for various applications. Novel anion-exchange membranes combining the advantages of densely functionalization architecture and crosslinking structure were fabricated by Q. Ge et al. [13]. A high ratio of hydroxide conductivity to water swelling suggests that these membranes have high potential for application in fuel cells. The analysis of some membrane properties affecting their overall performances is made in Refs. [14,15]. F. Luo et al. [14] examined trans-membrane potential across ion exchange membranes in order to evaluate their possible power efficiency, when applied in a "reverse electrodialysis heat engine". The impact of different ion compositions of a salt solution containing NaHCO3, Na2CO3, and NH4Cl electrolytes was examined. The performance of another membrane process, important in the recovery of fertilizers from wastewaters was studied by O. Rybalkina et al. [15]. Phosphorus transport through anion-exchange membranes in the course of electrodialysis of NaH2PO4 solution was investigated. It was shown that when H2PO4—ions enter the membrane, a part of these anions dissociates, hence parasitic transport of H+ ions occurs in the depleted solution, which essentially reduces the current efficiency of the process.

The analysis of interesting and diverse applications of artificial membranes is reported in Refs. [16-19]. An ionic-liquid-driven supported liquid membrane system was applied by J. Li et al. [16] for preparing a special kind of phosphors, which were characterized by good luminescent properties. The team of L. Bazinet for the first time realized simultaneous separation of peptides from salmon protein hydrolysate by three ultrafiltration membranes stacked in an electrodialysis system [17]. A thorough study of this green and ultra-selective process is presented. Another application concerning processing whey was investigated in references [18, 19]. G. Dufton et al. [18] used a special mode of electrodialysis where Pulsed Electric Fields (PEFs) are applied. It was found that the PEF mode, in which current pulses alternated with pauses of zero current, can increase the degree of both demineralization and deacidification of the whey, as well as reduce membrane scaling. Another kind of electric current pulses, called Pulsed Electrodialysis Reversal (PER), was applied by A. Merkel and A. Ashrafi [19] for demineralization of acid whey. They alternated relatively long direct current pulses with short reverse-polarity pulses to decrease the fouling onto membrane surface during ED. It was found that the fouling on the diluate side of both cation and anion exchange membranes in PER regime was reduced compared to the conventional ED.

The issues connected with the hydrodynamic conditions of mass transfer in membrane systems are considered in references [18–20]. G. Battaglia et al. [18] studied the effect of a trans-membrane pressure (TMP), which may arise in membrane stacks for electrodialysis (water desalination) or reverse electrodialysis (energy production by salinity gradient), on solution flow and mass transfer. A cognitive and helpful review of the effect of profiling ion-exchange membranes on the properties of electrodialysis was presented by S. Pawlowski, J. Crespo and S. Velizarov [19]. They were very convincing when showing that there is exciting potential for improving membrane performance due to the enormous degree of freedom in creating new profile geometries on a membrane surface. Some problems of electrokinetic instability of solution adjacent to an ion-exchange membrane are considered by Magnico [20]. The electrokinetic behavior of cation-exchange resin particles arranged in a well-defined geometrical structure has been studied by the team of Z. Slouka [21]. The understanding of this effect of coupling between water and ion transport is of utmost importance for improving the performance of electro-driven membrane separation processes.

Reference

1. Kedem, O.; Katchalsky, A. Thermodynamic analysis of the permeability of biological membranes to non-electrolytes. *Biochim. Biophys. Acta* **1958**, *27*, 229–246, doi:10.1016/0006-3002(58)90330-5.

2. Katchalsky, A.; Kedem, O. Thermodynamics of flow processes in biological systems. *Biophys. J.* **1962**, *2*, 53–78, doi:10.1016/S0006-3495(62)86948-3.

3. Rubinstein, I.; Schur, A.; Zaltzman, B. Artifact of "breakthrough" osmosis: Comment on the local Spiegler-Kedem-Katchalsky equations with constant coefficients. *Sci. Rep.* **2021**, *11*, 5051, doi:10.1038/s41598-021-83404-9.

4. Trofimov, Y.A.; Krylov, N.A.; Efremov, R.G. Confined dynamics of water in transmembrane pore of TRPV1 ion channel. *Int. J. Mol. Sci.* **2019**, *20*, 4285, doi:10.3390/ijms20174285.

5. Tingey, M.; Mudumbi, K.C.; Schirmer, E.C.; Yang, W. Casting a wider net: differentiating between inner nuclear envelope and outer nuclear envelope transmembrane proteins. *Int. J. Mol. Sci.* **2019**, *20*, 5248, doi:10.3390/ijms20215248.

6. Yue, K.; Sun, X.; Tang, J.; Wei, Y.; Zhang, X. A simulations study on the interaction between pollutant nanoparticles and the pulmonary surfactant monolayer. *Int. J. Mol. Sci.* **2019**, *20*, 3281, doi:10.3390/ijms20133281.

7. Wang, B.-Y.; Zhang, N.; Li, Z.-Y.; Lang, Q.-L.; Yan, B.-H.; Liu, Y.; Zhang, Y. Selective separation of acetic and hexanoic acids across polymer inclusion membrane with ionic liquids as carrier. *Int. J. Mol. Sci.* **2019**, *20*, 3915, doi:10.3390/ijms20163915.

8. Tien, W.; Chen, K.; Huang, F.; Chiu, C. Effects of cholesterol on water permittivity of biomimetic ion pair amphiphile bilayers: Interplay between membrane bending and molecular packing. *Int. J. Mol. Sci.* **2019**, *20*, 3252, doi:10.3390/ijms20133252.

9. Aziz, S.B.; Hamsan, M.H.; Kadir, M.F.Z.; Karim, W.O.; Abdullah, R.M. Development of polymer blend electrolyte membranes based on chitosan: Dextran with high ion transport properties for EDLC application. *Int. J. Mol. Sci.* **2019**, *20*, 3369, doi:10.3390/ijms20133369.

10. Aziz, S.B.; Karim, W.O.; Brza, M.A.; Abdulwahid, R.T.; Saeed, S.R.; Al-Zangana, S.; Kadir, M.F.Z. Ion transport study in CS: POZ based polymer membrane electrolytes using Trukhan model. *Int. J. Mol. Sci.* **2019**, *20*, 5265, doi:10.3390/ijms20215265.

11. Jaleh, B.; Etivand, E.S.; Mohazzab, B.F.; Nasrollahzadeh, M.; Varma, R.S. Improving wettability: deposition of TiO2 nanoparticles on the O2 plasma activated polypropylene membrane. *Int. J. Mol. Sci.* **2019**, *20*, 3309, doi:10.3390/ijms20133309.

12. Abbasi, A.; Hosseini, S.; Somwangthanaroj, A.; Mohamad, A.A.; Kheawhom, S. Poly(2,6-dimethyl-1,4-phenylene oxide)-based hydroxide exchange separator membranes for zinc–air battery. *Int. J. Mol. Sci.* **2019**, *20*, 3678, doi:10.3390/ijms20153678. Abbasi, A.; Hosseini, S.; Somwangthanaroj, A.; Mohamad, A.A.; Kheawhom, S. Correction: Abbasi, A. et al. Poly(2,6-dimethyl-1,4-phenylene oxide)-Based Hydroxide Exchange Separator Membranes for Zinc-Air Battery. *Int. J. Mol. Sci.* **2019**, *20*, 3678. Reprinted from: *Int. J. Mol. Sci.* **2020**, *21*, 377. doi:10.3390/ijms21020377.

13. Ge, Q.; Zhu, X.; Yang, Z. Highly conductive and water-swelling resistant anion exchange membrane for alkaline fuel cells. *Int. J. Mol. Sci.* **2019**, *20*, 3470, doi:10.3390/ijms20143470.

14. Luo, F.; Wang, Y.; Sha, M.; Wei, Y. Correlations of ion composition and power efficiency in a reverse electrodialysis heat engine. *Int. J. Mol. Sci.* **2019**, *20*, 5860, doi:10.3390/ijms20235860.

15. Rybalkina, O.; Tsygurina, K.; Melnikova, E.; Mareev, S.; Moroz, I.; Nikonenko, V.; Pismenskaya, N. Partial fluxes of phosphoric acid anions through anion-exchange membranes in the course of NaH2PO4 solution electrodialysis. *Int. J. Mol. Sci.* **2019**, *20*, 3593, doi:10.3390/ijms20143593.

16. Li, J.; Dong, H.; Yang, F.; Sun, L.; Zhao, Z.; Bai, R.; Zhang, H. Simple preparation of LaPO4:Ce, Tb phosphors by an ionic-liquid-driven supported liquid membrane system. *Int. J. Mol. Sci.* **2019**, *20*, 3424, doi:10.3390/ijms20143424.

17. Henaux, L.; Thibodeau, J.; Pilon, G.; Gill, T.; Marette, A.; Bazinet, L. How charge and triple size-selective membrane separation of peptides from salmon protein hydrolysate orientate their biological response on ggucose uptake. *Int. J. Mol. Sci.* **2019**, *20*, 1939, doi:10.3390/ijms20081939.

18. Dufton, G.; Mikhaylin, S.; Gaaloul, S.; Bazinet, L. Positive impact of pulsed electric field on lactic acid removal, Demineralization and membrane scaling during acid whey electrodialysis. *Int. J. Mol. Sci.* **2019**, *20*, 797, doi:10.3390/ijms20040797.

19. Merkel, A.; Ashrafi, A.M. An investigation on the application of pulsed electrodialysis reversal in whey desalination. *Int. J. Mol. Sci.* **2019**, *20*, 1918, doi:10.3390/ijms20081918.

20. Battaglia, G.; Gurreri, L.; Farulla, G.A.; Cipollina, A.; Pirrotta, A.; Micale, G.; Ciofalo, M. Membrane deformation and its effects on flow and mass transfer in the electromembrane processes. *Int. J. Mol. Sci.* **2019**, *20*, 1840, doi:10.3390/ijms20081840.

21. Pawlowski, S.; Crespo, J.G.; Velizarov, S. Profiled ion exchange membranes: A comprehensible review. *Int. J. Mol. Sci.* **2019**, *20*, 165, doi:10.3390/ijms20010165.

22. Magnico, P. Electro-kinetic instability in a laminar boundary layer next to an ion exchange membrane. *Int. J. Mol. Sci.* **2019**, *20*, 2393, doi:10.3390/ijms20102393.

23. Vobecká, L.; Belloň, T.; Slouka, Z. Behavior of embedded cation-exchange particles in a DC electric field. *Int. J. Mol. Sci.* **2019**, *20*, 3579, doi:10.3390/ijms20143579.

Victor Nikonenko, Natalia Pismenskaya
Editors

International Journal of
Molecular Sciences

MDPI

Article

Confined Dynamics of Water in Transmembrane Pore of TRPV1 Ion Channel

Yury A. Trofimov [1,2,3], Nikolay A. Krylov [1,2] and Roman G. Efremov [1,2,4,*]

1 M.M. Shemyakin & Yu.A. Ovchinnikov Institute of Bioorganic Chemistry, Russian Academy of Sciences, Miklukho-Maklaya Street, 16/10, 117997 Moscow, Russia
2 National Research University Higher School of Economics, Myasnitskaya ul. 20, 101000 Moscow, Russia
3 National Research Nuclear University Moscow Engineering Physics Institute, Kashirskoe Shosse, 31, 115409 Moscow, Russia
4 Moscow Institute of Physics and Technology (State University), Dolgoprudny, 141701 Moscow, Russia
* Correspondence: r-efremov@yandex.ru; Tel.: +7-903-743-16-56

Received: 6 August 2019; Accepted: 29 August 2019; Published: 1 September 2019

Abstract: Solvation effects play a key role in chemical and biological processes. The microscopic properties of water near molecular surfaces are radically different from those in the bulk. Furthermore, the behavior of water in confined volumes of a nanometer scale, including transmembrane pores of ion channels, is especially nontrivial. Knowledge at the molecular level of structural and dynamic parameters of water in such systems is necessary to understand the mechanisms of ion channels functioning. In this work, the results of molecular dynamics (MD) simulations of water in the pore and selectivity filter domains of TRPV1 (Transient Receptor Potential Vanilloid type 1) membrane channel are considered. These domains represent nanoscale volumes with strongly amphiphilic walls, where physical behavior of water radically differs from that of free hydration (e.g., at protein interfaces) or in the bulk. Inside the pore and filter domains, water reveals a very heterogeneous spatial distribution and unusual dynamics: It forms compact areas localized near polar groups of particular residues. Residence time of water molecules in such areas is at least 1.5 to 3 times larger than that observed for similar groups at the protein surface. Presumably, these water "blobs" play an important role in the functional activity of TRPV1. In particular, they take part in hydration of the hydrophobic TRPV1 pore by localizing up to six waters near the so-called "lower gate" of the channel and reducing by this way the free energy barrier for ion and water transport. Although the channel is formed by four identical protein subunits, which are symmetrically packed in the initial experimental 3D structure, in the course of MD simulations, hydration of the same amino acid residues of individual subunits may differ significantly. This greatly affects the microscopic picture of the distribution of water in the channel and, potentially, the mechanism of its functioning. Therefore, reconstruction of the full picture of TRPV1 channel solvation requires thorough atomistic simulations and analysis. It is important that the naturally occurring porous volumes, like ion-conducting protein domains, reveal much more sophisticated and fine-tuned regulation of solvation than, e.g., artificially designed carbon nanotubes.

Keywords: anomalous water diffusion; computer simulations; ion channel gating; molecular dynamics; nano-size water pore; physico-chemical properties of confined water; protein-water interactions; TRPV1 channel permeability for water; water dynamics; water H-bonding

1. Introduction

The behavior of water near molecular surfaces is known to be critically important in chemistry and biology [1,2]. Thus, in a protein hydration shell, it presumably plays a sufficient role in many biophysical processes and defines pivotal protein properties, like protein folding (water expulsion from a hydrophobic core mediates the rate of the late stage folding [1]); ligand binding (water molecules can affect the selectivity of a binding site [2]); protein aggregation (accelerated water dynamics speeds up the growth of aggregates by facilitating the binding of new peptide monomers [3]); thermal stability (strong protein–water interactions prevent unfolding of thermophilic proteins [4]), and so on. Nontrivial dynamics of water was also reported on the water/membrane interface and inside lipid bilayers [5]. These effects were associated with the functioning of cell membranes, for example, with their permeation for water and other substances, adsorption of proteins, protein–protein interactions in membranes, etc.

Experimental studies and atomistic simulations showed that water molecules in a solute hydration shell (e.g., proteins, membranes) have retarded dynamics compared to the bulk water by a factor ≈2 to 6 [6]. Such a slowdown can be explained by the extended molecular jump mechanism. It describes dynamics of water molecules as fast large-amplitude rotational jumps occurring due to the exchange of hydrogen bonded (hb) partners, and a relatively slow diffusion process occurring between the jumps [7]. The typical time constant of molecular rotational motions and hb lifetimes is about 2 ps [8] for bulk water. In the jump model, two factors affect water dynamics in solute hydration shell. The first one is the excluded volume effect that is caused by reducing the number of probable hb partners for water molecules due to the water–solute interface topology. The typical slowdown factor of water in this case is less than two at the local convex solute sites and exceeds two at the concave sites [9]. The second factor is water–solute hb interactions. This can change the rotational dynamics of water depending on the type (hb donor or acceptor) and strength of these bonds in comparison with the water–water hb [10]. The largest deceleration effect is induced by strong hb acceptors, such as carboxylates, whose slowdown factor also exceeds two.

Both factors have a strong influence on the dynamics of water confined in artificial and natural nanoscale channels, where the aforementioned "water braking" is much more pronounced. Thus, at the wall of a relatively broad (ø 24 Å) hydrophilic silica channel, water rotational dynamics are lengthened by roughly two orders of magnitude compared with the bulk phase [11]. These dynamics assume a broad distribution of rate constants that arises from the spatial heterogeneity of the channel surface [12]. While in the internal core region of the channel, nearly bulk-like homogenous dynamics are observed.

In narrow channels (e.g., carbon nanotubes), water molecules are often arranged in a single file and move as a unit due to strong hb interactions with each other. Water rotational dynamics in such channels are slowed down to the order of several nanoseconds [13]. For porous microstructures, like perturbed channels [14] and water permeable membranes [15], the interaction between the channel wall and water strongly effects the water hb–network (and water dynamics) due to the confinement and affinity of channels or membrane polar groups. Water permeability through such structures critically depends on their molecular design. For example, water transfer resistance of aromatic polyamide membranes can be reduced by enhancing the interfacial hydrophilicity and the interior hydrophobicity of their structures [15].

Besides the above said artificial chemically synthetized nano objects, it is very important to understand the detailed water behavior in the naturally occurring ones—e.g., in transmembrane (TM) protein ion channels. For the latter, spatial arrangement of hydrophobic/hydrophilic properties of the pores is often critical for their functionality. Thus, hydrophilic residues in the pore of aquaporins (a family of proteins, which provide rapid and highly selective conduction of water and other small molecules across the cellular membrane) lower the energy barrier for water permeation by offering replacement interactions to water molecules in order to compensate the loss of water–water hydrogen bonds [16]. An opposite example of this hydrophobic influence on the channel's activity is the so-called hydrophobic gating. In this case, a hydrophobic constriction in a channel pore mediates an energetic barrier to water and ion permeation by means of partial hydration/dehydration of the pore [17,18].

In this work, the effects of the biological pore on water dynamics were studied using the results of molecular dynamics (MD) simulations of water in confined conditions of the pore and filter domains of the TRPV1 ion channel. Vanilloid receptor TRPV1 (Transient Receptor Potential Vanilloid type 1) is a non-selective cation channel that accomplishes transport of ions and water molecules through a cellular membrane in response to capsaicin, temperatures above 42 °C, and other physico-chemical stimuli [19]. TRPV1 is preferentially expressed in neurons of the peripheral neuron system responsible for sensation to heat, hot spice, pain, itch, and so on [19,20].

TRPV1 is composed of four identical subunits, which form a pathway for ions and water molecules in a lipid bilayer [21–23]. This pathway has two "bottlenecks": The so-called upper gate that is formed by a selectivity filter created by short unstructured segments at the extracellular side of the channel and the lower gate located in the bundle composed of four crossing TM helices S6 disposed closer to the cytoplasmic side. The narrowest place of the upper gate is formed by carbonyl oxygen atoms of the residue Glu643 (the distance between the atoms of diagonally located subunits is 7.6 Å in the open state [22]). The lower gate bottleneck is formed by side chains of Ile679 (7.6 Å in the open state [22]).

The cavity between the gates (the pore) has mostly hydrophobic walls [24]. However, few polar groups here are exposed to solvent: Backbone carbonyls of some residues, hydroxyl group of Thr671, and polar groups of Asn676. Such an organization of the TRPV1 pore represents a rather unusual natural nano object—strongly amphiphilic confined volume with presumably flexible walls formed by four TRPV1 subunits stabilized by the membrane environment. It is reasonable to assume that the dynamic behavior of water may drastically differ from that in bulk water as well as on the protein or membrane interfaces.

2. Results

To investigate the dynamics of confined water in TRPV1, the most spatially limited protein domains were chosen: The pore segments of helices S6 between the upper and lower gates and the selectivity filter (residues 670–680 and 642–646, respectively, Figure 1). For comparison, three additional domains were considered: Pore vestibule, the less confined volume, which is formed by the segments of helices S6 under the lower gate (residues 681–692); TRP helix lying on the cytoplasmic side of the membrane and strongly exposed to the bulk water (residues 693–711); loops in the extracellular entrance of the channel—they represent protein regions, which are mostly exposed to bulk water in TRPV1 (residues 604–625).

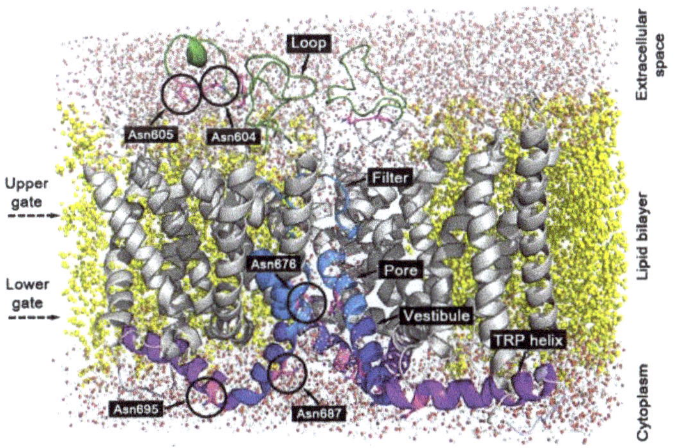

Figure 1. Ribbon representation of TM segments of TRPV1 embedded into hydrated lipid bilayer (three protein subunits are shown). The protein domains considered in the analysis of water dynamics are colored: green—extracellular loops (residues 604–625), blue—pore segment of the helices S6 (residues 670–680), and selectivity filter (residues 642–646), dark blue—vestibule segment of the helices S6 (residues 681–692), purple—TRP helices (residues 693–711), gray—other protein parts. Asparagine residues of TRPV1 further mentioned in the text are marked by black circles. Yellow—lipid bilayer, red and white dots—water molecules. Dashed arrows mark the levels of the upper and the lower gates.

2.1. Translational Dynamics of Water along the Pore Axis

Figure 2 shows the pore radius (R), water linear density (ϱ_{ln}) for all water molecules (red curve) and for molecules bound to protein polar groups via hb (green curve), and translational dynamics coefficient (D, see Materials and Methods). All these profiles were calculated in a cylinder (ø30 × 100 Å) oriented along the pore axis. The profiles were averaged over four MD trajectories. Minima of the curves, R(z) and $\varrho_{\text{ln}}(z)$, in the regions of z = −5 . . . 0 Å and 10 . . . 15 Å correspond to the narrowest zones of the pore: The lower and upper gates. It can be seen that water in the confined protein domains has about 1.5 to 3 times slower translational dynamics in comparison with the bulk (z < −30 Å and z > 50 Å). Above and below the gates, the mobility of water molecules increases as the channel expands. The maximum of D in the pore domain (about 0.4 Å^2/ps at z = 3 . . . 5 Å) corresponds to the maximum of the pore radius (*R* = 3.5 . . . 4.5 Å). In this region, there is a large cavity located between the gates, where about a half of the water molecules do not form hb with the protein. Furthermore, the maximum of D(z) is probably related to the formation of a bulk-like phase similar to the core water in silica nanotube in [11]. An opposite scenario can be seen in the range of z = 15 . . . 30 Å. Very slow water dynamics (D ≈ 0.2 Å^2/ps) take place in spite of a relatively large pore radius and water density. Such a slowdown is caused by the unstructured protein regions, which evenly fill about half of the volume in this region and intense water–protein hb interactions (about 75% of water molecules are bound to protein).

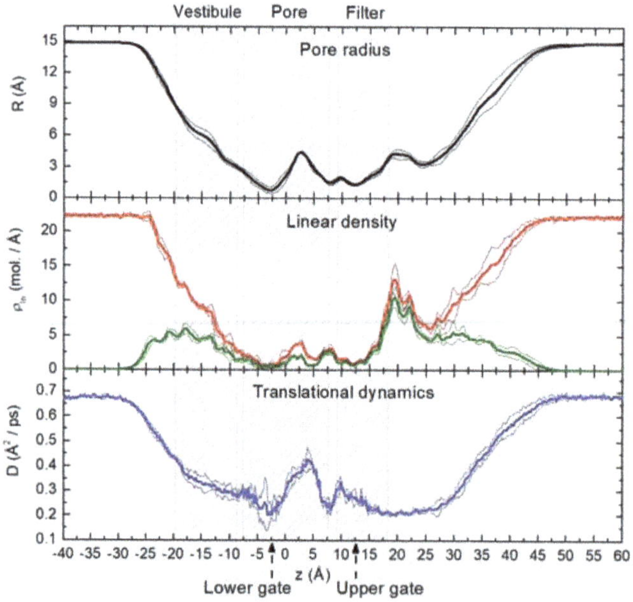

Figure 2. Pore radius (*R*), linear density (ρ_{ln}), and translational dynamics coefficient of water (*D*) along the pore axis (z). Linear density of all water molecules and waters bound to protein polar groups via hb are shown with red and green, respectively. Thick and thin curves display average values (calculated over four MD trajectories) and standard deviations, respectively. Hatched areas indicate vestibule, pore, and filter domains of the protein. TRP helix and loop domains are out of the boundaries of the volume under consideration (cylinder with the radius of 15 Å). Dashed arrows mark the location of the upper and the lower gates.

2.2. Spatial Distribution of Water in the Pore

Water distribution around particular protein groups (atoms) can be investigated with the radial distribution function (g(r)) [25]. This function for water oxygen atoms calculated around polar (and charged) and nonpolar groups of the pore and filter domains is shown in Figure 3. For analysis, the groups with coordination number ≥ 0.2 were chosen. It can be seen that around polar groups there is a pronounced peak within the radius of 3.5 Å. Because nonpolar groups reveal a broader peak shifted by 1 Å from the center, they form a low density region in their 3.5-Å neighborhood. The difference between the water radial distribution around polar and nonpolar groups can be explained by the water–protein hb interactions, which preferably retain water molecules near the polar groups.

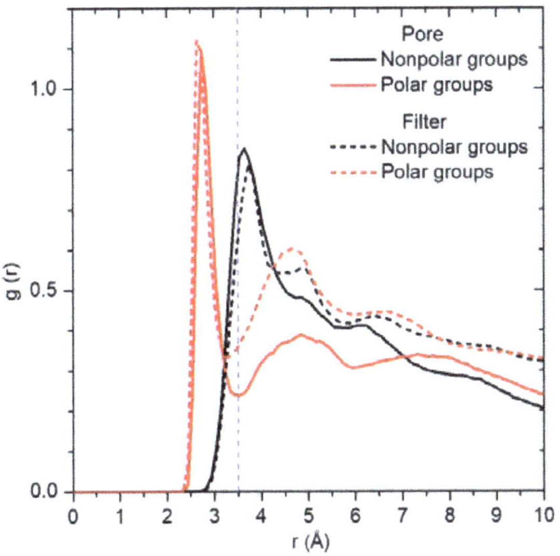

Figure 3. Radial distribution functions of water oxygens around polar and nonpolar groups of the pore and filter domains. The most hydrated groups of Gly643, Met644, Gly645, Asp646 (filter) and Tyr671, Asn676, Ile679, Ala680 (pore) were taken for calculation. $r = 0$ corresponds to the position of the central atom of a group (oxygen or nitrogen for polar groups and carbon for nonpolar groups). Dashed vertical line shows the radius of 3.5 Å around the protein groups.

This water molecules' localization can be seen in Figure 4, which shows the averaged spatial density distribution of water, protein, and water–protein hb. Isosurfaces corresponding to densities of (ϱ) 0.03 (semitransparent green) and 0.1 mol/Å3 (opaque dark green) are indicated. The former corresponds roughly to the bulk water density. The second shows the areas of anomalously high water density (water localization areas). High-density areas are preferably disposed near the protein polar groups, which manifest themselves via water–protein hb interactions (pink areas in Figure 4).

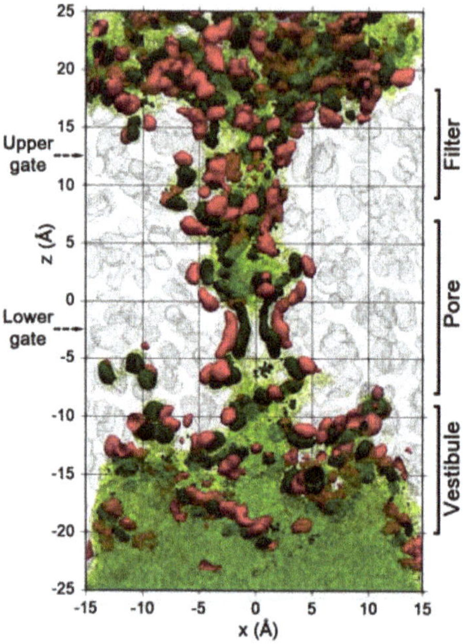

Figure 4. Spatial density distribution of protein atoms, water molecules, and protein–water hb in the cylinder (ø30 × 50 Å) along the pore axis averaged over one of the trajectories. Gray—isosurface of protein density $\rho = 0.1$ atom/Å^3 (part of the protein is removed for clarity), dark green—water with $\rho = 0.1$ mol/Å^3, green semi-transparent—water with density similar to the bulk water $\rho = 0.03$ mol/Å^3, pink—water-protein hb with $\rho = 0.1$ bound/Å^3. Boundaries of the vestibule, pore, and filter domains are shown on the right. Dashed arrows mark the levels of the upper and the lower gates.

A very heterogeneous water spatial distribution with a high abundance of water localization areas can be seen in the pore and filter domains, while in the less confined vestibule there is a more homogeneous water distribution. This picture corresponds to the linear density of water along the pore axis in Figure 2, where about 70% of the 60 water molecules in the pore and filter domains are bound to protein polar groups, while in the vestibule domain, only 50% of the 96 waters are bound. Moreover, the contribution of bonded molecules decreases to 36% in the bottom part of the vestibule ($z = -20$ Å).

2.3. *Water Residence Time in the Hydration Shells of Polar and Nonpolar Protein Groups*

The distributions of water residence time (τ_{res}, see Materials and Methods) were obtained separately for polar and nonpolar groups of different protein domains (Figure 5a). Nonpolar groups of the loop, vestibule, and TRP domains showed similar distributions with a narrow peak in the range of short residence times (1–5 ps), which indicates fast water dynamics, and a long tail of larger residence times that corresponds to slowed down dynamics. Polar groups of the same domains do not have the peak of fast dynamics, while the tail decreases slowly compared to that observed for nonpolar groups. Since the pore and filter domains possess a smaller number of hydrated groups, their distributions were merged. In this case, the fast dynamics peak in the distribution for nonpolar groups is less pronounced and extends to 8 ps. In the case of polar groups, the distribution demonstrates slower water dynamics; its left edge is shifted relative to the less confined domains by 5 to 8 ps to the larger residence times' direction.

In Figure 5b, the same τ_{res} distributions are shown in a box chart representation. The boxes report interquartile ranges of the τ_{res} distributions, with the line and number in each box showing the median

τ_{res}. For the loop, vestibule, and TRP domains, water molecules show a 3 to 4 times longer residence time near polar groups compared with nonpolar groups (median τ_{res} 16–28 ps and 5–7 ps, respectively). For the pore and filter domains, the similar ratio of τ_{res} occurs (12 and 42 ps, respectively). Furthermore, water residence times in these domains are 1.5 to 3 times longer than in other domains.

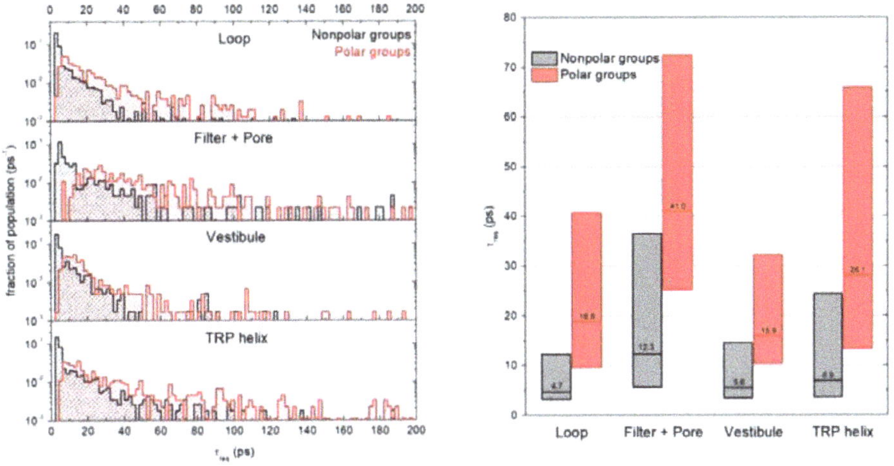

(a) (b)

Figure 5. The distributions of the water residence time in hydration shells of polar and nonpolar groups of different TRPV1 domains in a histogram (**a**) and box chart (**b**) representation. Red—polar (and charged) groups, gray—nonpolar groups. Boxes represent interquartile ranges of the corresponding histograms; line and number in each box is the median τ_{res} of the corresponding histogram.

Another way to compare the dynamics of confined water in a protein hydration shell is to calculate the water residence time for the same protein group located in different protein regions. The carbonyl oxygen atom of the asparagine residue side chain (OD1 group) was chosen for this purpose. Figure 6 shows water survival time correlation functions and τ_{res} calculated for the OD1 group of the next residues: Asn604 and Asn605 of the loop domain, Asn676 located in the pore, Asn687 in the vestibule, and Asn695 at the TRP helix (see Figure 1). Asn-OD1 groups of the loop, vestibule, and TRP domains reveal similar values of correlation function parameters: Coordination number (N_{α} (t = 0), see Materials and Methods) in the range of 2.0 to 2.6 and τ_{res} in the range of 11 to 16 ps. Asn676 located in the confined volume of the pore shows less hydration $N_{\alpha}(t = 0) = 1.6$ and slower water dynamics τ_{res} = 58 ps in spite of the same solvent exposure of its polar OD1 group.

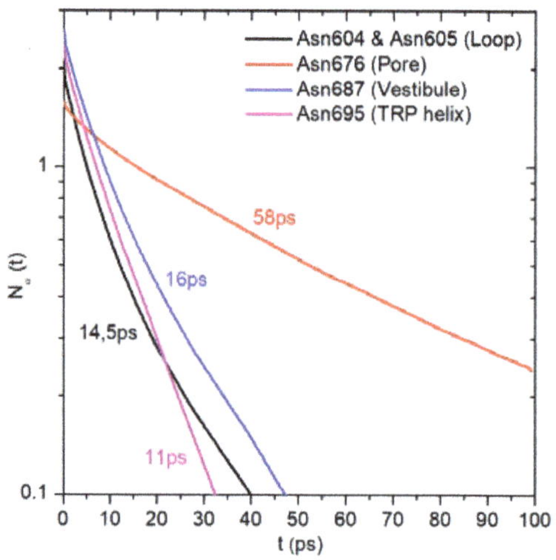

Figure 6. Water survival time correlation functions calculated for the carbonyl oxygen atom of the side chain of asparagine residues (OD1 group) located in different protein regions. The correlation functions were averaged over four identical protein subunits and over four MD trajectories. Corresponding values of τ_{res} are shown near the curves.

Averaging over the protein subunits and MD trajectories broadly used in this study is a convenient instrument for the comparison of large protein domains. However, a great variety of properties for the same protein sites or groups should be noted. Figure 7 shows the water survival time correlation functions calculated separately for the Asn676-OD1 group of four protein subunits and four MD trajectories. Despite the fact that they are the same groups of identical subunits, the parameters of water dynamics near the groups vary considerably: τ_{res} from 16 to 128 ps, coordination number from 1.1 to 2.2. Moreover, the atoms of Asn676-OD1 are fully dehydrated in two cases.

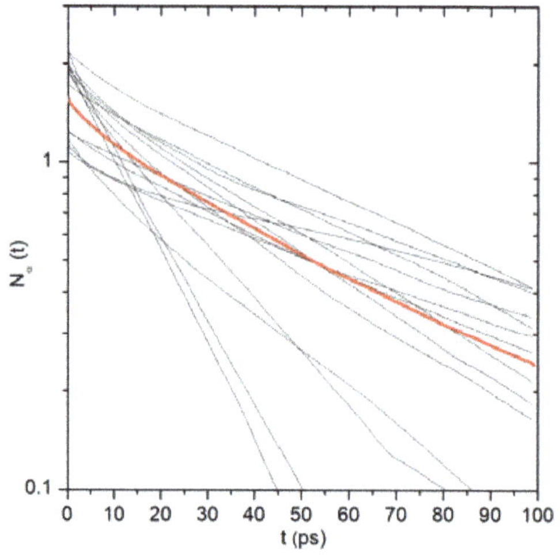

Figure 7. Water survival time correlation functions calculated separately for the Asn676-OD1 group of four protein subunits and four MD trajectories (black curves). Red curve is the average of the black curves.

The observed variety of the water dynamics is the consequence of two facts that we believe are not the artifacts of MD simulation but represent the natural property of the protein under study. The first one is the asymmetrical structure of TRPV1 that arises from the initial symmetric experimental model due to the flexibility of large protein domains in the course of MD simulations [24]. This is reflected in the asymmetrical spatial distribution of water in the channel (Figure 4). The second one is the flexibility of the side chains of individual residues. These two factors are responsible for heterogeneity of the microenvironment of particular protein groups and affect the water dynamics in their hydration shells.

3. Discussion and Conclusions

In this work, we employed MD simulations to study the atomistic resolution dynamics of water molecules confined in the transmembrane ion/water conducting pathway of the temperature sensor protein—TRPV1—channel. The main objective was to explore in detail the behavior of water in a rather unusual nano volume with varying geometry and complex distribution of physico-chemical properties of the internal walls of the pore. Firstly, the solvent-accessible space represents a bottle-like reservoir with two, instead of just one, bottlenecks, linked at the extremities to funnels—vestibules of the channel. The radius of the channel pathway changes from less than one to 4 to 7 Å along the membrane normal (Figure 2). Secondly, the walls of the inner channel are highly hydrophobic in the central part and contain a number of polar protein groups exposed to the pore to varying degrees [24]. Finally, the unique nature of the pore domain is also determined by asymmetrical disposition of the TRPV1 subunits and the highly dynamic character of particular protein polar and nonpolar groups lining the pore. Altogether, this makes the pore in TRPV1 a very interesting object to study the confined water inside—this naturally-occurring system drastically differs from the chemically synthesized artificial porous nano objects, like carbon nanotubes (CNTs), nano slits, and so on.

As expected, the water in the channel showed a noticeable slowdown, which is caused by the limited space (and, therefore, excluded volume effect) and the presence of protein groups capturing waters via hb interactions. Although these effects are commonly observed in any limited nano reservoir, the picture presented here for TRPV1 has some intriguing specific features. Thus, the water "braking" near polar groups of the confined protein domains, like the pore and selectivity filter of TRPV1, leads to an interesting effect of water localization at the particular sites of the protein (Figure 4). These areas (or zones with anomalously high water density) preferably occupy the protein regions, where most of the waters (about 70%) are bound to the protein polar groups. Apparently, such "water blobs" play an important role in the functional activity of TRPV1. In particular, Asn676-OD1 polar groups of four protein subunits may take part in the hydration of the hydrophobic TRPV1 pore by capturing up to six waters near the lower gate of the channel. This, in turn, reduces the free energy barrier for ion and water transport [26].

The effect of water slowdown near the confined protein domains (pore and filter) is evident from the analysis of residence time (τ_{res}) distributions obtained for all polar and nonpolar groups (Figure 5). In these domains, the τ_{res} values for water are 1.5 to 3 times larger in comparison with other protein regions. Meanwhile, the influence of water–protein hb interactions on water dynamics can be seen in Figure 5b: The residence time of water near polar groups is 3 to 4 times longer compared to nonpolar groups for all considered protein regions.

It is interesting to note that the τ_{res} distributions for nonpolar groups of the loop, vestibule, and TRP domains look similar to those obtained by Sterpone et al. [27] for water reorientation dynamics within a lysozyme hydration shell. Both sets of distributions reveal a narrow peak that corresponds to relatively fast water dynamics, and a long tail of slow water (Figure 5a). The authors explain the peak by the contribution of water molecules located near hydrophobic and hb donor protein groups, while the tail corresponds to water located near the groups that are confined in pockets and clefts on the protein surface. Most of the latter are hb acceptor groups. Such an assumption agrees with our observation that the tails of the distributions of τ_{res} for polar groups decrease more slowly than for nonpolar groups (Figure 5a). In other words, the slowest water molecules are mainly localized near polar than near nonpolar groups. However, earlier, we showed that waters usually form stronger H-bonds as acceptors than proton donors [28]. So, we suggest, that the slowest water can be a donor or acceptor of hb depending on the type of a particular protein group.

MD simulations also show that the water dynamics in the hydration shell of a particular protein polar group critically depend on the group location (i.e., on its microenvironment). For the polar group OD1 of Asn676 that is confined in the TRPV1 pore, the water τ_{res} value is 3 to 5 times larger than that for the same groups in other Asn residues located in less-confined conditions of the extracellular loop, protein surface (TRP helix), and pore vestibule (Figures 1 and 6). This is caused by thte slowed down dynamics of water molecules near the spatially restrained groups. Since the solvation properties (e.g., expressed in terms of the water-accessible surface, etc.) are very similar for the considered groups, the observed changes in water dynamics can only be explained by the excluded volume effect, which has a much greater impact in confined conditions [27].

The presented dynamic picture of water enclosed in the nano-scale TM pore domain of the ion channel TRPV1 radically differs from those observed earlier in artificial nano objects (see, e.g., [11,13,29] and references therein). Thus, in the latter cases, confined water forms either single-file ordered chains in narrow CNTs and small reverse micelles, various n-gonal ice nanotubes in wider CNTs, or two-dimensional clusters in slit-pore spaces. None of these phenomena were found in the biomolecular system under study. Such unique properties of water in nano spaces formed by membrane proteins (ion channels, receptors, and others) can be explained by the rather flexible walls of the pores and their finely tuned amphiphilic surface "portraits". In the first case, even a slight instant asymmetry in the packing of protein subunits (including thermal fluctuations of non-covalently bound chains) can lead to serious changes of the geometry of the water "nano pool". Furthermore, fast conformational dynamics of multiple polar/nonpolar protein groups inside the pore induces a prominent effect on the spatial distribution of the high-density water sites—the example of the residue Asn676 clearly demonstrates this. It should be noted that even such a complex picture is too simplified, since cations (Ca^{2+}, Na^{+}, K^{+}), which are the key players in channel functioning, were not considered here. Their appearance will certainly affect the dynamics of water in the pore domain of TRPV1, but this topic is beyond the scope of the present work.

4. Materials and Methods

4.1. System and Molecular Dynamics Protocol

The studied system was taken from our previous work [24]. It consisted of the TM-segment of TRPV1 (residues 427–719) in the open state (PDB structure 3J5Q22) embedded into a fully hydrated lipid bilayer with a composition similar to the neuronal membrane: 50% of palmitoyloleoylphosphatidylcholine (POPC), 25% of palmitoyloleoylphosphatidylethanolamine (POPE), and 25% of cholesterol molecules (Figure 1). Chloride ions were added to the system for restoring electroneutrality.

Molecular dynamics (MD) simulations were performed using the GROMACS 2018.5 package [30], Amber99sd-ildn force field [31], and TIP3P water model [32]. Simulations were carried out with an integration time of 2 fs, imposed 3D periodic boundary conditions, and constant temperature (310 K) and pressure (1 bar). For electrostatic interactions, the particle-mesh Ewald summation was used (real space cutoff 10 Å and 1.2 Å grid with fourth-order spline interpolation). A twin-range (10/12 Å) spherical cutoff function was employed to treat van der Waals interactions. Four starting configurations of the system were taken from one of the 500-ns MD trajectories calculated earlier [24]. These configurations were heated to 310 K during the 400-ps MD run with fixed positions of the protein heavy atoms. Then, four unrestrained MD simulations were carried out with a length of 2 ns and the time step between the stored states of 0.25 ps.

4.2. MD Data Analysis

Translational dynamics of water molecules along the pore axis (z) were characterized by the coefficient, D, that was named the translational dynamics coefficient and calculated as:

$$D(z) = \frac{1}{6T}\sum_{t=0}^{T}\frac{1}{N_w(z,t)}\sum_{i=0}^{N_w(z,t)}\Delta R_i^2(z,t) \quad (1)$$

where T is the trajectory length (2 ns); N_w(z,t) is the number of water molecules in a layer from z to z + dz at a moment t (dz = 0.25 Å); ΔR_i^2(z,t) is the square displacement over time of 1 ps of the molecules, which are located from z to z + dz at a moment t. Parameter D for bulk water numerically equals its diffusion coefficient (about 0.65 Å^2/ps for TIP3P water at 310 K) and, at the same time, it shows the local heterogeneity of water dynamics along the pore.

Channel radius along the pore axis (*R*) was calculated as:

$$R(z) = \sqrt{< V(z) > /\pi dz} \quad (2)$$

where <V(z)> is the MD-averaged solvent-accessible volume of a layer from z to z + dz (dz = 0.25 Å). That is, the volume, where the center of a sphere with a radius of 1.4 Å can be placed without overlapping with the protein surface.

To compare the water dynamics near polar and nonpolar groups in the protein domains with various confinement conditions, the residence time of water molecules in the hydration shells of protein groups (τ_{res}) was calculated. For this, a survival time correlation function was used [25]:

$$N_\alpha(t) = \frac{1}{T}\sum_{j=1}^{N_w}\sum_{t'=0}^{T} p_{\alpha,j}(t', t'+t; t_0) \quad (3)$$

Here, T is the trajectory length (2 ns). N_w is the total number of water molecules in the system. $p_{\alpha,j}$(t′, t′ + t; t_0) is the binary function, which equals 1 if the water molecule, j, continuously stays in the hydration shell of a group, α, during the time interval from t′ to t′ + t or leaves the shell during this interval, but returns for the time no longer than t_0 = 2 ps. Otherwise, the function equals zero. The radius of hydration shell was taken as 3.5 Å for all protein groups. The value N_α (*t* = 0) gives the mean hydration shell occupancy (coordination number) of the group, α.

The calculated N_α was fitted by a double exponential function:

$$n(t) = n_f e^{-t/\tau_f} + n_s e^{-t/\tau_s} \quad (4)$$

where τ_f and τ_s are the fast and the slow decay constants. Only groups with N_α (*t* = 0) ≥ 0.2 were taken for fitting. The latter was performed on the time interval from 0.25 to 100 ps or up to the time when N_α reaches the value of 0.1 for rapidly decaying functions. The residence time of water molecules was calculated for each group as:

$$\tau_{res} = \langle t \rangle = \frac{\int_0^\infty t n(t)dt}{\int_0^\infty n(t)dt} = \frac{n_f\tau_f^2 + n_s\tau_s^2}{n_f\tau_f + n_s\tau_s} \quad (5)$$

In Figures 2 and 4, origin of the pore axis (z) was taken as the pore center of mass calculated over Cα atoms of residues 642, 643, 644, 645, 671, 675, 676, 679, 680, 683, 686, and 687—those are located in the helices S6 and the selectivity filter. The following geometric criteria were used to define water–protein hb: Donor (D) – acceptor (A) atoms distance ≤ 3.5 Å, the angle D-H-A lies in the range 180° ± 30°.

Author Contributions: Conceptualization, Y.A.T. and R.G.E.; methodology, Y.T; software, N.A.K.; investigation, Y.A.T. and N.A.K.; data curation, Y.A.T. and N.A.K.; writing—original draft preparation, Y.A.T.; writing—review and editing, N.A.K. and R.G.E.; visualization, Y.A.T.; supervision, R.G.E.; project administration, R.G.E.; funding acquisition, R.G.E.

Funding: This research was supported by the Russian Science Foundation, grant #19-74-30014. MD simulations with the time step 0.25 ps and corresponding data analysis were supported by the Russian Foundation for Basic Research (grant #19-04-00350). Supercomputer calculations were sponsored in the framework of the Basic Research Program at the National Research University Higher School of Economics and Russian Academic Excellence Project '5-100'.

Acknowledgments: We thank Pavel Volynsky for 500-ns long MD trajectory of membrane bound TRPV1, which provided us with starting configurations for further MD simulations The latter were partially carried out using the equipment operated by the IBCH core facility (CKP IBCH, supported by the Russian Ministry of Education and Science, grant RFMEFI62117X0018). Access to computational facilities of the Supercomputer Center "Polytechnical" at the St. Petersburg Polytechnic University and Joint Supercomputer Center RAS (Moscow) is greatly appreciated.

Conflicts of Interest: The authors declare no conflict of interest.

Abbreviations

CNT	Carbon nanotube
hb	Hydrogen bond
MD	Molecular dynamics
TM	Transmembrane
TRPV1	Transient Receptor Potential Vanilloid type 1

References

1. Levy, Y.; Onuchic, J.N. Water mediation in protein folding and molecular recognition. *Annu. Rev. Biophys. Biomol. Struct.* **2006**, *35*, 389–415. [CrossRef] [PubMed]
2. Ball, P. Water as an active constituent in cell biology. *Chem. Rev.* **2008**, *108*, 74–108. [CrossRef] [PubMed]
3. Khatua, P.; Bandyopadhyay, S. Dynamical crossover of water confined within the amphiphilic nanocores of aggregated amyloid β peptides. *Phys. Chem. Chem. Phys.* **2018**, *20*, 14835–14845. [CrossRef] [PubMed]
4. Rahaman, O.; Melchionna, S.; Laage, D.; Sterpone, F. The effect of protein composition on hydration dynamics. *Phys. Chem. Chem. Phys.* **2013**, *15*, 3570–3576. [CrossRef] [PubMed]
5. Krylov, N.A.; Pentkovsky, V.M.; Efremov, R.G. Nontrivial behavior of water in the vicinity and inside lipid bilayers as probed by molecular dynamics simulations. *ACS Nano* **2013**, *7*, 9428–9442. [CrossRef]
6. Laage, D.; Elsaesser, T.; Hynes, J.T. Water dynamics in the hydration shells of biomolecules. *Cham. Rev.* **2017**, *117*, 10694–10725. [CrossRef]
7. Laage, D.; Hynes, J.T. A molecular jump mechanism of water reorientation. *Science* **2006**, *310*, 832–835. [CrossRef] [PubMed]
8. Laage, D.; Elsaesser, T.; Hynes, J.T. Perspective: Structure and ultrafast dynamics of biomolecular hydration shells. *Struct. Dyn.* **2017**, *4*. [CrossRef] [PubMed]
9. Laage, D.; Stirnemann, G.; Hynes, J.T. Why water reorientation slows without iceberg formation around hydrophobic solutes. *J. Phys. Chem. B* **2009**, *113*, 2428–2435. [CrossRef] [PubMed]
10. Sterpone, F.; Stirnemann, G.; Hynes, J.T.; Laage, D. Water hydrogen-bond dynamics around amino acids: The key role of hydrophilic hydrogen-bond acceptor groups. *J. Phys. Chem. B* **2010**, *114*, 2083–2089. [CrossRef]
11. Laage, D.; Thompson, W.H. Reorientation dynamics of nanoconfined water: Power-law decay, hydrogen-bond jumps, and test of a two-state model. *J. Chem. Phys.* **2012**, *136*. [CrossRef]
12. Fogarty, A.C.; Duboue-Dijon, E.; Laage, D.; Thompson, W.H. Origins of the non-exponential reorientation dynamics of nanoconfined water. *J. Chem. Phys.* **2014**, *14*. [CrossRef] [PubMed]
13. Chakrabort, Y.S.; Kumar, H.; Dasgupta, C.; Maiti, P.K. Confined water: Structure, dynamics and thermodynamics. *Acc. Chem. Res.* **2017**, *50*, 2139–2146. [CrossRef] [PubMed]
14. Murail, S.; Vasiliu, T.; Neamtu, A.; Barboiu, M.; Sterpone, F.; Baaden, M. Water permeation across artificial I-quartet membrane channels: From structure to disorder. *Faraday Discuss* **2018**, *209*, 125–148. [CrossRef] [PubMed]

15. Zhang, N.; Chen, S.; Yang, B.; Huo, J.; Zhang, X.; Bao, J.; Ruan, X.; He, G. Effect of hydrogen-bonding interaction on the arrangement and dynamics of water confined in a polyamide membrane: A molecular dynamics simulation. *J. Phys. Chem. B* **2018**, *122*, 4719–4728. [CrossRef] [PubMed]
16. Hub, J.S.; Grubmüller, H.; Groot, B.L. Dynamics and energetics of permeation through aquaporins. What do we learn from molecular dynamics simulations? *Handb. Exp. Pharmacol.* **2009**, *190*, 57–76. [CrossRef]
17. Areal, P.; Sansom, M.S.P.; Tucker, S.J. Hydrophobic gaiting in ion channels. *J. Mol. Biol.* **2015**, *427*, 121–130. [CrossRef] [PubMed]
18. Rao, S.; Lynch, C.I.; Klesse, G.; Oakley, G.E.; Stansfeld, P.J.; Tuncker, S.J.; Sansom, M.S.P. Water and hydrophobic gates in ion channels and nanopores. *Faraday Discuss* **2018**, *209*, 231–247. [CrossRef] [PubMed]
19. Clapham, D.E. TRP channels as cellular sensors. *Nature* **2003**, *426*, 517–524. [CrossRef] [PubMed]
20. Moore, C.; Gupta, R.; Jordt, S.E.; Chen, Y.; Liedtke, W.B. Regulation of pain and itch by TRP channels. *Neurosci. Bull.* **2018**, *34*, 120–142. [CrossRef]
21. Liao, M.; Cao, E.; Julius, D.; Cheng, Y. Structure of the TRPV1 ion channel determined by electron cryo-microscopy. *Nature* **2013**, *504*, 107–112. [CrossRef] [PubMed]
22. Cao, E.; Liao, M.; Cheng, Y.; Julius, D. TRPV1 structures in distinct conformations reveal activation mechanisms. *Nature* **2013**, *504*, 113–118. [CrossRef] [PubMed]
23. Gao, Y.; Cao, E.; Julius, D.; Cheng, Y. TRPV1 structures in nanodiscs reveal mechanisms of ligand and lipid action. *Nature* **2016**, *534*, 347–351. [CrossRef] [PubMed]
24. Chugunov, A.O.; Volynsky, P.E.; Krylov, N.A.; Nolde, D.E.; Efremov, R.G. Temperature-sensitive gating of TRPV1 channel as probed by atomistic simulations of its trans- and juxtamembrane domains. *Sci. Rep.* **2016**, *6*, 1–16. [CrossRef]
25. Hua, L.; Huang, X.; Zhou, R.; Berne, B.J. Dynamics of water confined in the interdomain region of a multidomain protein. *J. Phys. Chem. B* **2006**, *110*, 3704–3711. [CrossRef] [PubMed]
26. Kasimova, M.; Yazici, A.; Yudin, Y.; Granata, D.; Klein, M.L.; Rohacs, T.; Carnevale, V. Ion Channel Sensing: Are Fluctuations the Crux of the Matter? *J. Phys. Chem. Lett.* **2018**, *9*, 1260–1264. [CrossRef] [PubMed]
27. Sterpone, F.; Stirnemann, G.; Laage, D. Magnitude and molecular origin of water slowdown next to a protein. *J. Am. Chem. Soc.* **2012**, *134*, 4116–4119. [CrossRef] [PubMed]
28. Efremov, R.G. Dielectric-dependent strength of interlipid hydrogen bonding in biomembranes: Model case study. *J. Chem. Inf. Mod.* **2019**, *59*, 2765–2775. [CrossRef]
29. Fayer, M.D.; Levinger, N.E. Analysis of water in confined geometries and at interfaces. *Annu. Rev. Anal. Chem.* **2010**, *3*, 89–107. [CrossRef]
30. Hess, B.; Kutzner, C.; van der Spoel, D.; Lindahl, E. GROMACS 4: Algorithms for Highly Efficient, Load-Balanced, and Scalable Molecular Simulation. *J. Chem. Theory Comput.* **2008**, *4*, 435–447. [CrossRef]
31. Lindorff-Larsen, K.; Piana, S.; Palmo, K.; Maragakis, P.; Klepeis, J.L.; Dror, R.O.; Shaw, D.E. Improved side-chain torsion potentials for the Amber ff99SB protein force field. *Proteins* **2010**, *78*, 1950–1958. [CrossRef]
32. Jorgensen, W.L.; Chandrasekhar, J.; Madura, D. Comparison of simple potential functions for simulating liquid water. *J. Chem. Phys.* **1983**, *79*, 926–935. [CrossRef]

International Journal of
Molecular Sciences

MDPI

Review

Casting a Wider Net: Differentiating between Inner Nuclear Envelope and Outer Nuclear Envelope Transmembrane Proteins

Mark Tingey [1,†], Krishna C. Mudumbi [2,3,†], Eric C. Schirmer [4] and Weidong Yang [1,*]

1 Department of Biology, Temple University, Philadelphia, PA 19121, USA; Mark.Tingey@temple.edu
2 Department of Pharmacology, Yale University School of Medicine, New Haven, CT 06510, USA; Krishna.Mudumbi@yale.edu
3 Yale Cancer Biology Institute, Yale University, New Haven, CT 06516, USA
4 Wellcome Centre for Cell Biology, University of Edinburgh, Edinburgh EH8 3BF, UK; E.Schirmer@ed.ac.uk
* Correspondence: Weidong.Yang@temple.edu
† These authors contributed equally to this work.

Received: 27 September 2019; Accepted: 18 October 2019; Published: 23 October 2019

Abstract: The nuclear envelope (NE) surrounds the nucleus with a double membrane in eukaryotic cells. The double membranes are embedded with proteins that are synthesized on the endoplasmic reticulum and often destined specifically for either the outer nuclear membrane (ONM) or the inner nuclear membrane (INM). These nuclear envelope transmembrane proteins (NETs) play important roles in cellular function and participate in transcription, epigenetics, splicing, DNA replication, genome architecture, nuclear structure, nuclear stability, nuclear organization, and nuclear positioning. These vital functions are dependent upon both the correct localization and relative concentrations of NETs on the appropriate membrane of the NE. It is, therefore, important to understand the distribution and abundance of NETs on the NE. This review will evaluate the current tools and methodologies available to address this important topic.

Keywords: NETs; inner nuclear membrane; outer nuclear membrane; nuclear envelope

1. Introduction

The eukaryotic nuclear membrane consists of two separate lipid bilayers, the inner nuclear membrane (INM) and the outer nuclear membrane (ONM) that are separated by a perinuclear space of approximately 30–50 nm [1,2]. Both INM and ONM contain unique sets of nuclear envelope transmembrane proteins (NETs) that must target to their respective compartments after synthesis in the endoplasmic reticulum (ER). The ONM is contiguous with the ER and fuses with the INM at sites where nuclear pore complexes (NPCs) are inserted, often called the pore membrane [3]. NPCs are megadalton complexes built of more than 30 nucleoporin proteins that regulate directed transport of proteins and RNA in and out of the nucleus through their central channel. Along with the NE, the NPC provides a barrier against the free diffusion of large molecules into the nucleus. In the NPC, this barrier in the central channel with a narrowest waist of ~50 nm is formed by intrinsically disordered phenylalanine-glycine (FG) motifs on one thirds of nucleoporins [4,5]. In addition to the central channel, NPCs could also have ~10-nm peripheral channels between their core protein mass and the membrane [6,7]. Though these channels are not well characterized, it has been proposed that these peripheral channels facilitate the transit of INM NETs to their functional sites on the inner face of the NE.

It is critical for the cell to allow INM NETs into the nucleus while excluding ONM NETs from entering as many NETs have important functions on their designated membrane. These functions can range from cell and nuclear migration to connecting the rest of the cell to the genome to regulating genome functions. Both INM and ONM NETs provide structure to their respective membranes, for example, many INM NETs bind the lamina intermediate filament network underlying the INM while the ONM contains NETs that connect to all three major cytoplasmic filament systems [8–15]. There are also connections across the lumen of the NE between INM and ONM NETs that form the linker of nucleoskeleton and cytoskeleton (LINC) complex, which is central to mechanosignal transduction regulating the genome, cell and nuclear mechanical stability, and providing mechanical connections needed for cell and nuclear migration [16,17]. Accordingly, many INM NETs interact with DNA, chromatin proteins, chromatin-remodeling enzymes, transcription factors, transcriptional repressors, and even splicing factors [18–23]. INM proteins also contribute significantly to 3D spatial genome organization, which is a major factor in fine-tuned regulation of the genome during tissue development. Disruption of this complex interactome can result in a number of pathological conditions, often termed as nuclear envelopathies or laminopathies, most of which are highly tissue-specific, thus underscoring the importance of INM protein function in development [24–26].

In order to better understand the complex interactome of the nuclear envelope and these diseases, it is becoming increasingly apparent that differentiating between INM or ONM position is a critical question. At the simplest level, understanding the function of a NET requires also knowing whether it connects the NE to the cytoplasm or to the genome, thus knowing the INM/ONM distribution answers this question. This distribution is generally not absolutely binary because NETs are first synthesized in the ER and so will never be 100% in the INM, especially if NETs freely diffuse in both directions between the ONM and INM until they find a binding partner. Thus, a very high INM:ONM ratio likely indicates a more directed mechanism for translocation that might be receptor mediated, similar to NLS-mediated directional transport of soluble proteins through the central channels of the NPCs. This ratio information can also be important to gain insights about functioning of different pools of NETs because several NETs have multiple cellular locations and being able to isolate and distinguish if they perform different functions in these locations requires understanding also what controls their targeting. For example, a subunit of the plasma membrane Na,K-ATPase was separately shown by immunogold electron microscopy to have a non-mitochondrial pool in the inner nuclear membrane and this pool functions as a co-regulator of transcription [27]. Particularly in these cases more information about the INM:ONM ratio and translocation rates can help direct research efforts towards different types of mechanisms ranging from post-translational modifications that might create a cryptic and novel transport sequence to different splice variants with different targeting sequences. This latter is particularly relevant considering that NETs such as Lap2 have at least half a dozen splice variants that have never been carefully compared for their INM:ONM ratios or translocation and a recent study indicated that NETs in general have more splice variants and particularly more tissue-specific splice variants than non-NE proteins [28]. It is noteworthy that the proteins with tissue-specific splice variants includes both transmembrane and non-transmembrane nucleoporins that make up the core structure of the NPC.

Due to the close proximity of the ONM and INM, electron microscopy (EM) remains the only unequivocal method of determining a NET's location. However, this only provides a snapshot of potential NET locations and, by extension, their potential binding partners and functions. This exposes a critical need for assays capable of distinguishing between INM and ONM proteins accurately in vivo. Therefore, to truly better understand the involvement of NETs in cellular functions, it is critical to develop new methods that are capable of distinguishing the localization of NETs between INM and ONM with spatial and temporal accuracy. In recent years, many existing technologies have been repurposed to interrogate this question. However, many of these technologies suffer from their own shortcomings. This is true for imaging techniques, biochemical methods, and bioinformatic approaches. Within this review, we evaluate several cutting-edge technologies within the context of interrogating the location of NETs on the nuclear envelope.

2. Determining NET Location

2.1. Determining the Spatial Location of NETs on the NE

Determining the membrane distribution of NETs remains a difficult question to answer due to the close proximity of the INM and ONM as well as the transient nature of many NETs. An important caveat to all the research on this question is that most work relies upon exogenously expressed tagged proteins and one of the core hypotheses to how NETs get to the INM is that they diffuse laterally in the membrane and remain in the INM by binding lamins or chromatin. If correct, this "lateral-diffusion retention" hypothesis would limit binding sites in the INM to the levels of the binding partner; thus, exogenous overexpression of a NET that had limiting amounts of the INM binding partner would result in large pools of the NET moving freely in the INM, ONM, and ER. At the same time, many NETs have a plethora of splice variants. Lap2β has at least 6 different splice variants [29] and many tissue-specific variants of NETs that have not yet been cloned or characterized are indicated [28]. Correspondingly, staining with antibodies cannot distinguish splice variants. Due to the complexity and limitations of this problem, a variety of technologies have been repurposed or developed to evaluate a NET's localization.

2.1.1. Electron Microscopy

Often considered the gold standard for determining the localization of NETs, immunogold-label electron microscopy makes use of small particles of gold bound to an antibody to generate extremely high-resolution images. Where the gold nanoparticle is present, a dark spot will be present on the image. While small, ~1 nm, particles of gold are possible to generate and label with, it becomes difficult to differentiate between organic material and the gold label [30]. Furthermore, due to the primary and secondary antibody labeling method, the gold particle will typically be 15–30 nm away from the molecule of interest, which can cause issues with distinguishing between ONM and INM localizations [30]. The potentially large distance between the gold particle and the antigen when using primary and secondary antibodies can be circumvented through the use of gold-labeled nanobodies [31,32]. Finally, due to the high density of the gold particle it will still be the strongest signal even when not in the same plane as the stained membrane; thus sectioning may cut the NE at an angle that makes it appear that a gold particle on one side of the membrane is actually on the other. Hence, a reasonable percentage of particles might appear to be in the lumen despite that the region of the protein being labeled is 100% cytoplasmic or nucleoplasmic and luminal-appearing particles cannot be determined for INM or ONM localization.

With this difficulty in mind, the diameter of the gold nanoparticle is of critical importance. Until the development of scanning electron microscopy (SEM) instruments equipped with field emission guns, the limited resolution of SEM required the use of gold nanoparticles larger than 15 nm. However, modern SEM allows for resolution similar to that of transmission electron microscopy (TEM), approximately 0.5 nm [33].

The extreme localization precisions achievable through immunogold-label electron microscopy has been used to provide direct evidence of the localization of a small subset of NETs [34–36]. However, this approach is ponderous, expensive, and only viable in fixed cells; thereby limiting the efficacy and feasibility of this approach. In addition, the aforementioned requirement that the samples be fixed, limits the viability of this technique as it pertains to translocation of NETs.

2.1.2. Differential Membrane Permeabilization

Historically it was thought that, as most INM proteins associate with the nuclear lamina, the resistance of a protein to a pre-fixation detergent extraction indicated INM localization. It has since been shown that several ONM proteins interact with cytoskeletal filaments and so this is no longer a clear determination of INM localization; however, a variant involving differential detergent extraction can still provide information on INM/ONM localization. This method takes advantage of the fact that digitonin preferentially pokes holes in membranes containing cholesterol. While the cell membrane contains large amounts of cholesterol, the ER and its contiguous ONM appear to contain very little cholesterol [37]. Thus, one can fix cells and then preferentially permeabilize the plasma membrane with digitonin and stain with antibodies [38]. Proteins on the ONM will stain strongly, but the nuclear membrane will prevent the antibodies accessing the INM thereby preventing staining [39,40]. Some proteins will have partial pools in both the INM and ONM, therefore a separate staining of cells permeabilized with Triton X-100 will show the full staining with the antibodies. If this stain shows a difference in the relative staining patterns, it can be inferred that there are pools in both membranes. Similarly, if Green Fluorescence Protein (GFP)-tagged proteins are exogenously expressed, the relative GFP signal intensity in different membranes compared to the antibody staining can be compared to determine at least a population in the INM [41,42]. This approach can also be used to determine membrane topology if mapped antibodies are used. This is because the luminal domains in both the ER and nuclear envelope will be protected in the digitonin-permeabilized cells. These approaches are still used, but trap and super resolution approaches are more in favor due to the difficulties involved in titrating the proper amount of digitonin to use, as it is possible with high levels, or lengths of digitonin treatment, to also poke holes in the nuclear membrane.

2.1.3. Rapamycin Trapping

One rather clever approach to surmounting diffraction limitations is the rapamycin trap. This method is based upon the forced protein dimerization technique first reported by Chen et al. [43] which exploits the binding affinity of rapamycin. Rapamycin binds to both the 12 kDa FK506 binding protein (FKBP12) and the 100 amino acid domain of the mammalian target of rapamycin (mTOR) protein, also known as FKBP-rapamycin binding domain (FRB). This system has been used to tackle the challenge of NET localization by engineering two chimeric proteins: (1) NETs of interest expressing an FRB domain and a fluorescent tag. (2) A "trap" protein limited to the nucleus consisting of a complimentary nuclear localization signal (NLS) bearing glutathione s-transferase (GST) sequence tagged with an FKBP12 sequence and a different fluorophore from the aforementioned NET chimera (Figure 1a). In the absence of rapamycin, fluorescence microscopy will show the fluorescently-tagged NET in a ring around the periphery of the nucleus whether it is in the INM or the ONM while the soluble FKBP12-tagged GST-NLS will diffuse equally throughout the nucleus (Figure 1b). Upon rapamycin treatment, the membrane bound FRB-tagged NET will associate with rapamycin, which in turn will recruit the soluble FKBP12-tagged protein only if the NET is in the INM. Such an association will result in a distinct redistribution of the fluorophore-tagged GST-NLS protein to a ring-like staining at the nuclear periphery that can be visualized by fluorescence microscopy (Figure 1c). In contrast, if the NET is restricted to the ONM the nuclear-restricted FKBP12-tagged protein will not move to the nuclear periphery, but remain diffuse in the nucleoplasm [44,45]. This method provides a very clear condition, which when met, provides very strong support for the protein being present on the INM. However, this methodology only evaluates the presence of a protein on the INM and does not distinguish whether a separate pool can reside simultaneously in the ONM. Furthermore, the addition of two tags, the fluorophore and the FKBP12 sequence, introduces a greater possibility of error in this system, allowing for the possibility that in vivo wild-type interactions may be significantly different from what is observed. Nonetheless, there are many strengths to this system, including that it can be adapted in many ways to address related questions. For example, if the nuclear trap protein has a lamin-binding site that keeps it stably at the INM, then a NET that freely diffuses with different subcellular pools can be followed for its dynamics until it gets trapped at the INM after rapamycin is added.

This approach has been used to test requirements for targeting a heterologous reporter to the INM in a study supporting the lateral-diffusion retention hypothesis that found an ATP requirement for translocation of the reporter to the INM [46]. However, it was also used in a study supporting the lateral-diffusion retention hypothesis that argued against an ATP requirement for translocation [47].

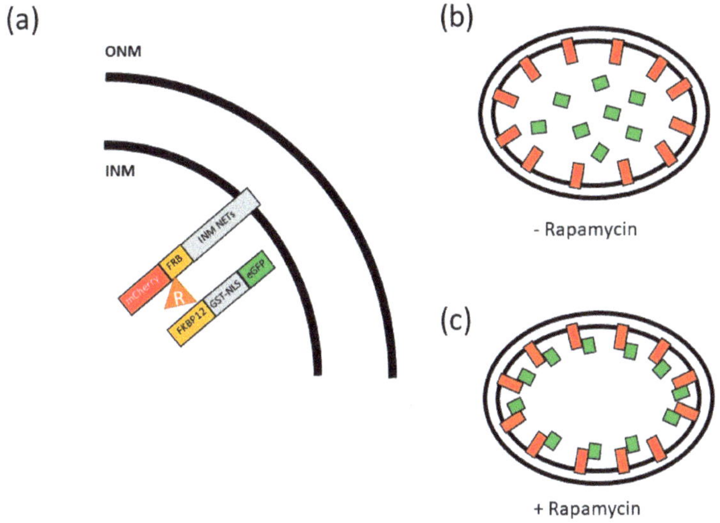

Figure 1. A rapamycin trap for evaluating the presence of nuclear envelope transmembrane proteins (NETs) on the INM: (**a**) A membrane bound NETs tagged with a fluorophore and FRB will dimerize with another fluorescently tagged protein with an 12 kDa FK506 binding protein (FKBP12) domain in the presence of rapamycin, represented here as a triangle. (**b**) In the absence of rapamycin, the soluble Green Fluorescent Protein (GFP) labeled protein (green blocks) will not dimerize with the membrane bound NETs (red blocks) and diffuse throughout the nucleus. (**c**) In the presence of rapamycin the soluble GFP labeled protein (green blocks) will dimerize with the membrane bound NETs (red blocks) and localize to the nuclear envelope.

2.1.4. Split GFP

Another proximity-based interaction system utilizes Superfolder GFP, which is capable of being split asymmetrically into two parts, GFP_{11} (3 kD) and GFP_{1-10} (24 kD). Individually, these two constructs do not fluoresce. However, they can be reconstituted into a fluorescing GFP (27 kD) when the two pieces are expressed within the same cellular compartment and associate (Figure 2a) [48,49]. This assay was adapted to identify the localization of NETs on the NE of *Saccharomyces cerevisiae* by Smoyer and colleagues [50]. To accomplish this, they created reporter proteins by fusing a yeast-codon optimized GFP_{11} and mCherry to a nuclear reporter (GFP_{11}-mCherry-Pus1), ONM/ER surface reporter (GFP_{11}-mCherry-Scs2TM), ER lumen reporter (mCherry-Scs2TM-GFP_{11}), and a cytoplasmic reporter (GFP_{11}-mCherry-Hxk1) (Figure 2e). NETs were then selected and fused with the complimentary GFP_{1-10} fragment. Each NET-GFP_{1-10} construct was then expressed with each of the reporters individually and observed in the green fluorescence channel. Since the location of the reporters is well established, if the nuclear protein fused to GFP_{1-10}, was present in the same compartment as the reporter protein, fluorescence in the green channel would be detected, and thus, the localization of the NET in question could be determined (Figure 2c,d). To confirm the accuracy of the system, control proteins were generated which localized to specific regions, thereby confirming the location of the experimental constructs (Figure 2e). Split GFP is a powerful and elegant system for qualitatively determining if a NET localizes to the INM. However, the system does not account for NETs with a dual role that may be present on both the INM and ONM as it does not allow for derivation of information about ONM quantity and proportion to the INM. Furthermore, similar to the rapamycin trap, split GFP carries with it the potential that adding several tags may alter the behavior of NETs, which may lead investigators to reach erroneous conclusions. Thus, despite being an elegant and straightforward approach to verifying the presence of NETs on the INM, the split GFP system is limited in its capacity to provide further information about the translocation and proportion of INM NETs.

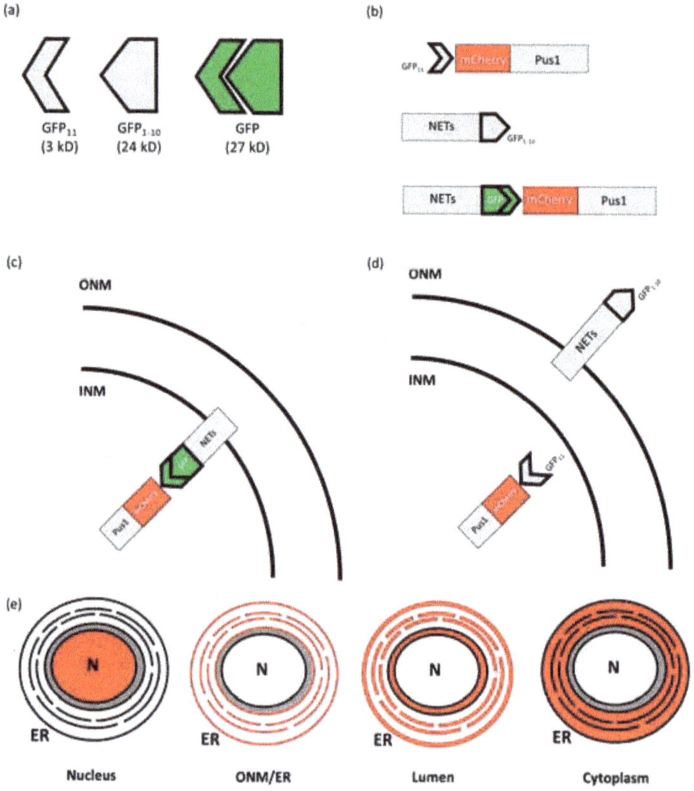

Figure 2. A conceptual representation of the split GFP system as it is used to identify the position of INM proteins: (**a**) Superfolder GFP can be split into two non-fluorescent components, which can also recombine into a fluorescently functional GFP. (**b**) A representation of the soluble nuclear yeast protein Pus1 tagged with mCherry and GFP_{11}, a NET of interest tagged with the complimentary GFP_{1-10}, the interaction between GFP_{1-10}-tagged NET and the GFP_{11} reporter resulting in green fluorescence. (**c**,**d**) Representations of how the localization of NETs tagged with GFP_{1-10} and reporter proteins tagged with GFP_{11} produce green fluorescence (**c**) or fail to do so (**d**). (**e**) A representation of control proteins tagged with mCherry and GFP_{11} in the nucleus, the outer nuclear membrane (ONM) and endoplasmic reticulum (ER), the lumen, and the cytoplasm of the cell in the absence of GFP_{1-10} fused NETs of interest.

2.1.5. MIET

Recently, Metal-Induced Energy Transfer (MIET), a technique that relies upon the principals of non-radiative electromagnetic energy transfer, was used to probe the distribution of NETs on the ONM and INM [51]. This method is similar to the more commonly used technique of Förster resonance energy transfer and fluorescence lifetime imaging microscopy (FRET/FLIM). Here, donor fluorophores close to a metal surface interact with surface plasmons and transfer their energy to the metal thereby reducing their fluorescence lifetime (τ) in a direct relationship with their distance from the metal surface. This technique can work over the range of about 150 nm, and can therefore help determine the location of proteins on the part of the NE that is close to the bottom of a cell and near the metal surface [51,52].

This technique has recently been utilized to generate a topography of the nuclear envelope. This was done by tagging landmark proteins Lap2β on the INM, and NUP358 for the ONM and then using MIET to localize the landmark proteins. MIET boasts an impressive axial localization of 2.5 nm, which allows for a very accurate differentiation between the INM and ONM, as well as a very accurate measurement of the basal region of the perinuclear space [51]. While no specific experiments have been published using MIET to differentiate membrane location of NETs, we propose that this technology could potentially be utilized to accurately determine the membrane location of NETs. MIET is a very interesting concept with many potential applications, however it is not without limitations. The temporal resolution of this technique is currently too low to provide real-time mobility measurements of NETs. Furthermore, the lateral resolution of this technique is still diffraction limited and therefore unable to determine the distribution of NETs on the ONM and INM. However, this may be overcome by employing Single-Molecule Localization Methods (SMLM) to improve the lateral precision.

2.1.6. Ensemble FRAP

Since the initial breakthrough experiments in the 1970s, fluorescence recovery after photobleaching (FRAP) has become an essential tool to determine the mobility and diffusion of transmembrane proteins embedded in lipid bilayers [53–55]. Since then, FRAP has been used on many membranous structures, including the plasma membrane [56–58], ER [59], and NE [34]. FRAP allows one to distinguish the diffusivity of molecules on a membrane as well as the fraction of immobile molecules that are unable to diffuse due to interactions with other macromolecules. For most NETs tested, however, the immobile fraction was sufficiently high that FRAP was principally measuring the translocation of protein accumulated in the ER into the INM instead of mobility of protein within the NE. While this is straightforward on single-membrane structures such as the plasma membrane, it becomes more convoluted on the NE due to the ~40 nm distance of the ONM and INM. FRAP relies on diffraction limited imaging which does not have the resolution to distinguish between the two membranes. Therefore, any information about NET diffusion coefficients or immobile fractions inherently is an average of behavior of the specific protein in question on both of the membranes.

2.1.7. Airyscan Confocal Microscopy and Differential Labeling

The expertise required for super-resolution or electron microscopy may be untenable for many research labs, therefore, Airyscan confocal microscopy is an attractive alternative to these other techniques due to its accessibility and comparative affordability. The Airyscan confocal microscope's principle of operation makes use of multiple extremely sensitive GaAsP detectors for a single illumination point. The detector consists of 32 detector elements arranged in a compound eye fashion. As the image is scanned, each detector records a portion of the whole. The resultant images are then concatenated and a point spread function (PSF) is generated. This PSF allows for a sub-diffraction limit image to be generated with a lateral resolution of approximately 140 nm and an axial resolution of approximately 350 nm [60–62]. This resolution is not sufficient to differentiate between INM and ONM. However, the addition of differential labeling overcomes this weakness by enabling investigators to create a landmark on the INM or ONM.

Previous publications have identified landmark proteins which localize to the ONM or the INM respectively. Labeling these known protein markers with a fluorescent protein (i.e., mCherry) and NETs of interest with a different color fluorophore (i.e., eGFP) allows one to study colocalization. This approach was used to great effect by Groves et al. to better understand how NETs are targeted to the INM in plants. In this study, ER tail-anchored proteins were tagged with an NLS and GFP and compared to the localization of calnexin-mCherry, a well characterized protein located exclusively on the ER and ONM. Since the Airyscan microscopy method does not have enough resolution to visually distinguish between the ONM and INM, line scans of the NE were used to determine co-localization of the two proteins (Figure 3a,b). However, even line scans are limited by the overall resolution of the system, therefore several statistical analyses had to performed on the line scan results before providing satisfying conclusions [63].

While this approach is relatively simple and easy to perform, it unfortunately has several pitfalls that can drastically affect the results. First, overexpression of NETs can often result in their mislocalization on the NE due to the leaky nature in which proteins are regulated by the NPC [64]. This is especially detrimental if the ONM marker, calnexin, is found on the INM as the line scan will show false colocalizations. Second, there is an inherent amount of uncertainty associated with fitting a line scan to determine the peaks. The resultant error makes using these values to colocalize proteins precarious. Line scans are a one-dimensional analysis method and do not provide information regarding diffusion or relative enrichment on the ONM or INM.

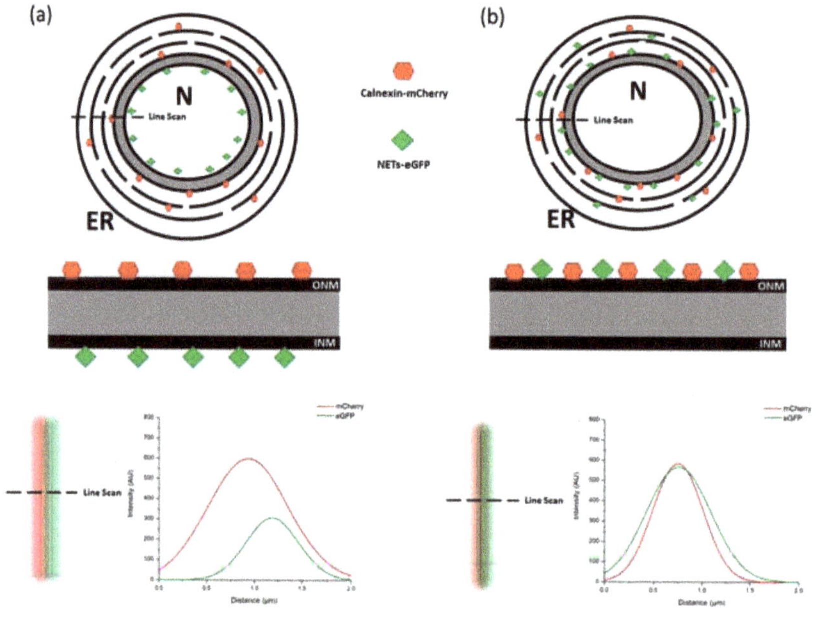

Figure 3. A model of differential staining using Airyscan confocal microscopy. (**a**) Calnexin tagged with mCherry (red) localizes exclusively to the ONM and ER. While the NETs of interest tagged with GFP (green) is enriched at the INM. A line scan is performed and a line profile is generated indicating that the two fluorophores do not co-localize. (**b**) A line scan of NETs tagged with GFP (green) that are enriched at the ONM and ER do co-localize with calnexin.

2.1.8. Super-Resolution Microscopy

As was discussed previously, the diffraction limit of light microscopy is approximately 250 nm. This limit can be overcome by using super-resolution (SR) microscopy techniques, which are theoretically capable of providing an image resolution between 100 to 20 nm. So far, three classes of SR techniques have been applied to imaging NETs: Structured illumination microscopy (SIM), stimulated emission depletion (STED) microscopy, and single-molecule localization microscopy (SMLM). While all three of these techniques can break the diffraction barrier of light, they vary drastically in approach. SIM relies on a diffraction pattern or grating placed in front of the excitation laser beam. These patterns are then moved and rotated several times to produce a Moiré effect allowing one to discern high frequency signals relating to fine cellular structures, reaching resolutions of ~100 nm laterally and ~250 nm axially. This technique was originally used to image the localization of the nuclear lamina and the INM protein Lap2β in relation to either an NPC protein of the nuclear basket that protrudes into the nucleoplasm by

~50 nm or an NPC protein of the cytoplasmic filaments that similarly protrude into the cytoplasm [65]. Thus, the INM or ONM localization was determined based on whether a protein was closer to one or the other NPC protein and the NPC proteins were in fact separated by well over 100 nm distance. This approach was subsequently expanded to study the localization and interactions between INM proteins, ONM proteins, and the cytoskeleton [66], as well as to systematically analyze the localization of 21 novel NETs identified by proteomics [41,67,68]. While this technique can be performed on live cells, the need to take thousands of images makes the image acquisition rate far too slow to detect the fast dynamics of transmembrane protein diffusion on the NE. It is important to remember when applying this technique that the NPC proteins used as landmarks are penetrating into the nucleoplasm or cytoplasm by ~50 nm. Thus, if the protein being interrogated has its tag or epitope being recognized by an antibody in the lumen or near the membrane on the nucleoplasmic side as opposed to similarly penetrating into the nucleoplasm, a clear distinction on its localization may not be possible.

Improving significantly upon the resolution of SIM is STED microscopy, which uses a depletion laser that is overlapped with an excitation laser. This depletion laser depletes the excited state of fluorophores on the outer edges of the excitation laser targets so that only the inner most region of the excitation laser will excite fluorophores to emit photons [69]. In other words, this method reduces the effective PSF of the excitation laser below the diffraction limit of light, giving lateral resolution of ~50 nm and an axial resolution of ~150 nm in 3D STED applications [70]. Using this technique in combination with FRAP, Giannios and colleagues examined the localization and mobility of lamin B receptor (LBR) and concluded that the mobility of LBR is greatly affected by the interfaces between ER tubules and the ONM as well as discrete LBR microdomains. Mobility was primarily determined by using FRAP/FLIP techniques, whereas localization was determined by fixing the cells and performing STED, which, naturally, does not preserve the dynamics of a live cell system.

Finally, SMLM takes advantage of the photophysical properties of fluorescent probes and relies on the blinking or 'on/off' switching of the probes. The two most commonly SMLM methods are stochastic optical reconstruction microscopy (STORM) and photoactivated localization microscopy (PALM), both which rely on the blinking of fluorophores, but differ in that STORM relies on the blinking of organic dyes, whereas PALM uses the photoactivation or photoswitching of genetically modified fluorescent proteins. These blinking events spatially and temporally separate fluorophores, allowing each active emitter to be distinctly observed and localized with mathematical functions to find the centroid. Thousands of subsets of active emitters are imaged and then reconstructed to recreate the original image. This powerful technique has a very high localization precision (~20 nm laterally and ~50 axially), however, it is time consuming and not ideal for capturing the dynamic movement of NETs on the NE. As such this methodology has mostly been used to localize different lamin subtypes within the nuclear lamina in fixed cells [71,72]. It is noteworthy, however, that application of multiple of the above-listed SR approaches to this problem of lamin subtype localization yielded somewhat differing results.

Recently, however, a technique was developed harnessing both SMLM and FRAP—named single-molecule fluorescence recovery after photobleaching (smFRAP)—to try and capture the dynamic movement and distribution of NETs on both the ONM and INM [73]. By photobleaching a small spot on the NE, the local concentration of fluorescently functional EGFP tagged NETs was brought to near zero, and the recovery of new, fluorescently functional, molecules on the ONM or INM was recorded using SMLM. Photobleaching and observing only a small spot at the equator of the NE (~0.5 μm), the ONM and INM could be treated as two parallel membranes with no overlap due to curvature. Then, by exploiting the high lateral localization precision of SMLM, the recorded events on either of the two membranes could be easily separated to determine the distribution of NETs on the respective membranes. Furthermore, observing such a small location allows for the use of a fast frame rate, preserving the natural dynamics of a live system while having the fast acquisition speed to provide information about translocation rates for NETs. However, a drawback of this system is that only a small region is imaged and used to represent the entire NE, which is not an isotropic structure. To overcome this, results measured from multiple NEs have been averaged to represent the final outcomes [66].

2.2. Determining the Translocation Rate of NETs

The membrane distribution of NETs can sometimes provide insight into the deeper question of how NETs import or export through the NPC and their translocation rate. While several of the aforementioned technologies are capable of providing a qualitative relative ratio of proteins on the ONM to proteins on the INM and many early studies did very elegant work to gain insights into the routes and mechanisms of translocation, in vivo direct measurement of translocation rate of NETs still remains both desirable and challengeable. However, some, when applied creatively and with mathematical modeling, have given insights into this question.

2.2.1. Mathematical Modeling of NETs Translocation Rate

One of the first studies to try to get at this question creatively used the rapamycin trap described above. In this case the fusion construct carrying the FKBP was fused to the lamin binding domain of Lap2β and, as this would stay at the nuclear periphery, it was used as a trap to capture freely diffusing FRB fused to pyruvate kinase, the Lap2β transmembrane domain, and GFP. Thus, when rapamycin was added the GFP signal would begin to accumulate at the NE and the rate of this accumulation could be inferred to reflect the translocation rate [46]. However, this system still would have a background from the pool of the FRB construct already in the INM, but not yet trapped, and that in the ONM. This was improved upon slightly in a subsequent study that used a combination of data from FRAP, photoactivatable GFP, and immunogold EM to model translocation and estimate translocation rates for a variety of NETs with different characteristics in terms of length of the nucleoplasmic region, number of transmembrane spans and the presence or absence of an NLS [36]. One of the particularly interesting findings in this paper was that it indicated that earlier FRAP studies that purported to estimate mobility in the NE were most likely mostly measuring translocation. They found that photoactivation of GFP-NETs in the ER resulted in GFP signal accumulating in the NE with kinetics similar to FRAP studies while photoactivation of these GFP-NETs in the NE resulted in much slower mobility, thus indicating that most of the protein observed at the NE for these particular NETs was immobile and therefore likely tethered in the INM. Further using data from all three experimental approaches they were able to model the translocation and infer the percent immobile fraction and translocation rates of NETs NET55, Lap2β, and Lap1-L, respectively at 50–70 s, 60–80 s, and 70–140 s.

Another study attempted to determine a translocation coefficient for just the lamin B receptor (LBR) using a slightly different approach. LBR is a well characterized INM membrane bound protein, with domains binding the B-type lamins, as well as Histones H3 and H4 and heterochromatin protein 1 (HP1). Thus it has many nucleoplasmic tethering sites consistent with the lateral-diffusion retention model of translocation and, in fact, it was the first NET studied by Soullam and Worman when presenting this hypothesis [74,75]. In addition, LBR contains 3 NLS sequences, consistent with the receptor-mediated model of translocation [76]. LBR is also a good substrate to test because, having 8 transmembrane spans, it is unlikely to be sustainable outside the membrane while it is possible that C-terminal anchored single-span transmembrane proteins never get inserted into the membrane. This study tried to determine the kinetics of LBR protein translocation using inducible expression of the LBR reporter fused to another protein making it too long to translocate through the peripheral NPC channels [77]. This reporter also had a protease site between the LBR and fusion so that before the addition of the protease, the LBR construct would be constrained to the ER/ONM. However, upon the addition of a protease, the fusion protein would be cleaved from the LBR, thus allowing it to freely localize to the INM, which is then observed over time to determine translocation kinetics and requirements. Using the data accumulated from the inducible cleavage reporter and FRAP, they were able to estimate the binding time of LBR to its nuclear binding partners as 0 to 4 min. Furthermore, they used the surface area of the NE and ER, the degradation of proteins on the ER/ONM and INM, as well as the kinetics of the inducible cleavage to determine the kinetics of LBR import. Finally, it was estimated that diffusion from the ER to the NE occurs in 5 to 15 s, and it was assumed that LBR was at a concentration of 1 μM (2.6×10^6 molecule). Using these parameters, Boni and colleagues calculated the translocation rate of LBR to be 4.6 molecules per minute per NPC.

2.2.2. Experimental determination of NETs Translocation Rate

A recent study by Mudumbi and colleagues used a new approach, combining smFRAP and ensemble FRAP, to directly determine the translocation rate of LBR. Using smFRAP, they were able to clearly distinguish the diffusion coefficient of LBR on the ONM and INM, along with the fraction of LBR distribution on the ONM and INM. In addition, the group used ensemble FRAP measurements of LBR to determine its mobile and immobile fractions. Finally, using previously published data about the number of LBR molecules and NPCs typically expressed in HeLa cells [78,79], they calculated the translocation rate of LBR to be approximately 5.4 molecules per minute per NPC. This method of calculating translocation rate is possible for all NETs so long as a relatively accurate quantification of total NET molecules is available. Here, the authors were able to take advantage of the high resolution of SMLM to directly determine distribution of LBR on both membranes as well as its diffusion coefficient on both membranes to calculate translocation rate, without relying on numerous theoretical assumptions.

3. Conclusions and Future Directions

Several microscopy techniques are able to provide the nanometer level spatial resolution required to distinguish protein localization on the ONM and INM, however they suffer from a small field of view or lack of appropriate temporal resolution. It is clear from a survey of recent literature that advances are quickly being made in the fields of microscopy and dye development that are pushing the boundaries of both temporal and spatial resolution in live cells. The application of these new technologies, together with approaches described above for measuring translocation rates from single molecules, will play an important role in addressing many outstanding questions in the area of nuclear envelope biology. For example, do different transmembrane NET 'cargos' have different transport kinetics? Furthermore, how might this reflect on their using distinct transport mechanisms? The application of these advances together with FRET, as shown above, and FFS (fluorescence fluctuation spectroscopy), and FCCS (fluorescence cross-correlation spectroscopy) approaches will enable further determination of which, if any, nucleoporins interact with NET cargos and if transportins might facilitate the transport process though the peripheral NPC channels. Moreover, these approaches might be also used to determine if—a totally new conceptual hypothesis—NETs could possibly themselves act as transport receptors for soluble cargos translocating through the peripheral channels of the NPCs. This could be an important backup mechanism, for example, when central channel transport is blocked by pathogens and would be consistent with the notion that the peripheral channels were actually the original mode of transport when NPCs presumably evolved from COP proteins of the ER [80] and before they acquired FGs and a directed transport mechanism. It is noteworthy that NLS deletion did not block lamin nuclear import [81] and there are actually many nuclear proteins for which an NLS has not been identified that make the possibility of multiple transport mechanisms through the peripheral channels the more likely.

One critical area is lacking on the technology side: many high-resolution biology techniques involve high laser power which causes the photobleaching of fluorophores and increases phototoxicity in the cell. This both may affect in vivo cellular dynamics and is incompatible with the need for single molecule tracking in some of the approaches discussed above. As such, there is high demand for further development of high-resolution techniques that are also gentle on the cell. This is particularly important for studying transport due to the need to simultaneously track the NET and its cargo while also following other molecules to obtain positional reference information against the NPC structure. It is also important to use these developing approaches together with CRISPR technologies to maintain endogenous expression levels and smaller tags to limit mistargeting: This is particularly important in working with the nuclear envelope because light has to pass through the whole cell to get to the nucleus and thus large amounts of out-of-focus light from inappropriate protein pools will yield poor signal-to-noise ratios. Harnessing these new methods in conjunction with the creative biological techniques presented here will help shed light in this area of nuclear biology that is poorly understood, but functionally extremely important as indicated by the many nuclear envelope-linked diseases [82,83].

One critical area is lacking on the biology side: this work will have to go hand-in-hand with developments in finding and annotating new splice variants of NETs and using these technologies that distinguish outer and inner nuclear membranes to directly compare each of these variants. One of the first NETs to be identified was Lap2, which has between 6 and 7 annotated splice variants [29,84]; however, for Lap1 which was identified at the same time, there were three bands identified by monoclonal antibodies on Western blot. Despite being discovered nearly 30 years ago, cDNAs encoding all three splice variants have yet to be annotated. A recent analysis of RNA-Seq data showed that genes encoding nuclear envelope proteins are enriched for splice variants compared to the rest of the genome as well as identifying many new tissue-specific splice variants [28]. Accordingly, it is not surprising that other studies are popping up occasionally showing important functions for novel splice variants of NETs, such as three SUN1 splice variants that each differently contribute to directional cell migration [85]. In light of the tissue-specificity of pathology in many identified nuclear envelope-linked diseases and that widely expressed proteins have mostly been implicated in these disorders [82,83], it is likely that tissue-specific splice variants that may use distinct transport pathways or exhibit distinct translocation kinetics could play a role in the disease pathomechanisms.

Funding: The project was supported by grants from the National Institutes of Health (NIH GM097037, GM116204 and RGM122552 to W.Y.). E.C.S. is supported by Medical Research Council grant MR/R018073 and Muscular Dystrophy UK grant 18GRO-PG24-0248. K.C.M. is supported by the National Institutes of Health (NIH R01-CA198164 awarded to Mark A. Lemmon at Yale University).

Conflicts of Interest: The authors declare no conflict of interest. The funders had no role in the design of the study; in the collection, analyses, or interpretation of data; in the writing of the manuscript, or in the decision to publish the results.

Abbreviations

NPC	Nuclear pore complex
NETs	Nuclear envelope transmembrane proteins
INM	Inner nuclear membrane
ONM	Outer nuclear membrane
ER	Endoplasmic Reticulum
FKBP12	12 kDa FK506 binding protein
FRB	FKBP12 rapamycin binding domain
mTOR	Mammalian target of rapamycin
NLS	Nuclear localization signal
GST	Glutathione s-transferase
SR	Super-Resolution
SIM	Structured Illumination Microscopy
FRAP	Fluorescence Recovery After Photobleaching
STORM	Stochastic-Optical-Reconstruction-Microscopy
PALM	Photo-Activated-Localization-Microscopy
STED	Stimulated emission depletion
SMLM	Single-Molecule Localization Microscopy
smFRAP	Single-Molecule Fluorescence Recovery After Photobleaching
TEM	Transmission electron microscopy
SEM	Scanning electron microscopy
MIET	Metal-Induced Energy Transfer
FRET	Förster Resonance Energy Transfer

References

1. Franke, W.W.; Scheer, U.; Krohne, G.; Jarasch, E.-D. The nuclear envelope and the architecture of the nuclear periphery. *J. Cell Biol.* **1981**, *91*, 39s–50s. [CrossRef] [PubMed]
2. Feldherr, C.M.; Akin, D. The permeability of the nuclear envelope in dividing and nondividing cell cultures. *J. Cell Biol.* **1990**, *111*, 1–8. [CrossRef] [PubMed]
3. Gerace, L.; Burke, B. Functional organization of the nuclear envelope. *Annu. Rev. Cell Biol.* **1988**, *4*, 335–374. [CrossRef] [PubMed]
4. Frey, S.; Richter, R.P.; Görlich, D. FG-rich repeats of nuclear pore proteins form a three-dimensional meshwork with hydrogel-like properties. *Science* **2006**, *314*, 815–817. [CrossRef] [PubMed]
5. Lim, R.Y.; Fahrenkrog, B.; Köser, J.; Schwarz-Herion, K.; Deng, J.; Aebi, U. Nanomechanical basis of selective gating by the nuclear pore complex. *Science* **2007**, *318*, 640–643. [CrossRef]
6. Hinshaw, J.E.; Carragher, B.O.; Milligan, R.A. Architecture and design of the nuclear pore complex. *Cell* **1992**, *69*, 1133–1141. [CrossRef]
7. Reichelt, R.; Holzenburg, A.; Buhle, E.; Jarnik, M.; Engel, A.; Aebi, U. Correlation between structure and mass distribution of the nuclear pore complex and of distinct pore complex components. *J. Cell Biol.* **1990**, *110*, 883–894. [CrossRef]
8. Arib, G.; Akhtar, A. Multiple facets of nuclear periphery in gene expression control. *Curr. Opin. Cell Biol.* **2011**, *23*, 346–353. [CrossRef]
9. Burns, L.T.; Wente, S.R. Trafficking to uncharted territory of the nuclear envelope. *Curr. Opin. Cell Biol.* **2012**, *24*, 341–349. [CrossRef]
10. de Las Heras, J.I.; Meinke, P.; Batrakou, D.G.; Srsen, V.; Zuleger, N.; Kerr, A.R.; Schirmer, E.C. Tissue specificity in the nuclear envelope supports its functional complexity. *Nucleus* **2013**, *4*, 460–477. [CrossRef]
11. Gruenbaum, Y.; Margalit, A.; Goldman, R.D.; Shumaker, D.K.; Wilson, K.L. The nuclear lamina comes of age. *Nat. Rev. Mol. Cell Biol.* **2005**, *6*, 21–31. [CrossRef] [PubMed]
12. Heessen, S.; Fornerod, M. The inner nuclear envelope as a transcription factor resting place. *EMBO Rep.* **2007**, *8*, 914–919. [CrossRef] [PubMed]
13. Hetzer, M.W.; Wente, S.R. Border Control at the Nucleus: Biogenesis and Organization of the Nuclear Membrane and Pore Complexes. *Dev. Cell* **2009**, *17*, 606–616. [CrossRef] [PubMed]
14. Wilson, K.L.; Foisner, R. Lamin-binding Proteins. *Cold Spring Harb. Perspect. Biol.* **2010**, *2*, a000554. [CrossRef] [PubMed]
15. Zuleger, N.; Korfali, N.; Schirmer, E.C. Inner nuclear membrane protein transport is mediated by multiple mechanisms. *Biochem. Soc. Trans.* **2008**, *36*, 1373–1377. [CrossRef] [PubMed]
16. Crisp, M.; Liu, Q.; Roux, K.; Rattner, J.; Shanahan, C.; Burke, B.; Stahl, P.D.; Hodzic, D. Coupling of the nucleus and cytoplasm: Role of the LINC complex. *J. Cell Biol.* **2006**, *172*, 41–53. [CrossRef]
17. Östlund, C.; Folker, E.S.; Choi, J.C.; Gomes, E.R.; Gundersen, G.G.; Worman, H.J. Dynamics and molecular interactions of linker of nucleoskeleton and cytoskeleton (LINC) complex proteins. *J. Cell Sci.* **2009**, *122*, 4099–4108. [CrossRef]
18. Burke, B.; Stewart, C.L. Functional architecture of the cell's nucleus in development, aging, and disease. *Curr. Top. Dev. Biol.* **2014**, *109*, 1–52. [CrossRef]
19. Dauer, W.T.; Worman, H.J. The Nuclear Envelope as a Signaling Node in Development and Disease. *Dev. Cell* **2009**, *17*, 626–638. [CrossRef]
20. Davidson, P.M.; Lammerding, J. Broken nuclei—Lamins, nuclear mechanics, and disease. *Trends Cell Biol.* **2014**, *24*, 247–256. [CrossRef]
21. Mendez-Lopez, I.; Worman, H.J. Inner nuclear membrane proteins: Impact on human disease. *Chromosoma* **2012**, *121*, 153–167. [CrossRef] [PubMed]
22. Schreiber, K.H.; Kennedy, B.K. When Lamins Go Bad: Nuclear Structure and Disease. *Cell* **2013**, *152*, 1365–1375. [CrossRef] [PubMed]
23. Worman, H.J.; Dauer, W.T. The nuclear envelope: An intriguing focal point for neurogenetic disease. *Neurotherapeutics* **2014**, *11*, 764–772. [CrossRef] [PubMed]
24. Chi, Y.H.; Chen, Z.J.; Jeang, K.T. The nuclear envelopathies and human diseases. *J. Biomed. Sci.* **2009**, *16*, 96. [CrossRef]

25. Maraldi, N.M.; Lattanzi, G.; Capanni, C.; Columbaro, M.; Merlini, L.; Mattioli, E.; Sabatelli, P.; Squarzoni, S.; Manzoli, F.A. Nuclear envelope proteins and chromatin arrangement: A pathogenic mechanism for laminopathies. *Eur. J. Histochem.* **2006**, *50*, 1–8.
26. Schirmer, E.C.; Gerace, L. The nuclear membrane proteome: Extending the envelope. *Trends Biochem. Sci.* **2005**, *30*, 551–558. [CrossRef]
27. Pestov, N.B.; Ahmad, N.; Korneenko, T.V.; Zhao, H.; Radkov, R.; Schaer, D.; Roy, S.; Bibert, S.; Geering, K.; Modyanov, N.N. Evolution of Na, K-ATPase βm-subunit into a coregulator of transcription in placental mammals. *Proc. Natl. Acad. Sci. USA* **2007**, *104*, 11215–11220. [CrossRef]
28. Capitanchik, C.; Dixon, C.R.; Swanson, S.K.; Florens, L.; Kerr, A.R.; Schirmer, E.C. Analysis of RNA-Seq datasets reveals enrichment of tissue-specific splice variants for nuclear envelope proteins. *Nucleus* **2018**, *9*, 410–430. [CrossRef]
29. Berger, R.; Theodor, L.; Shoham, J.; Gokkel, E.; Brok-Simoni, F.; Avraham, K.B.; Copeland, N.G.; Jenkins, N.A.; Rechavi, G.; Simon, A.J. The characterization and localization of the mouse thymopoietin/lamina-associated polypeptide 2 gene and its alternatively spliced products. *Genome Res.* **1996**, *6*, 361–370. [CrossRef]
30. Hermann, R.; Walther, P.; Müller, M. Immunogold labeling in scanning electron microscopy. *Histochem. Cell Biol.* **1996**, *106*, 31–39. [CrossRef]
31. Pleiner, T.; Bates, M.; Trakhanov, S.; Lee, C.-T.; Schliep, J.E.; Chug, H.; Böhning, M.; Stark, H.; Urlaub, H.; Görlich, D. Nanobodies: Site-specific labeling for super-resolution imaging, rapid epitope-mapping and native protein complex isolation. *Elife* **2015**, *4*, e11349. [CrossRef] [PubMed]
32. Kijanka, M.; van Donselaar, E.G.; Müller, W.H.; Dorresteijn, B.; Popov-Čeleketić, D.; El Khattabi, M.; Verrips, C.T.; van Bergen En Henegouwen, P.M.P.; Post, J.A. A novel immuno-gold labeling protocol for nanobody-based detection of HER2 in breast cancer cells using immuno-electron microscopy. *J. Struct. Biol.* **2017**, *199*, 1–11. [CrossRef] [PubMed]
33. Goldberg, M.W.; Fiserova, J. Immunogold labelling for scanning electron microscopy. *Methods Mol. Biol.* **2010**, *657*, 297–313. [CrossRef] [PubMed]
34. Ellenberg, J.; Siggia, E.D.; Moreira, J.E.; Smith, C.L.; Presley, J.F.; Worman, H.J.; Lippincott-Schwartz, J. Nuclear membrane dynamics and reassembly in living cells: Targeting of an inner nuclear membrane protein in interphase and mitosis. *J. Cell Biol.* **1997**, *138*, 1193–1206. [CrossRef]
35. Wilhelmsen, K.; Litjens, S.H.; Kuikman, I.; Tshimbalanga, N.; Janssen, H.; van den Bout, I.; Raymond, K.; Sonnenberg, A. Nesprin-3, a novel outer nuclear membrane protein, associates with the cytoskeletal linker protein plectin. *J. Cell Biol.* **2005**, *171*, 799–810. [CrossRef]
36. Zuleger, N.; Kelly, D.A.; Richardson, A.C.; Kerr, A.R.; Goldberg, M.W.; Goryachev, A.B.; Schirmer, E.C. System analysis shows distinct mechanisms and common principles of nuclear envelope protein dynamics. *J. Cell Biol.* **2011**, *193*, 109–123. [CrossRef]
37. Ginsbach, C.; Fahimi, H. Labeling of cholesterol with filipin in cellular membranes of parenchymatous organs. *Histochemistry* **1987**, *86*, 241–248. [CrossRef]
38. Adam, S.A.; Marr, R.S.; Gerace, L. Nuclear protein import in permeabilized mammalian cells requires soluble cytoplasmic factors. *J. Cell Biol.* **1990**, *111*, 807–816. [CrossRef]
39. Worman, H.J.; Evans, C.D.; Blobel, G. The lamin B receptor of the nuclear envelope inner membrane: A polytopic protein with eight potential transmembrane domains. *J. Cell Biol.* **1990**, *111*, 1535–1542. [CrossRef]
40. Worman, H.J.; Yuan, J.; Blobel, G.; Georgatos, S.D. A lamin B receptor in the nuclear envelope. *Proc. Natl. Acad. Sci. USA* **1988**, *85*, 8531–8534. [CrossRef]
41. Malik, P.; Korfali, N.; Srsen, V.; Lazou, V.; Batrakou, D.G.; Zuleger, N.; Kavanagh, D.M.; Wilkie, G.S.; Goldberg, M.W.; Schirmer, E.C. Cell-specific and lamin-dependent targeting of novel transmembrane proteins in the nuclear envelope. *Cell Mol. Life Sci.* **2010**, *67*, 1353–1369. [CrossRef] [PubMed]
42. Schirmer, E.C.; Florens, L.; Guan, T.; Yates, J.R.; Gerace, L. Nuclear membrane proteins with potential disease links found by subtractive proteomics. *Science* **2003**, *301*, 1380–1382. [CrossRef] [PubMed]
43. Chen, J.; Zheng, X.F.; Brown, E.J.; Schreiber, S.L. Identification of an 11-kDa FKBP12-rapamycin-binding domain within the 289-kDa FKBP12-rapamycin-associated protein and characterization of a critical serine residue. *Proc. Natl. Acad. Sci. USA* **1995**, *92*, 4947–4951. [CrossRef]
44. Blenski, M.; Kehlenbach, R.H. Targeting of LRRC59 to the Endoplasmic Reticulum and the Inner Nuclear Membrane. *Int. J. Mol. Sci.* **2019**, *20*, 334. [CrossRef] [PubMed]

45. Pfaff, J.; Rivera Monroy, J.; Jamieson, C.; Rajanala, K.; Vilardi, F.; Schwappach, B.; Kehlenbach, R.H. Emery-Dreifuss muscular dystrophy mutations impair TRC40-mediated targeting of emerin to the inner nuclear membrane. *J. Cell Sci.* **2016**, *129*, 502–516. [CrossRef]
46. Ohba, T.; Schirmer, E.C.; Nishimoto, T.; Gerace, L. Energy-and temperature-dependent transport of integral proteins to the inner nuclear membrane via the nuclear pore. *J. Cell Biol.* **2004**, *167*, 1051–1062. [CrossRef]
47. Ungricht, R.; Klann, M.; Horvath, P.; Kutay, U. Diffusion and retention are major determinants of protein targeting to the inner nuclear membrane. *J. Cell Biol.* **2015**, *209*, 687–704. [CrossRef]
48. Cabantous, S.; Terwilliger, T.C.; Waldo, G.S. Protein tagging and detection with engineered self-assembling fragments of green fluorescent protein. *Nat. Biotechnol.* **2005**, *23*, 102–107. [CrossRef]
49. Cabantous, S.; Waldo, G.S. In vivo and in vitro protein solubility assays using split GFP. *Nat. Methods* **2006**, *3*, 845–854. [CrossRef]
50. Smoyer, C.J.; Katta, S.S.; Gardner, J.M.; Stoltz, L.; McCroskey, S.; Bradford, W.D.; McClain, M.; Smith, S.E.; Slaughter, B.D.; Unruh, J.R.; et al. Analysis of membrane proteins localizing to the inner nuclear envelope in living cells. *J. Cell Biol.* **2016**, *215*, 575–590. [CrossRef]
51. Chizhik, A.M.; Ruhlandt, D.; Pfaff, J.; Karedla, N.; Chizhik, A.I.; Gregor, I.; Kehlenbach, R.H.; Enderlein, J. Three-Dimensional Reconstruction of Nuclear Envelope Architecture Using Dual-Color Metal-Induced Energy Transfer Imaging. *ACS Nano* **2017**, *11*, 11839–11846. [CrossRef] [PubMed]
52. Chizhik, A.I.; Rother, J.; Gregor, I.; Janshoff, A.; Enderlein, J. Metal-induced energy transfer for live cell nanoscopy. *Nat. Photonics* **2014**, *8*, 124–127. [CrossRef]
53. Axelrod, D.; Koppel, D.; Schlessinger, J.; Elson, E.; Webb, W.W. Mobility measurement by analysis of fluorescence photobleaching recovery kinetics. *Biophys. J.* **1976**, *16*, 1055–1069. [CrossRef]
54. Axelrod, D.; Ravdin, P.; Koppel, D.; Schlessinger, J.; Webb, W.; Elson, E.; Podleski, T. Lateral motion of fluorescently labeled acetylcholine receptors in membranes of developing muscle fibers. *Proc. Natl. Acad. Sci. USA* **1976**, *73*, 4594–4598. [CrossRef] [PubMed]
55. Edidin, M.; Zagyansky, Y.; Lardner, T. Measurement of membrane protein lateral diffusion in single cells. *Science* **1976**, *191*, 466–468. [CrossRef] [PubMed]
56. Schlessinger, J.; Elson, E.; Webb, W.; Yahara, I.; Rutishauser, U.; Edelman, G. Receptor diffusion on cell surfaces modulated by locally bound concanavalin A. *Proc. Natl. Acad. Sci. USA* **1977**, *74*, 1110–1114. [CrossRef]
57. Schlessinger, J.; Koppel, D.; Axelrod, D.; Jacobson, K.; Webb, W.; Elson, E. Lateral transport on cell membranes: Mobility of concanavalin A receptors on myoblasts. *Proc. Natl. Acad. Sci. USA* **1976**, *73*, 2409–2413. [CrossRef]
58. Schlessinger, J.; Shechter, Y.; Willingham, M.C.; Pastan, I. Direct visualization of binding, aggregation, and internalization of insulin and epidermal growth factor on living fibroblastic cells. *Proc. Natl. Acad. Sci. USA* **1978**, *75*, 2659–2663. [CrossRef]
59. Nehls, S.; Snapp, E.L.; Cole, N.B.; Zaal, K.J.; Kenworthy, A.K.; Roberts, T.H.; Ellenberg, J.; Presley, J.F.; Siggia, E.; Lippincott-Schwartz, J. Dynamics and retention of misfolded proteins in native ER membranes. *Nat. Cell Biol.* **2000**, *2*, 288–295. [CrossRef]
60. Huff, J.; Bergter, A.; Birkenbeil, J.; Kleppe, I.; Engelmann, R.; Krzic, U. The new 2D Superresolution mode for ZEISS Airyscan. *Nat. Methods* **2017**, *14*, 1223. [CrossRef]
61. Korobchevskaya, K.; Lagerholm, B.; Colin-York, H.; Fritzsche, M. Exploring the Potential of Airyscan Microscopy for Live Cell Imaging. *Photonics* **2017**, *4*, 41. [CrossRef]
62. Weisshart, K. The Basic Principle of Airyscanning. *Zeiss Technol. Note* **2014**, *22*, 8.
63. Groves, N.R.; McKenna, J.F.; Evans, D.E.; Graumann, K.; Meier, I. A nuclear localization signal targets tail-anchored membrane proteins to the inner nuclear envelope in plants. *J. Cell Sci.* **2019**, *132*. [CrossRef] [PubMed]
64. Popken, P.; Ghavami, A.; Onck, P.R.; Poolman, B.; Veenhoff, L.M. Size-dependent leak of soluble and membrane proteins through the yeast nuclear pore complex. *Mol. Biol. Cell* **2015**, *26*, 1386–1394. [CrossRef] [PubMed]
65. Schermelleh, L.; Carlton, P.M.; Haase, S.; Shao, L.; Winoto, L.; Kner, P.; Burke, B.; Cardoso, M.C.; Agard, D.A.; Gustafsson, M.G. Subdiffraction multicolor imaging of the nuclear periphery with 3D structured illumination microscopy. *Science* **2008**, *320*, 1332–1336. [CrossRef]

66. Horn, H.F.; Kim, D.I.; Wright, G.D.; Wong, E.S.M.; Stewart, C.L.; Burke, B.; Roux, K.J. A mammalian KASH domain protein coupling meiotic chromosomes to the cytoskeleton. *J. Cell Biol.* **2013**, *202*, 1023–1039. [CrossRef]
67. Korfali, N.; Wilkie, G.S.; Swanson, S.K.; Srsen, V.; Batrakou, D.G.; Fairley, E.A.; Malik, P.; Zuleger, N.; Goncharevich, A.; de las Heras, J. The leukocyte nuclear envelope proteome varies with cell activation and contains novel transmembrane proteins that affect genome architecture. *Mol. Cell. Proteom.* **2010**, *9*, 2571–2585. [CrossRef]
68. Wilkie, G.S.; Korfali, N.; Swanson, S.K.; Malik, P.; Srsen, V.; Batrakou, D.G.; de las Heras, J.; Zuleger, N.; Kerr, A.R.; Florens, L.; et al. Several novel nuclear envelope transmembrane proteins identified in skeletal muscle have cytoskeletal associations. *Mol. Cell Proteomics* **2011**, *10*, M110.003129. [CrossRef]
69. Hell, S.W.; Wichmann, J. Breaking the diffraction resolution limit by stimulated emission: Stimulated-emission-depletion fluorescence microscopy. *Opt. Lett.* **1994**, *19*, 780–782. [CrossRef]
70. Wildanger, D.; Medda, R.; Kastrup, L.; Hell, S. A compact STED microscope providing 3D nanoscale resolution. *J. Microsc.* **2009**, *236*, 35–43. [CrossRef]
71. Galland, R.; Grenci, G.; Aravind, A.; Viasnoff, V.; Studer, V.; Sibarita, J.-B. 3D high-and super-resolution imaging using single-objective SPIM. *Nat. Methods* **2015**, *12*, 641–644. [CrossRef] [PubMed]
72. Xie, W.; Chojnowski, A.; Boudier, T.; Lim, J.S.; Ahmed, S.; Ser, Z.; Stewart, C.; Burke, B. A-type lamins form distinct filamentous networks with differential nuclear pore complex associations. *Curr. Biol.* **2016**, *26*, 2651–2658. [CrossRef] [PubMed]
73. Mudumbi, K.C.; Schirmer, E.C.; Yang, W. Single-point single-molecule FRAP distinguishes inner and outer nuclear membrane protein distribution. *Nat. Commun.* **2016**, *7*, 12562. [CrossRef] [PubMed]
74. Soullam, B.; Worman, H.J. The amino-terminal domain of the lamin B receptor is a nuclear envelope targeting signal. *J. Cell Biol.* **1993**, *120*, 1093–1100. [CrossRef] [PubMed]
75. Soullam, B.; Worman, H.J. Signals and structural features involved in integral membrane protein targeting to the inner nuclear membrane. *J. Cell Biol.* **1995**, *130*, 15–27. [CrossRef] [PubMed]
76. King, M.C.; Lusk, C.; Blobel, G. Karyopherin-mediated import of integral inner nuclear membrane proteins. *Nature* **2006**, *442*, 1003–1007. [CrossRef]
77. Boni, A.; Politi, A.Z.; Strnad, P.; Xiang, W.; Hossain, M.J.; Ellenberg, J. Live imaging and modeling of inner nuclear membrane targeting reveals its molecular requirements in mammalian cells. *J. Cell Biol.* **2015**, *209*, 705–720. [CrossRef]
78. Maul, G.G.; Deaven, L. Quantitative determination of nuclear pore complexes in cycling cells with differing DNA content. *J. Cell Biol.* **1977**, *73*, 748–760. [CrossRef]
79. Schwanhäusser, B.; Busse, D.; Li, N.; Dittmar, G.; Schuchhardt, J.; Wolf, J.; Chen, W.; Selbach, M. Global quantification of mammalian gene expression control. *Nature* **2011**, *473*, 337–342. [CrossRef]
80. Devos, D.; Dokudovskaya, S.; Alber, F.; Williams, R.; Chait, B.T.; Sali, A.; Rout, M.P. Components of coated vesicles and nuclear pore complexes share a common molecular architecture. *PLoS Biol.* **2004**, *2*, e380. [CrossRef]
81. Mical, T.I.; Luther, P.W.; Monteiro, M.J. Intracellular assembly and sorting of intermediate filament proteins: Role of the 42 amino acid lamin insert. *Exp. Cell Res.* **2004**, *295*, 183–193. [CrossRef] [PubMed]
82. Nicolas, H.A.; Akimenko, M.-A.; Tesson, F. Cellular and Animal Models of Striated Muscle Laminopathies. *Cells* **2019**, *8*, 291. [CrossRef] [PubMed]
83. Worman, H.J.; Schirmer, E.C. Nuclear membrane diversity: Underlying tissue-specific pathologies in disease? *Curr. Opin. Cell Biol.* **2015**, *34*, 101–112. [CrossRef] [PubMed]
84. Harris, C.A.; Andryuk, P.J.; Cline, S.W.; Mathew, S.; Siekierka, J.J.; Goldstein, G. Structure and mapping of the human thymopoietin (TMPO) gene and relationship of human TMPO β to rat lamin-associated polypeptide 2. *Genomics* **1995**, *28*, 198–205. [CrossRef] [PubMed]
85. Nishioka, Y.; Imaizumi, H.; Imada, J.; Katahira, J.; Matsuura, N.; Hieda, M. SUN1 splice variants, SUN1_888, SUN1_785, and predominant SUN1_916, variably function in directional cell migration. *Nucleus* **2016**, *7*, 572–584. [CrossRef]

International Journal of
Molecular Sciences

MDPI

Article

A Simulation Study on the Interaction Between Pollutant Nanoparticles and the Pulmonary Surfactant Monolayer

Kai Yue *, Xiaochen Sun, Jue Tang, Yiang Wei and Xinxin Zhang

School of Energy and Environmental Engineering, University of Science and Technology Beijing, Beijing 100083, China

* Correspondence: yuekai@ustb.edu.cn

Received: 15 June 2019; Accepted: 1 July 2019; Published: 4 July 2019

Abstract: A good understanding of the mechanism of interaction between inhaled pollutant nanoparticles (NPs) and the pulmonary surfactant monolayer is useful to study the impact of fine particulate matter on human health. In this work, we established coarse-grained models of four representative NPs with different hydrophilicity properties in the air (i.e., $CaSO_4$, C, SiO_2, and $C_6H_{14}O_2$ NPs) and the pulmonary surfactant monolayer. Molecular dynamic simulations of the interaction during exhalation and inhalation breathing states were performed. The effects of NP hydrophilicity levels, NP structural properties, and cholesterol content in the monolayer on the behaviors of NP embedment or the transmembrane were analyzed by calculating the changes in potential energy, NP displacement, monolayer orderliness, and surface tension. Results showed that NPs can inhibit the ability of the monolayer to adjust surface tension. For all breathing states, the hydrophobic C NP cannot translocate across the monolayer and had the greatest influence on the structural properties of the monolayer, whereas the strongly hydrophilic SiO_2 and $C_6H_{14}O_2$ NPs can cross the monolayer with little impact. The semi-hydrophilic $CaSO_4$ NP can penetrate the monolayer only during the inhalation breathing state. The hydrophilic flaky NP shows the best penetration ability, followed by the rod-shaped NP and spherical NP in turn. An increase in cholesterol content of the monolayer led to improved orderliness and decreased fluidity of the membrane system due to enhanced intermolecular forces. Consequently, difficulty in crossing the monolayer increased for the NPs.

Keywords: air-pollutant nanoparticle; coarse-grained model; interaction; molecular dynamics simulation; pulmonary surfactant monolayer

1. Introduction

Studies show that high concentrations of fine particulate matter (PM2.5) is the underlying cause of haze. The surfaces of PM2.5 particles can carry many harmful organic and inorganic molecules, and the particles can readily enter the body via inhalation and ingestion primarily through the respiratory tract. The inhaled particles may deposit near the alveoli and interact with the pulmonary surfactant monolayer, which is used to regulate the surface tension of the lungs and, thus, maintain the stability of alveoli [1,2]. This interaction and particle invasion may affect the normal function of alveoli [3,4] and even cause acute lung injury [5], cardiovascular disease, inflammation of the respiratory tract, and hematological toxicity [6]. Meanwhile, nanoparticles (NPs) have significant advantages over normal bulk materials owing to their unique optical, photothermal, and electromagnetic properties. NPs can easily cross biological barriers and access specific intracellular locations, thereby enabling many potential biomedical applications such as sensing and imaging, drug delivery, and hyperthermia. Accordingly, NPs with membrane-penetration abilities and high delivery efficiencies should be designed.

The interaction between the pulmonary surfactant monolayer and inhalable particles is the most primitive bio-nano process. Understanding this interaction at the molecular level is one of the most important steps in studying the impact of NPs on the human body and can also provide guidance for NP-based clinical applications. Some experiments and simulations have been carried out, and many advances have been made [7–10]. The physicochemical properties of NPs reportedly play an important role in influencing their interaction with the pulmonary surfactant monolayer. Experimental results show that NP hydrophobicity increases their retention in the monolayer [11], and large NPs have a greater degree of damage to the monolayer [12]. Cholesterol can enhance the interaction between the monolayer and drug and, thus, inhibit drug penetration [13] when the cholesterol content in the monolayer reaches 20%. Some simulations have been performed using molecular dynamics methods, which are widely used to compute biomolecular interactions on micro/nano scales by determining the positions and velocities of particles in terms of the laws of classical physics [14]. The results of analysis of the equilibrium position of NPs, potential of mean force, and order parameters indicate that the structural or morphological properties of the monolayer are obviously affected by the structural and surface physicochemical properties of NPs [15–19]. However, most simulation studies have been carried out based on the generalized coarse-grained (CG) NP models and the CG models of a pure dipalmitoylphosphatidylcholine (DPPC) monolayer or cholesterol-free monolayer. The pulmonary surfactant monolayer comprises phospholipids and proteins, wherein approximately 90% of the monolayer is phospholipids and 10% is proteins. The main component of phospholipids is DPPC with a small amount of neutral lipids (5%~10%), and cholesterol constitutes the major fraction of neutral lipids [20]. Proteins include four surfactant proteins (i.e., SP-A, SP-B, SP-C, and SP-D). Among them, SP-A and SP-D are macromolecular proteins dominating the metabolism and immunoregulation of the monolayer [21], whereas SP-B and SP-C are small molecular proteins dominating the stability of their surface-sustaining monolayer [22].

In the present study, we used molecular dynamics simulations to investigate the interaction mechanism of NPs with the pulmonary surfactant monolayer and the behaviors of NP translocation across the monolayer at exhalation and inhalation breathing states for the representative atmospheric particulate matters. The coarse-grained (CG) model of DPPC containing cholesterol molecules and the surfactant-specific proteins SP-B and SP-C were established, and a qualitative comparison of the simulation results and experimental data from literature was carried out. On this basis, we comprehensively analyzed the influence of crucial factors affecting NPs and the monolayer such as NP type, shape, size, hydrophilicity/hydrophobicity, and cholesterol content on NP interactions and translocation.

2. Results and Discussion

To investigate the interaction mechanism of four kinds of representative NPs with the pulmonary surfactant monolayer during natural breathing patterns, we analyzed the mean-square displacement (MSD) of the NPs, centroid-to-centroid distance between the NP and monolayer, order parameter of the phospholipid molecule, and surface area per lipid molecule in the interaction. The MSD is used to evaluate the movement of a given particle, indicating its diffusion rate [23]. It is calculated with the following formula:

$$MSD = R(t) = \{|\vec{r}(t) - \vec{r}(0)|^2\}, \quad (1)$$

where $\vec{r}(t)$ is used to represent the position of a particle at time t.

The order parameter P_2 is defined as [24]

$$P_2 = \frac{1}{2}(3\cos^2\theta - 1), \quad (2)$$

where θ represents the angle between the bond linking two CG beads and the z-axis, and these two beads are DPPC C3B and C4B.

2.1. Interactions of Different Nanoparticles (NPs) with the Pulmonary Surfactant Monolayer in Exhalation and Inhalation Breathing States

Figure 1a,b show the snapshots of interactions among the different NPs and the pulmonary surfactant monolayer during compression and expansion, respectively. Water molecules are not shown in the figures to emphasize NP movement. We can observe that C and $CaSO_4$ NPs were embedded into the monolayer at the exhalation breathing state, whereas $C_6H_{14}O_2$ and SiO_2 NPs penetrated the monolayer. Compared with $CaSO_4$, C was almost completely wrapped by the monolayer and caused the monolayer to bulge, prominently, toward to the water phase, generating a large monolayer curvature. Thus, the interaction of C can inflict great disturbance to the monolayer structure. Figure 1b indicates that $C_6H_{14}O_2$ and SiO_2 NPs still translocated across the pulmonary surfactant monolayer throughout the inhalation breathing process as they did throughout the exhalation breathing process. However, unlike in the exhalation breathing state, $CaSO_4$ penetrated the monolayer, although C NP still could not cross and was embedded into the monolayer.

The MSDs of the NPs and the centroid-to-centroid distances between the NPs and the monolayer during compression and expansion were calculated, respectively, as shown in Figure 1c–f. In the exhalation breathing state, the MSD of SiO_2 was larger than those of the other three NPs, and the MSD of C was higher than that of $CaSO_4$ NP, indicating that the diffusion rate of the SiO_2 NP was the fastest, and C moved a longer distance than $CaSO_4$. Figure 1e also shows that C and $CaSO_4$ NPs reached a relatively equilibrated state within approximately 10 ns, whereas $C_6H_{14}O_2$ and SiO_2 NPs took a longer time (i.e., around 45 ns). The initial centroid-to-centroid distance between the NP and the monolayer at time $t = 0$ was set to 3.2 nm. The distance of the centroid position between the $CaSO_4$ NP and the monolayer at the equilibrium state was about 1.52 nm, and C was 0.27 nm away from the monolayer, meaning that C was embedded in the monolayer more deeply than $CaSO_4$. The final positions of $C_6H_{14}O_2$ and SiO_2 NPs were all approximately −3.3 nm away from the monolayer, with an average thickness of 2.08 nm, indicating that both NPs completely translocated across the monolayer. Compared with the compression process, the MSDs of all NPs during expansion were larger, which meant that all NPs could move faster, and their diffusion abilities were stronger than those in the exhalation breathing state. Another difference was that the MSD of $CaSO_4$ was larger than that of C, although SiO_2 showed the largest MSD. Figure 1f shows that $C_6H_{14}O_2$ and SiO_2 NPs reached equilibrium within approximately 40 ns, which was shorter than that in the exhalation breathing state. Notably, the final position of the C NP was about 0.88 nm away from the centroid of the monolayer in the inhalation breathing state, indicating the C NP was embedded into the monolayer more deeply in the exhalation breathing state. These analysis results of MSD and the centroid-to-centroid distance confirmed our previous findings obtained from the molecular dynamics simulation snapshots.

The differences in hydrophilicity/hydrophobicity of the NPs explained the difference among the interaction behaviors. C NPs are typical hydrophobic NPs, and $CaSO_4$ NP is semi-hydrophilic. $C_6H_{14}O_2$ NPs and SiO_2 NPs are strongly hydrophilic NPs, and SiO_2 NPs are more hydrophilic.

The strong hydrophilicity of $C_6H_{14}O_2$ and SiO_2 NPs led to the strong attractive forces between the NPs and the hydrophilic head groups of DPPC molecules or between the NPs and the water molecules on the other side of the monolayer, which enabled the NPs to overcome the energy barrier and translocate across the monolayer. The uncharged and hydrophobic C NP was attracted by the hydrophobic tail groups of DPPC molecules, and its interaction with the hydrophilic head group was repulsive at the same time. Consequently, C cannot translocate across the monolayer, and its deep encapsulation inside the monolayer caused monolayer deformation.

For the semi-hydrophilic $CaSO_4$ NP in the exhalation breathing state, the hydrophobic interaction between the NPs and hydrophobic apolar tail groups of DPPC molecules were not intensive. The interactions between the NP and hydrophilic head groups and water molecules were attractive but not sufficiently large for NP penetration, only causing the NP to stay in the surface region of the monolayer. In the inhalation breathing state, the reason for $CaSO_4$ translocation across the monolayer may be that the relatively loose arrangement of lipid molecules enabled the attractive interaction between the NP and the monolayer to overcome the energy barrier, resulting in NP penetration through the expanded monolayer. We further elucidated this aspect by analyzing the surface area per lipid molecule and the energy component of the simulation system, as discussed in the following sections.

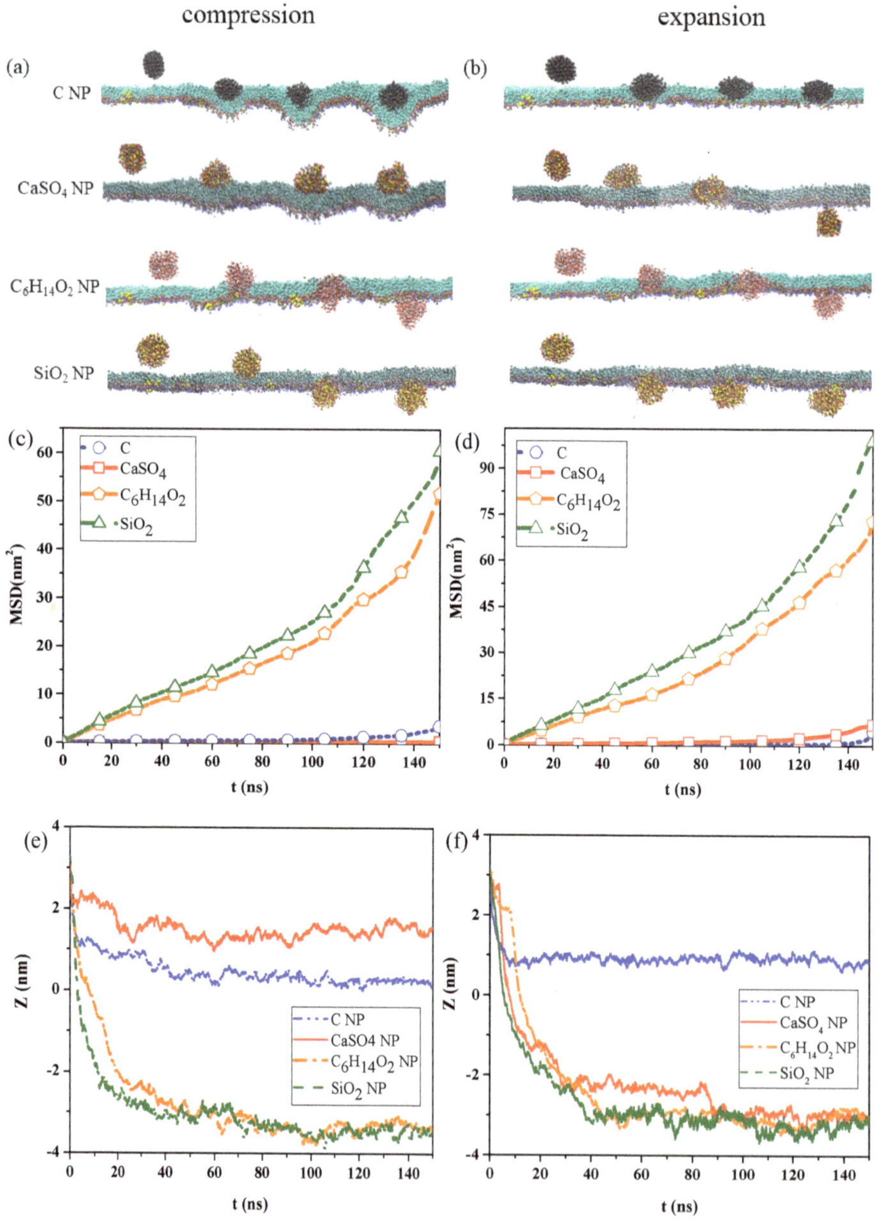

Figure 1. Interaction between nanoparticles (NPs) and the pulmonary surfactant monolayer. Snapshots of the interaction in the (**a**) exhalation and (**b**) inhalation breathing states; mean-square displacements (MSDs) of the NPs in (**c**) exhalation and (**d**) inhalation breathing states; centroid-to-centroid distance between the NPs and monolayer in (**e**) exhalation and (**f**) inhalation breathing states.

We then calculated the order parameters of lipid molecules in the pure monolayer system and NP monolayer systems over time, as well as the surface area per lipid molecule, to analyze the effect of interaction of NPs with the monolayer on the structural properties of the monolayer. Figure 2a,b shows the changes in order parameters and surface area per lipid molecule in the exhalation and inhalation breathing states, respectively. A larger order parameter represented better consistency between the bonding and arrangement directions of the phospholipid monolayer, indicating better orderliness of the membrane lipid molecules. Figure 2a shows that the order parameters of lipid molecules in the systems containing $C_6H_{14}O_2$ and SiO_2 were almost the same as those in the system without NPs, either in exhalation or at inhalation breathing states, all of which were approximately 0.458 or 0.285 (Figure 2a,b) at equilibrium, respectively. Thus, $C_6H_{14}O_2$ and SiO_2 penetrations had a negligible effect on the orderliness of lipid molecules after NP membrane translocation. The order parameters for the system containing $CaSO_4$ NPs were 0.452 and 0.285 in exhalation or at inhalation breathing states, respectively, but they were 0.413 and 0.204 for C, indicating that the penetration of $CaSO_4$ had little effect on the monolayer structure in the inhalation breathing state. The presence of C most obviously affected the monolayer structure and led to decreased orderliness in the monolayer.

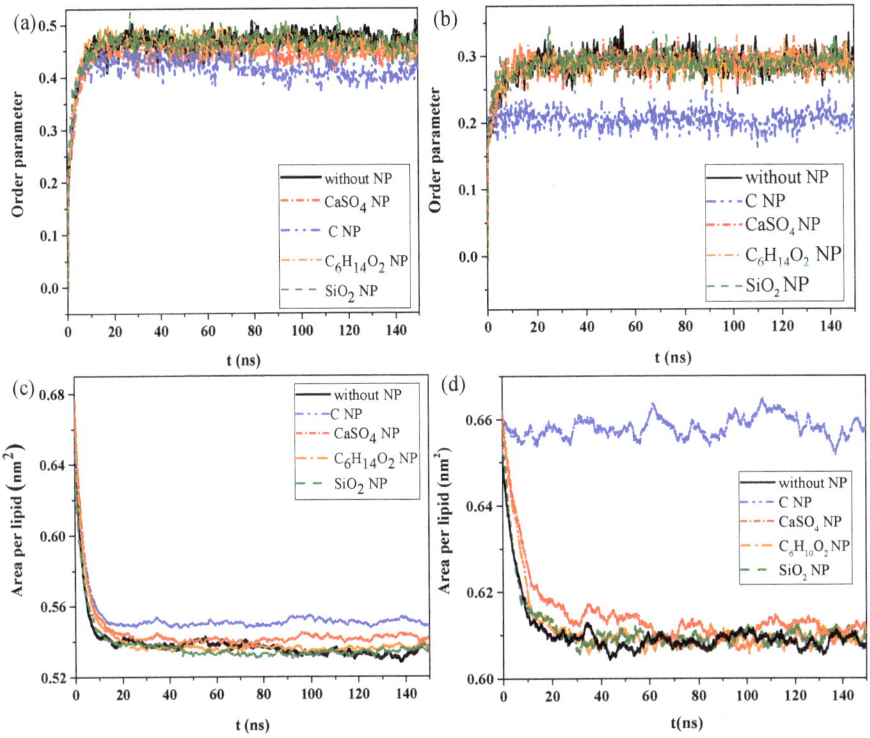

Figure 2. Order parameters of monolayer molecules in the (**a**) exhalation and (**b**) inhalation breathing states. Surface area per lipid molecule in the (**c**) exhalation and (**d**) inhalation breathing states.

Figure 2c,d shows that the surface areas per lipid molecule in the systems without NPs and with $C_6H_{14}O_2$ and SiO_2 NPs were all about 0.53 nm^2 after equilibrium was reached. In other words, the effect of $C_6H_{14}O_2$ and SiO_2 on the monolayer structure was almost negligible. The surface areas per lipid molecule of the system containing C was about 0.552 and 0.657 nm^2 in exhalation or inhalation breathing states, respectively. These values were noticeably larger than those of other systems because the hydrophobicity of C NPs produced remarkable monolayer curvature and a large change in area.

Analysis results of surface areas per lipid molecule were consistent with those of the order parameters. In general, the difference in interactions with the monolayer between SiO_2 and $C_6H_{14}O_2$ was not obvious because the differences in their hydrophilicity levels was not large.

2.2. Analyses of System Energy and Surface Pressure of the Monolayer

We calculated the total energy of different systems to compare the interactions of different NPs with the pulmonary surfactant monolayer in inhalation and exhalation breathing states and to further understand the mechanism underlying such interactions. All simulations in this study were performed in the NPT ensemble, and the total kinetic energy of the system remained unchanged considering the constant temperature under practical conditions. Thus, the change in total energy was equal to the change in potential energy.

As shown in Figure 3a, for the systems with the same NP, the total energy of the system in the inhalation breathing state was lower than that in the exhalation breathing state (i.e., the orderliness of the lipid molecules was more orderly, and the entropy of the system was lower). For all NPs, the total energies in both breathing states decreased after interacting with the monolayer, indicating that the interaction was irreversible. During the interaction process, the potential energy was converted to kinetic energy and was then dissipated by the simulation system to keep the total kinetic energy constant, so the potential energy of the system decreased. The NPs could not enter the air side again without applying external disturbance after it was embedded into the monolayer or crossed the membrane. Figure 3b shows the difference in total energy between the NP monolayer system and the pure monolayer system, which includes the energy of the NP, the interaction energy between the NP and monolayer, and the interaction energy between the NP and water molecules. The change in this difference can reflect the change in NP potential energy, owing to the constant kinetic energy in the NPT ensemble. We also found SiO_2 had higher potential than C, indicating a larger interaction of SiO_2 with the lipid and water molecules and a greater likelihood of NP crossing the monolayer. The potential energy of SiO_2 in the inhalation breathing state was larger than that in the exhalation breathing state, indicating that the NP was subject to greater force and could more easily penetrate the monolayer.

The attractive interaction between the hydrophobic NP and hydrophobic groups of the monolayer was the dominant driving force acting on the NP during the process of NP movement and embedment, which had the same direction as the NP motion. Accordingly, the potential energy continually decreased until it eventually stabilized at a position where it was wrapped and a potential well formed. Conversely, hydrophilic NPs were subjected to a repulsive force, owing to the existence of a hydrophobic barrier produced by the lipid monolayer. The NP overcame the potential barrier and finally penetrated the monolayer when the hydrophilic interaction between the NP and water molecules was sufficiently strong. Otherwise, the NP was stopped by the potential barrier and embedded into the monolayer.

A stable, low surface tension of the pulmonary surfactant monolayer can maintain the stability of lung alveolars [1,20], but the surface tension may be affected by the interaction between the NPs and the monolayer. We calculated the surface pressure and pressure–area isotherm of the monolayer, which was commonly used to analyze the functional properties of the monolayer. The surface pressure π is given by $\pi = \gamma_w - \gamma$, where γ_w and γ represent the surface tension of pure water and the monolayer, respectively.

Figure 3c shows the changes in surface pressure of the monolayer in the exhalation and inhalation breathing states for the pure monolayer and NP monolayer systems after reaching the equilibrium state. Compared with the surface tension of the monolayer in the system without NPs, all interactions of different NPs with the monolayer resulted in decreased surface tension of the monolayer to a certain extent. The effects of all interactions in the exhalation breathing state were more remarkable than those in the inhalation breathing state, in which C showed the most pronounced effect on the surface tension of the monolayer. C was embedded into the monolayer because of its hydrophobic interaction with the hydrophobic tail chain of DPPC molecules and its interaction with the hydrophilic head groups of the lipid. This phenomenon resulted in decreased free area available for the lipid molecules and increased surface density of the monolayer. Thus, the surface tension of the monolayer was reduced and the surface pressure increased. For $CaSO_4$, although it was also embedded in the monolayer in the exhalation breathing state, the effect of the interaction on the surface pressure of the monolayer was still smaller than that of C because it had a smaller embedding degree, as shown in Figure 1a. In the inhalation breathing state, the effect of the interaction of $CaSO_4$ with the monolayer was very slight, similar to those of $C_6H_{14}O_2$ and SiO_2 NPs, to their complete membrane translocation. These findings indicated that the breathing state played an important role in the interaction of NPs with the monolayer.

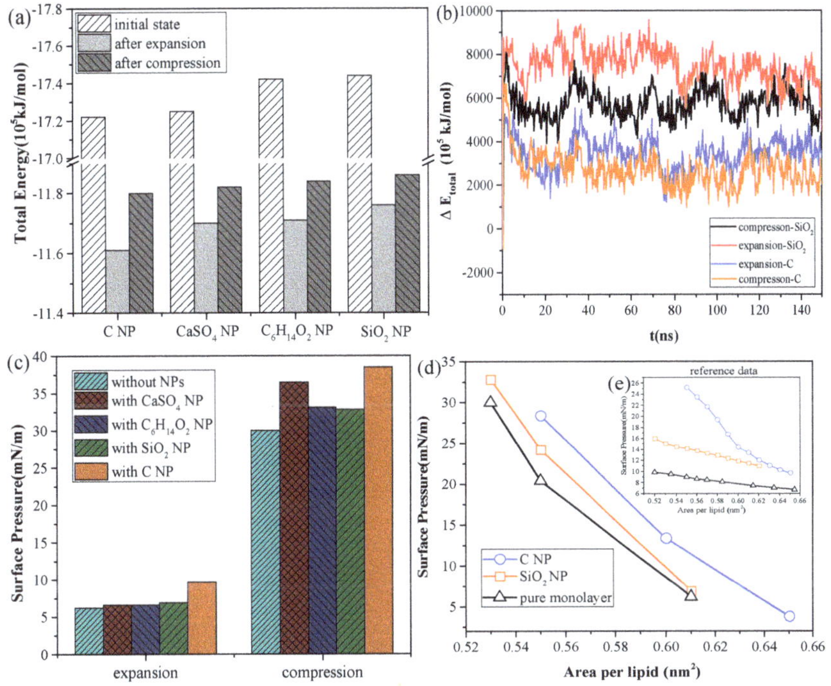

Figure 3. Analyses of system energy and surface pressure for the systems containing a pure monolayer or an NP monolayer: (**a**) total energy of different systems after interaction, (**b**) ΔE_{total} (i.e., the difference in total energy between the NP monolayer system and the pure monolayer system), (**c**) surface pressure of the monolayer in exhalation and inhalation breathing states, and (**d**) surface pressure–area isotherms of the monolayers from the simulation and Ref. [25].

We compared the calculated surface pressure–area isothermal (π-A) of the monolayer for the strongly hydrophilic SiO_2 NP and the hydrophobic C NP with experimental data obtained from Ref. [25] (Figure 3d). Qualitative agreement was observed between the simulation and experimental results (i.e., strongly hydrophilic NPs had a smaller effect on the surface tension of the monolayer than the hydrophobic NPs). This finding was due to ability of the strongly hydrophilic NPs to translocate across the monolayer and not stay in it at equilibrium. The reason for the quantitative difference in simulation and experimental results was that the NPs initially mixed with DPPC for compression in the experiment, whereas the NPs were initially located on the air-side in the simulation and took more time to move to and interact with the monolayer.

2.3. Effect of Cholestrol Content of the Monolayer and NP Structural Properties on the Interaction

Cholesterol is reportedly present in the lipid membrane, at 5%–10% by mass, and constitutes the major fraction of neutral lipids [3]. To investigate the effect of cholesterol content on the interaction of NPs with the pulmonary surfactant monolayer, we established CG models of the pulmonary membrane with 0%, 5%, and 10% cholesterol content, respectively, and simulated its interaction with C, $CaSO_4$, or SiO_2 NPs. We found that all kinds of NPs moved slower with increased cholesterol content, and that C could not penetrate all three kinds of the monolayers at the inhalation breathing state, whereas SiO_2 could cross all these monolayers.

However, the calculation results of the centroid–centroid distance of the NP with the monolayer and the MSD of the lipid molecules (Figure 4a,b) revealed that $CaSO_4$ can translocate only across the monolayer with 0% and 5% cholesterol content, and the MSD of the lipid molecules prominently decreased in the system with 10% cholesterol content. We then analyzed the order parameters of the lipid molecules of the monolayer with 0%, 5%, and 10% cholesterol contents, respectively, in the systems without NPs and with $CaSO_4$ (Figure 4c,d). An increase in cholesterol content of the monolayer led to an increase in the order parameters of the monolayer. The fluidity of lipid molecules is usually related to their structural ordering and is manifested in the tail of phospholipid molecules [26]. Our results indicated that the cholesterol content remarkably influenced the fluidity of the pulmonary surfactant monolayer and that increased cholesterol content led to a more orderly arrangement of lipid molecules. This ordering enhanced the packing of the hydrophobic portion of the lipid molecule and, thus, reduced the fluidity of the pulmonary surfactant monolayer. Subsequently, the NPs encountered increased difficulty in penetrating the monolayer. $CaSO_4$ could cross the monolayers with 0% and 5% cholesterol content, so little difference in order parameter between the pure monolayer and $CaSO_4$ NP monolayer systems was observed. For the lipid molecules of the monolayer with 10% cholesterol content, the order parameters of the two systems were 0.40 and 0.42, respectively. This finding indicated that the difference was not very remarkable because $CaSO_4$ stayed only on the surface region of the monolayer, and it inflicted slight disturbance to the monolayer structure because the centroid–centroid distance between the NP with 3 nm diameter and the monolayer with 2 nm thickness was only 2.23 nm.

We analyzed the effects of NP structural properties on the interaction between NPs and the pulmonary surfactant monolayer, taking the effects of size and shape into account. Figure 5a shows the centroid-to-centroid distances between a hydrophilic and spherical NP without charge and surfactant monolayer and the corresponding snapshots at equilibrium under different NP size conditions, respectively. Smaller NPs took shorter times to cross the monolayer and reach the equilibrium state, indicating that the difficulty of NP membrane translocation was positively correlated with NP size. Notably, all NPs attached onto the hydrophilic side of the monolayer after membrane translocation, and no detachment from the monolayer occurred due to the attractive interaction of the hydrophilic head groups of the monolayer with the NPs. Although all hydrophilic NPs can successfully cross the monolayer, the NPs did not stay in a certain position and remain fixed, and the smaller NPs were more easily affected by thermal fluctuations and had a larger fluctuation range.

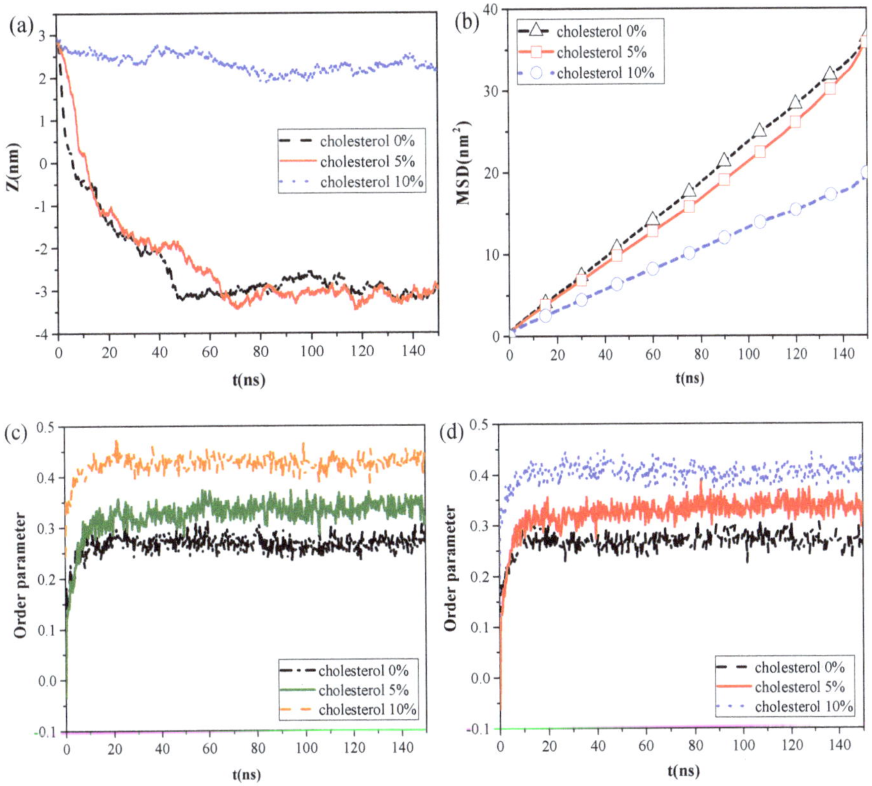

Figure 4. Effect of cholesterol content of the monolayer: (**a**) centroid–centroid distance between the $CaSO_4$ NP and the monolayer, (**b**) MSDs of lipid molecules in the $CaSO_4$ NP monolayer system, and (**c**,**d**) order parameters for the systems without NPs and with the $CaSO_4$ NP.

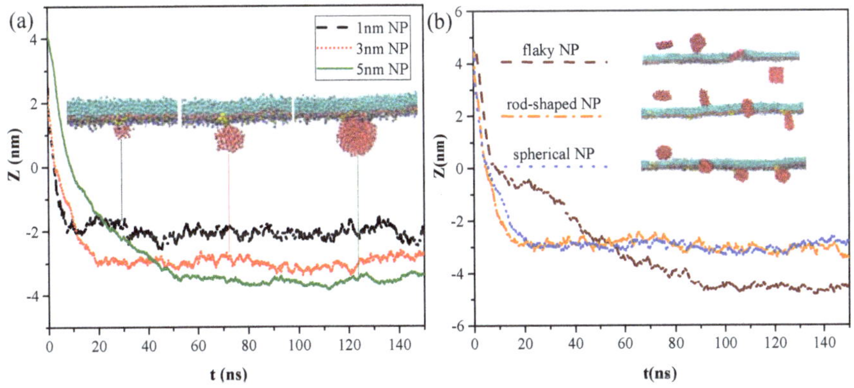

Figure 5. Snapshot of transmembrane and centroid-to-centroid distances of NPs with different (**a**) NP sizes and (**b**) NP shapes.

Three types of neutral hydrophilic NPs with three common shapes for NPs in the air, namely, spherical (diameter d = 3 nm), rod-shaped (diameter d = 3 nm and height h = 3 nm), and flaky (size d = 3 nm and thickness δ = 0.5 nm), were modeled to analyze the effect of NP shape on the interaction.

As shown in Figure 5b, all three types of NPs can translocate across the monolayer. The difference was that the spherical and rod-shaped NPs still adsorbed onto the monolayer at equilibrium, whereas the flaky NP penetrated through and completely separated from the monolayer, entering into the water phase. During NP translocation, we observed that the angles between the monolayer and rod-shaped or flaky NPs were continuously adjusted according to the changes in their relative position and the variation in their interactions. The rod-shaped NP contacted and penetrated the monolayer with the long axis perpendicular to the plane of the monolayer from the initial parallel placement. Similarly, the flaky NP spontaneously moved and rotated from the initial placement in parallel with the monolayer to the vertical placement with a minimum contact area made with the monolayer. After the NP was embedded into the monolayer, the NP eventually detached from the monolayer because attachment of the NP to hydrophilic head groups of the monolayer cannot counteract the attractive force between the NP and water molecules due to minimal contact.

To quantify the difference in penetration behaviors of NPs with different shapes, we calculated the centroid-to-centroid distance between NPs and the monolayer, and we found that rod-shaped NPs took less time (t = 17 ns) to cross the monolayer and reach the equilibrium state than spherical NPs (t = 21 ns), although all of them eventually attached onto the monolayer and could not detach from it. The flaky NP spent more time (t = 50 ns) to penetrate the pulmonary surfactant monolayer, and equilibrium was reached within about 90 ns, where it then completely entered the water phase and was surrounded by water molecules.

3. Methods

All simulations were carried out using the MARTINI [27,28] CG force field, in which several heavy atoms were incorporated into one bead to decrease computational cost as well as allow larger length scale and longer time-period simulations compared with all-atom models. The coarse-graining procedure would reduce the resolution of the system and might cause some key information at the atomic scale to be lost [29]. In our simulation, the model of pulmonary monolayer was established using the standard coarse-grained MARTINI force field, and the coarse-graining of NPs was performed by using one bead to represent one atom for reducing the loss of information.

The CG model of the DPPC pulmonary surfactant monolayer containing cholesterol and surfactant-specific proteins (i.e., SP-B and SP-C) was established using GROMACS [30] and Visual Molecular Dynamics [31], as shown in Figure 6a–d. Given that organic carbon, elemental carbon, SO_4^{2-}, and NO_3^- are the most abundant chemical components of PM2.5 [32], we selected C, $C_6H_{14}O_2$,and $CaSO_4$ NPs to represent elemental carbon, alcoholic content, and secondary sulfate particulates discharged from automobile exhaust, coal combustion, and biomass burning, and we selected SiO_2 to represent the main inorganic component of wind-blown soils. The CG models of these four kinds of NPs were constructed using GROMACS and MATERIAL STUDIO (Figure 6e). Relevant information about bond length and bond angle of the NPs was obtained by implementing the MSI2LMP tool packaged with LAMMPS software. To ensure rigidity of the NPs during simulation, the equilibrium distance among beads was set to 0.47 nm, and the force constant was set to 1250 kJ/mol.

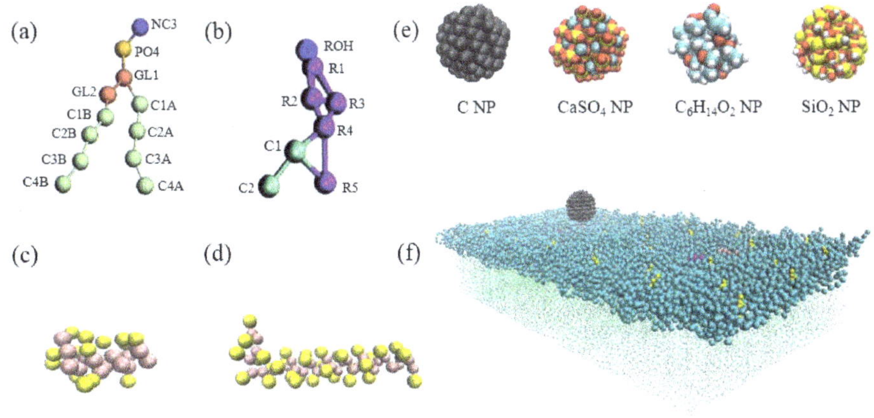

Figure 6. Coarse-grained (CG) models of (**a**) the dipalmitoylphosphatidylcholine (DPPC) molecule; (**b**) cholesterol molecule; (**c**) SP-B protein; (**d**) SP-C protein; (**e**) C, $CaSO_4$, $C_6H_{14}O_2$, and SiO_2 NPs; and (**f**) NP (cyan)- pulmonary surfactant monolayer with DPPC (blue), cholesterol (yellow), SP-B (purple), SP-C (pink), and water molecules (green).

The model of the pulmonary surfactant monolayer consisted of 1086 CG DPPC molecules, 34 CG cholesterol molecules, and 2 CG surfactant proteins (SP-B and SP-C). They were placed in a 20 × 40 × 80 nm cubic simulation box containing 46634 CG water molecules and one NP. The monolayer was placed on the liquid–air interface in the z-direction, and the NP was initially located in the air above the monolayer. The general shape of the NP was spherical, and its diameter was approximately 3 nm. The simulation box was stretched along the z-direction to prevent water molecules in the system from running through the periodic boundary to the airside of the monolayer, as shown in Figure 6f.

In the simulation, a cutoff distance of 1.2 nm was set for van der Waals interactions, and the Lennard–Jones potential was smoothly shifted to zero in the range of 0.9–1.2 nm to reduce cutoff noise. For electrostatic interactions, the Coulomb potential had a cutoff distance of 1.0 nm and a smooth shift to zero over the range of 0–1.0 nm. An isobaric–isothermal (NPT) ensemble with periodic boundary conditions was used for simulations. The temperature of the system was controlled at 310 K using the Berendsen thermostatting method, and semi-anisotropic coupling was used by the Berendsen barostat with a coupling constant of 3.0 ps and a compressibility of 4.5×10^{-5} bar^{-1} in the *x*–*y* plane and 0 bar^{-1} on the z-axis. Given that the respiratory scale is several orders of magnitude larger than the nanoscale of the molecular simulation study [33,34], a large lateral pressure was applied in the simulation to achieve the surface tension of the pulmonary surfactant monolayer in the respiration process, which was 30% higher and lower than that at initial state for the exhalation (compression) and inhalation (expansion) breathing states, respectively. We implemented postprocessing and visualization through VMD, and all simulations were performed using the Gromacs 4.5.4 simulation package. The simulation time of the system was set to 150 ns.

4. Conclusions

This work investigated the interaction mechanism between the representative pollutant NPs and the pulmonary surfactant monolayer using molecular dynamics simulations based on established CG models. By analyzing the snapshot, NP MSD and centroid distance, the order parameter and area of the lipid molecules, system energy, and surface pressure of the monolayer, we concluded that the strongly hydrophilic SiO_2 and $C_6H_{14}O_2$ NPs can cross the monolayer in either exhalation or inhalation breathing states and had little effect on the monolayer structure at equilibrium, whereas NP penetration of the semi-hydrophilic $CaSO_4$ NP occurred only in the inhalation breathing state with a slight influence on monolayer orderliness. No obvious difference was observed in NP membrane translocation and effect on the monolayer between SiO_2 and $C_6H_{14}O_2$ NPs because there was a slight difference in hydrophilicity levels.

The hydrophobic C NP had the most pronounced effect on monolayer orderliness because it could not cross and embedded only into the monolayer during both breathing states. The difficulty in NP membrane translocation for the hydrophilic NPs was positively correlated with the cholesterol content in the monolayer because of the change in fluidity caused by cholesterol (e.g., the semi-hydrophilic $CaSO_4$ NP could not penetrate the monolayer with 10% cholesterol content in the inhalation breathing state). The flaky shape had the advantage of membrane translocation for the hydrophilic NPs over the spherical and rod-shape ones, and only the hydrophilic flaky NP could detach from the monolayer and enter the water phase completely. The interaction of the hydrophilic SiO_2 NP had a relatively slight effect on the surface tension of the monolayer, whereas that of the hydrophobic C NP could obviously decrease the surface tension. These findings indicated that NP deposition could affect its ability to regulate interfacial tension. Our simulations can provide information on the effect of NP hydrophilicity, cholesterol content of the monolayer, and NP structural properties on the interactions of air-pollutant NPs with the pulmonary surfactant monolayer, and the knowledge can be useful for drug-delivery design.

Author Contributions: Conceptualization, K.Y.; methodology, K.Y.; X.S. and J.T.; software, X.S.; J.T. and Y.W.; validation, K.Y.; X.S. and Y.W.; formal analysis, K.Y.; X.S. and Y.W.; investigation, K.Y.; X.S. and J.T.; resources, X.S. and J.T.; data curation, X.S. and J.T.; writing—original draft preparation, K.Y.; X.S. and Y.W.; writing—review and editing, K.Y.; X.S. and Y.W.; project administration, K.Y. and X.Z.; funding acquisition, K.Y. and X.Z.

Funding: This work was financially supported by the National Natural Science Foundation of China (Grant No. 51890891 and No. 51890894) and the Beijing Engineering Research Center for Energy Saving and Environmental Protection.

Conflicts of Interest: The authors declare no conflict of interest.

References

1. Zasadzinski, J.A.; Ding, J.; Warriner, H.E. The physics and physiology of lung surfactants. *Curr. Opin. Colloid Interface Sci.* **2001**, *6*, 506–513. [CrossRef]
2. Zuo, Y.Y.; Veldhuizen, R.A.W.; Neumann, A.W. Current perspectives in pulmonary surfactant—Inhibition, enhancement and evaluation. *Biochim. Biophys. Acta Biomembr.* **2008**, *1778*, 1947–1977. [CrossRef] [PubMed]
3. Bakshi, M.S.; Zhao, L.; Smith, R. Metal Nanoparticle Pollutants Interfere with Pulmonary Surfactant Function In Vitro. *Biophys. J.* **2008**, *94*, 855–868. [CrossRef] [PubMed]
4. Salvador-Morales, C.; Townsend, P.; Flahaut, E.; Vénien-Bryan, C.; Vlandas, A.; Green, M.L.H.; Sim, R.B. Binding of pulmonary surfactant proteins to carbon nanotubes; potential for damage to lung immune defense mechanisms. *Carbon* **2007**, *45*, 607–617. [CrossRef]
5. Li, C.; Liu, H.; Sun, Y.; Wang, H.; Guo, F.; Rao, S.; Deng, J.; Zhang, Y.; Miao, Y.; Guo, C. PAMAM nanoparticles promote acute lung injury by inducing autophagic cell death through the Akt-TSC2-mTOR signaling pathway. *J. Mol. Cell Biol.* **2009**, *1*, 37. [CrossRef] [PubMed]
6. Buzea, C.; Pacheco, I.I.; Robbie, K. Nanomaterials and nanoparticles: Sources and toxicity. *Biointerphases* **2007**, *2*, 17–71. [CrossRef]
7. Harishchandra, R.K.; Saleem, M.; Galla, H.J. Nanoparticle interaction with model lung surfactant monolayers. *J. R. Soc. Interface* **2010**, *7* (Suppl. 1), S15–S26. [CrossRef]
8. Schleh, C.; Hohlfeld, J.M. Interaction of nanoparticles with the pulmonary surfactant system. *Inhal. Toxicol.* **2009**, *21* (Suppl. 1), 7. [CrossRef]
9. Ye, X.; Hao, C.; Yang, J.; Sun, R. Influence of modified silica nanoparticles on phase behavior and structure properties of DPPC monolayers. *Colloids Surf. B Biointerfaces* **2018**, *172*, 480–486. [CrossRef]
10. Guzmán, E.; Liggieri, L.; Santini, E.; Ferrari, M.; Ravera, F. Influence of silica nanoparticles on phase behavior and structural properties of DPPC—Palmitic acid Langmuir monolayers. *Colloids Surf A Physicochem. Eng. Asp.* **2012**, *413*, 280–287. [CrossRef]
11. Valle, R.P.; Huang, C.L.; Loo, J.S.C. Increasing hydrophobicity of nanoparticles intensifieslung surfactant film inhibition and particle retention. *ACS Sustain. Chem. Eng.* **2014**, *2*, 1574–1580. [CrossRef]

12. Dwivedi, M.V.; Harishchandra, R.K.; Koshkina, O.; Maskos, M.; Galla, H.J. Size Influences the effect of hydrophobic nanoparticles on lung surfactant model systems. *Biophys. J.* **2014**, *106*, 289. [CrossRef] [PubMed]
13. Zhao, L.; Feng, S.S. Effects of cholesterol component on molecular interactions between paclitaxel and phospholipid within the lipid monolayer at the air–water interface. *J. Colloid Interface Sci.* **2006**, *300*, 314–326. [CrossRef] [PubMed]
14. Dror, R.O.; Dirks, R.M.; Grossman, J.P.; Xu, H.; Shaw, D.E. Biomolecular simulation: A Computational Microscope for Molecular Biology. *Annu. Rev. Biophys.* **2012**, *41*, 429–452. [CrossRef] [PubMed]
15. Adhangale, P.S.; Gaver, D.P. Equation of state for a coarse-grained DPPC monolayer at the air/water interface. *Mol. Phys.* **2006**, *104*, 3011–3019. [CrossRef] [PubMed]
16. Hu, G.; Jiao, B.; Shi, X.; Valle, R.P.; Fan, Q.; Zuo, Y.Y. Physicochemical Properties of Nanoparticles Regulate Translocation across Pulmonary Surfactant Monolayer and Formation of Lipoprotein Corona. *ACS Nano* **2013**, *7*, 10525–10533. [CrossRef] [PubMed]
17. Chen, P.; Zhang, Z.; Xing, J.; Gu, N.; Ji, M. Physicochemical properties of nanoparticles affect translocation across pulmonary surfactant monolayer. *Mol. Phys.* **2017**, *115*, 3143–3154. [CrossRef]
18. Chen, P.; Zhang, Z.; Gu, N.; Xing, J. Effect of the surface charge density of nanoparticles on their translocation across pulmonary surfactant monolayer: A molecular dynamics simulation. *Mol. Simul.* **2017**, *44*, 85–93. [CrossRef]
19. Chiu, C.C.; Shinoda, W.; Devane, R.H.; Nielsen, S.O. Effects of spherical fullerene nanoparticles on a dipalmitoyl phosphatidylcholine lipid monolayer: A coarse grain molecular dynamics approach. *Soft Matter* **2012**, *8*, 9610. [CrossRef]
20. Goerke, J. Pulmonary surfactant: Functions and molecular composition. *Biochim. Biophys. Acta* **1998**, *1408*, 79–89. [CrossRef]
21. Reid, K.B. Functional Roles of the Lung Surfactant Proteins SP-A and SP-D in Innate Immunity. *Immunobiology* **1998**, *199*, 200–207. [CrossRef]
22. Oosterlaken-Dijksterhuis, M.A.; Haagsman, H.P.; van Golde, L.M.G.; Demel, R.A. Characterization of lipid insertion into monomolecular layers mediated by lung surfactant proteins SP-B or SP-C. *Biochemistry* **1991**, *30*, 10965–10971. [CrossRef] [PubMed]
23. Khajeh, A.; Modarress, H. The influence of cholesterol on interactions and dynamics of ibuprofen in a lipid bilayer. *Biochim. Biophys. Acta* **2014**, *1838*, 2431–2438. [CrossRef] [PubMed]
24. Feller, S.E.; Venable, R.M.; Pastor, R.W. Computer Simulation of a DPPC Phospholipid Bilayer: Structural Changes as a Function of Molecular Surface Area. *Langmuir* **1997**, *13*, 6555–6561. [CrossRef]
25. Guzmán, E.; Liggieri, L.; Santini, E.; Ferrari, M.; Ravera, F. Effect of Hydrophilic and Hydrophobic Nanoparticles on the Surface Pressure Response of DPPC Monolayers. *J. Phys. Chem. C* **2017**, *115*, 21715–21722. [CrossRef]
26. Mhashal, A.R.; Sudip, R.; Jie, Z. Effect of Gold Nanoparticle on Structure and Fluidity of Lipid Membrane. *PLoS ONE* **2014**, *9*, e114152. [CrossRef] [PubMed]
27. Monticelli, L.; Kandasamy, S.K.; Periole, X.; Larson, R.G.; Tieleman, D.P.; Marrink, S.J. The MARTINI Coarse-Grained Force Field: Extension to Proteins. *J. Chem. Theory Comput.* **2008**, *4*, 819. [CrossRef] [PubMed]
28. Marrink, S.J.; Risselada, H.J.; Yefimov, S.; Tieleman, D.P.; de Vries, A.H. The MARTINI Force Field: Coarse Grained Model for Biomolecular Simulations. *J. Phys. Chem. B* **2007**, *111*, 7812–7814. [CrossRef]
29. Saunders, M.G.; Voth, G.A. Coarse-Graining Methods for Computational Biology. *Annu. Rev. Biophys.* **2013**, *42*, 73–93. [CrossRef]
30. Hess, B.; Kutzner, C.; van der Spoel, D.; Lindahl, E. GROMACS 4: Algorithms for highly efficient, load-balanced, and scalable molecular simulation. *J. Chem. Theory. Comput.* **2008**, *4*, 435–447. [CrossRef]
31. Humphrey, W.; Dalke, A.; Schulten, K. VMD: Visual molecular dynamics. *J. Mol. Graph.* **1996**, *14*, 33–38. [CrossRef]
32. Cao, J.J.; Shen, Z.X.; Chow, J.C.; Watson, J.G.; Lee, S.H.; Tie, X.U.; Ho, K.I.; Wang, G.E.; Han, Y.O. Winter and Summer PM2.5 Chemical Compositions in Fourteen Chinese Cities. *J. Air Waste Manag. Assoc.* **2012**, *62*, 1214–1226. [PubMed]
33. Baoukina, S.; Mendez, V.E.; Tieleman, D.P. Molecular view of phase coexistence in lipid monolayers. *J. Am. Chem. Soc.* **2012**, *134*, 17543–17553. [CrossRef] [PubMed]

34. Baoukina, S.; Monticelli, L.; Marrink, S.J.; Tieleman, D.P. Pressure–Area Isotherm of a Lipid Monolayer from Molecular Dynamics Simulations. *Langmuir* **2007**, *23*, 12617–12623. [CrossRef] [PubMed]

International Journal of
Molecular Sciences

MDPI

Article

Selective Separation of Acetic and Hexanoic Acids across Polymer Inclusion Membrane with Ionic Liquids as Carrier

Bao-Ying Wang [1,4], Na Zhang [1,3], Zhen-Yu Li [1,3], Qiao-Lin Lang [1], Bing-Hua Yan [1], Yang Liu [2,*] and Yang Zhang [1,*]

1 Waste Valorization and Water Reuse Group, Qingdao Institute of Bioenergy and Bioprocess Technology, Chinese Academy of Sciences, 189 Songling Road, Laoshan District, Qingdao 266101, China
2 College of Environment and Safety Engineering, Qingdao University of Science and Technology, 53 Zhengzhou Road, Qingdao 266042, China
3 College of Environmental Science and Engineering, Ocean University of China, 238 Songling Road, Laoshan District, Qingdao 266100, China
4 University of Chinese Academy of Sciences, Beijing 100049, China
* Correspondence: liuyangqust2019@163.com (Y.L.); zhangyang@qibebt.ac.cn (Y.Z.); Tel.: +86-532-80662728 (Y.L.); +86-532-80662729 (Y.Z.)

Received: 29 June 2019; Accepted: 8 August 2019; Published: 12 August 2019

Abstract: This paper first reports on the selective separation of volatile fatty acids (VFAs) (acetic and hexanoic acids) using polymer inclusion membranes (PIMs) containing quaternary ammonium and phosphonium ionic liquids (ILs) as the carrier. The affecting parameters such as IL content, VFA concentration, and the initial pH of the feed solution as well as the type and concentration of the stripping solution were investigated. PIMs performed a much higher selective separation performance toward hexanoic acid. The optimal PIM composed of 60 wt% quaternary ammonium IL with the permeability coefficients for acetic and hexanoic acid of 0.72 and 4.38 μm s^{-1}, respectively, was determined. The purity of hexanoic acid obtained in the stripping solution increased with an increase in the VFA concentration of the feed solution and decreasing HCl concentration of the stripping solution. The use of Na_2CO_3 as the stripping solution and the involvement of the electrodialysis process could dramatically enhance the transport efficiency of both VFAs, but the separation efficiency decreased sharply. Furthermore, a coordinating mechanism containing hydrogen bonding and ion exchange for VFA transport was demonstrated. The highest purity of hexanoic acid (89.3%) in the stripping solution demonstrated that this PIM technology has good prospects for the separation and recovery of VFAs from aqueous solutions.

Keywords: polymer inclusion membrane; ionic liquids; volatile fatty acids (VFAs); acetic acids; hexanoic acids

1. Introduction

The conversion of organic residual waste into platform chemicals though anaerobic microbial fermentation is considered to be a promising alternative route to replace the petroleum based production of chemicals [1]. Anaerobic microbial fermentation can produce volatile fatty acids (VFAs), which are short chain monocarboxylic acids consisting of six or fewer carbon atoms (e.g., acetic, propionic, butyric, valeric, and caproic (hexanoic) acids) [2]. These VFAs have a wide range of applications such as bioplastic production [3], bioenergy [4] as well as the biological removal of nutrients from wastewater [5]. However, the commercialization of VFA value-added chemicals via fermentation is challenging due to the relatively low VFA concentration in the fermentation broths and the complex fermentation composition [6]. Therefore, the separation and purification of these organic acids from fermentation broths have recently received considerable attention [7].

Different methods have been investigated like salt precipitation [8], solvent extraction [9], liquid membranes (LMs) [10], adsorption [11], microfiltration and/or nanofiltration [12], crystallization [13], and electrodialysis [14], etc. Torri et al. [15] introduced lipophilic amines based LMs for selective conversion of VFAs (acetic, propionic, and butyric acids) from anaerobic fermentation systems and proposed that these LMs had a higher affinity for longer carbon chain VFAs. Similar results were also discovered by Nuchnoi et al. [16], who used a supported liquid membrane (SLM) with tri-n-octyl phosphine oxide (TOPO) as a carrier to separate formic, acetic, propionic, and butyric acids. The results obtained exhibited an obvious difference in transport flux for four VFAs and butyric acid was the easiest to transport across the membrane, followed by propionic, acetic, and formic acids. These studies all demonstrated the feasibility of LMs for VFA selective separation. However, these LMs tended to lose the solvent to the water phases and their lack of stability may hinder their applications [17].

Polymer inclusion membrane (PIM) technology has drawn the considerable attention of many researchers in recent years in the separation of small organic compounds and metal ions from aqueous solutions [18–20]. PIMs are a novel type of polymer based liquid membrane where the carrier and plasticizer are incorporated into the entangled chains of the base polymer [21]. The base polymer plays a vital role in providing mechanical strength to the membranes and the carrier is responsible for binding with the interest species and transporting them across the membrane [22]. The plasticizer improves the elasticity, flexibility, and compatibility of the membrane components [23]. It should be pointed out that the majority of carriers in PIMs have plasticizing properties and there is no need to add additional plasticizer to the membrane composition [22–24]. The popularity of PIMs is mainly due to their excellent stability than that of other kinds of LMs such as bulk liquids (BLMs), emulsion liquids (ELMs), and supported liquid membranes (SLMs) [25]. Furthermore, PIMs can provide many other advantages in studies such as high selectivity, simple preparation, long term use, excellent stability and versatility, quick transport, and flexible design [23,26,27], and thus possess much potential.

In recent years, ionic liquids (ILs) have attracted much attention as a kind of environmentally friendly solvent. ILs are salts consisting of an organic cation and inorganic or organic anion with a low melting point [28]. ILs have exhibited many unique properties such as selectivity for specific ions, excellent ionic conductivity, non-flammability, electrochemical stability, high thermal stability, and negligible vapor pressure as well as extractability for different organic and inorganic compounds [29], and are therefore favored by many researchers. In addition, ILs as a carrier can also be employed as effective plasticizers [30].

ILs have been reported as a carrier or extractant in solvent extraction and LMs for the separation of VFAs and have shown a superior performance to conventional solvents in terms of extraction and transport efficiency [31–33]. Yang et al. [34], using Aliquat 336 (methyltri-n-octylammonium chloride) as an extractant to extract lactic, acetic, propionic and butyric acids, found that Aliquat 336 had more potential to extract VFAs than that of tri-n-octylamine (TOA) as an extractant. In fact, these solvents were found to possess the superiority of high selectivity for VFAs, suitable affinity strength for VFAs, and high biocompatibility toward microbial systems [15]. Furthermore, Aliquat 336 and phosphonium-based ionic liquids like trihexyltetradecylphosphonium chloride (Cyphos IL 101) have recently attracted considerable attention in PIM research for the transport of metal ions and small molecular species [35–38].

To the best of our knowledge, most of the literature on the use of PIM have focused on the transport of individual VFAs such as lactic acid [24,39], citric acid [40], succinic acid [41], humic acid [42], or of total VFAs (oxalic, tartaric, and lactic acids) [43] from feed solutions. However, the separation of different kinds of VFAs using PIMs has not been reported. Furthermore, although PIMs have many of the advantages as described above, its relatively lower initial flux values or permeability has always been a major challenge [44]. Therefore, attempts have been made by researchers to improve the properties of PIMs such as introducing crosslinking between components [45], applying novel carriers [46], employing nanoscale additives [47], and applying electric fields on both sides of the membrane [48]. Among them, the application of an electric field by combining the PIM and electrodialysis (ED) process seems to be a favorable and effective method [49].

In this study, the investigation of PIMs for VFA (acetic and hexanoic acids as examples) separation was performed. Two hydrophobic ionic liquids, Aliquat 336 and Cyphos IL101, were selected as carriers to synthesis PIMs. Cellulose triacetate (CTA) was selected as the base polymer due to its good mechanical properties and compatibility. The main parameters influencing the separation process such as the effect of composition (carrier type and content), the effect of feed components (pH and acetic and hexanoic acid concentration) as well as the stripping solution components (stripping solution type and concentration) were investigated. In addition, an integrated system combining electrodialysis with PIM to separate both acids were further explored to verify the performance and feasibility of the PIM in VFA separation during electrodialysis. It is believed that this work may offer a method for green and sustainable VFA separation processes.

2. Results and Discussion

2.1. Transport Mechanism

Acetic acid (pKa = 4.74) and hexanoic acid (pKa = 4.83) exist in two forms (i.e., dissociated and undissociated forms) in aqueous solutions, depending on the solution pH. When the pH < pKa, both acids are protonated and thus exist in undissociated forms. In contrast, dissociated forms are dominant when the pH > pKa [50]. Amine extractants such as TOA and tridodecylamine (TDDA) are typical carriers for carboxylic acids in liquid–liquid extraction and liquid membrane systems [51,52], and only protonated (undissociated) carboxylic acids can be extracted by these extractants through the hydrogen bonding mechanism [9,53]. Nevertheless, the quaternary ammonium Aliquat 336 can extract most of the dissociated and partially undissociated forms of acids because Aliquat 336 is composed of an organic cation associated with a chloride ion [34]. Therefore, coordinating mechanisms are coexist when using Aliquat 336 as the carrier. In conditions of an initial pH 6, the values of pH in the feed solution decreased gradually due to the reverse transport of HCl [54]; when HCl was used as the stripping solution, dissociated (pH > pKa) and undissociated (pH < pKa) forms were observed throughout the operation time. However, the pH of the feed solutions all measured above 6.0 during the experiment when Na_2CO_3 was used as the stripping solution. This means that the anion-exchange mechanism was dominant under this condition. In conditions of pH < pKa, the undissociated acids extracted by Aliquat 336 through the interfacial hydrogen bonding mechanism is known by [51]:

$$\left(R_4N^+Cl^-\right)_{(mem)} + HA_{(aq)} \leftrightarrow \left(R_4N^+Cl^-\right)HA_{(mem)} \quad (1)$$

where $R_4N^+Cl^-$ and HA represent Aliquat 336 and the acetic and hexanoic acid molecules, respectively.

In conditions of pH > pKa, the extraction reaction for the dissociated acid anions by anion-exchange mechanism is described as follows:

$$R_4N^+Cl^-{}_{(mem)} + A^-{}_{(aq)} \leftrightarrow R_4N^+A^-{}_{(mem)} + Cl^-{}_{(aq)} \quad (2)$$

The acids were transferred to the feed/membrane interface and interacted with the carrier to form ionic adducts (Reactions (1) and (2)). The transported compounds are transported through the PIM following a Fickian diffusion pattern [54]. Ultimately, the compounds dissociate immediately at the membrane/stripping interface, according to the reverse reaction of Reactions (1) and (2). The IL molecules return according to their concentration gradient. Compared with acetic acid (Kow of −0.31–0.17), the more hydrophobic hexanoic acid (Kow of 1.88–1.91) is, the easier it reacts with the hydrophobic ionic liquid, thus facilitating its transport.

2.2. SEM Analysis

Scanning electron microscopy (SEM) images of the prepared PIMs were recorded to observe the surface images of the membrane, as presented in Figure 1. From the SEM images, no pores, holes, or cracks could be observed on the membrane surface, which suggested the homogeneous nature of the prepared membranes. Compared to PIM4, PIM6 exhibited a rougher surface with multiple small particles on the surface, which may be related to the dispersibility of the ionic liquid Cyphos IL 101 in the membrane phase.

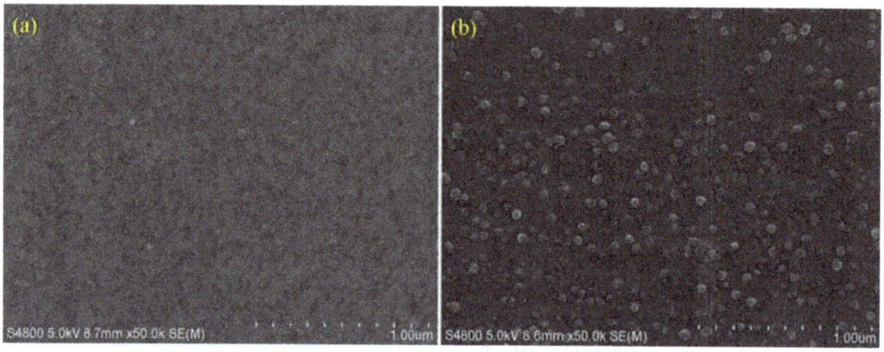

Figure 1. SEM micrographs of the PIMs. (**a**) PIM4; (**b**) PIM6.

2.3. Effect of Carrier Type and Transport Kinetics across the PIM

The nature of the carrier itself is closely related to the properties of the membrane. Therefore, PIMs with two types of ILs (i.e., Aliquat 336 and Cyphos IL 101) as the carrier for the separation of acetic and hexanoic acid were studied in order to investigate the effect of carrier type on the transport of both acids. The extraction rate of the acetic and hexanoic acid is shown in Figure 2a. It can be seen from Figure 2a that the extraction rate of hexanoic acid was much higher than that of acetic acid, regardless of the IL used. As with similar pKa (4.74 and 4.83 for acetic acid and hexanoic acid, respectively), the more hydrophobic the hexanoic acid (octanol–water partition coefficient (Kow) of 1.88–1.91) is, the easier it is to contact the hydrophobic ionic liquid than that of the acetic acid (Kow of −0.31–0.17), and thus easier to transport. This result was also consistent with the results of Torri et al. [15] and Aydin et al. [50], who confirmed that the longer (more lipophilic) the VFA alkyl chain, the higher its transport performance. In addition, the PIM containing Aliquat 336 as the carrier performed a higher extraction rate of hexanoic acid (71.9%) than Cyphos IL 101 as the carrier (56.2%) after 9 h of operation, while the difference in the extraction rate of acetic acid was not obvious. On one hand, this is because the molecular weight of Cyphos IL 101 (519.31) is higher than that of Aliquat 336 (404.17). Thus, the molar amount of ionic liquids in the Aliquat 336 membrane is higher under the same mass. On the other hand, studies have demonstrated that the extraction of carboxylic acids by most ILs in acidic solutions was dominated by hydrogen bonding. The central atom N of Aliquat 336 has an electronegativity of 3.05, which is higher than that of the central atom (P (2.19)) of Cyphos IL 101 [18]. It easily forms hydrogen bonds for Aliquat 336 due to the high electronegativity of N, which decreases the path of acid hopping transport by hydrogen bonding.

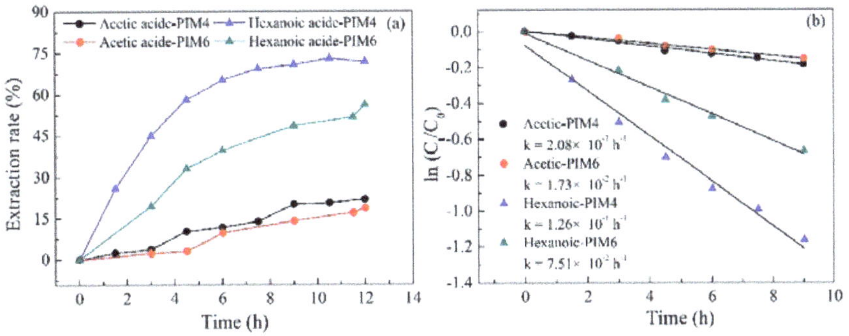

Figure 2. (**a**) Effect of the carrier type (Cyphos IL 101 and Aliquat 336) on the separation of acetic and hexanoic acids from their mixture and (**b**) kinetic plots for the transport of acetic and hexanoic acids across a PIM containing Cyphos IL 101 or Aliquat 336 as the carrier (PIMs: PIM4 and PIM6; Feed solution: 0.01 mol L^{-1} acetic acid + 0.01 mol L^{-1} hexanoic acid, pH = 6; Stripping solution: 0.1 mol L^{-1} HCl).

The plot of $ln(C_f/C_0)$ versus time for the transport of both acids through the PIM is shown in Figure 2b. Obviously, $ln(C_f/C_0)$ has a linear relationship ($R^2 > 0.98$) to time, indicating that the diffusion of acetic and hexanoic acids across the PIM follows the Fick's first-order kinetics. When the slope is greater, the transport of VFAs from the feed solution to the stripping solution is more rapid. From the slope, the value of the permeability coefficient (P) for hexanoic acids when using Aliquat 336 and Cyphos IL101 as the carrier can be calculated as 4.38 and 2.60 μm s^{-1}, respectively (Equation (7)), which were much higher than that of acetic acid (0.72 and 0.60 μm s^{-1}, respectively). These results demonstrate the potential of the Aliquat 336 based PIM developed in this study for the selective separation of acetic and hexanoic acids from aqueous solutions.

The purity of hexanoic acid in the stripping solution is exhibited in Figure 3. As can be seen from Figure 3, the purity of hexanoic acid when using the Aliquat 336 membrane was much higher than that of the Cyphos IL 101 membrane. The maximum purity of hexanoic acid attained 73.2% and 62.8%, respectively. Considering the excellent performance of the Aliquat 336 membrane during the experiment, IL Aliquat 336 was selected as the carrier for the following experiments.

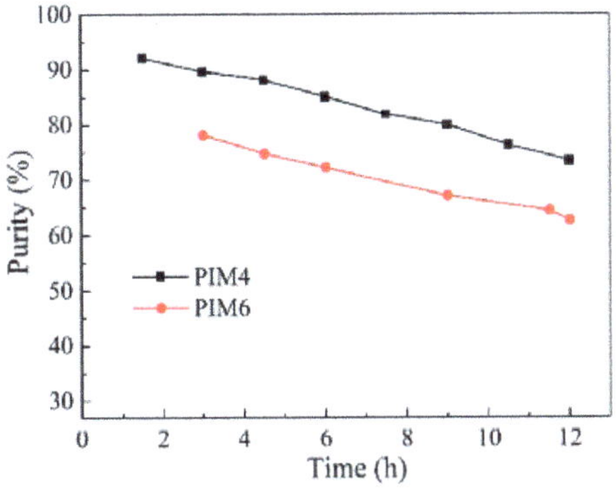

Figure 3. The purity of hexanoic acid in the stripping solution (PIMs: PIM4 and PIM6; Feed solution: 0.01 mol L^{-1} acetic acid +0.01 mol L^{-1} hexanoic acid, pH = 6; Stripping solution: 0.1 mol L^{-1} HCl).

2.4. Effect of Carrier Content

To investigate the effect of carrier content on the separation of acetic and hexanoic acids, membranes were prepared with different contents of Aliquat 336 (30, 40, 50, 60, and 70 wt%) while the total mass of the PIMs remained the same. The variation in the extraction rate of acetic and hexanoic acids as a function of Aliquat 336 content is presented in Figure 4. As shown in Figure 4, the extraction rates of acetic acid and hexanoic acid all increased with the increase of Aliquat 336 content in the membrane, which was due to the improvement in the number of available carriers [55]. Since the carrier also has a plasticizing effect in the PIMs, improving its content lowers the diffusive resistance of the membrane, thereby further increasing the extraction rate. In addition, the extraction rate of acetic acid was relatively lower than that of hexanoic acid under the same content of Aliquat 336. Moreover, the difference in extraction rate of hexanoic acid was more pronounced when using varied contents of Aliquat 336. A significant increase in the recovery rate of hexanoic acid from 29.2% to 71.6% was observed as the Aliquat 336 content improved from 40 wt% to 60 wt%, respectively, after 12 h operation. Further increases of carrier content showed no obvious change. This probably indicates that the Aliquat 336 was more evenly distributed in the membrane when the carrier content was 60 wt%, which was previously confirmed by the energy-dispersive X-ray analyzer (EDX) image [18]. Furthermore, the membrane became more viscous for contents higher than 60 wt%, which improved the membrane resistance [56].

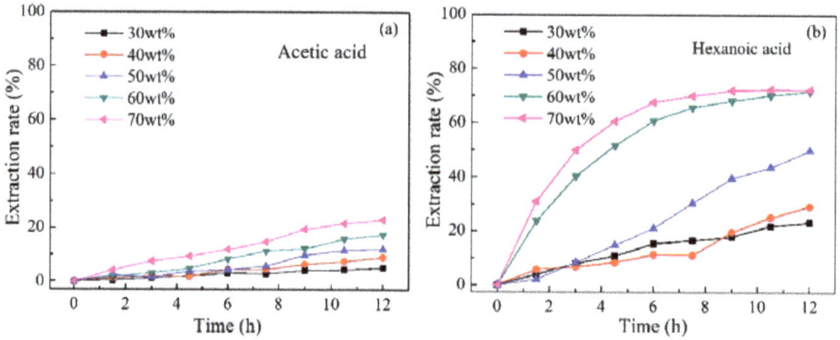

Figure 4. Effect of the Aliquat 336 content on the separation of (**a**) acetic acid and (**b**) hexanoic acid by PIMs (PIMs: PIM1, PIM2, PIM3, PIM4, and PIM5; Feed solution: 0.01 mol L^{-1} acetic acid +0.01 mol L^{-1} hexanoic acid, pH = 6; Stripping solution: 0.1 mol L^{-1} HCl).

Similarly, the permeability coefficient also increased with the carrier content and hexanoic acid showed a much higher value than that of acetic acid (Figure 5). When the content of Aliquat 336 increased from 40 wt% to 60 wt%, the permeability coefficient of hexanoic acid reached 1 μm s^{-1} and 5.39 μm s^{-1}, respectively. However, the permeability coefficients of acetic acid were all below 0.77 μm s^{-1}, although a higher permeability coefficient could be obtained when the ionic liquid content was 70 wt%. However, judging from the quality of the prepared membrane, the membrane became viscous when the carrier content was higher than 60 wt%. Therefore, a PIM containing 60 wt% Aliquat 366 as the carrier was selected for the subsequent experiments.

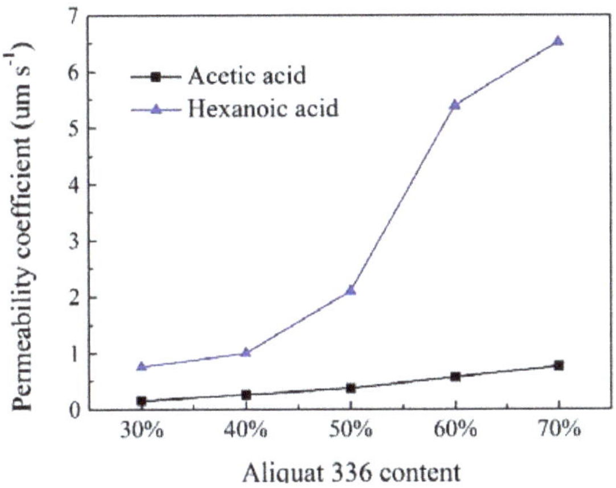

Figure 5. The permeability coefficient of acetic and hexanoic acid by PIMs with different Aliquat 336 content (PIMs: PIMs: PIM1, PIM2, PIM3, PIM4, and PIM5; Feed solution: 0.01 mol L^{-1} acetic acid +0.01 mol L^{-1} hexanoic acid, pH = 6; Stripping solution: 0.1 mol L^{-1} HCl).

2.5. *Effect of Initial pH in Feed Solution*

As is well-known, pH plays an important role in the extraction and separation of VFAs. Thus, it is essential to understand the effects of pH in the feed solution on the separation of acetic and hexanoic acids though the PIM. pH variation studies in the range of 2.0–10.0 were carried out and the results are illustrated in Figure 6 and Table 1. It can be seen from Figure 6 that the concentration of acetic acid and hexanoic acid in the feed solution decreased over time, regardless of the pH, and as expected, the hexanoic acid concentration decreased more rapidly. For acetic acid, the effect of changes in pH on its transport efficiency was not significant. For hexanoic acid, a relatively lower pH (pH = 4.0–5.0) seemed to be more beneficial to its transport. However, the transport efficiency slowed down under more acidic conditions (pH = 2). This may be due to competitive transport between hexanoic acid and HCl [54]. Studies have demonstrated that most amines extract organic acids from an aqueous solution by forming a hydrogen bond with the undissociated acid [53,57]. Since the concentration of undissociated acid greatly depends on the pH, the organic acid must be set at an acidic condition to be protonated. Nevertheless, a comparison of tertiary and quaternary amines (i.e., TOA and Aliquat 336) for the extraction of VFAs was reported by Yang et al. [34]. The authors discovered that quaternary amine Aliquat 336 could extract both dissociated and undissociated forms of acids, which was different to the tertiary amine TOA. Since Aliquat 336 is comprised of an organic cation associated with a chloride ion, the acid could be extracted by an anion-exchange mechanism when the acid is in a dissociated form (pH > pKa), while the coordinating mechanism by hydrogen bonding is favored when the acid is in an undissociated form (pH < pKa). The change in pH at different initial pH values during the experiment was monitored and the results are illustrated in Figure 7. As shown in Figure 7, the values of pH in the feed solution decreased gradually due to the reverse transport of HCl. In conditions of an initial pH 2.0–4.0, both acids existed in undissociated forms (pH < pKa), and the hydrogen bonding mechanism dominated under these conditions. However, in conditions of initial pH 5.0–10.0, the initial undissociated (pH < pKa) and subsequent dissociated (pH > pKa) forms coexisted throughout the operation time. Thus, acidic acid and hexanoic acid can be transported by both hydrogen bonding and anion-exchange mechanisms under these conditions. Therefore, both acetic and hexanoic acids can be transported though PIMs containing Aliquat 336 as the carrier under the initial pH of acidic and alkali conditions.

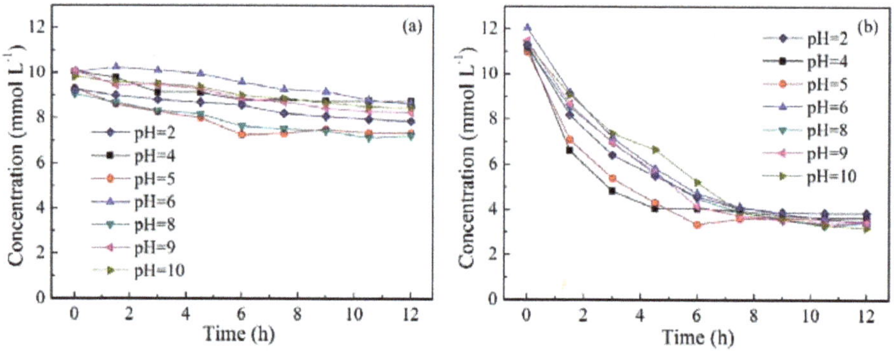

Figure 6. The changes in (**a**) acetic acid and (**b**) hexanoic acid concentrations in the feed solution with different initial pH (PIMs: PIM4; Feed solution: 0.01 mol L^{-1} acetic acid +0.01 mol L^{-1} hexanoic acid; Stripping solution: 0.1 mol L^{-1} HCl).

Table 1. Permeability coefficient (P) and purity at different pH values.

pH	**P (μm s^{-1})**		**Purity (%)**
	Acetic Acid	**Hexanoic Acid**	
2	0.42	5.63	76.2
4	0.42	6.67	77.4
5	0.42	6.25	77.1
6	0.83	5.63	73.5
8	0.63	5.00	75.9
9	0.42	5.63	76.5
10	0.42	5.63	77.9

PIMs: PIM4; Feed solution: 0.01 mol L^{-1} acetic acid +0.01 mol L^{-1} hexanoic acid; Stripping solution: 0.1 mol L^{-1} HCl. Operation time: 12 h.

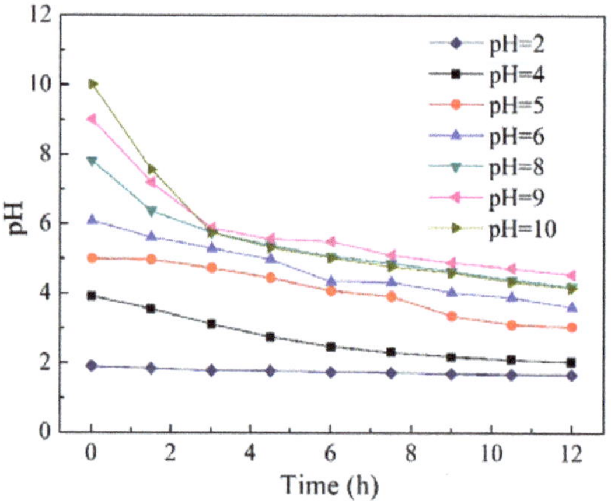

Figure 7. The change of pH in the feed solution at different initial pH values during the experiment (PIMs: PIM4; Feed solution: 0.01 mol L^{-1} acetic acid +0.01 mol L^{-1} hexanoic acid; Stripping solution: 0.1 mol L^{-1} HCl).

The permeability coefficient and purity of acetic and hexanoic acids at different initial pH values are exhibited in Table 1. It can be seen that the permeability coefficient of hexanoic acid was much higher than that of acetic acid. The maximum permeability coefficient of hexanoic acid was shown to have a pH around 4.0, and found to be 6.67 μm s^{-1}. Additionally, the purity of hexanoic acid was above 73.5% after 12 h operation. In general, the typical pH range of the fermentation broth in anaerobic acid fermentation varied. Hence, the results demonstrate that the Aliquat 336 based PIM developed in this study may have more potential for the selective separation of VFAs from aqueous solutions.

2.6. Effect of Initial Concentration in Feed Solution

To evaluate the influence of the initial acid concentration in the feed solution on the selective separation of acetic and hexanoic acids by PIM, experiments were performed by adjusting both acid concentrations from 0.01 mol L^{-1} to 0.05 mol L^{-1}. The acid concentration in the stripping solution during 12 h operation is illustrated in Figure 8. As can be seen from Figure 8, an increase in the initial concentration from 0.01 to 0.03 mol L^{-1} was beneficial for the transport of total acids across the membrane, resulting in the highest total acids concentration of 7.77 mmol L^{-1} at a concentration of 0.03 mol L^{-1} (Figure 8b). With a further increase in the initial concentration to 0.05 mol L^{-1}, the values of final concentration decreased to 6.16 mmol L^{-1}. The reason for the decrease at a higher initial concentration can be explained by the relatively small amount of carrier or the insufficient reaction sites in the membrane, leading to a low extraction rate at a sufficiently high acid concentration [58]. Additionally, as the initial acid concentration increased, the transport of acetic acid decreased. However, it should be noted that the recovery of hexanoic acid remained substantially invariable, which resulting in an improvement in the purity of hexanoic acid in the stripping solution. With an increase in the acid concentration from 0.01 mol L^{-1} to 0.05 mol L^{-1}, the purity of hexanoic acid varied from 73.5% to 89.3% after 12 h of operation. The reason for the increase in the purity of hexanoic acid at a higher concentration was due to the intensively competitive transport of both acids. Hexanoic acid with a longer alkyl chain is more hydrophobic, and is therefore more compatible with the hydrophobic IL Aliquat 336 in the membrane [34]. Hence, to achieve a higher acid purity in this system, improving the initial acid concentration may be a favorable choice.

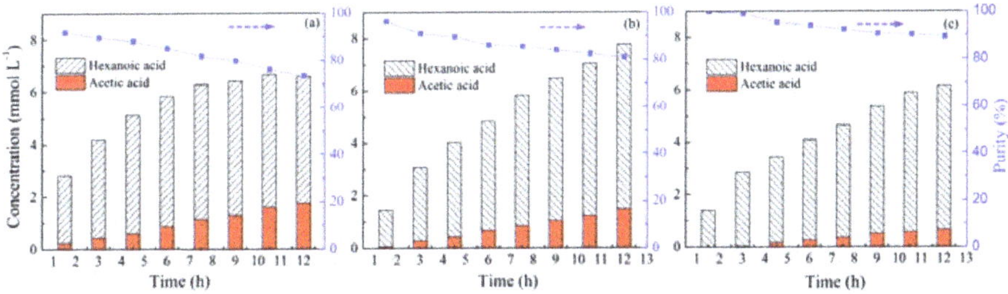

Figure 8. The changes in VFA concentration in the stripping solution with different initial concentrations and the purity of hexanoic acid in the stripping solution (PIMs: PIM4; Feed solution: (**a**) 0.01 mol L^{-1} acetic acid +0.01 mol L^{-1} hexanoic acid; (**b**) 0.03 M acetic acid +0.03 mol L^{-1} hexanoic acid; (**c**): 0.05 mol L^{-1} acetic acid +0.05 mol L^{-1} hexanoic acid, pH = 6; Stripping solution: 0.1 mol L^{-1} HCl).

2.7. Effect of Stripping Solution Type

The stripping solution plays a crucial role in the transport of acid ions through PIM because of the decomplexation reaction that occurs at the membrane/stripping interface. Therefore, two different solutions, namely hydrochloric acid and sodium carbonate, with the same concentration were utilized to detect their suitability as a stripping solution. It can be seen from Figure 9 that the difference between using these two types of stripping solution was obvious. Compared to HCl, using Na_2CO_3 as the stripping solution achieved a much higher transport efficiency of the total organic acids in the same time. The maximum concentration of using HCl and Na_2CO_3 reached 5.04 and 9.44 mmol L^{-1}, respectively. When Na_2CO_3 was used as the stripping solution, the pH of the feed solution all measured above 6.0 (pH > pKa) during the experiment. This means that the ion-exchange mechanism dominated under this condition. Therefore, a possible reason for the better transport performance may be due to the stronger electronegativity of CO_3^{2-} than that of chloride ions. In addition, Aliquat 336 is known to transport HCl as an HCl–Aliquat 336 complex [54], which indicates the reverse transport of HCl when utilizing HCl as the stripping solution. Thus, more acetic acid and hexanoic acid radical ions can be transported to the stripping solution when using Na_2CO_3 as the stripping solution. In addition, the pH difference between the two aqueous solution provides a large driving force for acid transport in the PIM system [51]. Although the amounts of acetic and hexanoic acids transported all increased when utilizing Na_2CO_3 as the stripping solution, the increase of acetic acid was more significant, which resulted in the reduction in hexanoic acid purity (Figure 9b). The optimal purity of hexanoic acid after 12 h of operation for HCl and Na_2CO_3 were 81.8% and 50.8%, respectively. These indicate that the selection of an appropriate stripping solution type is crucial for the PIM to achieve more efficient transport and selective separation.

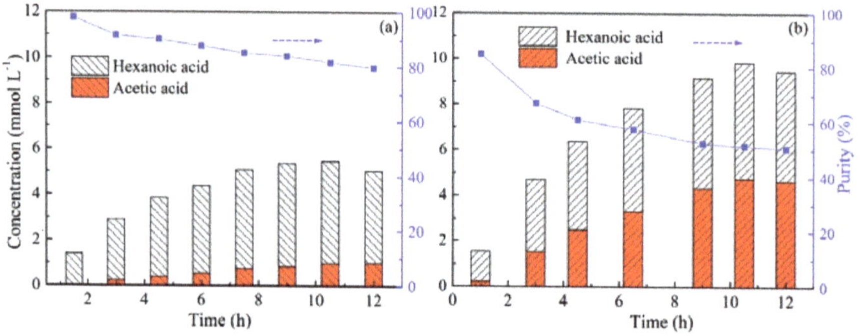

Figure 9. The changes in VFA concentration in the stripping solution with different types of stripping solution (PIMs: PIM4; Feed solution: 0.01 mol L^{-1} acetic acid +0.01 mol L^{-1} hexanoic acid, pH = 6; Stripping solution: (**a**) 0.05 mol L^{-1} HCl; (**b**) 0.05 mol L^{-1} Na_2CO_3).

2.8. Effect of Acid Concentration in the Stripping Solution

The type of stripping solution is also a significant parameter that influences the selective separation. To evaluate the influences of HCl concentration in the stripping solution on the transport of both acids, experiments were carried out by adjusting the HCl concentration from 0.05 to 0.2 mol L^{-1}. As shown in Figure 10, variation of the HCl concentration in the stripping solution did not have a dramatic effect on the total transport of both VFAs under these experimental conditions as per the given maximum concentration values, which were 5.03, 5.29, 5.69 and 6.14 mmol L^{-1}, respectively, for used HCl concentrations of 0.05, 0.1, 0.15, and 0.2 mol L^{-1}. The slight increase was due to the counter transport of protons that provided a driving force for the transport process [55]. However, the effect of HCl concentration on acetic acid was obvious. In particular, with an increase in the HCl concentration from 0.1 to 0.15 mol L^{-1}, the concentration of acetic acid in the stripping solution increased dramatically from 1.07 to 1.78 mmol L^{-1}, which resulted in the decrease of hexanoic acid purity in the stripping solution. As observed in Figure 11, with an increase in the HCl concentration, the purity values exhibited a decrease tendency. This is possibly related to the large quantitative transport of HCl in the reverse manner, which led to the reduced transport of hexanoic acid in the stripping solution [58]. In general, no discernible benefit was observed by improving the acid concentration, and it is desirable that the acid concentration should be as low as possible for cost and safety reasons. Thus, the optimum acid for back-extraction under transport conditions is 0.05 mol L^{-1}HCl.

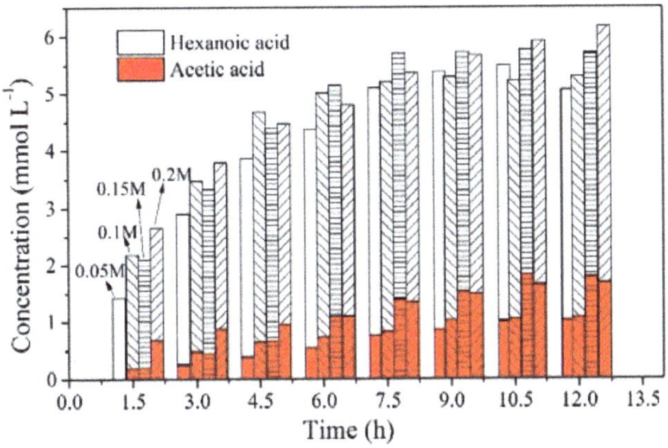

Figure 10. The changes in the acid concentration in the stripping solution with different HCl concentrations in the stripping solution (PIMs: PIM4; Feed solution: 0.01 mol L^{-1} acetic acid +0.01 mol L^{-1} hexanoic acid, pH = 6; Stripping solution: 0.05, 0.1, 0.15 and 0.2 mol L^{-1} HCl).

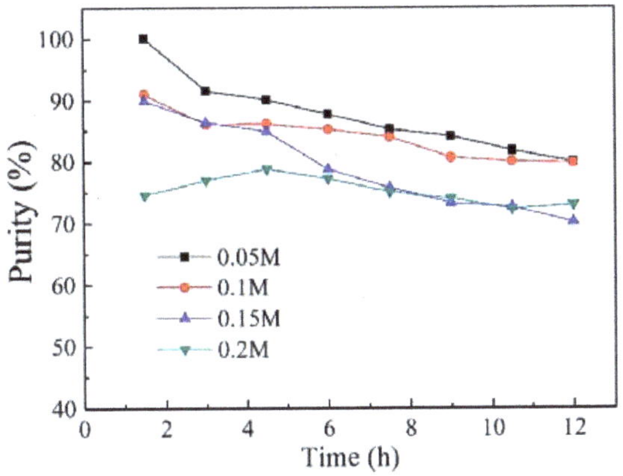

Figure 11. Effect of HCl concentration on the purity of hexanoic acid (PIMs: PIM4; Feed solution: 0.01 mol L^{-1} acetic acid +0.01 mol L^{-1} hexanoic acid, pH = 6; Stripping solution: 0.05, 0.1, 0.15, and 0.2 mol L^{-1} HCl).

2.9. Adsorption Experiment

To further investigate the transport mechanism of both acids within the PIM using Aliquat 336 as the carrier, an adsorption experiment of the membrane was conducted to verify the difference in transport efficiency between the two acids. As demonstrated in Figure 12, the adsorption amount of hexanoic acid rose rapidly in a short period of time, whereas only a small amount of acetic acid was adsorbed during the same time. The maximum adsorption capacity of hexanoic acid was 2.30 mmol g^{-1}, which was 10 times greater than that of acetic acid (2.30 mmol g^{-1}). This means that hydrophilic acetic acid finds it difficult to form ionic adducts with the carrier Aliquat 336 in the feed/membrane interface, resulting in a lower transport and recovery efficiency of acetic acid.

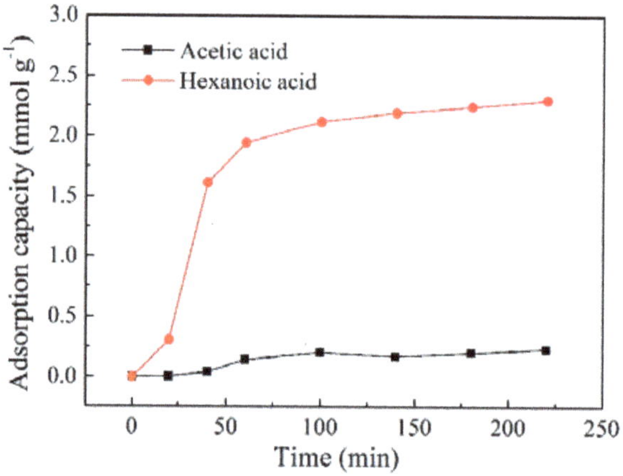

Figure 12. Adsorption capacity of PIM for acetic acid and hexanoic acid (PIMs: PIM4; Feed solution: 0.01 mol L^{-1} acetic acid +0.01 mol L^{-1} hexanoic acid, pH = 6).

2.10. Transport Experiment during Electrodialysis

In order to evaluate the feasibility of PIM in the transport and separation of both acids in electrodialysis, a transport experiment combing electrodialysis with the PIM was carried out under a current density of 5 mA cm^{-2}. As can be seen in Figure 13a, the transport efficiency was remarkably improved with an electric field was applied when compared with the obtained results without an electric field (Figure 8c). Under the electric field condition, the concentrations of acetic acid and hexanoic acid in the stripping solution reached 22.10 mmol L^{-1} and 21.55 mmol L^{-1}, respectively, after 3.5 h of operation. However, the concentration was only 1.38 and 7.22, respectively, after 12 h of operation without the application of an electric field (Figure 8c). This is because the transport of acid ions is accelerated when an electric field is applied. In addition, more acids may be dissociated and quickly transported across the membrane under the effect of an electric field. Nevertheless, according to the presented results, the concentration changes and recovery rates of both acids were very similar to each other (Figure 13a,b). This indicates that the application of an electric field was not beneficial to the selective separation of both acids. Moreover, the application of an electric field may overcome the binding resistance of more hydrophilic acetic acid to the carrier on the membrane, thus greatly increasing the transport efficiency. According to the investigation above, it can be concluded that applying PIM to electrodialysis can dramatically improve the efficiency of the transport of both acids, but is not conducive to their selective separation.

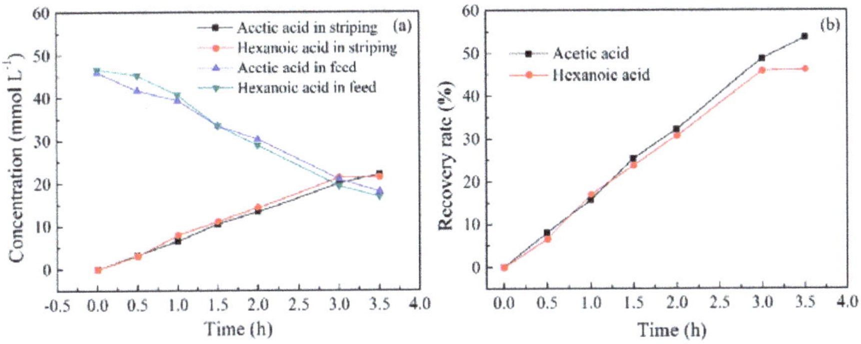

Figure 13. (**a**) The changes of acid concentration in the feed and stripping solution and (**b**) the recovery rate of both acids during electrodialysis (PIMs: PIM4; Feed solution: 0.05 mol L^{-1} acetic acid +0.05 mol L^{-1} hexanoic acid, pH = 6; Stripping solution: 0.1 mol L^{-1} HCl; Current density: 5 mA cm^{-2}).

3. Materials and Methods

3.1. Materials

Cyphos IL 101 (Merck Life Science Co., Ltd, Shanghai, China), Aliquat 336 (Alfa Aesar Chemical Co., Ltd, Beijing, China), CTA (cellulose triacetate) (Acros Organics, New Jersey, USA), and DCM (dichloromethane) (Sinopharm Chemical Reagent Co., Ltd, Shanghai, China) were used to cast the PIMs. Hexanoic acid was purchased from Merck Life Science Co., Ltd. (Shanghai, China). Acetic acid, HCl, H_2SO_4, NaOH, and Na_2CO_3 used in the experiments were of an analytical grade and purchased from Sinopharm Chemical Reagent Co., Ltd. (Shanghai, China). All chemicals were used without further purification. Deionized water was used to prepare all aqueous solutions. The commercial cation exchange membrane (CJMC-5) was obtained from Hefei Chemjoy Polymer Material Co. Ltd. (Hefei, China).

3.2. Membrane Preparation

PIMs with CTA as the base polymer, and Cyphos IL 101 or Aliquat 336 as the carrier were prepared using the solvent evaporation casting method, and the different compositions and proportions of PIMs studied are listed in Table 2. The composition of each component was varied while ensuring the same total amount of membrane matrix to ensure a consistent membrane thickness. The PIM components with a total mass of 1.1 g was dissolved in 20 mL of DCM solvent, followed by stirring for 2 h using a magnetic stirring bar at room temperature. The uniform casting solution was obtained after degassing for 15 min. The solutions were then poured onto flat glass (13 × 13 cm) placed horizontally, and covered with a watch glass to allow the slow evaporation of the DCM over 12 h. After the evaporation of DCM, the resulting PIMs were then carefully peeled off the glass plate and stored at low temperature for further experiments. The membrane thickness was determined by cutting the membrane in half diagonally and measuring the thickness of 20 points along the cut edge. To acquire knowledge on the surface morphology and characteristics of the membrane, the surface morphology of PIM4 and PIM6 were characterized using scanning electron microscopy (SEM) (Hitachi S-4800, Japan). The chemical structures of CTA and ionic liquids used in the experiment are shown in Figure 14.

Table 2. The information of the PIM composition and proportion.

Membrane Number	Base Polymer (wt%)	Carrier (wt%)		Total Weight (g)	Thickness (μm)
	CTA	Aliquat 336	Cyphos IL101		
1	70	30		1.1	58.1 ± 5.1
2	60	40		1.1	57.8 ± 6.2
3	50	50		1.1	58.6 ± 5.6
4	40	60		1.1	60.4 ± 5.2
5	30	70		1.1	59.6 ± 8.7
6	40		60	1.1	61.4 ± 4.6

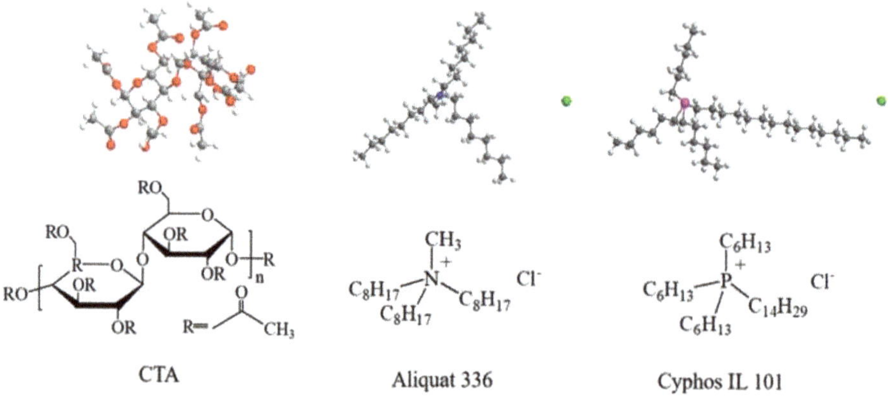

Figure 14. The chemical structures of the polymer and ionic liquids.

3.3. Transport Experiments

Transport experiments were conducted at approximately 25 °C in a device that comprised two equal transport cells (thickness, 1 cm) in contact through a PIM membrane with an exposed membrane area of 20 cm^2. The experimental apparatus is presented in detail in Figure 15a. The feed solution contained 250 mL of VFA (acetic and hexanoic acids) solution and the stripping solution contained 250 mL of Na_2CO_3 or HCl solution. The pH value in the feed solution was adjusted by the NaOH and HCl solution. The entire two cell solutions were driven by pumps at a certain flow rate of 300 mL min^{-1}.

Unlike the above transport experiments, the ED experiment was performed using the ED stack comprised of two titanium electrode boards coated with rare metals (ruthenium) and four cells (Figure 15b). Four cells were separated by one prepared PIM and two CJMC-5 commercial cation exchange membranes. The thickness of the cells and the effective membrane area was also 1 cm and 20 cm^2, respectively. The feed solution contained 250 mL 0.05 mol L^{-1} VFA (acetic and hexanoic acids) solution and the stripping solution contained 250 mL 0.1 mol L^{-1} HCl solution. Two electrode cells were pumped with the same concentration of 250 mL 0.1 mol L^{-1} H_2SO_4 solution. The constant current mode was supplied by a DC power supply (MCH-k305D, Shenzhen, China) with a current density of 5 mA cm^{-2}. Other experimental conditions were the same as the above experiment.

In all experiments, samples for VFA (acetic and hexanoic acids) analysis were taken periodically from both cells using a micropipette and a chemical determination was performed using a high performance liquid chromatography (HPLC) unit equipped with a UV detector (210 nm) (Agilent Corp., Santa Clara, CA, USA). An ion exchange column (Aminex HPX-87H) (300 × 7.8 mm, Bio-Rad, Hercules, CA, USA) was used at a temperature of 60 °C with a 5 mmol L^{-1} H_2SO_4 solution as an eluent at a flow rate of 0.6 mL min^{-1}.

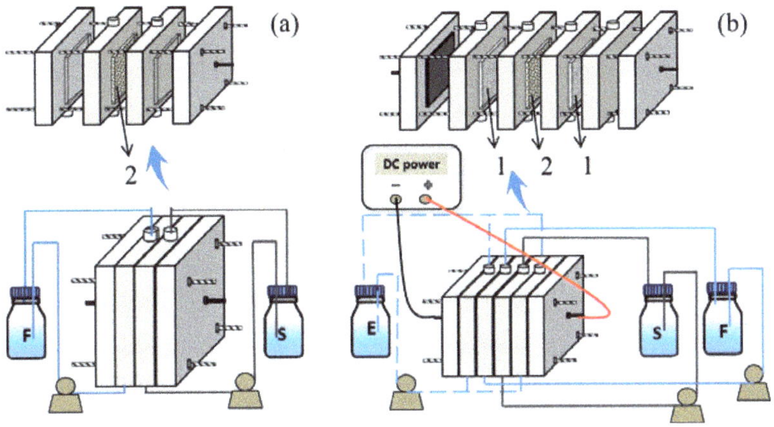

Figure 15. The experimental apparatus for the selective separation of VFAs (**a**) without the application of an electric field and (**b**) with the application of an electric field (1: Cation exchange membrane; 2: PIM; F: Feed solution; S: Stripping solution; E: Electrode solution).

3.4. Adsorption Capacity Test

The adsorption experiment was carried out using a CTA-60 wt% Aliquat 336 membrane to study its difference in the adsorption capacity of both acids. Membrane segments of an equal size were weighed and immersed into the feed solution containing 150 mL of 0.01 mol L^{-1} acetic and hexanoic acids, respectively. The solution was stirred using a constant temperature oscillator at 100 rpm and 1 mL samples were withdrawn at predetermined times, which were replaced with 1 mL of fresh aqueous feed solution for the purpose of keeping the aqueous feed solution volume constant. The samples were analyzed subsequently and the adsorption capacity of the membrane was obtained by calculation.

3.5. Data Processing

Adsorption capacity q_e (mol g^{-1}) of the membrane were calculated using Equation (3):

$$q_e = \frac{(C_i - C_e) \cdot V}{m} \tag{3}$$

where q_e is the equilibrium adsorption capacity for VFAs (mol g^{-1}). C_i and C_e are the concentrations of the initial and equilibrium VFA feed solutions (mol L^{-1}), respectively; V is the volume of the feed solution (L); and m is the measured mass of the PIM segments (g).

The extraction rate E (%) of the VFAs from the feed solution was calculated based on Equation (4):

$$E = \frac{C_0 - C_f}{C_0} \times 100\% \tag{4}$$

where C_0 is the concentration of VFAs in the feed solution after transport (mg L^{-1}), and C_f is the initial concentration of VFAs in the feed solution (mg L^{-1}).

The recovery rate R (%) of VFAs from the PIM was determined by Equation (5):

$$R = \frac{C_S}{C_0} \times 100\% \tag{5}$$

where C_s is the concentration of VFAs in the stripping solution after transport.

The kinetics of the transport process through the PIM can be described by a first order reaction with respect to VFA concentration.

$$\ln\left(\frac{C_f}{C_0}\right) = kt \quad (6)$$

where k is the transport process rate constant (s^{-1}), and t is the transport time (s).The relationship of $ln(C_f/C_0)$ vs. time was linear, which was verified by the high values of the correlation coefficients (R^2) ranging between 0.983 and 0.991.

The permeability coefficient P ($\mu m\ s^{-1}$) was determined via Equation (7):

$$P = \frac{V}{A} \times k \quad (7)$$

where V is the volume of the feed solution (m^3), and A is the effective area of the membrane (m^2).

The purity (%) of an individual VFA in the stripping solution was calculated using Equation (8):

$$\text{Purity} = \frac{C_s}{C_t} \times 100\% \quad (8)$$

where C_s is the concentration of an individual VFA in the stripping solution after transport (mol L^{-1}), and C_t is the concentration of total VFAs in the stripping solution (mol L^{-1}).

The average values of three parallel samples were utilized as the experimental data. The errors of all three experiments were all less than 5%. The obvious differences between each group was evaluated using Tukey's multiple range tests, with $P < 0.01$ as an obvious difference.

4. Conclusions

The transport and recovery of VFAs (acetic and hexanoic acids) though a PIM containing Aliquat 336 and Cyphos IL 101 as the carrier were investigated. The more hydrophobic hexanoic acid was much more selectively transported than that of acetic acid through PIMs. PIM with 60 wt% Aliquat 336 as the carrier was proven to have a more efficient separation performance. A larger initial concentration of VFAs tended to give a much higher purity of hexanoic acid in the stripping solution, which was 89.3 % at the maximum concentration of 0.05 mol L^{-1}. Furthermore, the use of Na_2CO_3 as a stripping solution achieved a much higher transport efficiency of the total VFAs when compared with HCl, whereas the purity of hexanoic acid in the stripping solution decreased from 81.8% to 50.8%. Additionally, the higher the concentration of HCl in the stripping solution, the lower the purity of the hexanoic acid. The efficient transport of the acid radical ions in an alkaline solution demonstrated that the transport mechanism was a coordinating mechanism of hydrogen bonding and ion-exchange. Moreover, the incorporation of electrodialysis and PIM resulted in a remarkable improvement in the transport efficiency of both acids, while selective separation was not achieved. The results in this study reveal that the PIM containing Aliquat 336 offers considerable potential for the selective separation of VFAs from aqueous solutions.

Author Contributions: Conceptualization, B.-Y.W. and N.Z.; Methodology, B.-Y.W., N.Z.; Software, B.-Y.W. and N.Z.; Validation, B.-Y.W. and N.Z.; Formal analysis, B.-Y.W. and N.Z.; Investigation, B.-Y.W., N.Z., and Z.-Y.L.; Resources, B.-Y.W., N.Z., Z.-Y.L., and Q.-L.L.; Data curation, B.-Y.W. and N.Z.; Writing—original draft preparation, B.-Y.W.; Writing—review and editing, Y.L. and Y.Z.; Visualization, B.-Y.W.; Supervision, Q.-L.L., B.-H.Y., Y.L., and Y.Z.; Project administration, Y.Z. and Y.L.; Funding acquisition, Y.Z. and Y.L.

Funding: This research was funded by the National Natural Science Foundation of China (51708542, 21878319), and the Shandong Provincial Natural Science Foundation (ZR2017BB062).

Acknowledgments: The authors would like to acknowledge the financial support of the National Natural Science Foundation of China (51708542, 21878319) and the Shandong Provincial Natural Science Foundation (ZR2017BB062).

Conflicts of Interest: The authors declare no conflict of interest.

References

1. Koller, M.; Marsalek, L.; Dias, M.M.D.S.; Braunegg, G. Producing microbial polyhydroxyalkanoate (PHA) biopolyesters in a sustainable manner. *New Biotechnol.* **2017**, *37*, 24–38. [CrossRef] [PubMed]
2. Lee, W.S.; Chua, A.S.M.; Yeoh, H.K.; Ngoh, G.C. A review of the production and applications of waste-derived volatile fatty acids. *Chem. Eng. J.* **2014**, *235*, 83–99. [CrossRef]
3. Cai, M.M.; Chua, H.; Zhao, Q.L.; Shirley, S.N.; Ren, J. Optimal production of polyhydroxyalkanoates (PHA) in activated sludge fed by volatile fatty acids (VFAs) generated from alkaline excess sludge fermentation. *Bioresour. Technol.* **2009**, *100*, 1399–1405.
4. Choi, J.D.R.; Chang, H.N.; Han, J.I. Performance of microbial fuel cell with volatile fatty acids from food wastes. *Biotechnol. Lett.* **2011**, *33*, 705–714. [CrossRef] [PubMed]
5. Zheng, X.O.; Chen, Y.G.; Liu, C.C. Waste activated sludge alkaline fermentation liquid as carbon source for biological nutrients removal in anaerobic followed by alternating aerobic-anoxic sequencing batch reactors. *Chin. J. Chem. Eng.* **2010**, *18*, 478–485. [CrossRef]
6. Pratt, S.; Liew, D.; Batstone, D.J.; Werker, A.G.; Morgan-Sagastume, F.; Lant, P.A. Inhibition by fatty acids during fermentation of pre-treated waste activated sludge. *J. Biotechnol.* **2012**, *159*, 38–43. [CrossRef]
7. Alkaya, E.; Kaptan, S.; Ozkan, L.; Uludag-Demirer, S.; Demirer, G.N. Recovery of acids from anaerobic acidification broth by liquid-liquid extraction. *Chemosphere* **2009**, *77*, 1137–1142. [CrossRef]
8. IJmker, H.M.; Gramblicka, M.; Kersten, S.R.A.; van der Ham, A.G.J.; Schuur, B. Acetic acid extraction from aqueous solutions using fatty acids. *Sep. Purif. Technol.* **2014**, *125*, 256–263. [CrossRef]
9. Van den Bruinhorst, A.; Raes, S.; Maesara, S.A.; Kroon, M.C.; Esteves, A.C.C.; Meuldijk, J. Hydrophobic eutectic mixtures as volatile fatty acid extractants. *Sep. Purif. Technol.* **2019**, *216*, 147–157. [CrossRef]
10. Kouki, N.; Tayeb, R.; Zarrougui, R.; Dhahbi, M. Transport of salicylic acid through supported liquid membrane based on ionic liquids. *Sep. Purif. Technol.* **2010**, *76*, 8–14. [CrossRef]
11. Li, Q.; Xing, J.M.; Li, W.L.; Liu, Q.F.; Su, Z.G. Separation of Succinic Acid from Fermentation Broth Using Weak Alkaline Anion Exchange Adsorbents. *Ind. Eng. Chem. Res.* **2009**, *48*, 3595–3599. [CrossRef]
12. Dey, P.; Linnanen, L.; Pal, P. Separation of lactic acid from fermentation broth by cross flow nanofiltration: Membrane characterization and transport modelling. *Desalination* **2012**, *288*, 47–57. [CrossRef]
13. Li, Q.; Wang, D.; Wu, Y.; Li, W.L.; Zhang, Y.J.; Xing, J.M.; Su, Z.G. One step recovery of succinic acid from fermentation broths by crystallization. *Sep. Purif. Technol.* **2010**, *72*, 294–300. [CrossRef]
14. Jones, R.J.; Massanet-Nicolau, J.; Guwy, A.; Premier, G.C.; Dinsdale, R.M.; Reilly, M. Removal and recovery of inhibitory volatile fatty acids from mixed acid fermentations by conventional electrodialysis. *Bioresour. Technol.* **2015**, *189*, 279–284. [CrossRef]
15. Torri, C.; Samori, C.; Ajao, V.; Baraldi, S.; Galletti, P.; Tagliavini, E. Pertraction of volatile fatty acids through biodiesel-based liquid membranes. *Chem. Eng. J.* **2019**, *366*, 254–263. [CrossRef]
16. Nuchnoi, P.; Yano, T.; Nishio, N.; Nagai, S. Extraction of volatile fatty-acids from diluted aqueous-solution using a supported liquid membrane. *J. Ferment. Bioeng.* **1987**, *65*, 301–310. [CrossRef]
17. Kaya, A.; Onac, C.; Alpoguz, H.K.; Yilmaz, A.; Atar, N. Removal of Cr(VI) through calixarene based polymer inclusion membrane from chrome plating bath water. *Chem. Eng. J.* **2016**, *283*, 141–149. [CrossRef]
18. Zhang, N.; Liu, Y.; Liu, R.; She, Z.L.; Tan, M.; Mao, D.B.; Fu, R.Q.; Zhang, Y. Polymer inclusion membrane (PIM) containing ionic liquid as a proton blocker to improve waste acid recovery efficiency in electrodialysis process. *J. Membr. Sci.* **2019**, *581*, 18–27. [CrossRef]
19. Garcia, A.; Alvarez, A.; Matamoros, V.; Salvadó, V.; Fontàs, C. Development of polymer inclusion membranes for the extraction of antibiotics from environmental waters. *Procedia Eng.* **2012**, *44*, 804–806. [CrossRef]
20. Garcia-Rodríguez, A.; Fontàs, C.; Matamoros, V.; Almeida, M.I.G.S.; Kolev, S.D. Development of a polymer inclusion membrane-based passive sampler for monitoring of sulfamethoxazole in natural waters. Minimizing the effect of the flow pattern of the aquatic system. *Microchem. J.* **2016**, *124*, 175–180. [CrossRef]
21. Ershad, M.; Almeida, M.I.G.S.; Spassov, T.G.; Cattrall, R.W.; Kolev, S.D. Polymer inclusion membranes (PIMs) containing purified dinonylnaphthalene sulfonic acid (DNNS): Performance and selectivity. *Sep. Purif. Technol.* **2018**, *195*, 446–452. [CrossRef]
22. Almeida, M.I.G.S.; Cattrall, R.W.; Kolev, S.D. Recent trends in extraction and transport of metal ions using polymer inclusion membranes (PIMs). *J. Membr. Sci.* **2012**, *415*, 9–23. [CrossRef]

23. Long, D.N.; Mornane, P.; Potter, I.D.; Perera, J.M.; Cattrall, R.W.; Kolev, S.D. Extraction and transport of metal ions and small organic compounds using polymer inclusion membranes (PIMs). *J. Membr. Sci.* **2006**, *281*, 7–41.
24. O'Rourke, M.; Cattrall, R.W.; Kolev, S.D.; Potter, I.D. The Extraction and Transport of Organic Molecules Using Polymer Inclusion Membranes. *Solvent Extr. Res. Dev.* **2009**, *16*, 1–12.
25. Baczyńska, M.; Słomka, Ż.; Rzelewska, M.; Waszak, M.; Nowicki, M.; Regel-Rosocka, M. Characterization of polymer inclusion membranes (PIM) containing phosphonium ionic liquids and their application for separation of Zn(II) from Fe(III). *J. Chem. Technol. Biotechnol.* **2018**, *93*, 1767–1777. [CrossRef]
26. O'Bryan, Y.; Cattrall, R.W.; Truong, Y.B.; Kyratzis, I.L.; Kolev, S.D. The use of poly(vinylidenefluoride-co-hexafluoropropylene) for the preparation of polymer inclusion membranes. Application to the extraction of thiocyanate. *J. Membr. Sci.* **2016**, *510*, 481–488. [CrossRef]
27. Zaheri, P.; Ghassabzadeh, H. Preparation of polymer inclusion membrane including mixture of D2EHPA and Cyanex272 for the extraction of Eu from nitrate media. *Chem. Pap.* **2017**, *71*, 1–9. [CrossRef]
28. Rynkowska, E.; Fatyeyeva, K.; Kujawski, W. Application of polymer-based membranes containing ionic liquids in membrane separation processes: A critical review. *Rev. Chem. Eng.* **2017**, *34*, 341–363. [CrossRef]
29. Baczyńska, M.; Waszak, M.; Nowicki, M.; Prządka, D.; Borysiak, S.; Regel-Rosocka, M. Characterization of polymer inclusion membranes (PIMs) containing phosphonium ionic liquids as Zn(II) carriers. *Ind. Eng. Chem. Res.* **2018**, *57*, 5070–5082. [CrossRef]
30. Darvishi, R.; Sabet, J.K.; Esfahany, M.N. Preparation and characterization of a novel calcium-conducting polymer inclusion membrane: Part, I. *Korean J. Chem. Eng.* **2018**, *35*, 1–13. [CrossRef]
31. Tonova, K.; Svinyarov, I.; Bogdanov, M.G. Hydrophobic 3-alkyl-1-methylimidazolium saccharinates as extractants for l-lactic acid recovery. *Sep. Purif. Technol.* **2014**, *125*, 239–246. [CrossRef]
32. Reyhanitash, E.; Zaalberg, B.; Kersten, S.R.A.; Schuur, B. Extraction of volatile fatty acids from fermented wastewater. *Sep. Purif. Technol.* **2016**, *161*, 61–68. [CrossRef]
33. Lopez, A.M.; Hestekin, J.A. Improved organic acid purification through wafer enhanced electrodeionization utilizing ionic liquids. *J. Membr. Sci.* **2015**, *493*, 200–205. [CrossRef]
34. Yang, S.T.; White, S.A.; Hsu, S.T. Extraction of carboxylic-acids with tertiary and quaternary amines-effect of pH. *Ind. Eng. Chem. Res.* **1991**, *30*, 1335–1342. [CrossRef]
35. Kagaya, S.; Ryokan, Y.; Cattrall, R.W.; Kolev, S.D. Stability studies of poly (vinyl chloride)-based polymer inclusion membranes containing Aliquat 336 as a carrier. *Sep. Purif. Technol.* **2012**, *101*, 69–75. [CrossRef]
36. Yıldız, Y.; Manzak, A.; Tutkun, O. Selective extraction of cobalt ions through polymer inclusion membrane containing Aliquat 336 as a carrier. *Desalin. Water Treat.* **2016**, *57*, 4616–4623. [CrossRef]
37. Litaiem, Y.; Dhahbi, M. Measurements and correlations of viscosity, conductivity and density of an hydrophobic ionic liquid (Aliquat 336) mixtures with a non-associated dipolar aprotic solvent (DMC). *J. Mol. Liq.* **2012**, *169*, 54–62. [CrossRef]
38. Yaftian, M.R.; Almeida, M.I.G.S.; Cattrall, R.W.; Kolev, S.D. Selective extraction of vanadium(V) from sulfate solutions into a polymer inclusion membrane composed of poly(vinylidenefluoride-co-hexafluoropropylene) and Cyphos® IL 101. *J. Membr. Sci.* **2018**, *545*, 57–65. [CrossRef]
39. Matsumoto, M.; Murakami, Y.; Minamidate, Y.; Kondo, K. Separation of lactic acid through polymer inclusion membranes containing ionic liquids. *Sep. Sci. Technol.* **2012**, *47*, 354–359. [CrossRef]
40. Gajewski, P.; Bogacki, M.B. Influence of alkyl chain length in 1-alkylimidazol on the citric acid transport rate across polymer inclusion membrane. *Sep. Sci. Technol.* **2012**, *47*, 1374–1382. [CrossRef]
41. Pratiwi, A.I.; Sato, T.; Matsumoto, M.; Kondo, K. Permeation mechanism of succinic acid through polymer inclusion membranes with ionic liquid Aliquat 336. *J. Chem. Eng. Jpn.* **2014**, *47*, 314–318. [CrossRef]
42. Manzak, A.; Kursun, C.; Yildiz, Y. Characterization of humic acid extracted from aqueous solutions with polymer inclusion membranes. *J. Taiwan Inst. Chem. Eng.* **2017**, *81*, 14–20. [CrossRef]
43. Przewozna, M.; Gajewski, P.; Michalak, N.; Bogacki, M.B.; Skrzypczak, A. Determination of the Percolation Threshold for the Oxalic, Tartaric, and Lactic Acids Transport through polymer inclusion membranes with 1-alkylimidazoles as a Carrier. *Sep. Sci. Technol.* **2014**, *49*, 1745–1755. [CrossRef]
44. Wang, D.; Hu, J.; Liu, D.; Chen, Q.; Jie, L. Selective transport and simultaneous separation of Cu(II), Zn(II) and Mg(II) using a dual polymer inclusion membrane system. *J. Membr. Sci.* **2017**, *524*, 205–213. [CrossRef]

45. O'Bryan, Y.; Truong, Y.B.; Cattrall, R.W.; Kyratzis, I.L.; Kolev, S.D. A new generation of highly stable and permeable polymer inclusion membranes (PIMs) with their carrier immobilized in a crosslinked semi-interpenetrating polymer network. Application to the transport of thiocyanate. *J. Membr. Sci.* **2017**, *529*, 55–62. [CrossRef]
46. Yoshida, W.; Baba, Y.; Kubota, F.; Kolev, S.D.; Goto, M. Selective transport of scandium(III) across polymer inclusion membranes with improved stability which contain an amic acid carrier. *J. Membr. Sci.* **2019**, *572*, 291–299. [CrossRef]
47. Kaya, A.; Onac, C.; Alpoguz, H.K.; Agarwal, S.; Gupta, V.K.; Atar, N.; Yilmaz, A. Reduced graphene oxide based a novel polymer inclusion membrane: Transport studies of Cr(VI). *J. Mol. Liq.* **2016**, *219*, 1124–1130. [CrossRef]
48. Meng, X.; Wang, C.; Ren, T.; Lei, W.; Wang, X. Electrodriven transport of chromium (VI) using 1-octanol/PVC in polymer inclusion membrane under low voltage. *Chem. Eng. J.* **2018**, *346*, 506–514. [CrossRef]
49. Sadyrbaeva, T.Z. Removal of chromium(VI) from aqueous solutions using a novel hybrid liquid membrane—Electrodialysis process. *Chem. Eng. Process.* **2016**, *99*, 183–191. [CrossRef]
50. Aydin, S.; Yesil, H.; Tugtas, A.E. Recovery of mixed volatile fatty acids from anaerobically fermented organic wastes by vapor permeation membrane contactors. *Bioresour. Technol.* **2018**, *250*, 548–555. [CrossRef]
51. Lee, S.C.; Hyun, K.S. Development of an emulsion liquid membrane system for separation of acetic acid from succinic acid. *J. Membr. Sci.* **2010**, *350*, 333–339. [CrossRef]
52. Yordanov, B.; Boyadzhiev, L. Pertraction of citric acid by means of emulsion liquid membranes. *J. Membr. Sci.* **2004**, *238*, 191–197. [CrossRef]
53. Schlosser, Š.; Marták, J.; Blahušiak, M. Specifc phenomena in carboxylic acids extraction by selected types of hydrophobic ionic liquids. *Chem. Pap.* **2018**, *72*, 567–584. [CrossRef]
54. Rao, R.V.S.; Sivakumar, P.; Natarajan, R.; Rao, P.R.V. Effect of Aliquat 336 concentration on transportation of hydrochloric acid across supported liquid membrane. *J. Radioanal. Nucl. Chem.* **2002**, *252*, 95–98.
55. Cai, C.Q.; Yang, F.; Zhao, Z.G.; Liao, Q.X.; Bai, R.X.; Guo, W.H.; Chen, P.; Zhang, Y.; Zhang, H. Promising transport and high-selective separation of Li(I) from Na(I) and K(I) by a functional polymer inclusion membrane (PIM) system. *J. Membr. Sci.* **2019**, *579*, 1–10. [CrossRef]
56. Kebiche-Senhadji, O.; Tingry, S.; Seta, P.; Benamor, M. Selective extraction of Cr(VI) over metallic species by polymer inclusion membrane (PIM) using anion (Aliquat 336) as carrier. *Desalination* **2010**, *258*, 59–65. [CrossRef]
57. Kertes, A.S.; King, C.J. Extraction chemistry of fermentation product carboxylic-acids. *Biotechnol. Bioeng.* **1986**, *28*, 269–282. [CrossRef]
58. Wang, Z.Y.; Sun, Y.; Tang, N.; Miao, C.L.; Wang, Y.T.; Tang, L.H.; Wang, S.X.; Yang, X.J. Simultaneous extraction and recovery of gold(I) from alkaline solutions using an environmentally benign polymer inclusion membrane with ionic liquid as the carrier. *Sep. Purif. Technol.* **2019**, *222*, 136–144. [CrossRef]

International Journal of
Molecular Sciences

MDPI

Article

Effects of Cholesterol on Water Permittivity of Biomimetic Ion Pair Amphiphile Bilayers: Interplay between Membrane Bending and Molecular Packing

Wu-jhao Tien [1,†], Kun-you Chen [1,†], Fong-yin Huang [1,†] and Chi-cheng Chiu [1,2,*,†]

1 Department of Chemical Engineering, National Cheng Kung University, Tainan 70101, Taiwan
2 Hierarchical Green-Energy Materials (Hi-GEM) Research Center, National Cheng Kung University, Tainan 70101, Taiwan
* Correspondence: ccchiu2@mail.ncku.edu.tw; Tel.: +886-6-2757575 (ext. 62659)
† These authors contributed equally to this work.

Received: 18 June 2019; Accepted: 28 June 2019; Published: 2 July 2019

Abstract: Ion pair amphiphile (IPA), a molecular complex composed of a pair of cationic and anionic amphiphiles, is an inexpensive phospholipid substitute to fabricate vesicles with various pharmaceutical applications. Modulating the physicochemical and permeation properties of IPA vesicles are important for carrier designs. Here, we applied molecular dynamics simulations to examine the cholesterol effects on the structures, mechanics, and water permittivity of hexadecyltrimethylammonium-dodecylsulfate (HTMA-DS) and dodecyltrimethylammonium- hexadecylsulfate (DTMA-HS) IPA bilayers. Structural and mechanical analyses indicate that both IPA systems are in gel phase at 298 K. Adding cholesterol induces alkyl chain ordering around the rigid sterol ring and increases the cavity density within the hydrophilic region of both IPA bilayers. Furthermore, the enhanced alkyl chain ordering and the membrane deformation energy induced by cholesterol increase the permeation free energy penalty. In contrast, cholesterol has minor effects on the water local diffusivities within IPA membranes. Overall, the cholesterol reduces the water permittivity of rigid IPA membranes due to the synergistic effects of increased alkyl chain ordering and enhanced membrane mechanical modulus. The results provide molecular insights into the effects of molecular packing and mechanical deformations on the water permittivity of biomimetic IPA membranes, which is critical for designing IPA vesicular carriers.

Keywords: biomimetic membrane; ion pair amphiphile; cholesterol; molecular dynamics; water permeation

1. Introduction

Phospholipid is an amphiphilic biomolecule and the major component of cell membranes. Phospholipids can self-assemble in vitro into lipid vesicles, also termed liposomes, which can carry both hydrophilic and hydrophobic substances within the hydrophilic core and the bilayer shell, respectively [1]. Liposomes have been widely applied in pharmaceutics to modify drug adsorption, prolong drug biological half-life, and reduce drug toxicity and metabolism [2–4]. Other than phospholipids, studies have demonstrated the formation of vesicles with various amphiphiles, including ionic or non-ionic surfactants, and polymers [5]. Kaler et al. first demonstrated the spontaneous vesicle formation from the mixture of hexadecyltrimethylammonium tosylate and sodium dodecylbenzene sulfonate, and the resulting vesicles are termed "catanionic vesicles" [6]. Further removing residual counter ions from a 1:1 molar mixture of cationic/anionic amphiphiles results in a molecular complex termed ion pair amphiphile, IPA, where the cationic and anionic head groups are held together via electrostatic attraction. A dicatenar IPA complex composed of a pair of single-chain ionic amphiphiles is therefore a pseudo-double-tailed bio-mimic to a zwitterionic phospholipid, e.g., phosphatidylcholine and phosphatidylethanolamine. As liposome substitutes, IPA vesicles have great potentials in cosmetics, transdermal delivery, and pharmaceutical applications [7,8].

Compared with liposomes, IPA vesicles have the advantage of high chemical stability against hydrolysis. Yet, IPA vesicles in general have lower physical and colloidal stability. Common strategies to improve the colloidal stability of catanionic vesicles include inter-vesicular and intra-vesicular modifications [7]. The inter-vesicular repulsion between catanionic vesicles, for instance, can be achieved by introducing additional charged double-tailed amphiphile into the dicatenar IPA vesicles [9–11]. In addition, the electrostatic repulsion therefore prohibits the vesicle collisions, improving the long term colloidal stability. The intra-vesicular modifications can be accomplished via introducing stabilizing additives such as cholesterol that strengthen the stability and the mechanical properties of the IPA vesicular bilayer [12–14]. Cholesterol is a common biological membrane additive known to alter the mechanical properties and fluidity of the lipid membranes. Previous simulation study by Kuo et al. demonstrated that cholesterol stabilizes alkyltrimethylammonium-alkylsulfate IPA bilayers in a similar manner as lipid bilayers [13]. Our preliminary simulation study on the same IPA membrane systems mixed with cholesterol further illustrated that cholesterol preferential interacts with anionic alkylsulfate, increasing the contribution of anionic component to the overall mechanical modulus [14,15].

At a given temperature, a lipid bilayer can exist in either gel (solid, S) or fluidic (or liquid disordered, Ld) phase. The S phase is a solid-like phase where most alkyl chains are aligned within the hydrophobic region of the bilayer; while in the Ld phase, lipid molecules have more disordered hydrophobic chains and can diffuse freely in lateral dimensions. For a Ld phase phospholipid bilayer, adding cholesterol can induce the ordering of neighboring alkyl chains by the rigid sterol ring, leading to an increased local membrane rigidity [16]. In contrast, adding cholesterol into a S phase lipid bilayer disrupt the packing of nearby hydrocarbon chains. Biomimetic IPA bilayers exhibit similar phase properties to the phospholipid bilayer systems [17,18]. Also, the cholesterol additives give rise to similar effects on the IPA bilayer's phase behavior and mechanical properties as the phospholipid bilayers [19]. Recent studies showed that high cholesterol content can stabilize the IPA vesicles, possibly due to the vanishing of the local phase separation [20]. These results demonstrate a close structural correlation and similar response to cholesterol between the biomimetic IPA and the lipid bilayer system.

A nonspecific diffusion for the substance permeating across the bilayer, also denoted as the passive membrane transport, is driven by the concentration gradient [21]. Theoretically, the permeation of a substance is characterized by the permeation coefficient P, which is related to the molecular flux J across a bilayer and the concentration gradient Δc [22,23]:

$$P = \frac{J}{\Delta c}. \tag{1}$$

A common permeation model for a lipid bilayer is the inhomogeneous solubility-diffusivity model, in which the permeability across a membrane with the thickness h is expressed as [23,24]:

$$\frac{1}{P} = \int_0^h \frac{1}{K(z)D(z)} dz \tag{2}$$

where $K(z)$ and $D(z)$ are the position-dependent partition coefficient and diffusion coefficient of the target substance. The model combined with computer simulation have been applied to study the rates and molecular mechanisms of the permeation of various molecules across bilayers [21,25–28]. The factors affecting the membrane permittivity include the composing lipids, the corresponding phases, and the packing characteristics, etc. In general, S phase lipid bilayers exhibit higher mechanical strength and lower membrane permittivity. Also, liposomes composed of saturated lipids with cholesterol exhibits lower membrane permittivity. Recent studies further showed that the fluctuations in the membrane conformation and the potential energy are critical to the permeation of hydrophilic molecules [29].

Passive transport is the mechanism for most of small neutral molecules and drug molecules. Hence, it is important to understand the process of passive permeation for medical and pharmaceutical applications. In this work, we applied molecular dynamics (MD) simulations to characterize the water permittivity of the biomimetic IPA membrane. Our early studies demonstrated the molecular effect of cholesterol on modulating the structural and mechanical properties of IPA membranes [14]. Here, we further examined on the mechanisms of cholesterol on modulating the IPA membrane permittivity form the structural, thermodynamic, and kinetic perspectives. The combined results provides important insights into IPA vesicle leakage stability and the membrane permeation of hydrophilic substance.

2. Results and Discussion

The target IPA series were the hexadecyltrimethylammonium-dodecylsulfate (HTMA-DS, $CH_3(CH_2)_{15}N(CH_3)_3^+$-$CH_3(CH_2)_{11}SO_4^-$) and dodecyltrimethylammonium-hexadecylsulfate (DTMA-HS, $CH_3(CH_2)_{11}N(CH_3)_3^+$-$CH_3(CH_2)_{15}SO_4^-$), which have inverse alkyl chain asymmetry as illustrated in Figure 1. All molecular dynamics (MD) simulations were conducted under the isothermal-isobaric condition at 1 bar and 298 K. According to early experimental and simulation works, both systems are in S phase at 298 K [15,17]. Different cholesterol (Chol) concentrations were introduced into IPA bilayers to investigate the molecular effects of cholesterol on modulating the water permittivity of the biomimetic IPA membrane. The molecular compositions for the two IPA-Chol bilayer systems utilized in the presented MD studies are listed in Table 1.

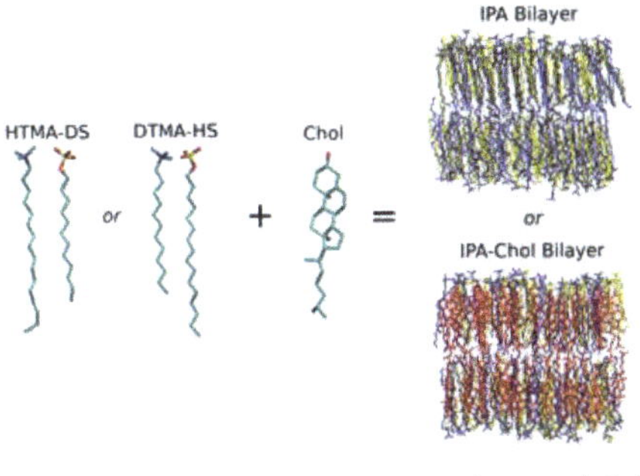

Figure 1. Molecular structures of HTMA-DS, DTMA-HS IPA complexes, and cholesterol. The representative bilayer structures of pure IPA and IPA-Chol systems are also shown, where the molecule color codes are: alkyltrimethylammonium in blue, alkylsulfate in yellow, and cholesterol in red. Graphics were generated using VMD package [30].

Table 1. Compositions of bilayer systems comprising HTMA-DS or DTMA-HS IPA, cholesterol (Chol) and water with the mole fraction of cholesterol (X_{Chol}) ranged from 0 to 0.5.

System	X_{Chol}	N_{IPA}	N_{Chol}	N_{water}
	0	128	0	
HTMA-DS	0.094	116	12	
or	0.203	102	26	3464
DTMA-HS	0.375	80	48	
	0.5	64	64	

2.1. Bilayer Structural and Mechanical Properties

It is known that cholesterol can alter the structural and mechanical properties of lipid and IPA bilayers. To characterize the effect of cholesterol on the S phase HTMA-DS and DTMA-HS IPA bilayer structures, we first analyzed the alkyl chain conformation and ordering via deuterium order parameter ($|S_{CD}|$) and gauche conformer fraction. In MD simulation, the deuterium order parameter, $|S_{CD}|$, can be evaluated as: [31,32]

$$S_{CD} = \frac{1}{2}\langle 3cos^2(\theta) - 1\rangle, \quad (3)$$

where θ denotes the angle between the C-H bond and the bilayer normal, and the angle brackets represent the ensemble average. According to the study by Chen et al., a Ld phase IPA bilayer has the $|S_{CD}|$ profile smaller than 0.3, and a S phase lipid bilayer has the one larger than 0.3 [15]. Meanwhile, the gauche fraction is evaluated as the fraction of gauche conformers along alkyl chains, where a gauche conformer is defined for the alkyl dihedral angle between −120 and 120 degrees. Smaller gauche fraction values indicate the higher chain ordering with more extended alkyl chains [33]. Chen et al. also determined the threshold gauche faction of 0.15 to distinguish Ld and S phase [15].

For both pure IPA bilayer systems, i.e., HTMA-DS and DTMA-HS systems, the $|S_{CD}|$ profiles have the plateau values of 0.45 and the gauche fraction profiles have the plateau values below 0.15 as illustrated in Figure 2. These results confirm that both pure HTMA-DS and DTMA-HS IPA bilayers are in S phase at 298 K. Upon the addition of Chol, the $|S_{CD}|$ values increase and the gauche fractions decrease with X_{chol} for the middle alkyl segments of all the surfactant species in both IPA systems. This suggests that the chain order is enhanced by the rigid sterol ring of cholesterol. Furthermore, the HTMA-DS-Chol system has slightly higher $|S_{CD}|$ and lower gauche fraction than the DTMA-HS-Chol system, particularly near the terminus of alkyl chains. This indicates that the HTMA-DS bilayer have higher alkyl chain ordering than the DTMA-HS system, due to the intrinsic alkyl chain mismatch between the alkyltrimethylammonium and alkylsulfate [15].

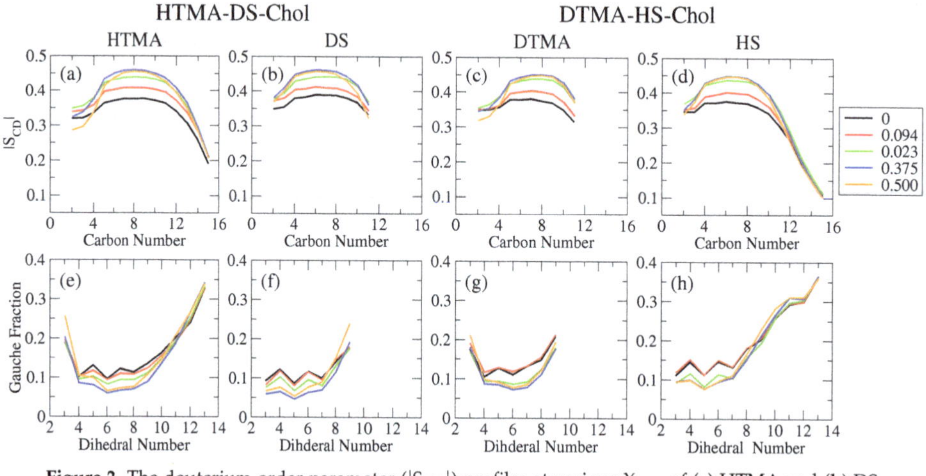

Figure 2. The deuterium order parameter ($|S_{CD}|$) profiles at various X_{chol} of (**a**) HTMA and (**b**) DS components for HTMA-DS-Chol systems and (**c**) DTMA and (**d**) HS components for DTMA-HS-Chol systems are shown in top panels. Also, the gauche fraction profiles at various X_{chol} along the alkyl chains of (**e**) HTMA and (**f**) DS components for HTMA-DS-Chol systems and (**g**) DTMA and (**h**) HS components for DTMA-HS-Chol systems are shown in bottom panels.

Several experimental and simulation studies showed that cholesterol can alter the molecular packing within a lipid membrane and modulate the water permeability of the membrane [34,35]. Here, we calculated the cavity density P_{cav} along the bilayer normal to probe the effects of cholesterol on the alkyl chain packing for HTMA-DS and DTMA-HS IPA membranes. The cavity density profile P_{cav} was evaluated using a uniform grid of the 0.5 Å gridsize [34]. Each bin was examined for any atom occupation to calculated the probability density at position z. As shown in Figure 3, the cavity density profiles for pure IPA systems peak at the bilayer center and is reduced to near 0.3 at z = 0.6–1.8 nm, similar to that for the lipid bilayers [34]. The low P_{cav} region for the HTMA-DS system is wider than the DTMA-HS membrane, which can be correlated with the higher alkyl chain ordering within the HTMA-DS bilayer. Further adding cholesterol reduces the P_{cav} values in the range of z = 0.6–1.4 nm, where the sterol ring are populated. In contrast, the P_{cav} in the region of z = 1.4–2.0 nm increases with the cholesterol concentration. These effects of cholesterol on the P_{cav} values become increasingly significant starting from X_{chol} = 0.203. This can be correlated with the variations on the alkyl chain ordering as shown by the $|S_{CD}|$ and gauche fraction analyses in Figure 2. The reduction of P_{cav} around the alkyl middle segments indicates a induced molecular packing by cholesterol which lowering the nearby free cavity. Similar effects have been reported on DPPC-Chol and sphingolipid membranes [34]. Yet, the increased P_{cav} near the hydrophilic region for IPA-Chol systems differ from the those reported for lipid membranes, in which cholesterol also reduces the P_{cav} near the hydrophilic region [34]. The reported lipid-cholesterol membrane studies were focusing on the lipid systems in Ld phase, compared with the S phase IPA bilayers in this work. According to previous studies on IPA systems, adding cholesterol into S phase IPA bilayers can increase the spacing between IPAs [13,14]. Thus, adding cholesterol disrupts the molecular packing within the hydrophilic region for both HTMA-DS and DTMA-HS bilayers, leading to increased cavity probability near the membrane surface.

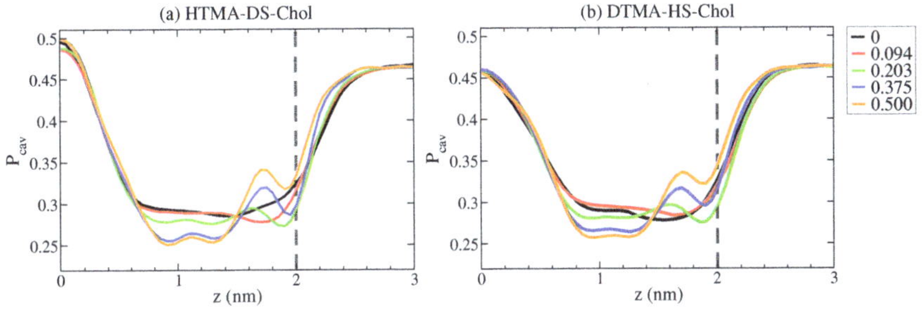

Figure 3. Cavity density profile $P_{cav}(z)$ of (**a**) HTMA-DS-Chol and (**b**) DTMA-HS-Chol bilayers at various X_{chol} with the standard deviations of less than 0.015. Error bars are not shown for clarity.

To characterize the effects of cholesterol on the mechanical properties of the IPA bilayers, we calculated the area compressibility modulus K_A, the molecular tilt modulus χ, and the bending modulus K_C. For the pure IPA bilayers, as shown in Figure 4, all three moduli (K_A, χ, and K_C) for HTMA-DS system are greater than those for the DTMA-HS system. This is attributed to the higher alkyl chain ordering within in the HTMA-DS system, as illustrated by the $|S_{CD}|$ and gauche fraction analyses in Figure 2. The K_A and χ values for both pure IPA systems are higher than the reported threshold values of $K_A = 700$ mN/m and $\chi = 13\ k_BT/\text{rad}^2$, respectively [15]. This also suggests that both HTMA-DS and DTMA-HS are in S phase at 298 K. After introducing cholesterol, the mechanical moduli for both IPA membranes increases. Such mechanical enhancement can be corresponded to the enhanced ordering of the middle alkyl chain segment induced by the rigid sterol ring of cholesterol. Around X_{chol} = 0.375, both K_A and χ are at their maxima while K_C dramatically increase for both IPA-Chol systems. This is due to the competing effects among the induced ordering at the middle alkyl segment, the disordering at the alkyl tails, and increased spacing within the hydrophilic regions induced by cholesterol addition, consistent with the results reported by Huang et al. [14]. Furthermore, the DTMA-HS-Chol bilayers generally have lower mechanical strength than the HTMA-DS-Chol systems. This can be attributed to the more disordered alkyl tail region within the DTMA-HS-Chol membranes.

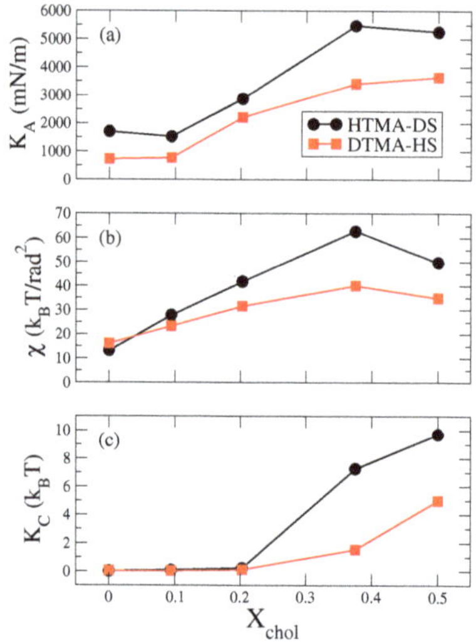

Figure 4. Three mechanical moduli for HTMA-DS-Chol (black circle) and DTMA-HS-Chol (red square) bilayers at various X_{chol}: (**a**) the area compressibility modulus K_A, (**b**) molecular tilt modulus χ, and (**c**) bending modulus K_C.

2.2. Free Energy of Water Crossing IPA-Chol Bilayer

According to the inhomogeneous solubility-diffusivity model as described in Equation (2), the water permittivity of the membrane depends on both the partition coefficient $K(z)$ and the diffusion coefficient $D(z)$ of the permeate water in the membrane [23,24]. The position-dependent partition coefficient $K(z)$ can be related to the free energy of permeation $\Delta G(z)$ as:

$$K(z) = e^{-\Delta G(z)/k_B T}. \tag{4}$$

where k_B and T denote the Boltzmann constant and the temperature, respectively. In this work, we calculated the free energy profiles for a water molecule across the HTMA-DS-Chol and DTMA-HS-Chol bilayer systems, as shown in Figure 5. As the permeate water goes from the bulk phase toward the membrane interior of both pure HTMA-DS and DTMA-HS bilayers, the free energy starts raising at around 2.2 nm, close to the average positions of the IPA head groups at 2 nm. The free energy keeps increasing as the water moves toward the center of the bilayer, which can be attributed to the membrane hydrophobic region. Please note that there exhibits a local free energy minimum near the membrane center. Such free energy local minimum is a common feature for lipid bilayers in S phase [28,36,37], compared with the free energy barrier with a narrow plateau near the core observed for lipid bilayers in Ld phase [34,38]. This again supports that both IPA membranes are in S phase at 298 K. The local free energy minimum observed for IPA bilayers is resulted from the void near the bilayer center as shown in Figure 3, which is originated from the alkyl chain mismatch between two IPA components [13,14]. The void space with locally lowered density provides the penetrating water molecule a relatively stable region in the membrane hydrophobic region. For pure IPA bilayers, a larger mismatch near the membrane core is observed in DTMA-HS bilayer [15]. As illustrated by the lower cavity density for the DTMA-HS system in Figure 3, the voids near the bilayer center are hence filled with less ordered DTMA-HS alkyl tails. This results in a wider and higher basin around the free energy minimum for the DTMA-HS system.

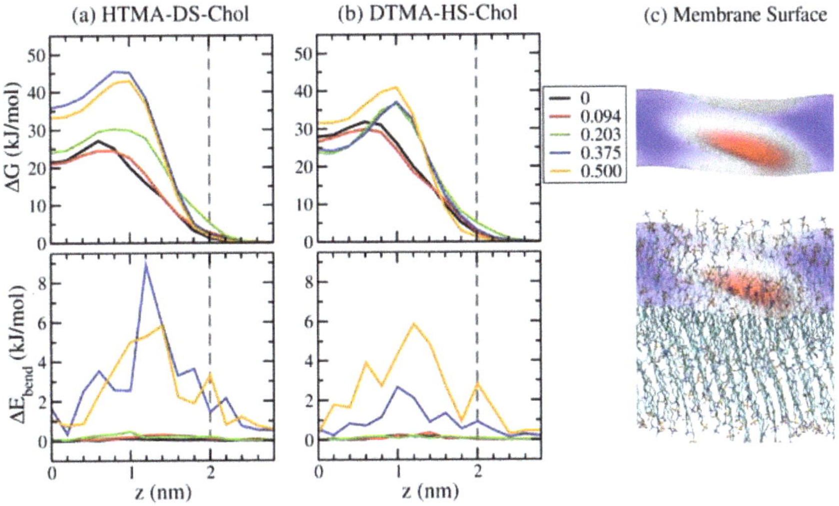

Figure 5. The free energy profiles (top panels) and the membrane bending energy profiles during water permeation for (**a**) HTMA-DS-Chol and (**b**) DTMA-HS-Chol bilayers at various X_{chol}. In addition (**c**) the representative membrane surface with maximum bending energy at z = 1.2 nm, and the superimposition of the surface with the bilayer structure where the permeant water oxygen is labeled with red sphere. Graphics were generated using VMD package [30].

With low amount of cholesterol addition (X_{chol} = 0.094), the free energy profiles for water permeation for both IPA systems are only slightly affected where the locations of the maxima remain but the barrier heights reduces 1–2 kJ/mol. When adding a few cholesterol into the S phase IPA membranes, the cholesterol only marginally increases the alkyl chain ordering shown in Figure 2. Meanwhile, low amount of cholesterol slightly increase the cavity density in the region around z = 1 nm, which corresponds to the minor reduction of free energy barriers for both HTMA-DS-Chol and DTMA-HS-Chol systems. With the cholesterol addition of $X_{chol} > 0.203$, the water permeation free energy for both IPA bilayers increases. This is attributed to the changes in the cavity density starting at X_{chol} = 0.203. The increased barriers observed for both IPA-Chol system at z = 1 nm can be related to the reduction cavity in the region of z = 0.6–1.4 nm induced by the ordering effect of the cholesterol sterol ring. Comparing the two IPA systems, both IPA systems have the permeation barrier of around 30 kJ/mol for $X_{chol} \leq 0.203$. Further increasing $X_{chol} > 0.375$, the HTMA-DS-Chol bilayers exhibits larger free energy barrier than the DTMA-HS-Chol bilayers, which is partly attributed to the more ordered structure in the HTMA-DS-Chol systems.

As the permeate water entering the membrane, it can cause local membrane deformation as demonstrated by the representative simulation snapshot in Figure 5. Such membrane deformation indicates that the membrane mechanical energy also affects the water permeation across a S phase IPA membrane. To characterize the contribution of the membrane deformation to the overall free energy, we evaluated the bending energy ΔE_{bend} using the Helfrich free energy via integration over the membrane surface [39]:

$$\Delta E_{bend} = \int \frac{K_C}{2}(c_1 + c_2)^2 \, dA, \tag{5}$$

where K_C is the bending modulus. The parameters c_1 and c_2 denote the two principal curvatures, and can be obtained from the membrane height field $h(x,y)$ defined as the positions of the bilayer headgroups in the normal direction [40]. Here, the membrane height field $h(x,y)$ was approximated as:

$$h(x,y) = \sum_{m=0}^{2}\sum_{n=0}^{2}\left(a_{m,n}sin(\frac{2m\pi}{L_x} + \frac{2n\pi}{L_y}) + b_{m,n}cos(\frac{2m\pi}{L_x} + \frac{2n\pi}{L_y})\right), \quad (6)$$

where (L_x, L_y) and (m, n) denote the system box size and the wave vectors in x and y dimensions, respectively; and $a_{m,n}$ and $b_{m,n}$ are the fitting parameters for the surface $h(x,y)$. The analytical expression of $h(x,y)$ thus allowed us to determine c_1 and c_2 for any (x,y) coordinate on the membrane surface.

As illustrated in Figure 5, the membrane bending energy profile for water permeation have the maximum at z = 1.2 nm, close to the free energy barrier at z = 1 nm. This suggests that the membrane deformation also provides additional water permeation barriers of the IPA membrane. As the X_{chol} increases, the bending energy barrier also raises due to the enhanced bending modulus K_C as shown in Figure 4. However, the HTMA-DS-Chol system with $X_{chol} = 0.375$, despite of its smaller K_C, has a higher bending energy than the $X_{chol} = 0.5$ system. This is because the high K_C for the $X_{chol} = 0.5$ reduces the deformation of the membrane, resulting in smaller surface curvatures and hence a lower bending energy barrier. Please note that in the region of z = 1.4–2.0 nm, the P_{cav} increases with X_{chol}, which should lead to decreased free energy. Yet, the variation of ΔE_{bend} is more dominate over the cavity effect when adding cholesterol, leading to overall increased free energy in the hydrophilic region. According to our early study on comparing IPA and phospholipid bilayers, IPA membranes have higher mechanical strength than lipid systems [18]. Hence, the effects of membrane deformation should be considered for the water permeation across most types of IPA bilayers. Comparing the two IPA-Chol systems, the membrane bending energy for HTMA-DS-Chol systems are greater than that for DTMA-HS-Chol systems when $X_{chol} > 0.375$, attributed to the higher K_C of the HTMA-DS-Chol systems. This also contributes to a higher permeation free energy of HTMA-DS-Chol membranes with high cholesterol content.

2.3. Water Permittivity of IPA-Chol Bilayer

Other than the permeation free energy $\Delta G(z)$, the position- dependent diffusion coefficient $D(z)$ is also an important parameter to evaluate water permittivity of IPA-Chol Bilayer. Here, we calculated the local diffusion coefficient using the force autocorrelation function [21]:

$$D(z) = \frac{(RT)^2}{\int_0^{\infty} \langle \Delta F_z(t)\Delta F_z(0)\rangle \, dt} \quad (7)$$

where R is the gas constant and $\Delta F_z(t) = F_z(t) - \langle F_z \rangle$ is the instantaneous force deviation from the mean force along the bilayer normal. As shown in Figure 6, the local diffusivity profiles for all IPA-Chol bilayers show plateau of lower $D(z)$ at around 1.4 to 2 nm. In this region, water molecule experiences both the hydrophobic repulsion from the carbon segments and the hydrophilic association from the polar groups, leading to slower dynamics, similar to that observed in the lipid systems [34]. In addition, the alkyl chain ordering altered by cholesterol addition does not significantly affect the local diffusivity profile near the hydrophobic region. As the permeate water entering the bilayer, the membrane deformation can decreases the nearby alkyl chain ordering, leading to increased local diffusivity. Such effects can be observed for HTMA-DS-Chol system in the range of z = 0.6–1.2 nm: the packing disruption induced by deformation enhances with increased X_{chol}, resulting in increased local diffusivity in the region. However, compared with the large dependency of the free energy profile on the cholesterol concentration, the local diffusivity profiles exhibit less obvious changes upon varying X_{chol}.

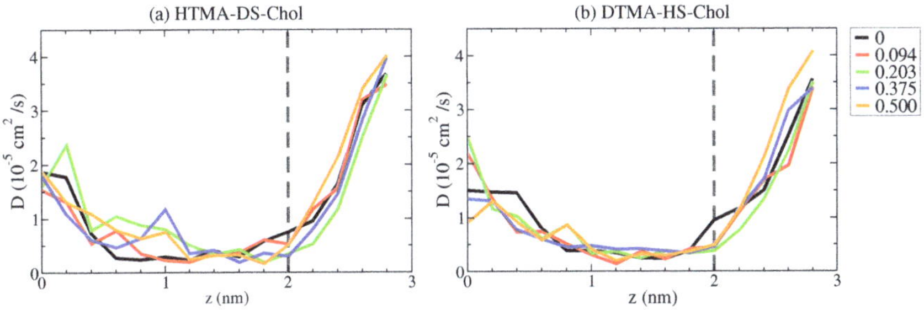

Figure 6. The local diffusivity profiles $D(z)$ of the permeant water for (**a**) HTMA-DS-Chol and (**b**) DTMA-HS-Chol bilayers at various X_{chol}.

With the results of free energy profiles and local diffusivity profiles shown in Figures 5 and 6, respectively, the local permeation resistance $R_{local}(z)$ can be calculated as [23]:

$$R_{local}(z) = \frac{1}{K(z)D(z)} = \frac{e^{\Delta G(z)/k_B T}}{D(z)}. \tag{8}$$

Figure 7 shows the local permeation resistance profile for the HTMA-DS-Chol and DTMA-HS-Chol systems. Comparing the free energy profiles and local diffusivity profiles, the overall variation in local diffusivity has little dependence on X_{chol}. Hence, the local resistance difference depends primarily on the variance in the free energy of permeation. For all the systems containing cholesterol, the region between 0.6 to 1.4 nm shows higher local resistance than other regions, corresponding to where cholesterol sterol rings reside. This suggests that the enhanced alkyl chain ordering and reduced cavity density induced by cholesterol addition decreases the water local permeation. In addition, with cholesterol addition, the increased permeation free energy induced by the enhanced mechanical modulus leads to the increased $R_{local}(z)$ in the hydrophilic region of $z = 1.4 - 2.0$ nm.

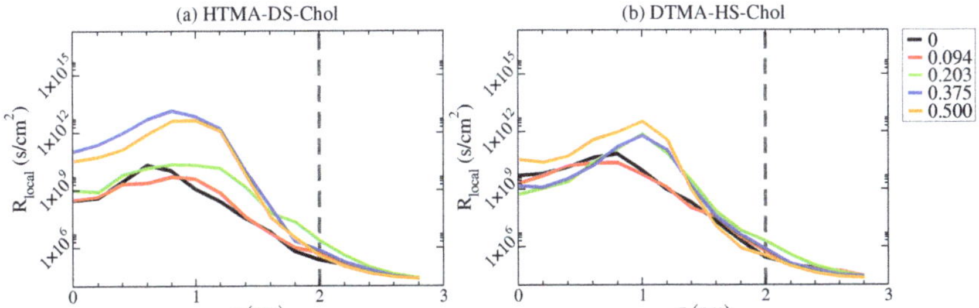

Figure 7. The local permeation resistance profiles $R_{local}(z)$ for (**a**) HTMA-DS-Chol and (**b**) DTMA-HS-Chol bilayers at various X_{chol}.

Integrating the values of local resistance along the permeation path yields the overall permeation resistance, or the reciprocal of permeability:

$$R = \frac{1}{P} = \int R_{local}(z) = \int \frac{e^{\Delta G(z)/k_B T}}{D(z)} \tag{9}$$

Figure 8 shows the resulting permeabilities for HTMA-DS-Chol and DTMA-HS-Chol bilayers. Compared with two IPA-Chol system, the DTMA-HS-Chol membrane has lower permeability, resulting form the lower cavity density near the bilayer center where the voids between two leaflet filled with less ordered alkyl tails. Since $R_{local}(z)$ primarily depends on the permeation free energy, the variations in permeability can be mainly interpreted by the free energy changes induced by cholesterol. For both types of IPA-Chol systems, the permeabilities increase at X_{chol} = 0.094 due to the slightly decreased free energy barrier. When further increasing X_{chol}, the permeabilities dramatically decrease 1-2 orders of magnitudes. Above X_{chol} = 0.375, the permeabilities for both IPA-Chol membranes become similar and decline to 2.3^{-6} cm/s for HTMA-DS-Chol and 4.6×10^{-6} cm/s for DTMA-HS-Chol at X_{chol} = 0.5. Such permeability reduction results from the synergistic effects of induced alkyl chain ordering and the enhanced membrane bending mechanics upon cholesterol addition. Please note that adding cholesterol into phospholipid bilayers may induced local phase seperation, resulting in a complex permittivity response [28]. Yet, no local phase domains were observed in this work, possibly due to the relatively small system size. In addition, the phase domain effect on the IPA membrane permittivity will be examined in the future.

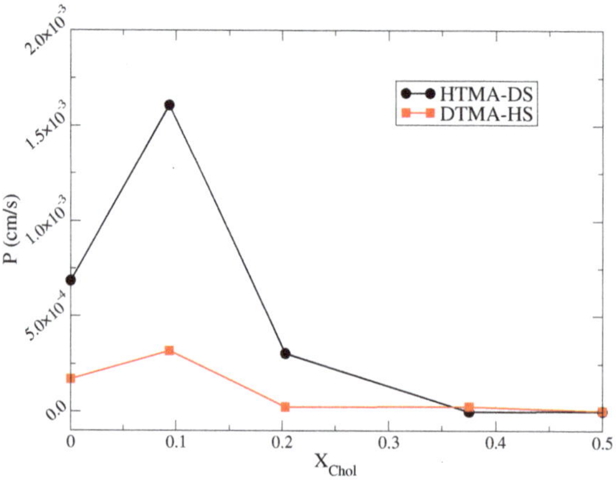

Figure 8. The water permittivity coefficients for HTMA-DS-Chol (black circle) and DTMA-HS-Chol (red square) bilayers as a function of cholesterol concentration X_{chol}.

3. Materials and Methods

3.1. Simulations Details

All molecular dynamics (MD) simulations were conducted with Gromacs 5.0.4 package with periodic boundary conditions applied in all three dimensions [41,42]. Each system was composed of 128 IPA and cholesterol with 64 molecules per leaflet and fully hydrated with 3464 water molecules. The initial configurations for the HTMA-DS-Chol and the DTMA-HS-Chol bilayers were constructed via Packmol [43]. All simulations were performed using the isothermal-isobaric (NPT) ensemble. Temperature and pressure were maintained at 298 K and 1 bar using the Nosé-Hoover and semi-isotropic Parrinello-Rahman algorithm, respectively [44–47]. The Lennard-Jones and short-range electrostatic potentials were cut off at 1.2 nm where the Lennard-Jones interactions were smoothly shifted to zero starting from 0.8 nm. Long-range electrostatic interactions were evaluated using Particle mesh Ewald (PME) method [48,49]. All bonds were constrained at their equilibrium length using the LINCS algorithm [50]. A 2 fs integration timestep was used to evaluate the equations of motions of atoms.

The CHARMM36 united-atom (C36-UA) force field parameters were applied for IPA and cholesterol molecules and the TIP3P model for water [51–53]. This force field combination have been used in various MD studies on alkyltrimethylammonium-alkylsulfate iPA and phospholipid bilayers [15,51]. Each bilayer system was first energy minimized via the steepest descent minimization algorithm, then equilibrated at 348 K and 1 bar for 40 ns to ensure the bilayer in Ld phase. The IPA bilayer system was then annealed from 348 K to 298 K with a 2.5 K/ns cooling rate, allowing the bilayer to naturally transition to S phase. After the annealing process, each simulation was first equilibrated at 298 K and 1 bar for 40 ns followed by a production run of 160 ns, in which system configurations were saved every 10 ps for analyses of the membrane structural and mechanical characteristics. All analyses were conduced with the in-house codes following the algorithms described in the main text.

3.2. Mechanical Modulus

To characterize the mechanical properties of the IPA-Chol bilayers, we calculated three different mechanical moduli, including the area compressibility modulus K_A, the molecular tilt modulus χ, and the bending modulus K_C. The area compressibility modulus K_A characterizes the bilayer resistance against the membrane lateral deformation and can be calculated from MD trajectories as [54,55]:

$$K_A = \frac{k_B T \langle A_{mol} \rangle}{N \langle \delta A_{mol}^2 \rangle}, \tag{10}$$

where k_B is the Boltzmann constant, T is the simulated temperature, $\langle A_{mol} \rangle$ is the average lateral area per molecule, $\langle \delta A_{mol}^2 \rangle$ is the variance of A_{mol}, and $N = 64$ denotes the number of the molecule per leaflet in our simulation. Meanwhile, the molecular tilt modulus, χ, characterizes the resistance against the alkyl chain tilting within the bilayer, and can be obtained via the quadratic fitting to the free energy profile of molecular tilting $F(\alpha)$ [56,57]. In addition, the tilting free energy profile $F(\alpha)$ was evaluated from the Boltzmann inversion of the normalized tilt angle distribution $P(\alpha)$:

$$F(\alpha) = -k_B T ln \left[\frac{P(\alpha)}{\sin(\alpha)} \right] = F(\alpha_0) + \frac{\chi}{2}(\alpha - \alpha_0)^2, \tag{11}$$

where α denotes the molecular tilt of the alkyl chain, $\sin(\alpha)$ is the normalizing Jacobian factor, and $F(\alpha_0)$ is the free energy minimum at the equilibrium tilt angle α_0. Here, α was defined as the angle between bilayer normal and the alkyl chain direction vector, i.e., the vector connecting the first and the second last carbons of the alkyl chain. Lastly, the bilayer bending modulus K_C characterizes the energetic costs of the membrane bending deformations. Here, we extracted K_C of IPA-Chol bilayers from MD simulations using the spectrum approach based on the modified Helfrich-Canham theory [40]. The bilayer deformation is quantified during the simulation via the height field $h(x, y)$ defined as the positions of the bilayer hydrophilic groups in the normal direction. After Fourier transform, the power spectrum of the height fluctuations is predicted as a function of the wavefactor q [58]:

$$\langle |h(q)|^2 \rangle = \frac{k_B T}{K_C q^4} + \frac{k_B T}{\chi q^2}, \tag{12}$$

which includes the contributions of membrane bending and lipid tilting to the overall deformations.

3.3. Permeation Free Energy Calculation

The permeation free energy profiles of water across IPA-Chol bilayers were evaluated with the constrained molecular dynamics as a function of the z distance between the permeant water and the membrane center of mass, z [59]. Each simulation was conducted by constraining at a fixed z value with a stiff harmonic spring constant of 10^5 kJ/mol/nm^2. A total of 15 constraint values were chosen every 0.2 nm in the range of 0 to 2.8 nm. For each z value, 10 different initial configurations were used to obtain the sufficient ensemble sampling. The 10 configurations were extracted every 1 to 10 ns from a 100 ns trajectory of NPT equilibrium simulation. In each extracted configuration, one water molecule was inserted randomly into free cavity at a desired z location. Then, an energy minimization was applied to eliminate the bad contacts between the inserted water and the surrounded atoms. With these 15 times 10 configurations, constrained MD runs were carried out for 500 ps each at fixed z locations under the canonical (NVT) ensemble to record the mean constraint forces $\langle f(z) \rangle$. The permeation free energy, $\Delta G(z)$ was then obtained from the mean constraint forces via thermodynamic integration:

$$\Delta G(z) = \int_0^{2.8 \text{ nm}} \langle f(z) \rangle dz. \tag{13}$$

Please note that the instantaneous constraint forces $f(z)$ were also used to calculate local diffusivity $D(z)$ via Equation (7).

4. Conclusions

An all-atom molecular dynamics simulation was applied to examine how cholesterol addition with X_{chol} = 0–0.5 affects the structural and permeation properties of the S phase biomimetic bilayers composed of hexadecyltrimethylammonium-dodecylsulfate (HTMA-DS) and dodecyltrimethylammonium-hexadecylsulfate (DTMA-HS). Simulation results showed that DTMA-HS-Chol systems have an overall smaller degree of chain ordering compared with HTMA-DS-Chol systems because of the greater intrinsic alkyl chain mismatch near the core region [14,15]. Cholesterol addition also enhances the membrane mechanical properties for both HTMA-DS and DTMA-HS systems, where the HTMA-DS-Chol bilayers have higher mechanical strengths owing to the more ordered alkyl chain packing. The combined effects of molecular packing and mechanical modulation cause the water permeation free energy barrier to slightly decrease at $X_{chol} = 0.094$ and to significantly increase when $X_{chol} \geq 0.203$ for both HTMA-DS and DTMA-HS systems. The analyses of membrane deformation energy further demonstrate that the enhanced mechanical strength induced by cholesterol can contribute additional energy costs for water permeations. In contrast, the local diffusivity is less affected by cholesterol addition.

Combining both free energy and local diffusivity data, we summarized the overall effects of cholesterol on the water permittivity of HTMA-DS and DTMA-HS membranes. With a low cholesterol amount of $X_{chol} = 0.094$, the water permittivity for both IPA systems slightly increases due to the slight reduction of the free energy barrier. When $X_{chol} \geq 0.203$, the synergistic effects of increased alkyl ordering and enhanced mechanical strength leads to a dramatical reduction of water permittivity of both HTMA-DS-Chol and DTMA-HS-Chol bilayers. Please note that the main phase transition temperatures for IPA systems are higher than the corresponding lipid systems, suggesting that the biomimetic IPA membranes are in S phase with higher mechanical strength under the physiological condition. Hence, the modulation of molecular packing and mechanical properties becomes important for controlling the permittivity of IPA membranes. Other than cholesterol addition, chemical penetration enhancers such as ethanol may also be applied to modulate the IPA membrane permittivity. In addition, the mechanical and free energy analyses used in this work can provide invaluable molecular insights into the mechanisms of modulating the rigid biomimetic membrane permittivity for various delivery systems.

Author Contributions: conceptualization, C.-c.C.; methodology, W.-j.T., F.-y.H., and C.-c.C.; software, W.-j.T., K.-y.C., F.-y.H., and C.-c.C.; validation, W.-j.T. and C.-c.C.; formal analysis, W.-j.T., K.-y.C., and C.-c.C.; investigation, W.-j.T., K.-y.C., F.-y.H., and C.-c.C.; resources, C.-c.C.; data curation, W.-j.T. and C.-c.C.; writing—original draft preparation, C.-c.C.; writing—review and editing, C.-c.C.; visualization, C.-c.C.; supervision, C.-c.C.; project administration, C.-c.C.; funding acquisition, C.-c.C.

Funding: This work was partially supported by the Hierarchical Green-Energy Materials (Hi-GEM) Research Center, from The Featured Areas Research Center Program within the framework of the Higher Education Sprout Projec tby the Ministry of Education (MOE) in Taiwan. The authors also acknowledge the financial supports by the Ministry of Science and Technology of Taiwan through Grant Nos. MOST 106-2221-E-006-199 and MOST 107-2221-E-006-102. The computational resource was partially supported by the high-performance cluster computing (HPCC) service from computer and network center of National Cheng Kung University.

Acknowledgments: The authors thank Chien-Hsiang Chang and Yu-Min Yang for fruitful discussions and experimental insights.

Conflicts of Interest: The authors declare no conflict of interest. The funders had no role in the design of the study; in the collection, analyses, or interpretation of data; in the writing of the manuscript, or in the decision to publish the results.

Abbreviations

The following abbreviations are used in this manuscript:

IPA	Ion Pair Amphiphile
MD	molecular dynamics
HTMA	hexadecyltrimethylammonium
DTMA	dodecyltrimethylammonium
HS	hexadecylsulfate
DS	dodecylsulfate
Chol	cholesterol

References

1. Lasic, D.D. Novel applications of liposomes. *Trends Biotechnol.* **1998**, *16*, 307–321. [CrossRef]
2. Moghimi, S.M.; Hunter, A.C.; Murray, J.C. Nanomedicine: Current status and future prospects. *FASEB J.* **2005**, *19*, 311–330. [CrossRef] [PubMed]
3. Yadav, D.; Sandeep, K.; Pandey, D.; Dutta, R.K. Liposomes for Drug Delivery. *J. Biotechnol. Biomater.* **2017**, *7*, 276. [CrossRef]
4. Zylberberg, C.; Matosevic, S. Pharmaceutical liposomal drug delivery: A review of new delivery systems and a look at the regulatory landscape. *Drug Deliv.* **2016**, *23*, 3319–3329. [CrossRef] [PubMed]
5. Soussan, E.; Cassel, S.; Blanzat, M.; Rico-Lattes, I. Drug Delivery by Soft Matter: Matrix and Vesicular Carriers. *Angew. Chem. Int. Ed. Engl.* **2009**, *48*, 274–288. [CrossRef] [PubMed]
6. Kaler, E.; Murthy, A.; Rodriguez, B.; Zasadzinski, J. Spontaneous vesicle formation in aqueous mixtures of single-tailed surfactants. *Science* **1989**, *245*, 1371–1374. [CrossRef] [PubMed]
7. Kuo, A.T.; Chang, C.H. Recent Strategies in the Development of Catanionic Vesicles. *J. Oleo Sci.* **2016**, *65*, 377–384. [CrossRef]
8. Dhawan, V.V.; Nagarsenker, M.S. Catanionic systems in nanotherapeutics—Biophysical aspects and novel trends in drug delivery applications. *J. Control. Release* **2017**, *266*, 331–345. [CrossRef] [PubMed]
9. Walker, S.A.; Zasadzinski, J.A. Electrostatic Control of Spontaneous Vesicle Aggregation. *Langmuir* **1997**, *13*, 5076–5081. [CrossRef]
10. Kuo, A.T.; Chang, C.H.; Shinoda, W. Molecular Dynamics Study of Catanionic Bilayers Composed of Ion Pair Amphiphile with Double-Tailed Cationic Surfactant. *Langmuir* **2012**, *28*, 8156–8164. [CrossRef]
11. Wu, C.J.; Kuo, A.T.; Lee, C.H.; Yang, Y.M.; Chang, C.H. Fabrication of positively charged catanionic vesicles from ion pair amphiphile with double-chained cationic surfactant. *Colloid Polym. Sci.* **2013**, *292*, 589–597. [CrossRef]
12. Lee, C.H.; Yang, Y.M.; Chang, C.H. Enhancing physical stability of positively charged catanionic vesicles in the presence of calcium chloride via cholesterol-induced fluidic bilayer characteristic. *Colloid Polym. Sci.* **2014**, *292*, 2519–2527. [CrossRef]
13. Kuo, A.T.; Chang, C.H. Cholesterol-induced condensing and disordering effects on a rigid catanionic bilayer: A molecular dynamics study. *Langmuir* **2014**, *30*, 55–62. [CrossRef] [PubMed]
14. Huang, F.Y.; Chiu, C.C. Interplay between alkyl chain asymmetry and cholesterol addition in the rigid ion pair amphiphile bilayer systems. *J. Chem. Phys.* **2017**, *146*, 035102. [CrossRef] [PubMed]
15. Chen, C.H.; Tian, C.A.; Chiu, C.C. The Effects of Alkyl Chain Combinations on the Structural and Mechanical Properties of Biomimetic Ion Pair Amphiphile Bilayers. *Bioengineering* **2017**, *4*, 84. [CrossRef] [PubMed]
16. De Meyer, F.J.M.; Benjamini, A.; Rodgers, J.M.; Misteli, Y.; Smit, B. Molecular Simulation of the DMPC-Cholesterol Phase Diagram. *J. Phys. Chem. B* **2010**, *114*, 10451–10461. [CrossRef] [PubMed]
17. Lee, W.H.; Tang, Y.L.; Chiu, T.C.; Yang, Y.M. Synthesis of Ion-Pair Amphiphiles and Calorimetric Study on the Gel to Liquid-Crystalline Phase Transition Behavior of Their Bilayers. *J. Chem. Eng. Data* **2015**, *60*, 1119–1125. [CrossRef]

18. Tian, C.A.; Chiu, C.C. Importance of Hydrophilic Groups on Modulating the Structural, Mechanical, and Interfacial Properties of Bilayers: A Comparative Molecular Dynamics Study of Phosphatidylcholine and Ion Pair Amphiphile Membranes. *Int. J. Mol. Sci.* **2018**, *19*, 1552. [CrossRef]
19. Liu, Y.S.; Wen, C.F.; Yang, Y.M. Cholesterol Effects on the Vesicular Membrane Rigidity and Drug Encapsulation Efficiency of Ethosome-Like Catanionic Vesicles. *Sci. Adv. Mater.* **2014**, *6*, 954–962. [CrossRef]
20. Chang, W.H.; Chuang, Y.T.; Yu, C.Y.; Chang, C.H.; Yang, Y.M. Effects of Sterol-Like Additives on Phase Transition Behavior of Ion-Pair Amphiphile Bilayers. *J. Oleo Sci.* **2017**, *66*, 1229–1238. [CrossRef]
21. Shinoda, W. Permeability across lipid membranes. *Biochim. Biophys. Acta Biomembr.* **2016**, *1858*, 2254–2265. [CrossRef] [PubMed]
22. Votapka, L.W.; Lee, C.T.; Amaro, R.E. Two Relations to Estimate Membrane Permeability Using Milestoning. *J. Phys. Chem. B* **2016**, *120*, 8606–8616. [CrossRef] [PubMed]
23. Marrink, S.J.; Berendsen, H.J.C. Simulation of water transport through a lipid membrane. *J. Phys. Chem.* **1994**, *98*, 4155–4168. [CrossRef]
24. Diamond, J.M.; Katz, Y. Interpretation of nonelectrolyte partition coefficients between dimyristoyl lecithin and water. *J. Membr. Biol.* **1974**, *17*, 121–154. [CrossRef] [PubMed]
25. Palaiokostas, M.; Ding, W.; Shahane, G.; Orsi, M. Effects of lipid composition on membrane permeation. *Soft Matter* **2018**, *14*, 8496–8508. [CrossRef] [PubMed]
26. De Vos, O.; Venable, R.M.; Van Hecke, T.; Hummer, G.; Pastor, R.W.; Ghysels, A. Membrane Permeability: Characteristic Times and Lengths for Oxygen and a Simulation-Based Test of the Inhomogeneous Solubility-Diffusion Model. *J. Chem. Theory Comput.* **2018**, *14*, 3811–3824. [CrossRef]
27. Lee, C.T.; Comer, J.; Herndon, C.; Leung, N.; Pavlova, A.; Swift, R.V.; Tung, C.; Rowley, C.N.; Amaro, R.E.; Chipot, C.; et al. Simulation-Based Approaches for Determining Membrane Permeability of Small Compounds. *J. Chem. Inf. Model.* **2016**, *56*, 721–733. [CrossRef] [PubMed]
28. Cordeiro, R.M. Molecular Structure and Permeability at the Interface between Phase-Separated Membrane Domains. *J. Phys. Chem. B* **2018**, *122*, 6954–6965. [CrossRef]
29. Qiao, B.; de la Cruz, M.O. Driving Force for Water Permeation Across Lipid Membranes. *J. Phys. Chem. Lett.* **2013**, *4*, 3233–3237. [CrossRef]
30. Humphrey, W.; Dalke, A.; Schulten, K. VMD: Visual molecular dynamics. *J. Mol. Graph.* **1996**, *14*, 33–38. [CrossRef]
31. Egberts, E.; Marrink, S.J.; Berendsen, H.J. Molecular dynamics simulation of a phospholipid membrane. *Eur. Biophys. J.* **1994**, *22*, 423–436. [CrossRef] [PubMed]
32. Douliez, J.P.; Léonard, A.; Dufourc, E.J. Restatement of order parameters in biomembranes: calculation of C-C bond order parameters from C-D quadrupolar splittings. *Biophys. J.* **1995**, *68*, 1727–1739. [CrossRef]
33. Pink, D.A.; Green, T.J.; Chapman, D. Raman scattering in bilayers of saturated phosphatidylcholines. Experiment and theory. *Biochemistry* **1980**, *19*, 349–356. [CrossRef] [PubMed]
34. Saito, H.; Shinoda, W. Cholesterol Effect on Water Permeability through DPPC and PSM Lipid Bilayers: A Molecular Dynamics Study. *J. Phys. Chem. B* **2011**, *115*, 15241–15250. [CrossRef]
35. Khajeh, A.; Modarress, H. Effect of cholesterol on behavior of 5-fluorouracil (5-FU) in a DMPC lipid bilayer, a molecular dynamics study. *Biophys. Chem.* **2014**, *187–188*, 43–50. [CrossRef]
36. Paloncýová, M.; DeVane, R.H.; Murch, B.P.; Berka, K.; Otyepka, M. Rationalization of Reduced Penetration of Drugs through Ceramide Gel Phase Membrane. *Langmuir* **2014**, *30*, 13942–13948. [CrossRef]
37. Hartkamp, R.; Moore, T.C.; Iacovella, C.R.; Thompson, M.A.; Bulsara, P.A.; Moore, D.J.; McCabe, C. Composition Dependence of Water Permeation Across Multicomponent Gel-Phase Bilayers. *J. Phys. Chem. B* **2018**, *122*, 3113–3123. [CrossRef]
38. Bemporad, D.; Essex, J.W.; Luttmann, C. Permeation of small molecules through a lipid bilayer: A computer simulation study. *J. Phys. Chem. B* **2004**, *108*, 4875–4884. [CrossRef]
39. Kawamoto, S.; Klein, M.L.; Shinoda, W. Coarse-grained molecular dynamics study of membrane fusion: Curvature effects on free energy barriers along the stalk mechanism. *J. Chem. Phys.* **2015**, *143*, 243112. [CrossRef]
40. Levine, Z.A.; Venable, R.M.; Watson, M.C.; Lerner, M.G.; Shea, J.E.; Pastor, R.W.; Brown, F.L.H. Determination of Biomembrane Bending Moduli in Fully Atomistic Simulations. *J. Am. Chem. Soc.* **2014**, *136*, 13582–13585. [CrossRef]

41. Hess, B.; Kutzner, C.; Van Der Spoel, D.; Lindahl, E. GROMACS 4: algorithms for highly efficient load-balanced, and scalable molecular simulation. *J. Chem. Theory Comput.* **2008**, *4*, 435–447. [CrossRef] [PubMed]
42. Pronk, S.; Páll, S.; Schulz, R.; Larsson, P.; Bjelkmar, P.; Apostolov, R.; Shirts, M.R.; Smith, J.C.; Kasson, P.M.; van der Spoel, D.; et al. GROMACS 4.5: A high-throughput and highly parallel open source molecular simulation toolkit. *Bioinformatics* **2013**, *29*, 845–854. [CrossRef] [PubMed]
43. Martínez, L.; Andrade, R.; Birgin, E.G.; Martínez, J.M. PACKMOL: A package for building initial configurations for molecular dynamics simulations. *J. Comput. Chem.* **2009**, *30*, 2157–2164. [CrossRef] [PubMed]
44. Nose, S. A molecular dynamics method for simulations in the canonical ensemble. *J. Chem. Phys.* **1984**, *81*, 511–519. [CrossRef]
45. Nose, S. A unified formulation of the constant temperature molecular dynamics methods. *J. Chem. Phys.* **1984**, *81*, 511–519. [CrossRef]
46. Hoover, W.G. Canonical dynamics: equilibrium phase-space distributions. *Phys. Rev.* **1985**, *31*, 1695–1697. [CrossRef]
47. Parrinello, M.; Rahman, A. Polymorphic transitions in single crystals: A new molecular dynamics. *J. Appl. Phys.* **1981**, *52*, 7182–7190. [CrossRef]
48. Darden, T.; York, D.; Pedersen, L. Particle mesh Ewald An Nlog(N) method for Ewald sums in large systems. *J. Chem. Phys.* **1993**, *98*, 10089–10092. [CrossRef]
49. Essmann, U.; Perera, L.; Berkowitz, M.L.; Darden, T.; Lee, H.; Pedersen, L.G. A smooth particle mesh Ewald method. *J. Chem. Phys.* **1995**, *103*, 8577–8596. [CrossRef]
50. Hess, B.; Bekker, H.; Berendsen, H.J.C.; Fraaije, J.G.E.M. LINCS: A linear constraint solver for molecular simulations. *J. Comput. Chem.* **1997**, *18*, 1469–1472. [CrossRef]
51. Lee, S.; Tran, A.; Allsopp, M.; Lim, J.B.; Hénin, J.; Klauda, J.B. CHARMM36 United Atom Chain Model for Lipids and Surfactants. *J. Phys. Chem. B* **2014**, *118*, 547–556. [CrossRef] [PubMed]
52. Klauda, J.B.; Venable, R.M.; Freites, J.A.; O'Connor, J.W.; Tobias, D.J.; Mondragon-Ramirez, C.; Vorobyov, I.; Mackerell, A.D.; Pastor, R.W. Update of the CHARMM All-Atom Additive Force Field for Lipids: Validation on Six Lipid Types. *J. Phys. Chem. B* **2010**, *114*, 7830–7843. [CrossRef] [PubMed]
53. Jorgensen, W.L.; Chandrasekhar, J.; Madura, J.D.; Impey, R.W.; Klein, M.L. Comparison of simple potential functions for simulating liquid water. *J. Chem. Phys.* **1983**, *79*, 926. [CrossRef]
54. Feller, S.E.; Pastor, R.W. Constant surface tension simulations of lipid bilayers: The sensitivity of surface areas and compressibilities. *J. Chem. Phys.* **1999**, *111*, 1281–1287. [CrossRef]
55. Shinoda, W.; Shinoda, K.; Baba, T.; Mikami, M. Molecular dynamics study of bipolar tetraether lipid membranes. *Biophys. J.* **2005**, *89*, 3195–3202. [CrossRef] [PubMed]
56. Khelashvili, G.; Pabst, G.; Harries, D. Cholesterol Orientation and Tilt Modulus in DMPC Bilayers. *J. Phys. Chem. B* **2010**, *114*, 7524–7534. [CrossRef] [PubMed]
57. Khelashvili, G.; Harries, D. How cholesterol tilt modulates the mechanical properties of saturated and unsaturated lipid membranes. *J. Phys. Chem. B* **2013**, *117*, 2411–2421. [CrossRef] [PubMed]
58. Safran, S.A. *Statistical Thermodynamics of Surfaces, Interfaces, and Membranes*, 1st ed.; CRC Press: Boca Raton, FL, USA, 2018.
59. Sprik, M.; Ciccotti, G. Free energy from constrained molecular dynamics. *J. Chem. Phys.* **1998**, *109*, 7737–7744. [CrossRef]

International Journal of
Molecular Sciences

MDPI

Article

Development of Polymer Blend Electrolyte Membranes Based on Chitosan: Dextran with High Ion Transport Properties for EDLC Application

Shujahadeen B. Aziz [1,2,*], Muhamad H. Hamsan [3], Mohd F. Z. Kadir [3], Wrya O. Karim [4] and Ranjdar M. Abdullah [1]

1 Advanced Polymeric Materials Research Lab., Department of Physics, College of Science, University of Sulaimani, Qlyasan Street, Sulaimani 46001, Kurdistan Regional Government, Iraq
2 Komar Research Center (KRC), Komar University of Science and Technology, Sulaimani 46001, Kurdistan Regional Government, Iraq
3 Centre for Foundation Studies in Science, University of Malaya, Kuala Lumpur 50603, Malaysia
4 Department of Chemistry, College of Science, University of Sulaimani, Qlyasan Street, Sulaimani 46001, Kurdistan Regional Government, Iraq
* Correspondence: shujahadeenaziz@gmail.com

Received: 20 May 2019; Accepted: 5 July 2019; Published: 9 July 2019

Abstract: Solid polymer blend electrolyte membranes (SPBEM) composed of chitosan and dextran with the incorporation of various amounts of lithium perchlorate ($LiClO_4$) were synthesized. The complexation of the polymer blend electrolytes with the salt was examined using FTIR spectroscopy and X-ray diffraction (XRD). The morphology of the SPBEs was also investigated using field emission scanning electron microscopy (FESEM). The ion transport behavior of the membrane films was measured using impedance spectroscopy. The membrane with highest $LiClO_4$ content was found to exhibit the highest conductivity of 5.16×10^{-3} S/cm. Ionic (t_i) and electronic (t_e) transference numbers for the highest conducting electrolyte were found to be 0.98 and 0.02, respectively. Electrochemical stability was estimated from linear sweep voltammetry and found to be up to ~2.3V for the Li^+ ion conducting electrolyte. The only existence of electrical double charging at the surface of electrodes was evidenced from the absence of peaks in cyclic voltammetry (CV) plot. The discharge slope was observed to be almost linear, confirming the capacitive behavior of the EDLC. The performance of synthesized EDLC was studied using CV and charge–discharge techniques. The highest specific capacitance was achieved to be 8.7 $F \cdot g^{-1}$ at 20th cycle. The efficiency (η) was observed to be at 92.8% and remained constant at 92.0% up to 100 cycles. The EDLC was considered to have a reasonable electrode-electrolyte contact, in which η exceeds 90.0%. It was determined that equivalent series resistance (R_{esr}) is quite low and varies from 150 to 180 Ω over the 100 cycles. Energy density (E_d) was found to be 1.21 $Wh \cdot kg^{-1}$ at the 1st cycle and then remained stable at 0.86 $Wh \cdot kg^{-1}$ up to 100 cycles. The interesting observation is that the value of P_d increases back to 685 $W \cdot kg^{-1}$ up to 80 cycles.

Keywords: biopolymer electrolyte membranes; XRD analysis; FTIR study; Morphology; Impedance study; EDLC fabrication

1. Introduction

Electrochemical capacitors are classified into three types: Pseudocapacitor, electrical double-layer capacitor (EDLC), and hybrid capacitor [1]. In pseudocapacitor, Faradaic process is involved as the energy storage mechanism with metal oxide and conducting polymer electrodes [2]. However, in EDLC, which is usually composed of two identical carbon-based electrodes, non-Faradaic process is envolved during the charge storage process, where the accumulation of charge at the double-layer occurs on the surface of carbon-based electrodes [3]. In the third type, i.e., in the hybrid capacitor, the energy storage mechanism is based on the combination of EDLC and pseudocapacitor, i.e., both Faradaic and non-Faradaic processes contribute. Among these capacitors, EDLC has shown to obtain through a straightforward fabrication process. Moreover, EDLC possesses valuable properties, such as long life cycle, high power density and light in weight [4]. In making such capacitor, activated carbon has been reported to be compatible with polymer electrolyte [5–7]. Several features of activated carbon, for instance, high surface area, high electrical conductivity, cost-effectiveness, and excellent chemical stability, make it preferable for EDLC applications [8].

Biopolymers are preferable over non-degradable synthetic polymers as host polymers in polymer electrolyte system because of their renewability, biocompatibility, and biodegradability [9,10]. They are usually extracted from natural resources, e.g., cellulose from plants [11], starches from potato, maize and cassava [12], chitosan from crustacean animals [13], carrageen from seaweed [14], and dextran from bacteria [15]. The last one is obtained by growing cultures of bacteria called *Leuconostocmesenteroides* in a medium filled with sucrose. Dextransucrase as an enzyme is excreted in the medium where excess sucrose had converted to dextran [16]. Dextran with 1,6-α-D-glucopyranosidic linkages is a biodegradable and a non-toxic polymer in the polymer backbone [17]. Regarding the backbone of dextran, it is clear to observe two main functional groups, such as hydroxyl and glycosidic bond, which possess electron lone pairs that contribute in the ionic conduction [18]. Herein, it is interesting that chitosan as one of the derivatives of amino polysaccharides called chitin produced from exoskeleton of crustaceans or insects. Both amine and hydroxyl functional groups enable chitosan to serve as one of ionic conductors [19].

Blending two or more polymers has motivated researchers to improve the characteristics of individual polymers. Hamsan et al. [20] have showed that potato starch-methylcellulose blend film is more amorphous in structure than both pure methylcellulose film and potato starch film. It is well-known that the amorphous region in polymer electrolyte is mainly responsible for ionic conduction [21]. As concluded in the report published by Tamilselvi and Hema [22], mechanical properties or structure stability of a material can be manipulated by polymer blending method. Blended polymer composites provide more sites for ionic complexation process to occur, which makes ionic conduction to be efficient than single polymer [23]. Based on our previous work [24], in 60 wt.% chitosan and 40 wt.% dextran blend system, the amorphous structure was shown to be dominant. Therefore, in this work, a similar percentage of chitosan and dextran has been blended along with incorporation of lithium perchlorate ($LiClO_4$). In addition to that, thehighest relatively conducting electrolyte has also been used as electrode separators in the EDLC.

2. Results and Discussion

2.1. Structural (XRD and FTIR) Analysis

XRD analysis was carried out on pure CS, CS:Dextran and CS:Dextran: $LiClO_4$ systems at the ambient temperature. In our previous work, it was shown that pure CS exhibits several crystalline peaks at 2θ = 15.1, 17.7, and 20.9° [25,26], as shown in Figure 1a, whereas Dextran exhibited two hollows at 2θ = 18 and 23° [27]. In the present work, the XRD pattern of CS:Dextran showed two hollows and no crystalline peaks can be observed, as can be seen in Figure 1b. These broad hollows indicate that a fully amorphous structure has been formed [28]. Earlier reports documented that a broad diffraction peaks corresponds to amorphous nature of the polymer electrolyte [29,30]. In the current study, it has been found that the intensity of the hump of CS:Dextrandecreases in the intensity with the addition of $LiClO_4$ salt and broad nature also increases, as shown in Figure 2a,b. It is clear that at 20 wt.% of the salt some new peaks appeared, which are due to polymer slat complexes rather than pure salt. Interestingly, at 40 wt.% of the added salt, these new peaks disappeared and the intensity of the hump decreased. The results obtained here confirm the amorphous nature of the polymer electrolytes. The amorphous nature provides greater ionic diffusivity and high ionic conductivity. No peaks corresponding to pure $LiClO_4$ appeared in the CS:Dextran blend polymer electrolyte, which indicates the complete dissociation of the dopant salt in the polymer blend matrix.

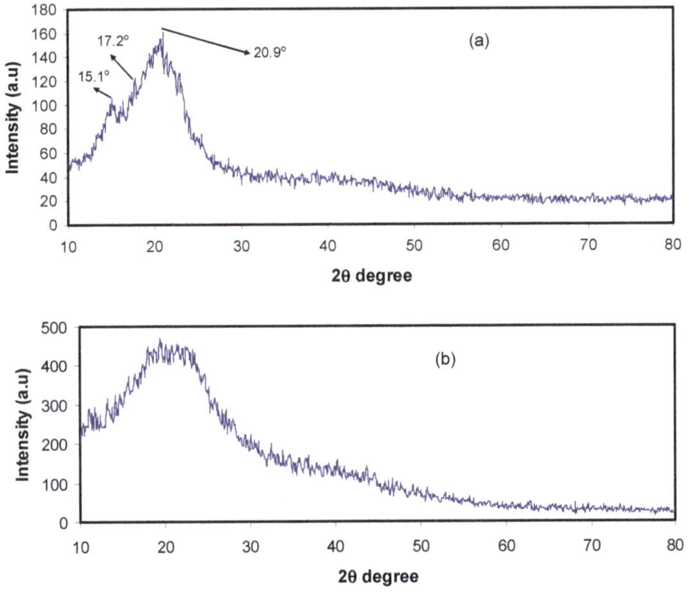

Figure 1. The XRD pattern for (**a**) pure CS and (**b**) CS:Dextran blend film.

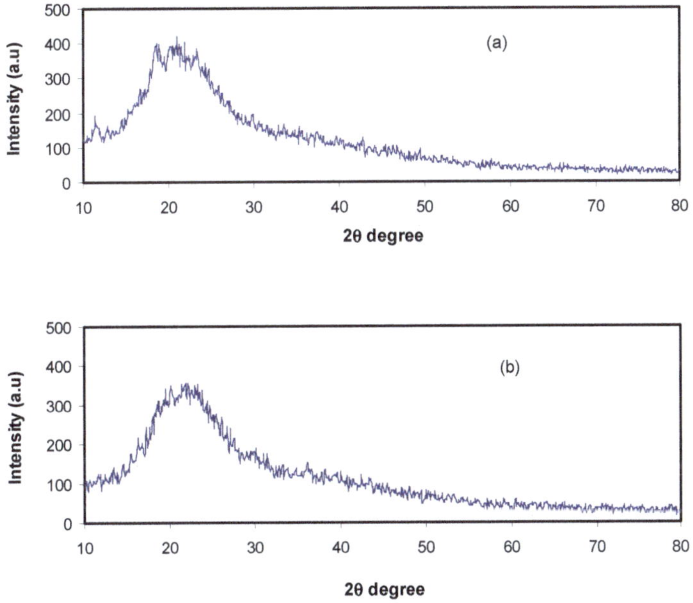

Figure 2. The XRD pattern for (**a**) CSDPB2 and (**b**) CSDPB4 blend electrolytes.

Figure 3a–c shows the FTIR spectra of the pure CS:Dextran and blend electrolyte films at three separate regions. Fourier transform infrared (FTIR) spectroscopy has been widely used by many researchers in dealing with the formation of polymer blends. FTIR spectroscopy provides insight into intermolecular interaction through analysis of FTIR spectra on the basis of stretching or bending vibrations of particular bonds. The main characteristic bands in the spectra of dextran were found at 1146 and 1021 cm^{-1}. The band peak at 1086 cm^{-1} is assigned to both valent vibrations of C–O–C bond and glycosidic bridge. The presence of a peak at 1021 cm^{-1} belongs to the high chain flexibility that present in dextran around the glycosidic bonds [31]. For CS:Dextran, the FT-IR spectrum exhibited the presence of OH groups that could be confirmed by the appearance of broad band with a maximum at 3351cm^{-1} [32,33]. In a comparison, the band peaked at 2906 cm^{-1} can be attributed to C–H stretching in dextran [32,33], since such band was not observed in the FTIR spectra of chitosan [34]. The peak at 1000 cm^{-1} characterizes the significant chain flexibility exist in dextran around the glycosidic bond [32]. The C–H bending usually appeared at 1450 cm^{-1}, whereas the broad band starts at 1158 cm^{-1}, indicating asymmetrical –C–O–C– stretching of the ring [33]. The sharp peak at 1009 cm^{-1} and small one at 1067 cm^{-1} is ascribed to the existence of C–O bands [34]. With increasing LiClO4 salt, these peaks are becomes distinguishable, as observed in our previous work for chitosan-based electrolyte. For C-H configuration in dextran, a characteristic peak appeared at 615 cm^{-1} [35]. For both stretching vibrations of the C–O–C bond and glycosides bridge a peak centered at 1155 cm^{-1}. Two peaks centered at 1651 cm^{-1} and 1554 cm^{-1} corresponding to carboxamide (O=C–NHR) and amine (NH2) bands, respectively [34]. It is interesting to observe that a shift occurred in the carboxamide (O=C–NHR) and amine (NH_2) bands which strongly confirms complexation between chitosan: Dxetran and the dopant salt. In fact, this attachment of cation salt to nitrogen and oxygen atoms can reduce the vibration intensity of the N–H or O=C–NHR bonds owing to the higher molecular weight after cation binding and eventually resulted in shifting and lowering in intensity [36]. More interesting observation is the incorporation of $LiClO_4$ salt into CS:Dextran resulting in a great change in the intensity of the bands. This change in intensity of these bands is strongly related to the alterations in the macromolecular order. These bands in the spectra of the complexes may result from more and less ordered structures [37].

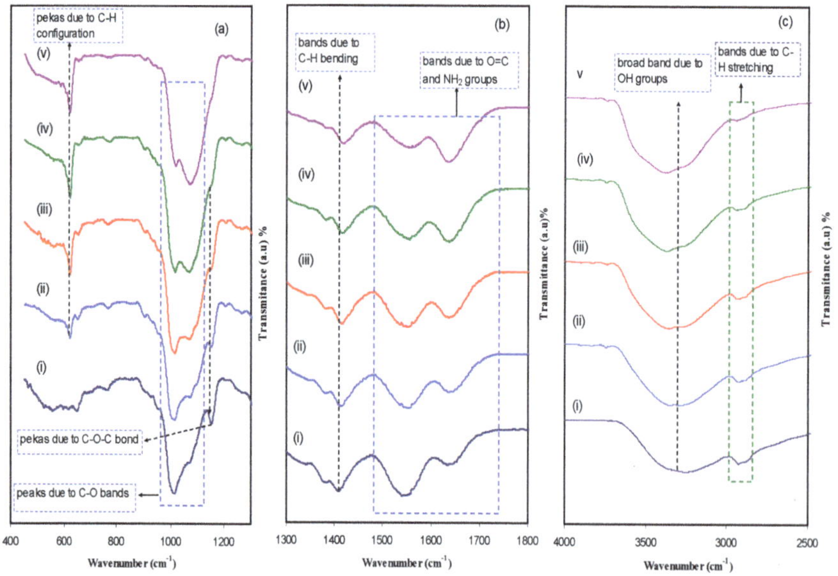

Figure 3. FTIR spectra of (i) CS:Dextran (pure film), (ii) CSDPB1, (iii) CSDPB2, (iv) CSDPB3, and (v) CSDPB4 in the region (**a**) 700 cm^{-1} to 1300 cm^{-1}, (**b**) 1400 cm^{-1} to 1800 cm^{-1}, and (**c**) 3000 cm^{-1} to 3800 cm^{-1}.

2.2. Morphological Study

Dealing with material surface is vital to understand the structural changes over a number of processes. Recent studies revealed that the morphology aspect in polymers provides some insights into the changes in structural or electrical properties [9,25,28]. Polymer family can be categorized in terms of crystallinity into crystallineand amorphous polymers. On the one hand, polymer crystals are characterized by compact assembly of stereo-regular chains and thereby exhibit high modulus and hardness, but weak toughness. On the other hand, amorphous polymers features are rubbery or glassy in behavior [38]. In our previous work, SEM technique was used to study the compatibility of salts with polar polymers [39]. Figure 4a–d shows the FESEM images for a number of the CS:Dextran systems incorporated with various weight percentage of $LiClO_4$ salt. The surface morphologies of the blend electrolyte samples are almost smooth and there are no protruded particles on the sample surfaces. This was observed in our previous work [40] where, as more salt is added, more particles protruded out of the surface. This indicates that polymer matrix capacity is limited to accommodate excess salt, which in turn led to salt recrystallization. It is apparent, as recrystallization proceeds, the amount of free ions lowers, which results in conductivity value decrement [41]. In fact, the smooth surface of the samples indicates that the complexation had taken place among the polymer blends and the incorporated $LiClO_4$ salt. The data results of the present work indicate that polymer blend fabrication is a novel approach and straightforward methodology with high DC conductivity. The occurrence of the extent of complexation of the dopant salt and CS:Dextran was realized from the FTIR study.

Figure 4. *Cont.*

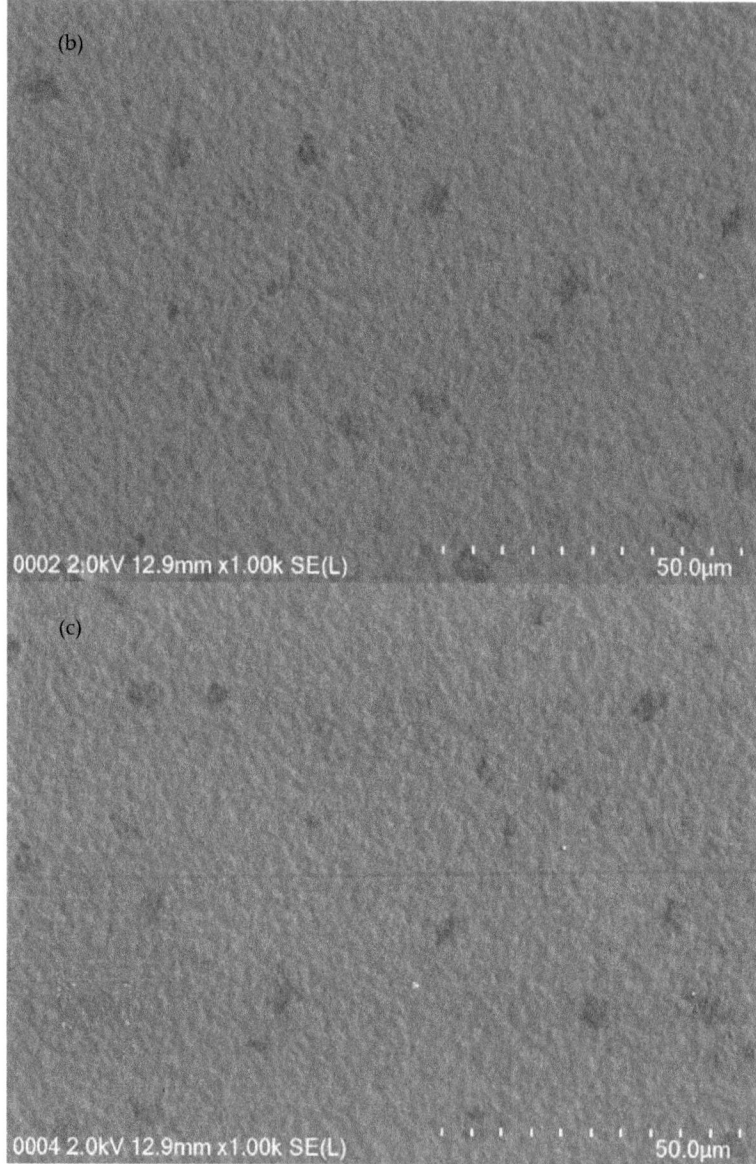

Figure 4. *Cont.*

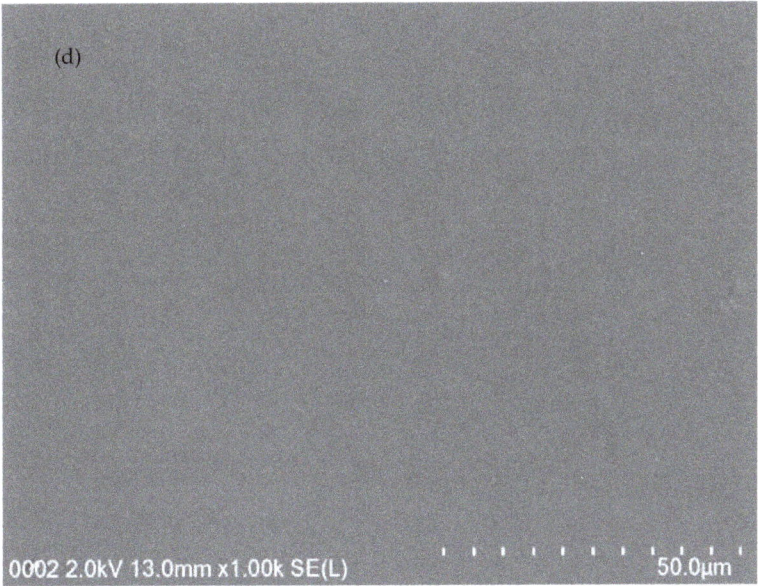

Figure 4. FESEM images for (**a**) CSDPB1, (**b**) CSDPB2, (**c**) CSDPB3, and (**d**) CSDPB4 blend electrolytes.

2.3. Impedance Study

A comparably new and powerful technique in the characterization of a number of the electrical properties of electrolyte materials and the interface region with electronically conducting electrodes is electrochemical impedance spectroscopy [28,42,43]. The impedance plot for CS:Dextran–$LiClO_4$ polymer blend electrolyte systems at room temperature are shown in Figure 5a–d. The complex impedance plots reveal two main distinct regions: The semicircle observed at the high frequency region, which is due to the bulk character of the electrolytes, and the linear region at the low frequency range, which is attributed to the blocking electrodes [42,44,45]. The membrane electrolytes carry ion carriers, and thus, ion diffusion occurs through the membrane as an AC electric field is applied, and consequently, ion accumulation builds up at the electro/electrolyte interface. Due to the electronic nature of the stainless steel electrodes, ions cannot cross the system, and thus, the real and imaginary parts of the impedance can be measured at various frequencies, which resulted in impedance plots. It is interesting that, at the intermediate frequencies, certainly at −45 ~ inclined lines indicate the occurrence of Warburg impedance as a consequence of diffusion of ions to the electrode surface [46]. Furthermore, the spike feature at the low frequency region is a characteristic of diffusion process [29]. It is apparent, in Figure 5 that, with increasing salt concentration, the bulk resistance (see the insets) decreased. The R_b value is determined by the point where the semicircle intersects the real axis (Zr). The equation below has been applied to determine the sample conductivity based on the R_b value and the sample dimensions:

$$\sigma_{dc} = \left(\frac{1}{R_b}\right) \times \left(\frac{t}{A}\right) \quad (1)$$

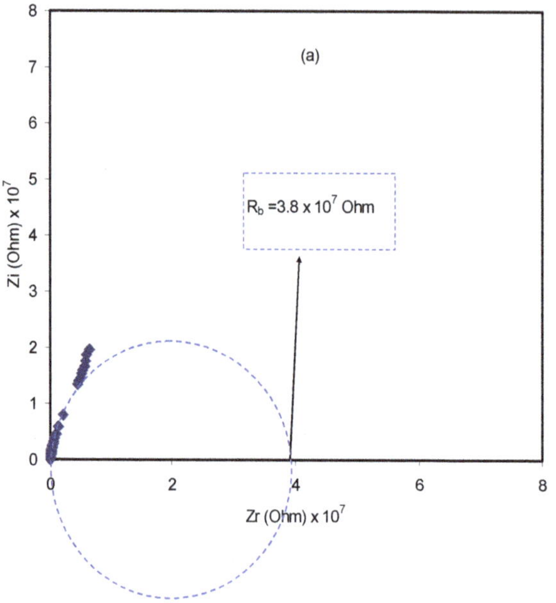

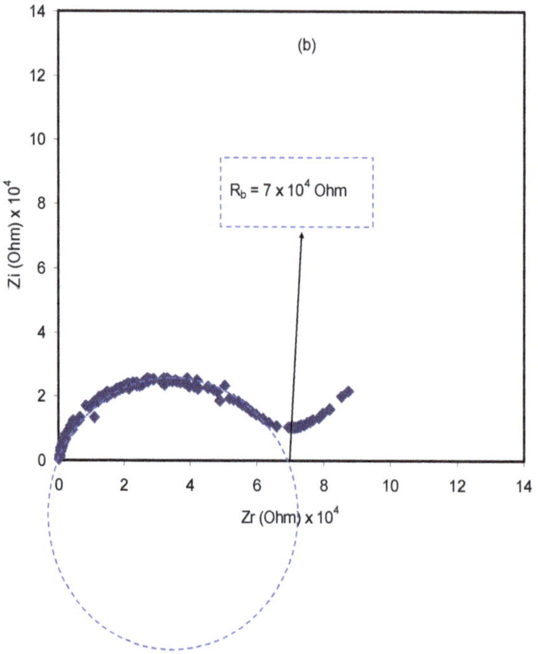

Figure 5. *Cont.*

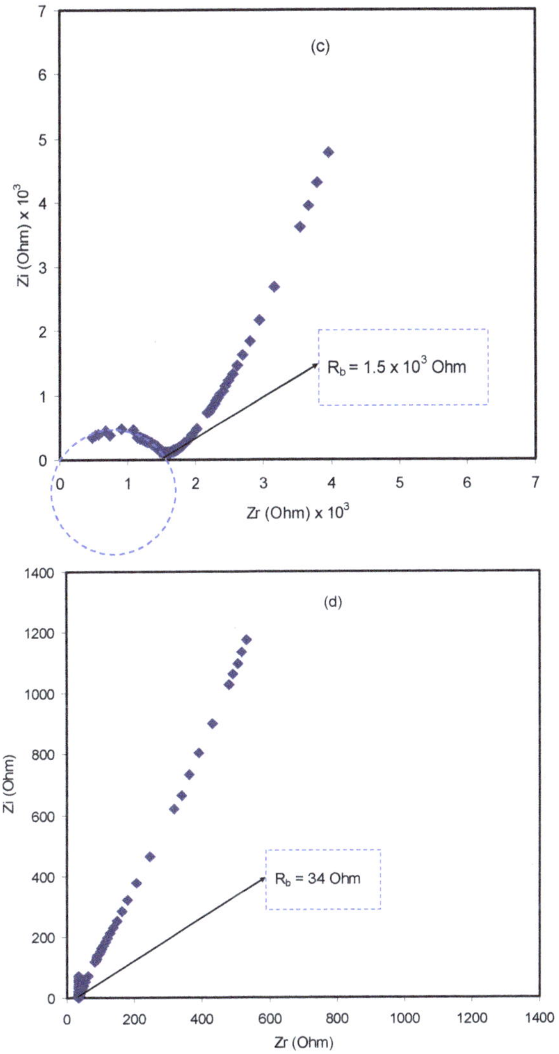

Figure 5. *Cont.*

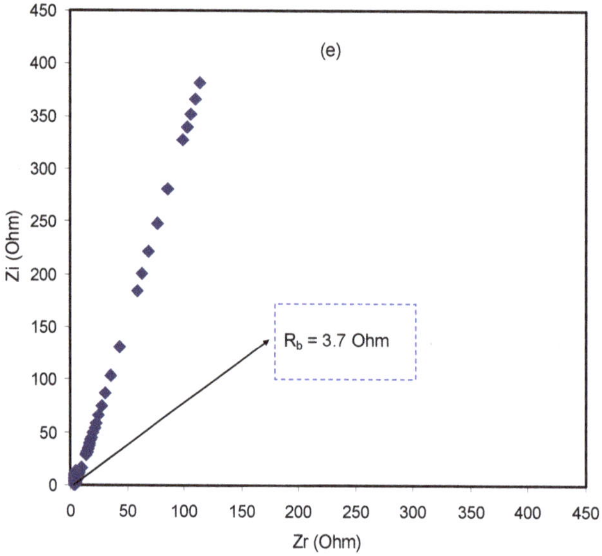

Figure 5. Experimental Impedance plots for (**a**) pure CS:Dextran film (**b**) CSDPB1, (**c**) CSDPB2, (**d**) CSDPB3, and (**e**) CSDPB4 blend electrolyte films.

In the above equation, the polymer electrolyte film thickness and the film surface area are, respectively, denoted by t and A. Table 1 shows the calculated DC conductivity for all the samples. The high DC conductivity of blend electrolytes is a guarantee for EDLC application. To get more information about the charge transfer resistance of the samples Bode plots were also studied. More insights about the electrical properties of the blend electrolyte samples can be grasped from the modeling of the impedance plots using electrical equivalent circuits (EECs). Through the modeling of impedance plots, it is possible to estimate the bulk resistance and circuit elements.

Table 1. DC conductivity for pure CS:Dex and blend electrolyte films at room temperature.

Sample Designation	DC Conductivity (S/cm)
CS:Dex	5.01×10^{-10}
CSDPB1	2.73×10^{-7}
CSDPB2	1.27×10^{-5}
CSDPB3	5.62×10^{-4}
CSDPB4	5.16×10^{-3}

Moreover, the results of Nyquist plots are further established by the consideration of Bode plot. From the electrochemical viewpoint, Bode plots are principally helpful in understanding the charge transfer process in electrolyte materials [47]. Figure 6 shows the Bode plots for the pure CS:Dextran blend film and blend electrolyte films at ambient temperature. Previous studies have demonstrated that three distinguished regions should be recognized from the Bode plots, which are capacitive, diffusion, and charge transfer regions [39,47–49]. Usually, the capacitive region (namely plateau region) can be observed at a very low frequency; from 10^{-2} to 100 Hz. However, this region could not be examined in this study, owing to the frequency limitations of the measuring equipment. As described in the impedance plots of Figure 5, the semicircle has been correlated to the ion transfer in amorphous phase of electrolytes and the tails were related to the contribution of Warburg or diffusion of ions, and therefore, their accumulation at the electrode/electrolyte interface [40,43,45]. The ion accumulation on both sides of the electrolyte membrane will produce electrical double layer capacitances. The results clearly indicated that, with increasing salt concentration, from 10 to 40 wt.%, the Warburg contribution (tail regions) has been increased, and therefore, the resistance decreased due to the large amount of carrier density. It is obvious from Figure 6a that pure CS:Dextran films show high charge transfer resistance. Obviously, the charge transfer resistance decreases with increasing salt concentrations as shown in Figure 6b,c. From Figures 5 and 6, it is clear that the sample incorporated with 40 wt.% of LiClO4 exhibits the lowest resistance and thus a high conductivity resulted.

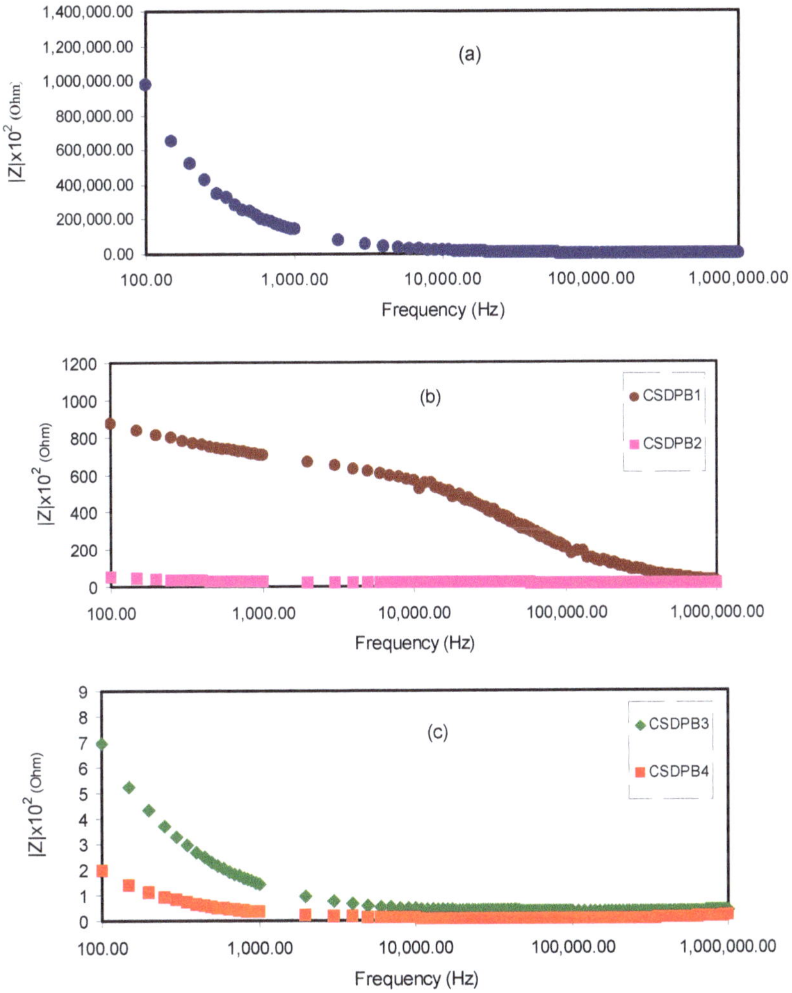

Figure 6. Bode plots for (**a**) pure CS:Dextran film (**b**) CSDPB1 and CSDPB2, and (**c**) CSDPB3, and CSDPB4 blend electrolyte films.

The EECs model can be usually utilized for fitting, i.e., it can be used in the analysis of impedance spectroscopy, since the model is straightforward, quick, and provides a complete picture of the system [50]. The acquired impedance plots can be interpreted with respect to the equivalent circuit including R_b for the charge carriers in the sample and two constant phase elements (CPE), as presented in the insets of Figure 7a. The high frequency region shows the combination of R_b and CPE, whilst the low frequency region exhibits CPE, i.e., the developed double layer capacitance between the electrodes and SPE. The abbreviated term CPE is more commonly used in the equivalent circuit rather than as an ideal capacitor in real system. This is due to the fact that the behavior of the actual SPE is varied from that of an ideal capacitor considered in a pure semicircular pattern [51]. As discussed above, Warburg impedance at −45 ~ inclined lines are definitely a consequence of diffusion of ions to the electrode surface. In this report, the depressed semicircle has been explained by CPE instead of a capacitor [52]. The impedance of Z_{CPE} can be written as [53,54]:

$$Z_{CPE} = \frac{\cos(\pi n/2)}{Y_m \omega^n} - j\frac{\sin(\pi n/2)}{Y_m \omega^n} \quad (2)$$

where Y_m refers to the CPE capacitance, ω is the angular frequency and n is associated to the deviation of the vertical axis of the plot in the complex impedance plots. Finally, the real (Z_r) and imaginary (Z_i) values of complex impedance (Z^*) related with the equivalent circuit (insets of Figure 7a) can be expressed as:

$$Z_r = R_s + \frac{R_1 + R_1^2 Y_1 \omega^{n1} \cos(\pi n_1/2)}{1 + 2R_1 Y_1 \omega^{n_1} \cos(\pi n_1/2) + R_1^2 Y_1^2 \omega^{2n_1}} + \frac{\cos(\pi n_2/2)}{Y_2 \omega^{n_2}} \quad (3)$$

$$Z_i = \frac{R_1^2 Y_1 \omega^{n_1} \sin(\pi n_1/2)}{1 + 2R_1 Y_1 \omega^{n_1} \cos(\pi n_1/2) + R_1^2 Y_1^2 \omega^{2n_1}} + \frac{\sin(\pi n_2/2)}{Y_2 \omega^{n_2}} \quad (4)$$

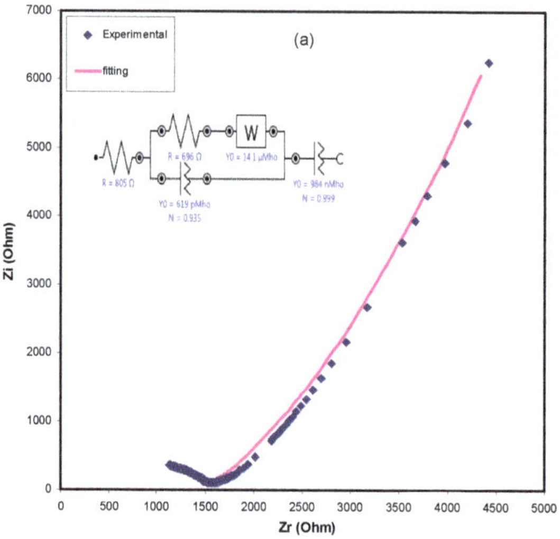

Figure 7. *Cont.*

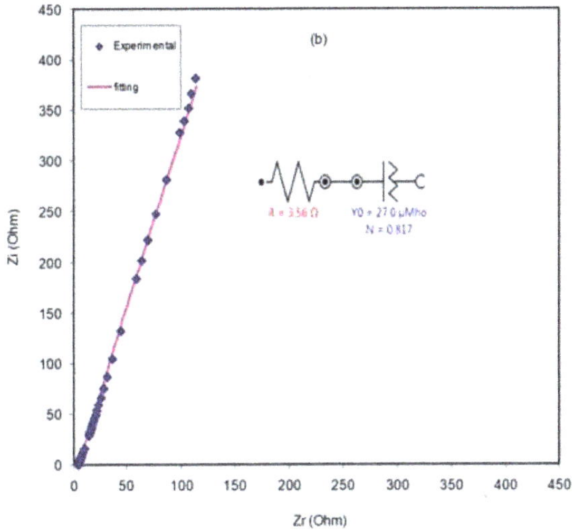

Figure 7. Experimental Impedance and fitting (EEC) plots for (**a**) CSDPB2 and (**b**) CSDPB4 blend electrolyte films.

All circuit element parameters that are used for fitting the experimental impedance plots for all the selected samples are presented in Table 2. These elements are exactly related to the parameters in the above equations, such that Y_o corresponds to Y_m, N corresponds to n_1 and n_2 according to CPE1 and CPE2 elements and R corresponds to R_1. In the Cole–Cole plot, the semicircle disappears at a certain high salt concentration of the salt (see Figure 7b), suggesting that only the resistive component of the polymer prevails [55]. In this case, the values of Z_r and Z_i associated to the EEC can be expressed as:

$$Z_r = R + \frac{\cos(\pi n/2)}{Y_m \omega^n} \quad (5)$$

$$Z_i = \frac{\sin(\pi n/2)}{Y_m \omega^n} \quad (6)$$

Table 2. The parameters of the circuit elements of the selected blend electrolyte membranes at ambient temperature.

Sample	R_1/Ohm	Y_1	n_1	R_s	Y_2/nMho	n_2
CSDPB2	696	619 pMho	0.9345	805	984	0.99
CSDPB4	3.56	27 μMho	0.817	-	-	-

2.4. *Transference Number Measurement (TNM) Study*

The ionic (t_i) and electronic (t_e) transference number can be gained from the ratio of steady-state current (I_{ss}) and initial current (I_i). The value of t_i can be calculated from the following equation [55]:

$$t_i = \frac{I_i - I_{ss}}{I_i} \quad (7)$$

Figure 8 shows the polarization plot of current versus time for the highest conducting electrolyte. The current at initial time is large due to migration of both ions and electrons. It is seen that the current is decreased rapidly prior to achieving a constant value of 0.4 μA, since only electron can pass through the stainless steel electrodes. At steady state, the cell is polarized as the current flow is stayed due to electron [56]. The constant current value indicates ionic conductor behavior of the electrolyte [57], where I_{ss} and I_i are observed to be at 0.4 and 27.8 μA, respectively. Therefore, t_i and t_e transference number for the electrolyte are found to be 0.98 and 0.02, respectively. This confirms the fact that ion is the dominant charge carrier in the electrolyte. Othman et al. [58] documented t_i values from 0.93 to 0.98 for poly methyl methacrylate (PMMA)-lithium trifluoro methane sulfonate ($LiCF_3SO_3$). Therefore, the high transference number may be correlated with the effect of polymer-ion and ion-ion interactions on the microscopic parameter.

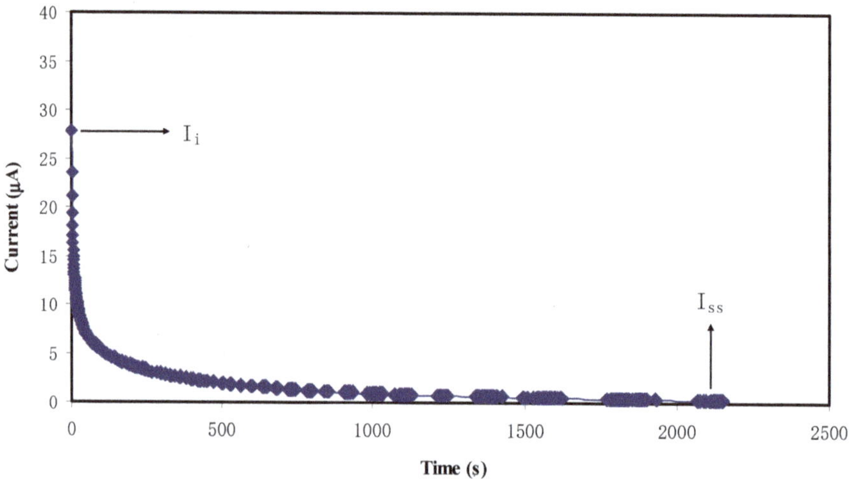

Figure 8. Polarization current versus time for the CSDPB4 blend electrolyte film.

2.5. LSV Analysis

Determination of potential stability is an imperative characteristic of the polymer electrolyte for energy storage device applications. Figure 9 shows the LSV plot of the highest conducting (CSDPB4) electrolyte. It is noticeable that there is no current flowing below 2.3 V, which indicates that there is no electrochemical reaction occurring below this potential window. The increase of current beyond 2.3 V is related to the decomposition of the polymer electrolyte, signifying the electrochemical reaction within the polymer electrolyte [59]. Monisha et al. [60] stated that the threshold voltage is that the current flows through the cells. Shukur et al. [52] reported a decomposition voltage at 2.10 V of lithium salt based biopolymer electrolyte. Thus, the potential stability, i.e., potential window of the relatively high conducting electrolyte in this work is suitable for energy storage device applications.

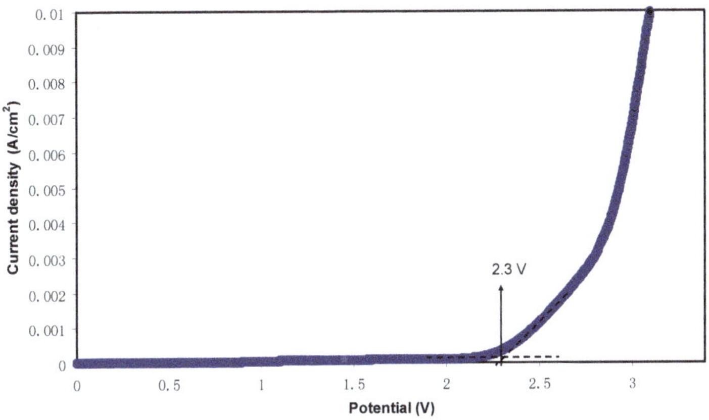

Figure 9. LSV plot for the highest conducting chitosan-dextran-$LiClO_4$ (CSDPB4) sample.

2.6. EDLC Study

CV and EDLC Characteristics

Figure 10 shows the cyclic voltammogram of the fabricated EDLC at a sweep rate of 10 $mV{\cdot}s^{-1}$. It can be observed that there is no peak within the potential range of 0 to 1.0 V. However, there is an electrical double layer, i.e., non-Faradaic current at the surface of the electrodes [61]. As stated previously, energy storage mechanism in EDLC goes via non-Faradaic, which means that no redox reaction process. The addition of salt in the electrolyte produces positively charged ion (i.e., cations) and negatively charged ions (i.e., anions). Once the EDLC is charged, cations and anions will migrate to negative and positive electrodes, respectively. At negative electrode, the induced electric field at the electrode attracts cations and repels anions, where the opposite action takes place at the positive electrode. The intense electric field holds ions from the electrolyte and electrons from the electrode. This is called as development of charge double-layer, where it stores the energy as potential energy [62]. The shape of CV in Figure 10 depicts rapid switching of ions at the electrode electrolyte interfaces as well the good capacitive behavior of electrodes. The internal resistance and electrode porosity resulted in current dependence of voltage and makes the shape of the CV plot a less perfect rectangular [63].

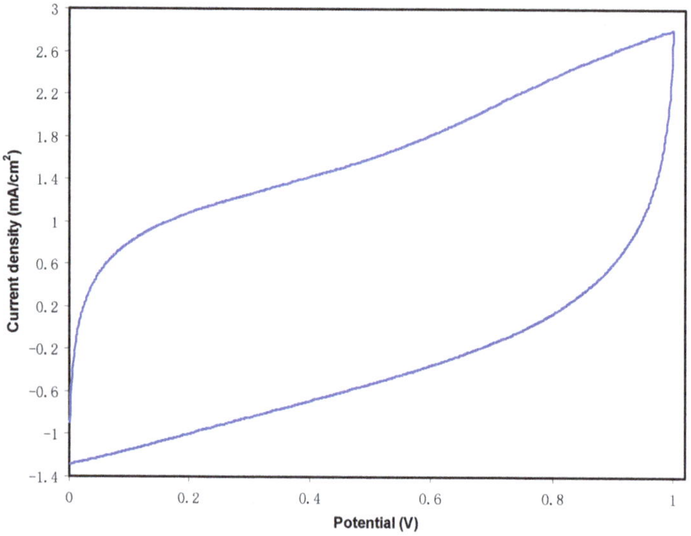

Figure 10. CV plot of the fabricated EDLC in the potential range of 0 V to 1 V.

The charge–discharge profiles of the fabricated EDLC are investigated through galvanostatic technique. Figure 11 shows the charge-discharge plot of the fabricated EDLC at 0.5 mA·cm^{-2} in the potential range of 0 to 1 V. The discharge slope is observed to be almost linear, which verifies the capacitive behavior in the EDLC [64]. As the slope of the discharge curve (s) is determined, the specific capacitance (C_s) can be calculated from the following equation:

$$C_S = \frac{i}{sm} \tag{8}$$

here, i is the constant current and m stands for active material mass, which is the mass of active carbon in this study. In Figure 12, C_s of the EDLC for 100 cycles can be seen, in which C_s of the EDLC at the 1st cycle is found to be 8.7 F·g^{-1}. Teoh et al. [65] have recorded a C_s of 7.1 F·g^{-1} for free plasticizer $LiClO_4$ based corn starch polymer electrolyte EDLC type capacitor. At 5th cycle, the C_s drops to 6.5 F·g^{-1} and remains constant in the range of 6.0 to 6.5 F·g^{-1}. This capacitance synthesized in the present work is of the great interest compared to the specific capacitance values of 2.6–3.0 and 1.7–2.1 F·g^{-1}, which have been recorded for the EDLC cells with the Mg- and Li-based PEO polymer electrolytes incorporated with ionic liquids [66]. Therefore, polymer blend electrolytes can be established as new materials in fabricating EDLC cells with high specific capacitances.

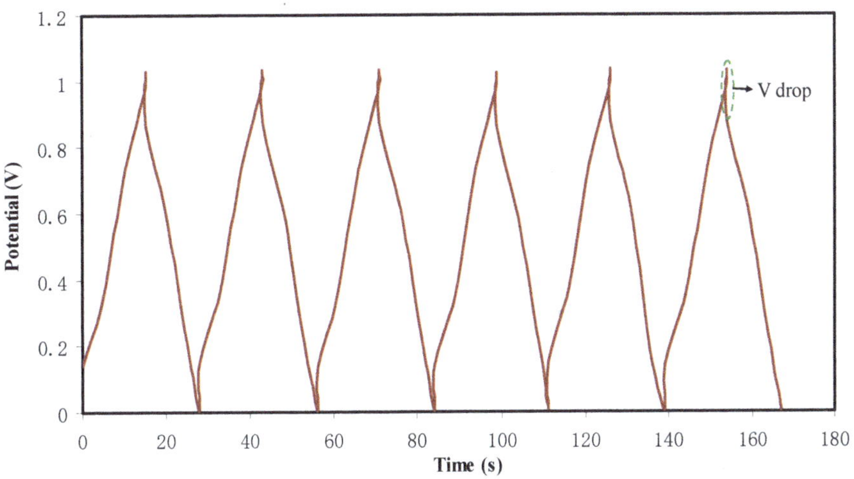

Figure 11. Charge-discharge profiles for the fabricated EDLC at 0.5 mA·cm^{-2}.

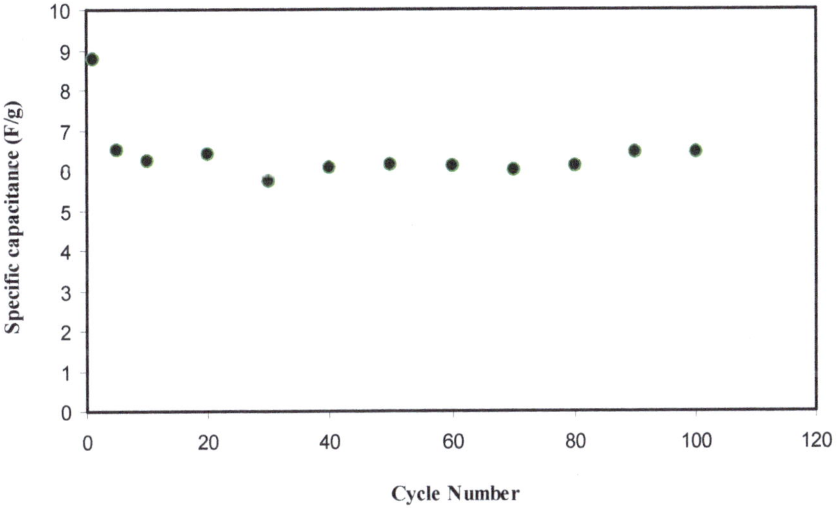

Figure 12. Specific capacitance of the fabricated EDLC for 100 cycles.

Coulombic efficiency (η) is another imperative parameter regarding the cycling stability of the EDLC where it can be calculated from the following equation:

$$\eta = \frac{t_d}{t_c} \times 100 \quad (9)$$

where discharge and charge time are denoted as t_d and t_c, respectively. Figure 13 shows the η of the EDLC of 100 cycles. The coulombic efficiency, η at the 1st cycle is found to be 20.6% and increases to 72.2% and 86.6% at 5th and 10th cycles, respectively. At 20th cycle, η is observed to be 92.8% and then lowered and remained constant at 92.0% up to 100 cycles. It is considered that the EDLC possesses plausible electrode-electrolyte contact as the η is beyond 90.0% [67].

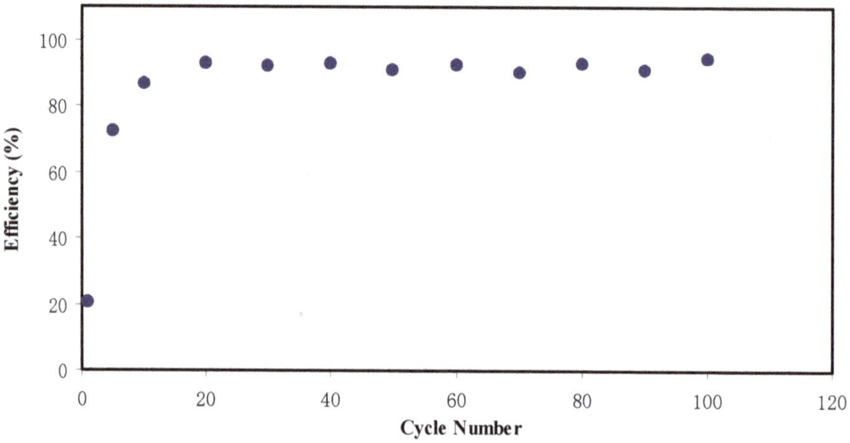

Figure 13. Cycling stability of the EDLC up to 100 cycles.

As observed in Figure 11, there are tiny potential drops (V_d) before the discharging process commences. This can be related to the existence of internal resistance in the EDLC, which is called equivalent series resistance (R_{esr}). This resistance R_{esr} of the EDLC can be obtained from the following equation:

$$R_{esr} = \frac{V_d}{i} \tag{10}$$

Figure 14 shows the R_{esr} of the EDLC for 100 cycles. It is determined that the R_{esr} varies from 150 to 180 Ω over the 100 cycles. The existence of internal resistance has been assumed to be due to various factors. The first one is from the electrolyte, where fast charge/discharge process cause free ions to recombine and reduce the ionic conductivity. Secondly, it is from current collectors, which in this case is the aluminum foil. Lastly, it is the gap between the electrolyte and electrode, where ions from the electrolyte and electrons from the carbon electrode form a charge double layer or called potential energy [68].

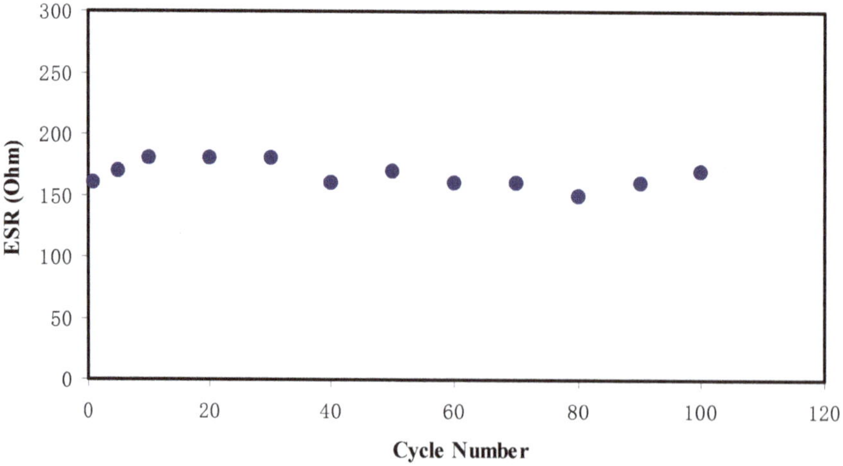

Figure 14. The pattern of equivalent series resistance of the EDLC for 100 cycles.

The energy density (E_d) and power density (P_d) of the EDLC can be expressed as:

$$E_d = \frac{C_s V}{2} \quad (11)$$

$$P_d = \frac{V^2}{4mR_{esr}} \quad (12)$$

where V is the applied voltage (1 V). In Figure 15, E_d is found to be 1.21 Wh·kg^{-1} at the 1st cycle and then lowered to 0.90 Wh·kg^{-1} at 5th cycle. E_d is then kept stable at 0.86 Wh·kg^{-1} up to 100 cycles. The E_d values of the EDLC are almost constant, which harmonized with the pattern of C_s. The stabilization study showed that the ions experience the same energy barrier during the migration towards the electrodes from 10th to 100th cycle of charge-discharge process [69]. The energy density achieved for the EDLC cell (0.86 Wh/Kg) in the present work is of great interest compared to that reported (0.3 Wh/Kg) for ionic liquid incorporated PEO based polymer electrolyte [66]. Figure 16 exhibits the P_d of the EDLC for 100 cycles. The value of P_d of the EDLC is found to be 643 W·kg^{-1} at the 1st cycle and dropped to 571 W·kg^{-1} at 10th cycle. The interesting observation is that the value of P_d increases back to 685 W·kg^{-1} up to 80cycles. In Figure 14, R_{esr} of the EDLC is also increased at 10th cycle and decreased until 80th cycle. At 90th and 100th cycles, the EDLC experiences reduction in P_d value. This could be related to the growth of ion aggregates/pairs and electrolyte depletion during rapid charge-discharge process. The main drawbacks of polymer electrolyte membranes are ion aggregates/pairs. In this case little ions diffuse to the electrode/electrolyte interface. Consequently, the accumulated ion in the forming double-layer, at the surface of the electrodes, is reduced, and thus, reducing the power density [61].

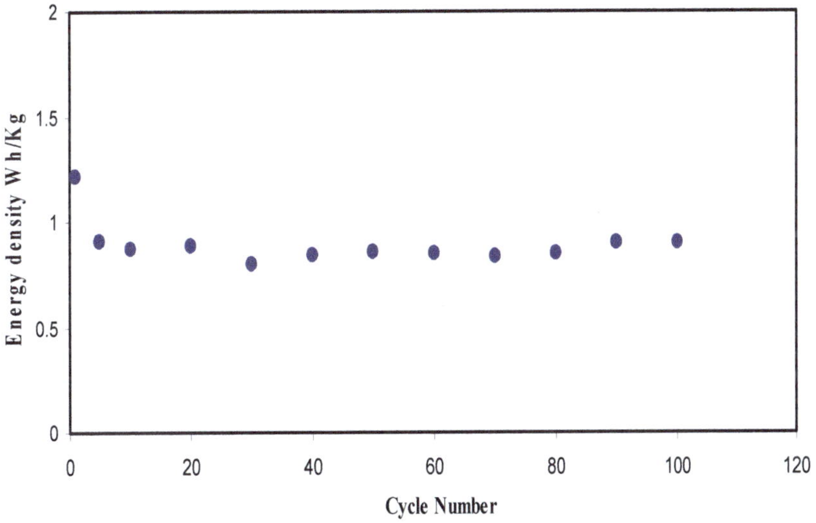

Figure 15. Energy density of the fabricated EDLC for 100 cycles.

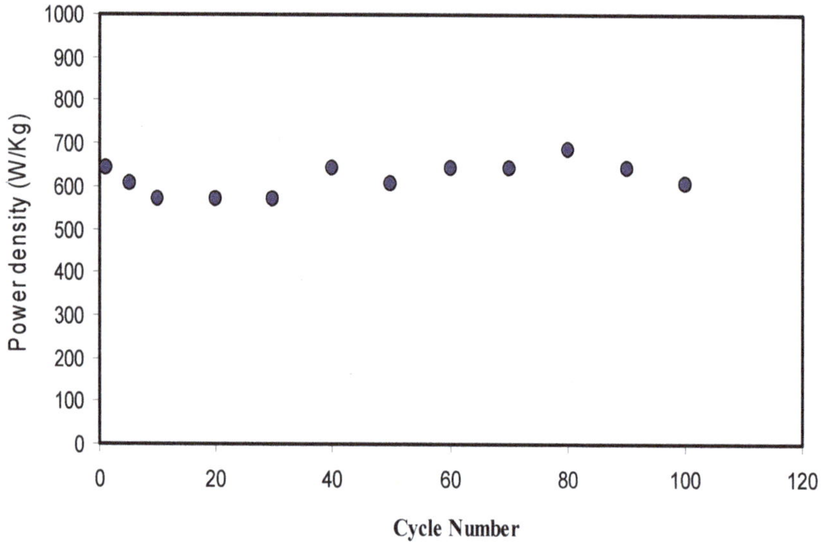

Figure 16. Power density of the fabricated EDLC for 100 cycles.

3. Experimental Method

3.1. Materials and Sample Preparation

High molecular weight chitosan (CS) (average molecular weight 310,000–375,000) and Dextran powder (average molecular weight 35,000–45,000) materials were used as the raw materials (Sigma-Aldrich, Warrington, PA, USA). For the fabrication of the polymer blending based on CS:Dextran, 60 wt.% chitosan and 40 wt.% dextran was dissolved separately in 50 mL of 1% acetic acid at room temperature for 90 min. Subsequently, these solutions then mixed and stirred for 3 h to gain a homogeneous blending solution. For the blended solution of CS:Dextran, various amounts of $LiClO_4$ ranging from 10 to 40 wt.% of $LiClO_4$ in steps of 10 was added separately with continuous stirring to prepare CS:Dextran: LiTf electrolytes. The polymer blend electrolytes were coded as CSDPB1, CSDPB2, CSDPB3, and CSDPB4 for CS:Dextran and incorporated with 10, 20, 30, and 40 wt.% of $LiClO_4$, respectively. After casting in different Petri dishes, the solutions were left to dry at room temperature for films to form. The films were transferred into a desiccator for further drying, which produces solvent-free films. The thickness of the SPBEM was from 0.0123 to 0.0131 cm.

3.2. TNM and LSV Measurements

V&A Instrument DP3003 digital DC power supply was employed to conduct the transference number (TNM) analysis via DC polarization method [70]. The cell was polarized at 30 mV and the DC current was monitored continuously as a function of time at room temperature. Stainless steel was used as electrodes for TNM due to its ion-blocking characteristic. Ionic and electronic transference numbers analyses are used to observe the contribution of ion and electron to the total conductivity and prove ionic conduction. The potential stability of the highest conducting electrolyte (CSDPB4) was studied using linear sweep voltammetry (LSV) (DY2300 Potentiostat) at sweep rate of 50 $mV{\cdot}s^{-1}$. For the LSV analysis, stainless steel was also used as counter, working and reference electrodes. For both LSV and TNM, the highest conducting electrolyte (CSDPB4) was placed in between two stainless steels in a Teflon conductivity holder, as shown in Figure 17.

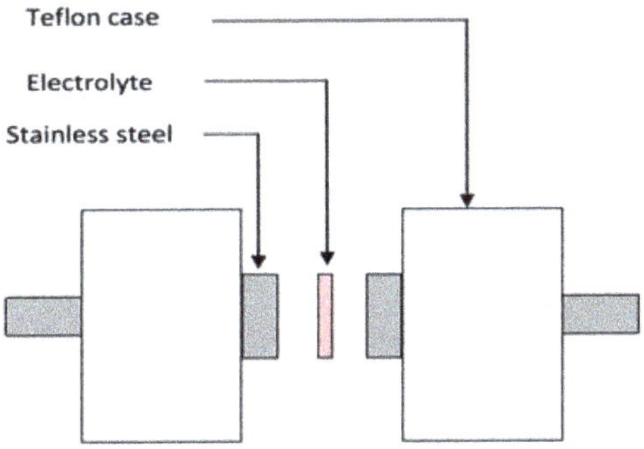

Figure 17. Schematic diagram of cell arrangement for LSV, TNM and impedance study.

3.3. EDLC Preparation

An amount of 0.50 g polyvinylidene fluoride (PVdF) was dissolved with stirring in 15 mL *N*-methyl pyrrolidone (NMP). On the other hand, activated carbon (3.25 g) and carbon black (0.25 g) powders were dry mixed. The dry mixing process was carried out by using a planetary ball miller (XQM-0.4). Six metal balls were inserted in the chamber along with the powders. The powders were mixed at rotational speed of 500 r/min for 15 min. The powders were then added to the obtained solution of PVdF/NMP and stirred for 90 min. The homogeneous solution was doctor bladed on aluminum foil and heated at 60 °C for a period of time. The electrodes were then stored in desiccator filled with silica gel. The thickness of the electrode was 0.01 cm. The highest conducting electrolyte was sandwiched between two carbon electrodes, which were cut into circle shape with area of 2.01 cm^2 and packed in CR2032 coin cells. The schematic diagram of the EDLC cell is shown in Figure 18. The galvanostatic charge-discharge characteristics of the EDLC were carried out using a battery cycler (Neware, Shenzhen, China) with a current density of 0.5 $mA \cdot cm^{-2}$. Digi-IVY DY2300 Potentiostat was used to conduct cyclic voltammetry (CV) of the EDLC at 10 $mV \cdot s^{-1}$ in the potential window of 0–1 V.

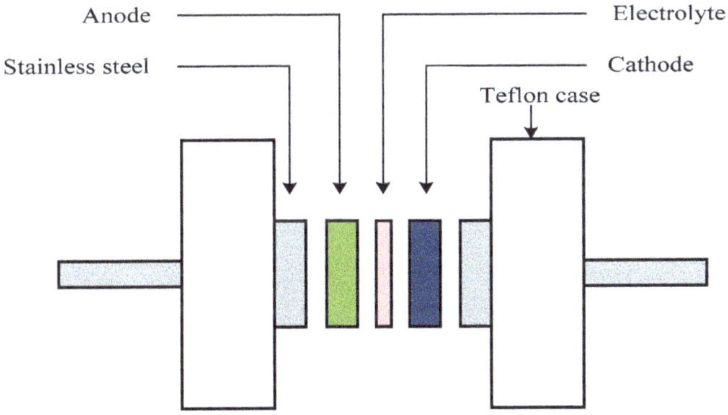

Figure 18. Diagram of the EDLC cell for CV and galvanostatic charge-discharge measurement.

3.4. Structural, Morphological, and Impedance Characterizations

Spotlight 400 Perkin-Elmer spectrometer was used in conducting Fourier transform infrared (FTIR) spectroscopy with a resolution of 1 cm^{-1} (450–4000 cm^{-1}). The surface of the electrolyte was analyzed via Hitachi SU8220 FESEM with 10K× magnification. For structural analysis, XRD pattern was acquired via D5000 X-ray diffractometer (1.5406 Å). The 2θ angle was continuously altered from 5° to 80° (resolution = 0.1°). In the mechanism study, HIOKI 3532–50 LCR HiTESTER was employed to analyze electrical impedance spectroscopy (EIS) measurements of the samples (50 Hz to 5 MHz). The cell arrangement for EIS is shown in Figure 17.

4. Conclusions

Solid polymer blend electrolytes (SPBE) composed of chitosan and dextran incorporated with various amounts of lithium perchlorate ($LiClO_4$) was prepared. The relatively highest ion conducting sample was utilized to fabricate EDLC supercapacitor. Non-crystalline behavior of the polymer blend electrolytes has been confirmed from XRD pattern. The FTIR emphasized the strong interaction between the constituents of polymer electrolyte. The relatively smooth surface morphology of the electrolyte was found to be an indication of compatible $LiClO_4$/polymer system. Faradaic process has shown to be absent and has definitely been noticed in the cyclic voltammetry. The conductivity of the samples was measured using impedance spectroscopy. The ionic (t_i) and electronic (t_e) transference number for the highest conducting electrolyte were found to be 0.98 and 0.02, respectively. The electrochemical stability window, as estimated from cyclic voltammetry, was found to be around 2.25 V for the Li^+ ion conducting electrolyte. The absence of a peak in CV plot indicating the presence of electrical double layer at the surface of the electrodes. The highest ion conducting samples was used to fabricate the EDLC supercapacitor. The discharge slope has been observed to be almost linear, which verified that the EDLC possesses the capacitive behavior. The performance of fabricated EDLC was also studied using cyclic voltammetry and charge–discharge techniques. The highest specific capacitance was achieved to be at 8.7 $F{\cdot}g^{-1}$. At 20th cycle, the efficiency (η) was observed at 92.8%, where it remained constant at 92.0% up to 100 cycles. It was found that the EDLC possesses good electrode-electrolyte contact as η is above 90.0%. Another key finding is that the R_{esr} was quite low and varied from 150 to 180 Ω over the 100 cycles. The energy density was quite high and equal to 1.21 $Wh{\cdot}kg^{-1}$ at the 1st cycle and then kept stable at 0.86 $Wh{\cdot}kg^{-1}$ up to 100 cycles. Lastly, the value of P_d was also found to increase, returning to 685 $W{\cdot}kg^{-1}$ up to 80 cycles.

Author Contributions: Conceptualization, Supervision, Methodology, Project Administration, Investigation, Writing—Original Draft Preparation, S.B.A.; Methodology, Investigation, Writing—Original Draft Preparation, Formal Analysis, M.H.H.; Writing—Review & Editing, Conceptualization, M.F.Z.K.; Writing—Review & Editing, Validation, W.O.K.; Writing—Review & Editing R.M.A.

Funding: This research was funded by Ministry of Higher Education and Scientific Research-Kurdish National Research Council (KNRC), Kurdistan Regional Government/Iraq. The financial support from the University of Sulaimani and Komar Research Center (KRC), Komar University of Science and Technology is greatly appreciated.

Acknowledgments: The authors gratefully acknowledge the Ministry of Higher Education and the Scientific Research, Kurdistan Regional Government/Iraq, University of Sulaimani, University of Malaya and Komar Research Center (KRC), Komar University of Science and Technology for supporting this research project.

Conflicts of Interest: The authors declare no conflicts of interest.

References

1. Iro, Z.S.; Subramani, C.; Dash, S.S. A Brief Review on Electrode Materials for Supercapacitor. *Int. J. Electrochem. Sci.* **2016**, *11*, 10628–10643. [CrossRef]
2. Kiamahalleh, M.V.; Zein, S.S.H.; Najafpour, G.; Sata, S.A.; Buniran, S. Multiwalled carbon nanotubes based nanocomposites for supercapacitors: A review of electrode materials. *Nano* **2012**, *7*, 1230002. [CrossRef]
3. Shobana, V.; Parthiban, P.; Balakrishnan, K. Lithium based battery-type cathode material for hybrid supercapacitor. *J. Chem. Pharm. Res.* **2015**, *7*, 207–212.
4. Kamarudin, K.H.; Hassan, M.; Isa, M.I.N. Lightweight and Flexible Solid-State EDLC based on Optimized CMC-NH4NO3 Solid Bio-Polymer Electrolyte. *ASM Sci. J. Spec. Issue* **2018**, *1*, 29–36.
5. Shukur, A.; Fadhlullah, M. Characterization of Ion Conducting Solid Biopolymer Electrolytes Based on Starch-Chitosan Blend and Application in Electrochemical Devices. Ph.D. Dissertation, University of Malaya, Wilayah Persekutuan Kuala Lumpur, Malaysia, 2015.

6. Kadir, M.F.Z.; Salleh, N.S.; Hamsan, M.H.; Aspanut, Z.; Majid, N.A.; Shukur, M.F. Biopolymeric electrolyte based on glycerolized methyl cellulose with NH4Br as proton source and potential application in EDLC. *Ionics* **2017**, *24*, 1651–1662. [CrossRef]
7. Aziz, S.B.; Hamsan, M.H.; Abdullah, R.M.; Kadir, M.F.Z. A Promising Polymer Blend Electrolytes Based on Chitosan: Methyl Cellulose for EDLC Application with High Specific Capacitance and Energy Density. *Molecules* **2019**. accpected for publication.
8. Wang, H.; Lin, J.; Shen, Z.X. Polyaniline (PANi) based electrode materials for energy storage and conversion. *J. Sci. Adv. Mater. Devices* **2016**, *1*, 225–255. [CrossRef]
9. Hamsan, M.H.; Aziz, B.; Shukur, M.F.; Kadir, M.F.Z. Protonic cell performance employing electrolytes based on plasticized methylcellulose-potato starch-NH4NO3. *Ionics* **2019**, *25*, 559–572. [CrossRef]
10. Nyuk, C.M.; Isa, M.I.N. Solid biopolymer electrolytes based on carboxymethyl cellulose for use in coin cell proton batteries. *J. Sustain. Sci. Manag.* **2018**, *2017*, 42–48.
11. Salleh, N.S.; Aziz, S.B.; Aspanut, Z.; Kadir, M.F.Z. Electrical impedance and conduction mechanism analysis of biopolymer electrolytes based on methyl cellulose doped with ammonium iodide. *Ionics* **2016**, *22*, 2157–2167. [CrossRef]
12. Hassan, M.F.; Azimi, N.S.N.; Kamarudin, K.H.; Sheng, C.K. Solid Polymer Electrolytes Based on Starch-Magnesium Sulphate: Study on Morphology and Electrical Conductivity. *ASM Sci. J. Spec. Issue* **2018**, *1*, 17–28.
13. Du, B.W.; Hu, S.Y.; Singh, R.; Tsai, T.T.; Lin, C.C.; Ko, F.U. Eco-Friendly and Biodegradable Biopolymer Chitosan/Y_2O_3 Composite Materials in Flexible Organic Thin-Film Transistors. *Materials* **2017**, *10*, 1026.
14. Moniha, V.; Alagar, M.; Selvasekarapandian, S.; Sundaresan, B.; Hemalatha, R.; Boopathi, G. Synthesis and characterization of bio-polymer electrolyte based on iota-carrageenan with ammonium thiocyanate and its applications. *J. Solid State Electrochem.* **2018**, *22*, 3209–3223. [CrossRef]
15. Netsopa, S.; Niamsanit, S.; Sakloetsakun, D.; Milintawisamai, N. Characterization and Rheological Behavior of Dextran from Weissella confusa R003. *Int. J. Polym. Sci.* **2018**, *2018*, 5790526. [CrossRef]
16. Sarwat, F.; Ahmed, N.; Aman, A.; Qader, S.A.U. Optimization of growth conditions for the isolation of dextran producing Leuconostoc spp. from indigenous food sources Pak. *J. Pharm. Sci.* **2013**, *26*, 793–797.
17. Barsbay, M.; Guner, A. Miscibility of dextran and poly(ethylene glycol) in solid state: Effect of the solvent choice, Carbohydr. *Polymers* **2007**, *69*, 214–223. [CrossRef]
18. Telegeev, G.; Kutsevo, N.; Chumachenko, V.; Naumenko, A.; Telegeeva, P.; Filipchenko, S.; Harahuts, Y. Dextran-Polyacrylamide as Matrices for Creation of Anticancer Nanocomposite. *Int. J. Polym. Sci.* **2017**, *2017*, 4929857. [CrossRef]
19. Misenan, M.S.M.; Isa, M.I.N.; Khiar, A.S.A. Electrical and structural studies of polymer electrolyte based on chitosan/methyl cellulose blend doped with BMIMTFSI. *Mater. Res. Express* **2018**, *5*, 055304. [CrossRef]
20. Hamsan, M.H.; Shukur, M.F.; Kadir, M.F.Z. The effect of NH4NO3 towards the conductivity enhancement and electrical behavior in methyl cellulose-starch blend based ionic conductors. *Ionics* **2017**, *23*, 1137–1154. [CrossRef]
21. Kharbachi, A.E.; Hu, Y.; Yoshida, K.; Vajeeston, P.; Kim, S.; Sørby, M.H.; Orimo, S.; Fjellvåg, H.; Hauback, B.C. Lithium ionic conduction in composites of Li(BH4)0.75I0.25 and amorphous 0.75Li2S0.25P2S5 for battery applications. *Electrochim. Acta* **2018**, *278*, 332–339. [CrossRef]
22. Tamilselvi, P.; Hema, M. Conductivity studies of LiCF3SO3 doped PVA: PVdF blend polymer electrolyte. *Physica B* **2014**, *437*, 53–57. [CrossRef]
23. Yusof, Y.M.; Shukur, M.F.; Illias, H.A.; Kadir, M.F.Z. Conductivity and electrical properties of corn starch–chitosan blend biopolymer electrolyte incorporated with ammonium iodide. *Phys. Scr.* **2014**, *89*, 035701–035711. [CrossRef]
24. Kadir, M.F.Z.; Hamsan, M.H. Green electrolytes based on dextran-chitosan blend and the effect of NH4SCN as proton provider on the electrical response studies. *Ionics* **2018**, *24*, 2379–2398. [CrossRef]
25. Aziz, S.B.; Abidin, Z.H.Z.; Kadir, M.F.Z. Innovative method to avoid the reduction of silver ions to silver nanoparticles in silver ion conducting based polymer electrolytes. *Phys. Scr.* **2015**, *90*, 035808. [CrossRef]
26. Aziz, S.B.; Kadir, M.F.Z.; Abidin, Z.H.Z. Structural, morphological and electrochemical impedance study of CS:LiTf based solid polymer electrolyte: Reformulated Arrhenius equation for ion transport study. *Int. J. Electrochem. Sci.* **2016**, *11*, 9228–9244. [CrossRef]

27. Hamsan, M.H.; Shukur, M.F.; Aziz, S.B.; Kadir, M.F.Z. Dextran from Leuconostoc mesenteroides-doped ammonium salt-based green polymer electrolyte. *Bull. Mater. Sci.* **2019**, *42*, 57. [CrossRef]
28. Aziz, S.B.; Abidin, Z.H.Z.; Arof, A.K. Effect of silver nanoparticles on the DC conductivity in chitosan–silver triflate polymer electrolyte. *Physica B* **2010**, *405*, 4429–4433. [CrossRef]
29. Malathi, J.; Kumaravadivel, M.; Brahmanandhan, G.M.; Hema, M.; Baskaran, R.; Selvasekarapandian, S. Structural, thermal and electrical properties of PVA–LiCF3SO3 polymer electrolyte. *J. Non-Cryst. Solids* **2010**, *356*, 2277–2281. [CrossRef]
30. Aziz, S.B. Role of dielectric constant on ion transport: Reformulated Arrhenius equation. *Adv. Mater. Sci. Eng.* **2016**, *2016*, 2527013. [CrossRef]
31. Shukla, R.; Shukla, S.; Bivolarski, V.; Iliev, I.; Ivanova, I.; Goyal, A. Structural Characterization of Insoluble Dextran Produced by Leuconostoc mesenteroides NRRL B-1149 in the Presence of Maltose. *Food Technol. Biotechnol.* **2011**, *49*, 291–296.
32. Vettori, M.H.P.B.; Franchetti, S.M.M.; Contiero, J. Structural characterization of a new dextran with a low degree of branching produced by Leuconostoc mesenteroides FT045B dextransucrase. *Carbohydr. Polym.* **2012**, *88*, 1440–1444. [CrossRef]
33. Dumitraşcu, M.; Meltzer, V.; Sima, E.; Vîrgolici, M.; Albu, M.G.; Ficai, A.; Moise, V.; Minea, R.; Vancea, C.; Scărişoreanu, A.; et al. Characterization of electron beam irradiated collagenpolyvinylpyrrolidone (PVP) and collagen-dextran (DEX) blends. *Dig. J. Nanomater. Biostruct.* **2011**, *6*, 1793–1803.
34. Aziz, S.B.; Abidin, Z.H.Z. Electrical conduction mechanism in solid polymer electrolytes: New concepts to arrhenius equation. *J. Soft Matter* **2013**, *2013*, 323868. [CrossRef]
35. Nikoli, G.S.; Caki, M.; Miti, Z.; Ili, B.; Premovic, P. Attenuated Total Reflectance–Fourier Transform Infrared Microspectroscopy of Copper(II) Complexes with Reduced Dextran Derivatives. *Russian J. Phys. Chem. A* **2009**, *83*, 1520–1525. [CrossRef]
36. Wei, D.; Sun, W.; Qian, W.; Ye, Y.; Ma, X. The synthesis of chitosan-based silver nanoparticles and their antibacterial activity. *Carbohydr. Res.* **2009**, *344*, 2375–2382. [CrossRef] [PubMed]
37. Mitić, Ž.; Cakić, M.; Nikolić, G. Fourier-Transform IR spectroscopic investigations of Cobalt(II)–dextran complexes by using D2O isotopic exchange. *Spectroscopy* **2010**, *24*, 269–275. [CrossRef]
38. Han, C.C.; Shi, W.; Jin, J. Morphology and Crystallization of Crystalline/Amorphous Polymer Blends. *Encyclopedia Polym. Compos.* **2013**, 1–19. [CrossRef]
39. Aziz, S.B.; Abdullah, R.M.; Kadir, M.F.Z.; Ahmed, H.M. Non suitability of silver ion conducting polymer electrolytes based on chitosan mediated by barium titanate ($BaTiO_3$) for electrochemical device applications. *Electrochim. Acta* **2019**, *296*, 494–507. [CrossRef]
40. Aziz, S.B.; Abdullah, O.G.; Rasheed, M.A.; Ahmed, H.M. Effect of high salt concentration (HSC) on structural, morphological, and electrical characteristics of chitosan based solid polymer electrolytes. *Polymers* **2017**, *9*, 187. [CrossRef]
41. Shukur, M.F.; Kadir, M.F.Z. Hydrogen ion conducting starch-chitosan blend based electrolyte for application in electrochemical devices. *Electrochim. Acta* **2015**, *158*, 152–165. [CrossRef]
42. Polu, A.R.; Kumar, R. AC impedance and dielectric spectroscopic studies of Mg2+ionconducting PVA–PEG blended polymer electrolytes. *Bull. Mater. Sci.* **2011**, *34*, 1063–1067. [CrossRef]
43. Aziz, S.B.; Woo, T.J.; Kadir, M.F.Z.; Ahmed, H.M. A conceptual review on polymer electrolytes and ion transport models. *J. Sci. Adv. Mater. Devices* **2018**, *3*, 1–17. [CrossRef]
44. Aziz, S.B.; Abidin, Z.H.Z.; Arof, A.K. Influence of silver ion reduction on electrical modulus parameters of solid polymer electrolyte based on chitosan-silver triflate electrolyte membrane. *ExpressPolym. Lett.* **2010**, *4*, 300–310. [CrossRef]
45. Aziz, S.B. The mixed contribution of ionic and electronic carriers to conductivity in chitosan based solid electrolytes mediated by CuNt Salt. *J. Inorg. Organomet. Polym.* **2018**, *28*, 1942. [CrossRef]
46. Hirankumar, G.; Selvasekarapandian, S.; Bhuvaneswari, M.S.; Baskaran, R.; Vijayakumar, M. AC Impedance Studies on Proton Conducting Polymer Electrolyte Complexes (PVA+CH3COONH 4). *Ionics* **2004**, *10*, 135. [CrossRef]
47. Aziz, S.B.; Faraj, M.G.; Abdullah, O.G. Impedance Spectroscopy as a Novel Approach to Probe the Phase Transition and Microstructures Existing in CS:PEO Based Blend Electrolytes. *Sci. Rep.* **2018**, *8*, 14308. [CrossRef] [PubMed]
48. Eftekhari, A. The mechanism of ultrafast supercapacitors. *J. Mater. Chem. A.* **2018**, *6*, 2866–2876. [CrossRef]

49. Eftekhari, A. Surface Diffusion and Adsorption in Supercapacitors. *ACS Sustain. Chem. Eng.* **2019**, *7*, 3692–3701. [CrossRef]
50. Pradhan, D.K.; Choudhary, P.; Samantaray, B.K.; Karan, N.K.; Katiyar, R.S. Effect of Plasticizer on Structural and Electrical Properties of Polymer Nanocompsoite Electrolytes. *Int. J. Electrochem. Sci.* **2007**, *2*, 861–871.
51. Mohapatra, S.R.; Thakur, A.K.; Choudhary, R.N.P. Effect of nanoscopic confinement on improvement in ion conduction and stability properties of an intercalated polymer nanocomposite electrolyte for energy storage applications. *J. Power Sources* **2009**, *191*, 601–613. [CrossRef]
52. Shukur, M.F.; Ithnin, R.; Kadir, M.F.Z. Electrical characterization of corn starch-LiOAc electrolytes and application in electrochemical double layer capacitor. *Electrochim. Acta* **2014**, *136*, 204–216. [CrossRef]
53. Aziz, S.; Abdullah, R.M. Crystalline and amorphous phase identification from the tanδ relaxation peaks and impedance plots in polymer blend electrolytes based on [CS: AgNt]x:PEO (x−1)(10 ≤ x ≤ 50). *Electrochim. Acta* **2018**, *285*, 30–46. [CrossRef]
54. Teo, L.P.; Buraidah, M.H.; Nor, A.F.M.; Majid, S.R. Conductivity and dielectric studies of Li2SnO3. *Ionics (Kiel)* **2012**, *18*, 655–665. [CrossRef]
55. Hema, M.; Selvasekarapandian, S.; Arunkumar, D.; Sakunthala, A.; Nithya, H.F.T.I.R. FTIR, XRD and ac impedance spectroscopic study on PVA based polymer electrolyte doped with NH4X (X = Cl, Br, I). *J. Non-Cryst. Solids* **2009**, *355*, 84–90. [CrossRef]
56. Kufian, M.Z.; Aziz, M.F.; Shukur, M.F.; Rahim, A.S.; Ariffin, N.E.; Shuhaimi, N.E.A.; Arof, A.K. PMMA-LiBOB gel electrolyte for application in lithium ion batteries. *Solid State Ionics* **2012**, *208*, 36–42. [CrossRef]
57. Diederichsen, K.M.; McShane, E.J.; McCloskey, B.D. McCloskey Promising Routes to a High Li+ Transference Number Electrolyte for Lithium Ion Batteries. *ACS Energy Lett.* **2017**, *2*, 2563–2575. [CrossRef]
58. Othman, L.; Isa, K.B.; Osman, Z.; Yahya, R. Ionic Conductivity, Morphology and Transport Number of Lithium Ions in PMMA Based Gel Polymer Electrolytes. *Defect Diffus. Forum* **2013**, *334–335*, 137–142. [CrossRef]
59. Sampathkumar, L.; Selvin, P.C.; Selvasekarapandian, S.; Perumal, P.; Chitra, R.; Muthukrishnan, M. Synthesis and characterization of biopolymer electrolyte based on tamarind seed polysaccharide, lithium perchlorate and ethylene carbonate for electrochemical applications. *Ionics* **2019**, *25*, 1067–1082. [CrossRef]
60. Monisha, S.; Mathavan, T.; Selvasekarapandian, S.; Benial, A.M.F.; Premalatha, M. Preparation and characterization of cellulose acetate and lithium nitrate for advanced electrochemical devices. *Ionics* **2017**, *23*, 2697–2706. [CrossRef]
61. Liew, C.W.; Ramesh, S. Electrical, structural, thermal and electrochemical properties of corn starch-based biopolymer electrolytes. *Carbohydr. Polym.* **2015**, *124*, 222–228. [CrossRef]
62. Kadir, M.F.Z.; Arof, A.K. Application of PVA–chitosan blend polymer electrolyte membrane in electrical double layer capacitor. *Mater. Res. Innov.* **2013**, *15*, 217–220. [CrossRef]
63. Bandaranayake, C.M.; Weerasinghe, W.A.D.; Vidanapathirana, K.P.; Perera, K.S. A Cyclic Voltammetry study of a gel polymer electrolyte based redox-capacitor. *Sri. Lankan J. Phys.* **2015**, *16*, 19–27. [CrossRef]
64. Teoh, K.H.; Liew, C.W.; Ramesh, S. Electric double layer capacitor based on activated carbon electrode and biodegradable composite polymer electrolyte. *Ionics* **2014**, *20*, 251–258.
65. Teoh, K.H.; Lim, C.S.; Liew, C.W.; Ramesh, S. Electric double-layer capacitors with corn starch-based biopolymer electrolytes incorporating silica as filler. *Ionics* **2015**, *21*, 2061–2068. [CrossRef]
66. Pandey, G.P.; Kumar, Y.; Hashmi, S.A. Ionic liquid incorporated PEO based polymer electrolyte for electrical double layer capacitors: A comparative study with lithium and magnesium systems. *Solid State Ionics* **2011**, *190*, 93–98. [CrossRef]
67. Lim, C.S.; Teoh, K.H.; Liew, C.W.; Ramesh, S. Capacitive behavior studies on electrical double layer capacitor using poly (vinyl alcohol)-lithium perchlorate based polymer electrolyte incorporated with TiO_2. *Mater. Chem. Phys.* **2014**, *143*, 661–667. [CrossRef]
68. Arof, A.K.; Kufian, M.Z.; Syukur, M.F.; Aziz, M.F.; Abdelrahman, A.E.; Majid, S.R. Electrical double layer capacitor using poly(methyl methacrylate)–C4BO8Li gel polymer electrolyte and carbonaceous material from shells of mata kucing (Dimocarpus longan) fruit. *Electrochim. Acta* **2012**, *74*, 39–45. [CrossRef]

69. Hamsan, M.H.; Shukur, M.F.; Kadir, M.F.Z. NH_4NO_3 as charge carrier contributor in glycerolized potato starch-methyl cellulose blend-based polymer electrolyte and the application in electrochemical double-layer capacitor. *Ionics* **2017**, *23*, 3429–3453. [CrossRef]
70. Shukur, M.F.; Ithnin, R.; Kadir, M.F.Z. Protonic transport analysis of starch-chitosan blend based electrolytes and application in electrochemical device. *Mol. Cryst. Liq. Cryst.* **2014**, *603*, 52–65. [CrossRef]

International Journal of
Molecular Sciences

MDPI

Article

Ion Transport Study in CS: POZ Based Polymer Membrane Electrolytes Using Trukhan Model

Shujahadeen B. Aziz [1,2,*], Wrya O. Karim [3], M. A. Brza [1,4], Rebar T. Abdulwahid [1,5], Salah Raza Saeed [6], Shakhawan Al-Zangana [7] and M. F. Z. Kadir [8]

1 Prof. Hameeds Advanced Polymeric Materials Research Lab., Department of Physics, College of Science, University of Sulaimani, Sulaimani 46001, Iraq; mohamad.brza@gmail.com (M.A.B.); rebar.abdulwahid@univsul.edu.iq (R.T.A.)
2 Komar Research Center (KRC), Komar University of Science and Technology, Sulaimani 46001, Iraq
3 Department of Chemistry, College of Science, University of Sulaimani, Sulaimani 46001, Iraq; wrya.karim@univsul.edu.iq
4 Department of Manufacturing and Materials Engineering, Faculty of Engineering, International Islamic University of Malaysia, Kuala Lumpur, Gombak 53100, Malaysia
5 Department of Physics, College of Education, University of Sulaimani, Sulaimani 46001, Iraq
6 Charmo Research Center, Charmo University, Sulaimani 46001, Iraq; salah1966sh@gmail.com
7 Department of Physics, College of Education, University of Garmian, Kalar 46021, Iraq; shakhawan.al-zangana@garmian.edu.krd
8 Centre for Foundation Studies in Science, University of Malaya, Kuala Lumpur 50603, Malaysia; mfzkadir@um.edu.my
* Correspondence: shujahadeenaziz@gmail.com

Received: 23 September 2019; Accepted: 22 October 2019; Published: 23 October 2019

Abstract: In this work, analysis of ion transport parameters of polymer blend electrolytes incorporated with magnesium trifluoromethanesulfonate ($Mg(CF_3SO_3)_2$) was carried out by employing the Trukhan model. A solution cast technique was used to obtain the polymer blend electrolytes composed of chitosan (CS) and poly (2-ethyl-2-oxazoline) (POZ). From X-ray diffraction (XRD) patterns, improvement in amorphous phase for the blend samples has been observed in comparison to the pure state of CS. From impedance plot, bulk resistance (R_b) was found to decrease with increasing temperature. Based on direct current (DC) conductivity (σ_{dc}) patterns, considerations on the ion transport models of Arrhenius and Vogel–Tammann–Fulcher (VTF) were given. Analysis of the dielectric properties was carried out at different temperatures and the obtained results were linked to the ion transport mechanism. It is demonstrated in the real part of electrical modulus that chitosan-salt systems are extremely capacitive. The asymmetric peak of the imaginary part (M_i) of electric modulus indicated that there is non-Debye type of relaxation for ions. From frequency dependence of dielectric loss (ε'') and the imaginary part (M_i) of electric modulus, suitable coupling among polymer segmental and ionic motions was identified. Two techniques were used to analyze the viscoelastic relaxation dynamic of ions. The Trukhan model was used to determine the diffusion coefficient (D) by using the frequency related to peak frequencies and loss tangent maximum heights ($\tan\delta_{max}$). The Einstein–Nernst equation was applied to determine the carrier number density (n) and mobility. The ion transport parameters, such as D, n and mobility (μ), at room temperature, were found to be 4×10^{-5} cm^2/s, 3.4×10^{15} cm^{-3}, and 1.2×10^{-4} cm^2/Vs, respectively. Finally, it was shown that an increase in temperature can also cause these parameters to increase.

Keywords: polymer blends; impedance study; dielectric properties; electric modulus study; loss tangent peaks; ion transport parameters; Trukhan model

1. Introduction

Solid polymer electrolytes (SPEs) are of particular relevance to devices that are technologically significant, such as batteries and fuel cells. To be scaled to an industrial level, one of the important criteria is achieving a large number of free ions that contribute to the net charge transport in liquid state materials [1]. Previously, it has been shown that polymers can be used as insulators and structural materials. Since 1975, more developments have been achieved that use polymers as ion conductors from the combination of suitable salts, i.e., enhancing their ionic conductivities. Manipulation of polymer electrolytes has been constantly carried out in order to improve three main characteristics, which are ionic conductivity, thermal stability, and mechanical strength [2]. In the period ahead, research and development on SPEs will show that they can readily replace the conventional organic sol–gel electrolytes in accordance with dimensional stability, electrochemical durability, processability, flexibility, safety, and long life [3]. Since the1970s, much attention has been given to the study of SPEs. The main motivation in this field is the pioneer works of Wright et al. and Armand et al., which were carried out on ionically conducting polymer-salt mixtures and polymer-salt complexes, respectively [4,5]. Chitosan as a derivative of chitin was used in the present study due to its invaluable properties, such as natural abundance and low-cost biopolymer. Chitosan can act as a host for ionic conduction as a result of its structural composition. It is known that a weakly alkaline polymer electrolyte can be produced as a consequence of dissolving a salt in the polymer matrix. Blending chitosan membranes can significantly improve the thermal/chemical stability while achieving a satisfactory mechanical strength. Moreover, the presence of hydroxyl and amino groups on the backbone structure enable chitosan membranes to have a relatively high level of hydrophilicity, which is essential for operation in polymer electrolyte membrane fuel cells [6,7]. It is widely recognized that a polymer that contains an electron donor group can have the ability to dissolve inorganic salts with low lattice energy through weak coordination bonds to form polymer electrolytes. Nearly all SPE membranes, which contain silver salts, were fabricated for the facilitated olefin transport and poly (2-ethyl-2-oxazoline) (POZ) was used as a polymer solvent to dissolve silver salts [8–10]. The monomer of POZ containing O and N atoms is responsible for complexation in POZ based electrolytes [11,12]. POZ can serve as an efficient host polymer in synthesizing polymer electrolytes because of the presence of an electron donor group in the backbone, which is very important for making polymer blends. The blending methodology of two or more polymers has inspired many researchers to improve the chemical and physical properties of individual polymers. It is clearly defined that ionic conduction in polymer electrolytes can be enhanced via increasing the amorphous region [13–16]. As previously confirmed, it is likely that the mechanical properties and/or structure stability of the material can be manipulated towards improvement through the polymer blending methodology [14–17]. The obtained blended polymer systems provide more sites for ionic complexation to occur, which facilitates ionic conduction compared to single polymer ones [14–18]. In the past, various ion transport mechanisms have been discussed [1]. Munar et al. [19] have found three parameters, such as ionic mobility, carrier density, and diffusion coefficient in lithium salts incorporated into PEs using a dielectric spectroscopy-based approach. In fact, the ion transport mechanism in polymer physics is not fully understood yet, and thus, the lack of information about ion transport has encouraged researchers to study ion conducting polymer electrolytes extensively [20–24].

In this work, analysis of ion transport parameters of chitosan (CS): POZ polymer blend electrolytes incorporated with magnesium trifluoromethanesulfonate ($Mg(CF_3SO_3)_2$) was carried out through the use of Trukhan model. This model can allow us to use the related values of peak frequencies and loss tangent maximum heights ($\tan\delta_{max}$) to calculate the ion transport parameters, such as mobility (μ), diffusion coefficient (D), and charge carrier number density (n). From the diffusion coefficient and direct current (DC) conductivity values at various temperatures, one can calculate mobility (μ) and charge carrier number density (n) [19]. Therefore, in the present work, the main objective isto estimate ion transport parameters in CS: POZ based SPEs using the Trukhan model.

2. Results and Discussion

2.1. XRD Study

Polymer materials are recognized as either being amorphous or semi-crystalline [25]. The crystallinity of polymer structure can be predicted only when a precise method is in place. This not only helps in identifying the molecular structure of a crystalline polymer, but also in comprehending and rationalizing the inherent features of polymeric materials that significantly affect crystal packing [26]. The XRD pattern of pure CS is shown in Figure 1. In the pure state of CS membrane, two peaks can be identified at 14.5° and 20.9°, owing to the crystalline domain of CS polymer. These two peaks are pertinent to two distinct kinds of crystals [27,28]. The first peak that occurs at 14.5° is related to crystals with a unit cell of a = 7.76, b = 10.91, c = 10.30 (Å), and β=90°. Here, the unit cell has two monomer units across the major chain axis. The second peak at 20.9° is related to the crystal within the chitosan membrane. The lattice parameters of the unit cell of crystal related to θ = 20.9°are as follows: a = 4.4, b = 10.0, c = 10.30 (Å), and β = 90° [29]. It can be noticed in Figure 1 that the wide X-ray peak is at approximately 2θ = 41°. It is evident in the previous studies that the wide peaks within the XRD patterns of pure CS polymers are created because of the inter-chain segment scattering as they are in the amorphous state [30]. It can be seen in Figure 2 that when POZ increases (15 wt.%), crystallinity of CS considerably decreases. The reason for this is the creation of hydrogen bonding (H-bonding) between CS and POZ matrices. The creation of hydrogen bonding among CS and POZ monomers disrupts the inter chain hydrogen bonding in the CS host polymer and thus results in less crystallinity. H-bonding is known to occur after the interaction of electron-deficient hydrogen with an area of high electron density. H-bonding can be demonstrated in the form of an intermolecular interaction as X–H ... Y, where X and Y indicate electronegative elements and Y contains one more solitary electron pair. This means that X and Y can be signified as F, O, and N atoms [31]. The molecular structure of POZ (refer to Scheme 1) makes it obvious that the POZ monomer contains O and N atoms, which allows it to form H-bonding [11,12]. The hydrogen-bonding is known to be a secondary force, which is weak in comparison to the primary bond within the molecules, such as covalent bonds and other polar bonds. However, this H-bonding is stronger compared to the vander Waals interaction [31]. It is evident from Scheme 1a that around three hydrogen bonds can be created among CS monomers (two -OH and one NH_2). This means that a chitosan monomer can form three hydrogen bonds at the same time. On the other hand, a POZ monomer has two sites for hydrogen bonding. Therefore, chitosan has strong inter-molecular hydrogen bonding compared to POZ. It is clear from Scheme 1b,c that three hydrogen bonds can be developed between CS and POZ monomers, and thus, the inter-molecular hydrogen bonds between CS monomers can be disrupted. Consequently, a decrease in crystallinity is expected in CS: POZ polymer blends as a result of the loss of intermolecular hydrogen bonds among CS monomers. It can be implied by the broadness of the X-ray peak of the CS: POZ blend that there is greater inter-chain spacing in the blend sample in comparison to that of the CS polymer. The extended inter-chain spacing in the blend makes it very easy to apply dipole reorientation to the field, which leads to a large dielectric constant in comparison to that of the neat polymers [30]. Greater information can be offered by the dielectric analysis regarding how the electrical properties are influenced by polymer blending.

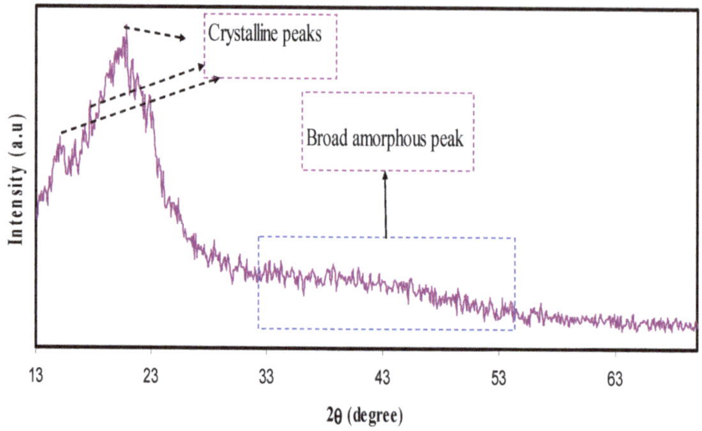

Figure 1. XRD pattern for pure chitosan (CS) film.

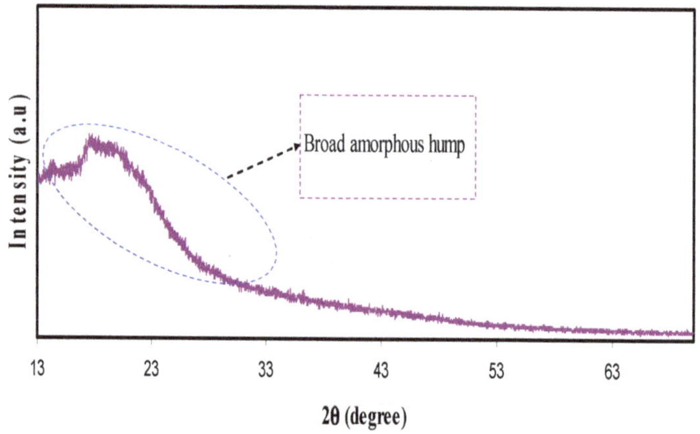

Figure 2. XRD pattern for chitosan: poly (2-ethyl-2-oxazoline) (POZ) polymer blend film.

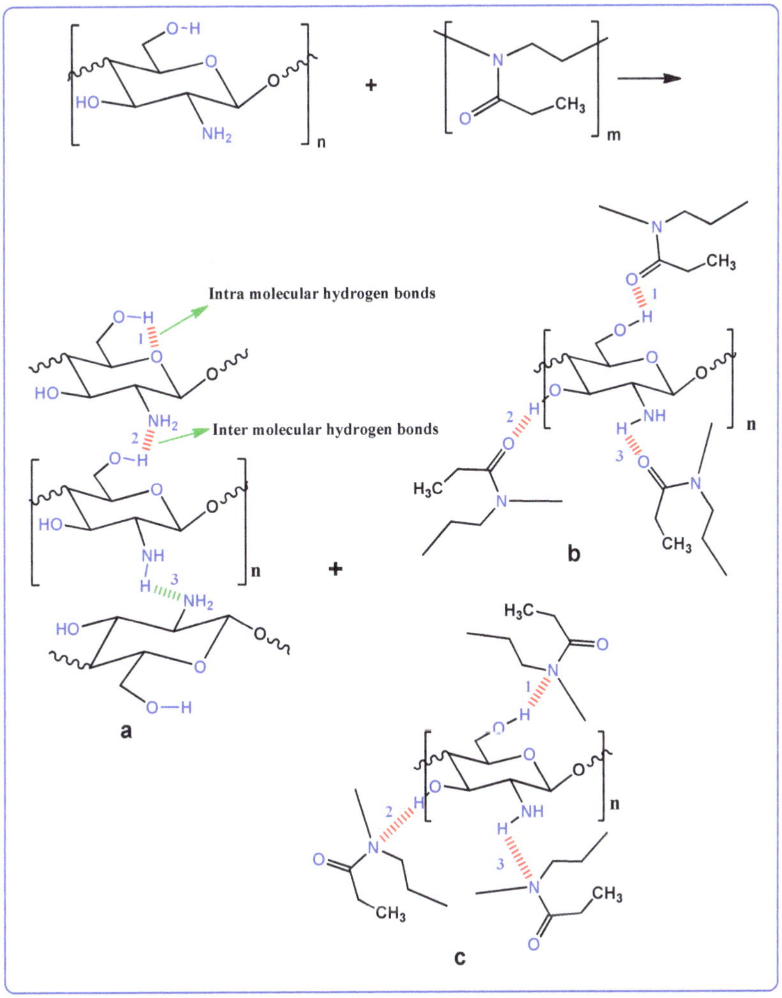

Scheme 1. Proposed hydrogen boding between POZ and CS polymers.

2.2. Impedance and DC Conductivity Study

It is vital to fully explain the charge transfer and ionic transport processes in composite materials from fundamental and technological viewpoints. The impedance spectroscopy is a versatile and a powerful tool in analyzing material properties and making establishment of structure–property correlations [32,33].

In Figure 3, the impedance plots (Z_i versus Z_r) for the CS: POZ: $Mg(CF_3SO_3)_2$ system is shown at different temperatures. It is clearly seen that there is a semicircle in the high frequency region and a linear line (i.e., spike) in the low frequency region. The physical meaning of the semicircle is that this bulk effect reflects a parallel combination of bulk resistance (R_b) and bulk capacitance of the polymer electrolytes. For the spike explanation, there is an effect of electrode blocking [34,35]. From the data analysis, R_b values were extracted from the intercept of high frequency semicircles with the real axis of the impedance plots. Moreover, there is the fact that the complex impedance is dominated by the ionic conductivity when the phase angle approaches zero [36]. Furthermore, the direct current (DC) conductivity (σ_{dc}) values were calculated from the following relation:

$$\sigma_{dc} = \left(\frac{1}{R_b}\right) \times \left(\frac{t}{A}\right) \quad (1)$$

where t and A are the thickness and area under study of the sample. In fact, three different cation transport modes have been reported theoretically; firstly, cation diffusion along the polymer chain, secondly, cooperative motion of the cation within the polymer chain, and thirdly, transferring of cations between different polymer chains. It is obvious that the attached ions cause the local polymer motion to slow down and accordingly, the binding of several functional groups to the ion reduces the number of torsional degrees of freedom within the polymer backbone. This all gives rise to the whole polymer-ion complex moving cooperatively [1].

Obviously, ionic conductivity is one of the key aspects of electrolyte materials to be addressed extensively. It is self-evident that the number of charge carriers and their mobility (i.e., size and electronegativity of an ion) are the two factors that govern conductivity. This is true for polymer electrolyte systems and the mathematical relationship can be described in the following equation [24]:

$$\sigma = \sum_i n_i q_i \mu_i \quad (2)$$

where n_i is the charge carrier's concentration (i.e., number of charge), q is the electron charge, and μ_i is the ions mobility (where i refers to the identity of the ion) [20]. It is clear from Equation (2) that the ionic conductivity (σ) increases as the charge carrier concentration (n) and/or the ionic species mobility increase in the system. Indeed, the ionic mobility is determined by the three main factors, electronegativity, size of ions, and interaction between polymers and ions. In comparison, the mobility of smaller ions (Li^+ and/or Mg^{2+}) is higher than that of larger ions (Na^+ and/or Zn^{2+}) in the polymer electrolyte [37]. The careful choice of SPEs in a certain application of an electrochemical device depends on the σ_{dc} value. Figure 4 shows the direct current(DC) conductivity as a function of 1000/T for the CS: POZ: MgTf solid polymer blend electrolyte film. Herein, the relationship between DC conductivity and temperature is explained in the form of the Arrhenius equation for linear parts (T≤ 60 °C) as given below:

$$\sigma_{dc}(T) = \sigma_0 \exp\left(-\frac{E_a}{K_B T}\right) \quad (3)$$

where σ_o is the pre-exponential factor, E_a is the activation energy, and k_B is the Boltzmann constant [38]. This equation shows that the motion of cations is not a result of the molecular motion of the polymer host. Thereby, when the temperature and ionic conductivity data strongly obey the Arrhenius relationship, the mechanism of cation transport can be correlated to the occurrence in ionic crystals, where an ion jumps to the nearest vacant site, causing the direct current (DC) ionic conductivity to reach its highest value [39]. Studies have confirmed that ion transport mainly occurs in the amorphous regions. However, while the conduction mechanisms are not fully understood yet, it is widely recognized that cations are interrelated to functional groups of the host polymer chains. It can move through re-coordination along the polymer backbone [40]. The activation energy was calculated and was found to be 0.612 eV for the linear region. The activation energy can be considered as an energy barrier that the ion needs to overcome for a successful jump between the sites [27]. It can be seen that at higher temperatures (T > 333 K) the pattern of DC conductivity vs 1000/T is almost curvature rather than linear behavior. In this case, the Vogel–Tammann–Fulcher (VTF) model is applicable to interpret the behavior of DC conductivity [41]. This is means that ion transport occurs with the help of polymer chain motion. To clarify, the glass transition temperature (T_g) of POZ is as low as 63 °C [11], whereas T_g for CS is relatively quite high and found to be above 200 °C [7,42]. Thus, the curvature of DC conductivity is related to the low T_g value of POZ. Consequently, blending CS with POZ can be interesting in developing new polymers with relatively high chain flexibility.

In view of the recent advances in polymer electrolytes, the chain in the polymer is folded in the form of cylindrical tunnels, where the cations are situated and coordinated with the functional groups [43,44]. These cylindrical tunnels offer a pathway for the cation movement. Therefore, there is two options for the segmental motion, either the ions are allowed to hop from one site to another site or they are given a pathway to move [45,46]. It is also understood that the ionic motion can take place through the transitional motion/hopping and dynamic segmental motion within the host polymer [46–48]. In this case, the pattern of DC conductivity vs 1000/T is almost linear as observed in the present work at a low temperature (T ≤ 333K). As the amorphous phase progressively expands at high temperature, the polymer chain gains faster internal motion, where bond rotation creates segmental motion. This favors the inter-chain and intra-chain ion hopping and thus ions move successively. Accordingly, this makes the conductivity of the polymer electrolyte to be relatively high [47,49,50]. It is also emphasized that the charge carrier density, n, depends mainly on both the dissociation energy (U) and dielectric permittivity (ε') of the polymer electrolyte as shown below [51–53]:

$$n = n_0 \exp\left(-\frac{U}{\varepsilon' K_B T}\right) \tag{4}$$

where k_B and T are the Boltzmann constant and temperature, respectively. This equation indicates that there is a correlation between dielectric permittivity (ε') and charge carriers. It is well-defined that the dielectric constant is equal to the ratio of the material capacitance (C) to the capacitance of the empty cell (C_o) ($\varepsilon' = C/C_o$). Indeed, an increase of dielectric constant causes an increaseof charge concentrations in the electrolyte. As mentioned in Equation (2), conductivity (σ) depends upon both the number of charge carriers (n) and the mobility of the ionic species in the system [54]. Furthermore, from Equation (4), the number of charge carriers (n) can be increased by increasing the dielectric constant, in other words, an increase of charge carriers leads to increasing dielectric constant. Overall, the conductivity increases with increasing of dielectric constant based on Equation (4). Here, one can conclude that the dielectric constant is an informative parameter in studying the conductivity behavior of polymer electrolytes [20]. From our discussion and interpretation regarding the ion transport and linkage between DC conductivity and dielectric constant ε', we can realize that ion transport is a complicated process in polymer electrolyte systems [55]. The lack of information about the cation transport mechanism in polymer electrolytes is thought to be one of the main obstacles in achieving a high conducting polymer electrolyte at room temperature [56,57]. Finally, it is clarified that the ion carrier concentration and segmental mobility are not the only factors that determine the conductivity behavior of polymer electrolytes, but also the dielectric constant and ion dissociation energy are of significant importance to ion transport. Study of the dielectric constant at various temperatures as can be seen in later sections and may support the above explanation.

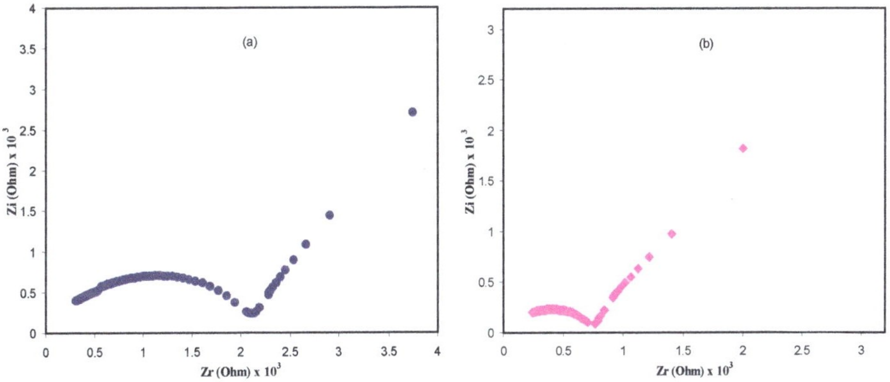

Figure 3. *Cont.*

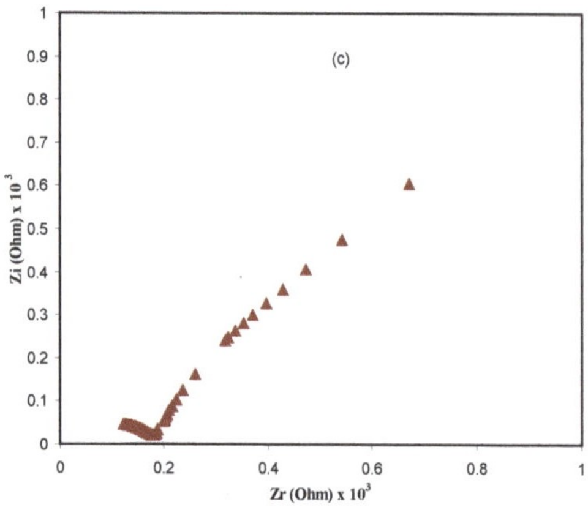

Figure 3. Electrical impedance plots for CS: POZ: MgTf solid polymer blend electrolyte film at (**a**) 303 K, (**b**) 323 K, and (**c**) 343 K.

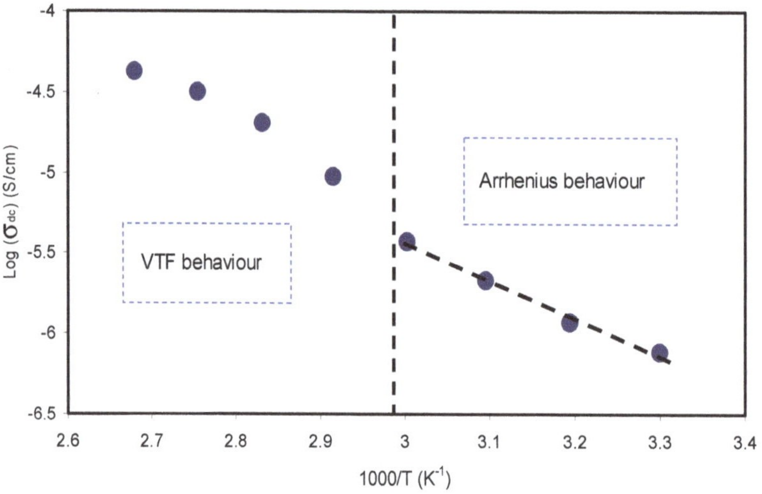

Figure 4. Direct current (DC) ionic conductivity versus 1000/T for CS: POZ: MgTf solid polymer blend electrolyte film.

2.3. Dielectric Properties and Electric Modulus Analysis

Figures 5 and 6 show the dielectric constant (ε′) and dielectric loss (ε′′) as a function of frequency, respectively. As the frequency increases, both ε′ and ε′′ decrease gradually to a minimum value, becoming almost constant at high frequencies. The obtained high values of both parameters in the low frequency region can be attributed to electrode polarization effect [20,58,59]. This effect is found to result from charge accumulation at the electrode/electrolyte interfacial region [60]. The constant values of both the dielectric constant and dielectric loss at high frequencies can result from the periodic reversal of the high electric field. Consequently, there is no charge accumulation at the interface. The absence of a more interesting observation for the dielectric relaxation peaks within the dielectric loss plot can be due to the mask of polymer relaxation segments by the DC ionic conductivity of charge carriers [22–24]. Studying temperature clearly showed that with increasing temperatures, the dielectric constant also increases. This can be clearly observed in Figure 5, which also supports our interpretations given in Section 3.1 for causes of increasing DC conductivity. Once the dielectric constant is obtained, it will be easy for researchers to identify the two following important phenomena: reduction of silver ions to metallic silver particles and electrical percolation in ion conducting solid polymer composites [21,58,61–63]. One can note from Figures 5 and 6 that both ε′ and ε′′ values recorded high values at high temperature. This can be explained in terms of the forces that bind the polymer chains together. Obviously, there is an existence of forces within polymer material matrices, which are generally classified into primary (intra-chain) and secondary (inter-chain) forces, that result in the stabilization of polymer structure [23,64]. On the one hand, the primary ones arise from the covalent bond formation (2.2–8.6 eV), which binds the chains of backbone atoms together. On the other hand, there are four types of secondary forces in polymers, which are dipole–dipole bonding (0.43–0.87 eV), hydrogen bonding (0.13–0.30 eV), induced interaction (0.07–0.13 eV), and dispersion interaction (0.002–0.09 eV). These forces possess relatively low dissociation energies; thereby, these forces are extremely susceptible to temperature change compared to their primary counterpart.

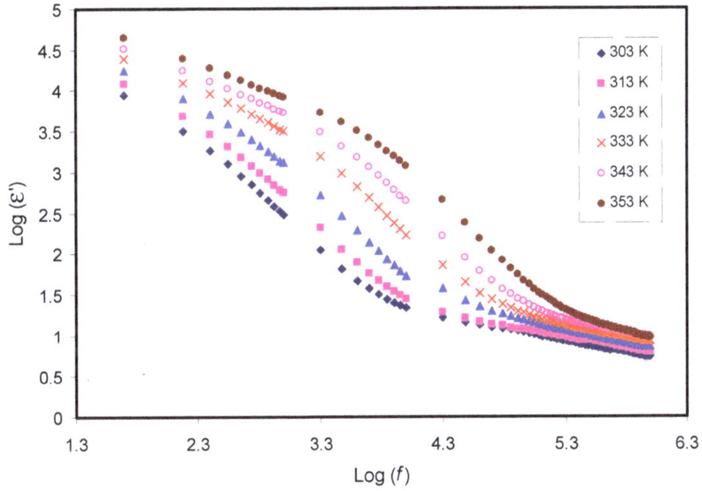

Figure 5. Frequency dependence of dielectric constant (ε′) at selected temperatures for the CS: POZ: MgTf system. Noticeably, the dielectric constant rises with temperature rise.

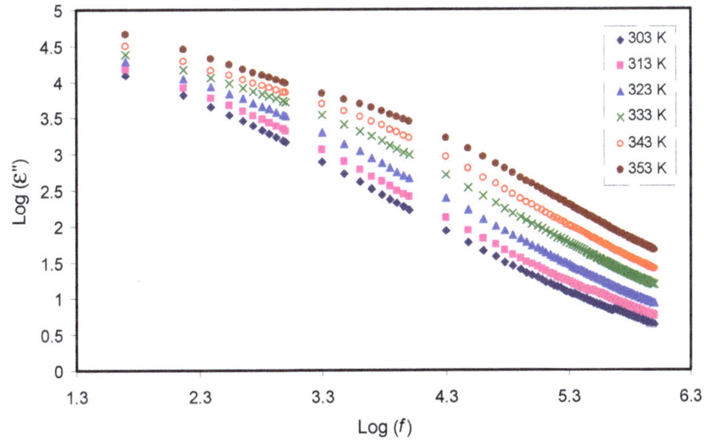

Figure 6. Frequency dependence of dielectric loss (ε′′) at selected temperatures for the CS: POZ: MgTf system. The dispersion in dielectric loss spectra rises with rising temperature.

The electric modulus is the inverse of electric permittivity and can be used to study the dielectric behavior of a polymer resulting from ion relaxation. In this work, suppression of charge accumulation near the electrode is related to the electrode polarization effects [52,65]. Figures 7 and 8 exhibit the real (M_r) and imaginary (M_i) components of electric modulus versus frequency, respectively. In contrast to the ε' spectra in Figure 5, detectable peaks, especially for curves at 303K and 313K resulting from conductivity relaxation, are seen in the M_i spectra as presented in Figure 8. It is observable that both M_r and M_i decrease towards low frequencies, indicating the negligibility of electrode polarization influence. In both plots, the curves show long tail features at low frequency at various temperatures that can be due to the suppression of low frequency electrodes/sample double layer effects, arising from the large capacitance establishment [66,67]. In contrast to the ε' spectra in Figure 5, detectable peaks, which result from conductivity relaxation, can be seen in the M_i spectra as shown in Figure 8. Previously, it had been confirmed that ε' parameter is always influenced by an ohmic conduction, i.e., DC conductivity, which leads to the hidden loss peaks in the ε' spectra [68]. A peak in the M_i spectra indicates the regions where the carrier can move a long distance (left of the peak) or where the carrier is confined (right of the peak) [69]. Viscoelastic relaxation or conductivity relaxation using Argand plots is the well-known method to explain the relaxation dynamic. Studying Argand plots at various temperatures is informative to address the nature of relaxation processes within the present polymer electrolytes.

Figure 9 reveals the Argand curves at room temperature. As can be seen from the figure, the diameter of the circles does not coincide with the real axis, which presents (M_r), and thus the ion transport occurs via the viscoelastic relaxation. The characteristics of the Argand plot are am incomplete half semicircle and arc-like appearance, which cannot be explained by the Debye model, i.e., single relaxation time [70]. Therefore, a distribution of relaxation time is necessary to interpret the experimental data, since more dipoles and ions are activated and contribute to the dielectric relaxation with increasing temperature [7,65,71]. Obviously, with increasing temperature, the Argand distributed data points shift towards the origin. This can be attributed to the increase of conductivity due to the increase of ionic mobility with temperature elevation, and accordingly caused both Z' and Z'' to decrease. Mohomed et al. [72] have explained that Argand plots (M_i vs. M_r) may have a complete semicircular shape and in this case, the relaxation could be due to the conductivity relaxation dynamic, i.e., pure ionic relaxation, that is, the polymer segmental motion and ion transport are completely separated or decoupled. On the other hand, if Argand plots (M_i vs. M_r) have an incomplete semicircular shape, the relaxation could be then attributed to viscoelastic relaxation. This implies that ion motion and polymer segmental motion are well coupled. From our Argand plots, the semicircular behavior with diameter well below the real axis can signify that the relaxation process has arisen from the viscoelastic relaxation dynamic. This is means that ion transport is coupled with the polymer segmental motion and thus no peaks can be seen in the dielectric loss spectra as shown in Figure 6. The obtained results can be interpreted on the basis of the fact that the ion transports in SPEs are enhanced by large scale segmental movements of the polymer chain.

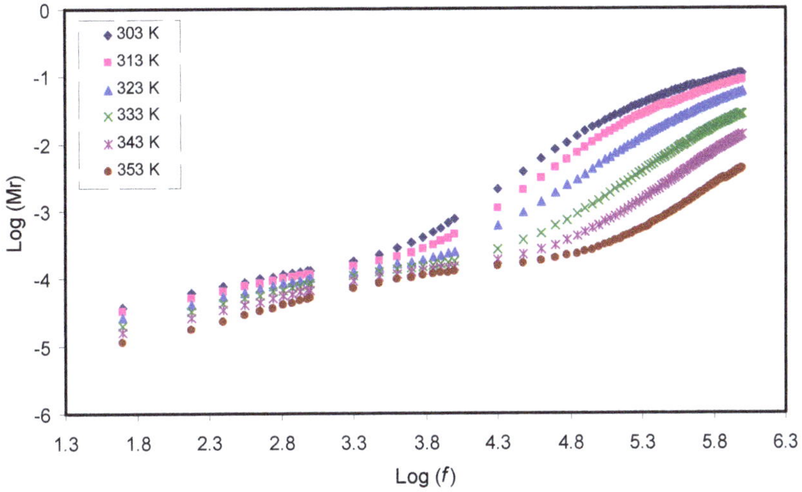

Figure 7. Frequency dependence of the real part (M_r) of M* for the CS: POZ: MgTfsystem at different temperatures.

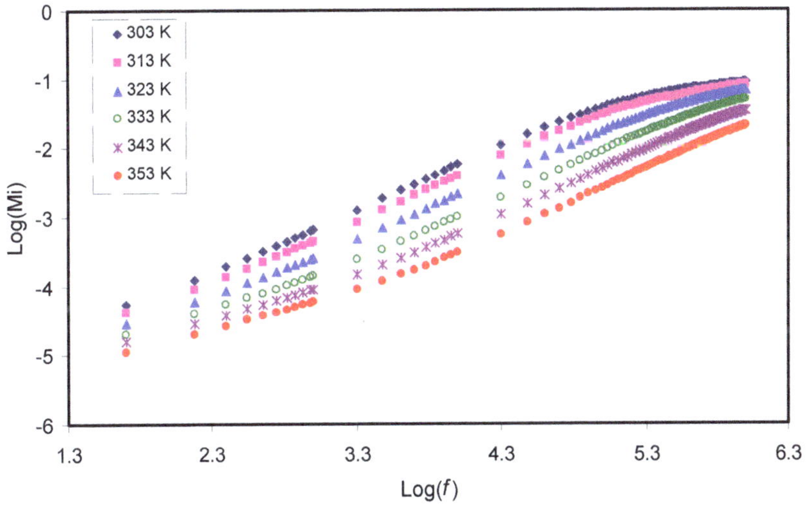

Figure 8. Frequency dependence of the imaginary part (M_i) of M* for the CS: POZ: MgTf systemat different temperatures.

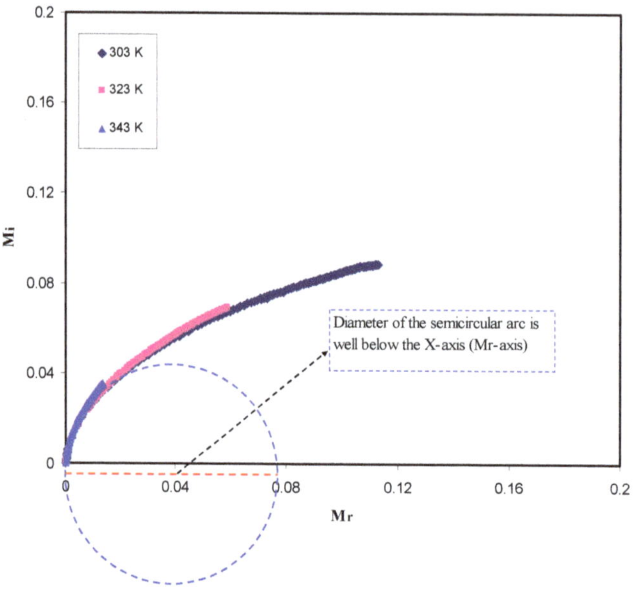

Figure 9. Argand plots for the CS: POZ: MgTf system at different temperatures.

2.4. *Tanδ Spectra and Ion Transport Study*

In polymer electrolyte systems with high electrical conductivity, the peaks at the low frequency relaxation can be masked due to permanent or induced dipoles by polarization relaxation of mobile charged species in the material [73]. To understand the relaxation process, the dielectric loss tangent (tanδ) as a function of frequency has been plotted as shown in Figure 10. The plot shape of tanδ can be interpreted in terms of Koop's phenomenological principle [74]. Thereby, tanδ increases with increasing frequency, reaching a maximum value, and thereafter decreases with further increasing frequency. On the one hand, at lower frequencies, where tanδ increases, theohmic component of the current increases more sharply than its capacitive component ($X_c = 1/2\pi fC$). On the other hand, at higher frequencies, where tanδ decreases, the ohmic component of current is obviously frequency independent and the capacitive component increases due to the high value of f and as a consequence, the value of X_c is small [74,75]. It is worth-noticing that the broad nature of the tanδ peaks suggests that the relaxation process is of non-Debye type [76]. The complex impedance relationship is $Z^* = R\text{-j}X_c$, where R is the real or resistor element and X_c is the capacitive element [62]. This function indicates that, at low frequency, the capacitive component becomes very high and as a result, most of the current passes through the resistor element. The tanδ relation ($\tan\delta = \varepsilon''/\varepsilon'$) shows the proportionality of tanδ to ε''. Also, from these relationships, one can obtain ε'', which is exactly in proportion to the real part (Z' or R) of the impedance function, through the following equation:

$$\varepsilon'' = \frac{Z'}{\omega C_0\left(Z'^2 + Z''^2\right)} \tag{5}$$

As the temperature is increased, the loss tanδ peaks shift to the higher frequency side. Figure 10 shows that the position and height of the peaks are increased with increasing the temperature. This is explained on the basis of the fact that the higher temperature makes the charge carrier movement easier and thus capable of relaxation at higher frequencies [77,78]. From all these discussions stated above, it can be realized that both the shape and intensity of tanδ peaks at various temperatures are completely related to the ion transport parameters, such as mobility and diffusion coefficient. From $\tan\delta_{max}$ and peak frequency, one can extract mobility (μ), carrier density (n), and diffusivity (D) as a function of temperature, which will be discussed in the next sections.

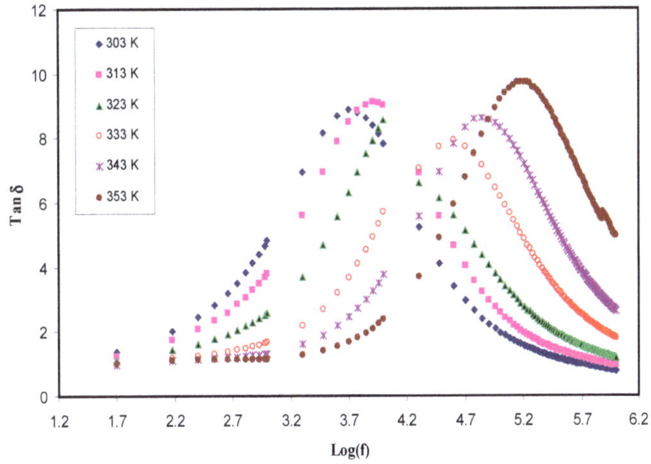

Figure 10. Tanδ plots for the CS: POZ: MgTfsystem at different temperatures.

In the analysis of the loss tangent, the shape of the tanδ plots is correlated to both the capacitive and resistive component of the solid electrolytes. Furthermore, it is explained that the shifting of tanδ peaks towards the high frequency side is associated with the thermally activated ions. These findings support our previous explanation for the DC conductivity pattern versus 1000/T. To calculate the charge density, mobility, and diffusion coefficient of the charge carriers, the Trukhan model was employed by analyzing the loss tangent data points. In this model, the diffusion coefficient of cations and anions are assumed to be equal and thus, a simple expression has been used to calculate the diffusion coefficient from the peak appearance in the loss tangent plots. The expression is shown as follows [19]:

$$D = \frac{2\pi f_{max} L^2}{32\ tan^3\delta_{max}} \quad (6)$$

where L is the sample thickness and all other terms have normal meanings.

Figure 11 shows a plot between diffusion coefficient and temperature. From the plot it is observed that the diffusion coefficient increases non-linearly with the temperature.

Another important parameter, which is in relation to conductivity, is the density of mobile ions. The density number of mobile ions (n) can be obtained from the well-known Einstein relation using the following equation [19]:

$$n = \frac{\sigma_{DC} K_B T}{De^2} = \left(\frac{\sigma_{DC} K_B T}{e^2}\right)\left[\frac{32\ tan^3\delta_{max}}{2\pi f_{max} L^2}\right] \quad (7)$$

where σ_{dc} is the DC conductivity and obtainable from impedance plots, K_B, T, e, and D have the usual meaning. From Equation (7), it is obvious that charge carrier density is proportional to cubic of $\tan\delta_{max}$ value and inversely proportional to the shifting of peak frequencies. It is also clarified that the $\tan\delta_{max}$ values do not change significantly as there is no considerable change in intensity values; as a consequence, the carrier density becomes almost constant.

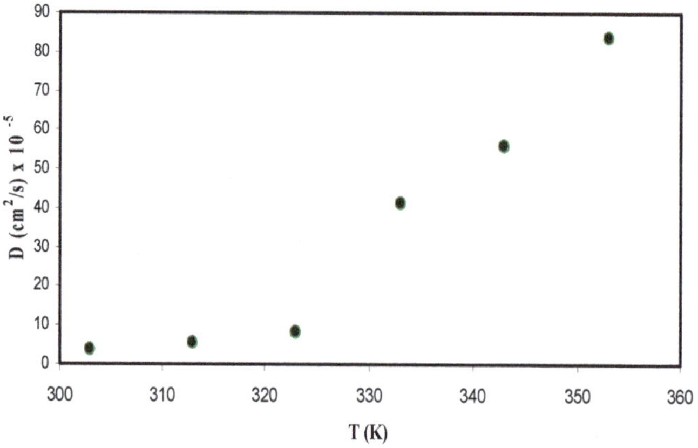

Figure 11. Diffusion coefficient of ions versus temperature for the CS: POZ: MgTfsystem.

Figure 12 shows charge density (n) as a function of temperature. It is seen that the n does not change as temperature increases. The Trukhan model allows an estimation of the diffusion coefficient (D), mobile ion concentration (n), and mobility (μ) from the value of ($\tan\delta_{max}$), where δ is a phase angle. The model is simple in analysis of the data points and there is no need for microscopic information about the distance between adjacent functional groups. However, any increase and decrease of $\tan\delta_{max}$ intensity directly affects the behavior of carrier density. Chandra et al. [79] have recorded that the value of n is found to be in the range of 10^{16} to 10^{18} cm^{-3} for PEO: PVP: $AgNO_3$ based SPE systems. Agrawal et al. [80] have studied the mobile ion concentration in the range of 10^{15} to 10^{16} cm^{-3} for hot press PEO: $AgNO_3$: SiO_2 nano-composite system. It is also documented that the conductivity is dependent on both the number of mobile ions and mobility [80]. In the present work, values of the carrier density of (1.5–4.7) × 10^{15} cm^{-3} were recorded and are in good agreement compared to those reported for polymer based solid electrolytes in the literature [79,80].

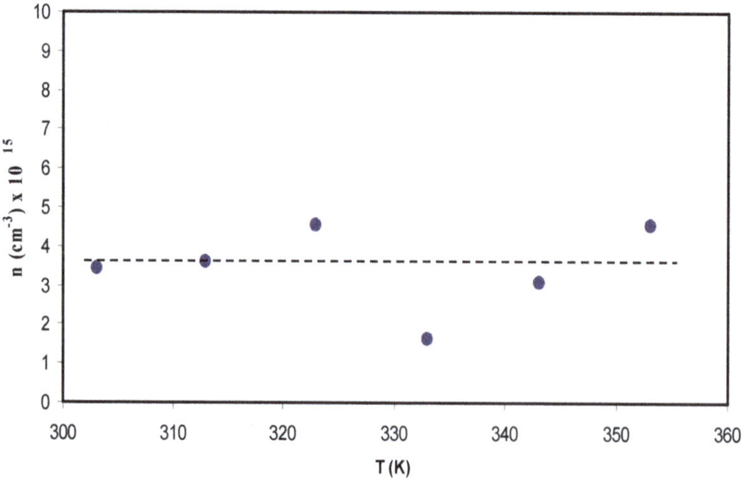

Figure 12. Charge carrier density (n) of ions versus temperature for the CS: POZ: MgTf system.

Ion mobility is also another parameter that can be calculated from the following equation:

$$\mu = \frac{\sigma_{DC}}{en} = \frac{De}{K_B T} = \left(\frac{e}{K_B T}\right)\left[\frac{2\pi f_{max} L^2}{32\ tan^3 \delta_{max}}\right] \quad (8)$$

where μ is the ionic mobility and all other terms have normal meanings.

Figure 13 shows the relation between temperature and ion mobility. The dependence of μ on temperature is similar to that in diffusivity. An increase in temperature results in an increase in the mobility, however, this increase does not linearly change. The concept of increasing the ion mobility is supported by the free volume model, in which the temperature increase leads to enlarged free volume in the amorphous phase, which helps ion mobility [75]. For example, the ion mobility values have been found to be 1.5 × 10^{-3} cm^2/Vs at room temperature, while it is about 2.8 × 10^{-2} cm^2/Vs at 353 K. Winie et al. [81] have also confirmed such a direct proportionality relationship between ion mobility and temperature. Several research groups have reported results on ion mobility. Majid and Arof [82] recorded the value of mobility (μ) in the range 10^{-8} to 10^{-6} $cm^2\ V^{-1}\ s^{-1}$. Agrawal et al. [80] reported a different value of 10^{-3} $cm^2\ V^{-1}\ s^{-1}$ for ionic mobility. Moreover, Arya and Sharma [83] reported ion mobility (μ) in the range 10^{-10} to 10^{-12} $cm^2\ V^{-1}\ s^{-1}$. More recently, Patla et al. [84] have also documented ion mobility (μ) in the range 1.8 × 10^{-4} to 9.5 × 10^{-11} $cm^2\ V^{-1}\ s^{-1}$ for polyvinylidene fluoride (PVDF) based polymer nano-composites incorporated with ammonium iodide (NH_4I) salt. From the results of the present work, it is clear that ion transport parameters can be easily estimated from the Trukhan model. It has been reported that polymers with optical absorption behaviour and high-performance optical properties are important in the application of solar cell devices [85–87]. Similarly, fabrication of polymer electrolytes with high ionic conductivity and improved ion transport properties are crucial in the application of electrochemical devices [14,16].

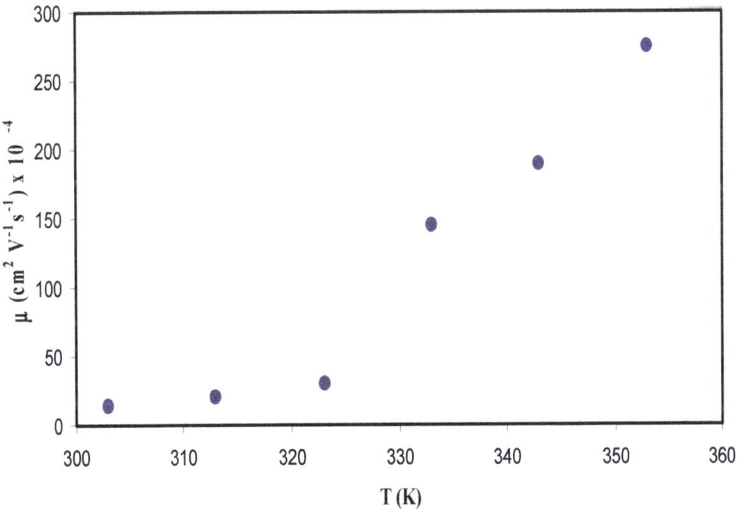

Figure 13. Ion mobility (μ) versus temperature for the CS: POZ: MgTf system.

3. Materials and Methods

3.1. Materials and Sample Preparation

The SPEs based on the CS: POZ blend was fabricated using a conventional solution cast technique. The procedure involves dissolution of 1 g of CS in 80 mL of 1% acetic acid. To prepare the polymer blend, 15 wt.% (0.1764 g) of poly (2-ethyl-2-oxazoline) (POZ) is dissolved in 20 mL of distilled water and added to the CS solution. This solution was continuously stirred with the aid of magnetic stirrer for several hours until a homogeneous viscous blend solution was achieved. Afterwards, 30 wt.% of $Mg(CF_3SO_3)_2$ salt was added to the CS: POZ solution to obtain an alkaline solution of CS: POZ: $Mg(CF_3SO_3)_2$ polymer blend electrolyte. Subsequently, the mixture was further stirred until a homogeneous solution was achieved. Finally, the solution was put into a Petri dish and then dried by leaving it to form a film at room temperature. The last step in this procedure was putting it into a desiccator to ensure dryness.

3.2. *Characterization Techniques*

X-Ray Diffraction (XRD) patterns were acquired using an Empyrean X-ray diffractometer, (PANalytical, Netherland) with operating current and voltage of 40 mA and 40 kV, respectively. The samples were irradiated with a beam of monochromatic CuKα X-radiation of wavelength λ = 1.5406 Å and the glancing angle of the X-ray diffraction was in the range of $5° \leq 2\theta \leq 80°$ withastepsizeof0.1°.

The acquisition of impedance spectra of the films was conducted using a HIOKI 3531 Z Hi-tester in the frequency range of 50 Hz to 1000 kHz and at various temperatures ranging from 303 K to 373 K. The prepared SPE films were cut into small discs of 2 cm in diameter and placed between two stainless steel electrodes under spring pressure. The cell was connected to a computer equipped with customized software to record both real (Z′) and imaginary (Z″) parts of the complex impedance (Z*) spectra. Consequently, from these data, complex permittivity (ε*) and complex electric modulus (M*) can be then estimated.

4. Conclusions

A solution cast technique was used to perform the integration of chitosan (CS) with poly (2-ethyl-2-oxazoline) (POZ). It was shown in the XRD spectra that there was an improvement in the amorphous phase within CS: POZ blend films when compared to pure CS. The broad peak in the CS: POZ blend films compared to the crystalline peaks of the CS polymer confirms this improvement. It can be concluded that a potential technique for studying the diffusion coefficient, mobility, and charge carrier number density in CS: POZ: MgTf solid polymer blend electrolyte film can be the Trukhan model. The calculations carried out in this technique employed peaks in loss tangent spectra. Temperature increases are found to cause a reduction in the bulk resistance and a rise in permittivity. There is also a change found in the position and height of loss tangent peaks verses frequency plots with temperature. The increase in temperatures causes the height of the peaks to shift positively. An Arrhenius type dependence on temperature was exhibited from the DC conductivity; hence, when temperature increases, the conductivity also improved. The increase in the amorphous phase is found to be the reason for such increment in conductivity. Emphasis has been laid on the role of association and disassociation of ionic motion and polymer segmental relaxation. Assessments of dielectric properties were carried out at different temperatures and their findings were related to the ion transport mechanism. It is demonstrated in the real part of the electrical modulus that chitosan-salt mechanisms are very capacitive. According to the asymmetric peak of the imaginary part (M_i) of the electric modulus, there is a non-Debye form of relaxation for chitosan-salt systems. It was shown by the frequency dependence of dielectric loss (ε'') and the imaginary part (M_i)of the electric modulus that there was suitable coupling among polymer segmental and ionic motions. Studies were carried out on the viscoelastic relaxation dynamic of the chitosan-based solid electrolyte using two techniques. The estimation of the diffusion parameter (D) was carried out through the Trukhan model by obtaining the frequency related to peak frequencies and loss tangent maximum heights ($\tan\delta_{max}$). The Einstein–Nernst equation was used to determine the carrier number density (n) and mobility. It was found that the ion transport parameters at room temperature are 4×10^{-5}cm^2/s, 3.4×10^{15} cm^{-3}, and 1.2×10^{-4} cm^2/Vs for D, n, and μ, respectively. It was shown that as the temperature increased; there was also an increase in the parameters.

Author Contributions: Conceptualization, S.B.A., S.R.S. and M.F.Z.K.; Investigation, S.B.A.; Methodology, S.B.A., M.A.B. and R.T.A.; Project administration, S.B.A.; Validation, W.O.K., R.T.A. and S.A.-Z.; Writing—Original Draft, S.B.A.; Writing—Review and Editing, W.O.K., M.A.B., R.T.A., S.R.S., S.A.-Z. and M.F.Z.K.

Funding: This research was funded by the Ministry of Higher Education and Scientific Research-Kurdish National Research Council (KNRC), Kurdistan Regional Government/Iraq. The financial support from the University of Sulaimani and Komar Research Center (KRC), Komar University of Science and Technology is greatly appreciated.

Acknowledgments: The authors gratefully acknowledge the financial support from the Kurdistan National Research Council (KNRC)-Ministry of Higher Education and Scientific Research-KRG. The authors also acknowledge the University of Sulaimani, and the Komar University of Science and Technology for providing the facility and financial support to carry out this work.

Conflicts of Interest: The authors declare no conflict of interest.

References

1. Diddens, D.; Heuer, A.; Borodin, O. Understanding the Lithium Transport within a Rouse-Based Model for a PEO/LiTFSI Polymer Electrolyte. *Macromolecules* **2010**, *43*, 2028–2036. [CrossRef]
2. Katherine, A.F.; Chiam-Wen, L.; Ramesh, S.; Ramesh, K.; Ramesh, S. Effect of ionic liquid 1-butyl-3-methylimidazolium bromide on ionic conductivity of poly(ethyl methacrylate) based polymer electrolytes. *Mater. Express* **2016**, *6*, 252–258.
3. Hema, M.; Selvasekarapandian, S.; Arunkumar, D.; Sakunthala, A.; Nithya, H. FTIR, XRD and ac impedance spectroscopic study on PVA based polymer electrolyte doped with NH4X (X = Cl, Br, I). *J. Non-Cryst. Solids* **2009**, *355*, 84–90. [CrossRef]
4. Fenton, D.E.; Parker, J.M.; Wright, P.V. Complexes of alkali metal ions with poly(ethylene oxide). *Polymer* **1973**. [CrossRef]
5. Farrington, G.C.; Briant, J.L. Fast Ion Transport in Solids. *Science* **1979**, *366*, 1371–1379. [CrossRef]
6. Wan, Y.; Peppley, B.; Creber, K.A.M.; Bui, V.T.; Halliop, E. Preliminary evaluation of an alkaline chitosan-based membrane fuel cell. *J. Power Sources* **2006**, *162*, 105–113. [CrossRef]
7. Aziz, S.B. Occurrence of electrical percolation threshold and observation of phase transition in chitosan(1 − x):AgIx (0.05 ≤ x ≤ 0.2)-based ion-conducting solid polymer composites. *Appl. Phys. A* **2016**, *122*, 706. [CrossRef]
8. Won, J.; Yoon, Y.; Kang, Y.S. Changes in Facilitated Transport Behavior of Silver Polymer Electrolytes by UV Irradiation. *Macromol. Res.* **2002**, *10*, 80–84. [CrossRef]
9. Kang, S.W.; Kim, J.H.; Won, J.; Char, K.; Kang, Y.S. Enhancement of facilitated olefin transport by amino acid insilver–polymer complex membranes. *Chem. Commun.* **2003**, 768–769. [CrossRef] [PubMed]
10. Kang, S.W.; Char, K.; Kim, J.H.; Kim, C.K.; Kang, Y.S. Control of Ionic Interactions in Silver Salt-Polymer Complexeswith Ionic Liquids: Implications for Facilitated Olefin Transport. *Chem. Mater.* **2006**, *18*, 1789–1794.
11. Ruiz-Rubio, L.; Alonso, M.L.; Pérez-Álvarez, L.; Alonso, R.M.; Vilas, J.L.; Khutoryanskiy, V.V. Formulation of Carbopol/Poly(2-ethyl-2-oxazoline)s Mucoadhesive Tablets for Buccal Delivery of Hydrocortisone. *Polymers* **2018**, *10*, 175. [CrossRef] [PubMed]
12. Moreadith, R.W.; Viegas, T.X.; Bentley, M.D.; Harris, J.M.; Fang, Z.; Yoon, K.; Dizman, B.; Weimer, R.; Rae, B.P.; Li, X.; et al. Clinical development of a poly(2-oxazoline) (POZ) polymer therapeutic for the treatment of Parkinson's disease–Proof of concept of POZ as a versatile polymer platform for drug development in multiple therapeutic indications. *Eur. Polym. J.* **2017**, *88*, 524–552. [CrossRef]
13. Kharbachi, A.E.; Hu, Y.; Yoshida, K.; Vajeeston, P.; Kim, S.; Sørby, M.H.; Orimo, S.; Fjellvåg, H.; Hauback, B.C. Lithium ionic conduction in composites of $Li(BH_4)_{0.75}I_{0.25}$ and amorphous $0.75Li_2S$ $0.25P_2S_5$ for battery applications. *Electrochim. Acta* **2018**, *278*, 332–339.
14. Aziz, S.B.; Hamsan, M.H.; Abdullah, R.M.; Kadir, M.F.Z. A Promising Polymer Blend Electrolytes Based on Chitosan: Methyl Cellulose for EDLC Application with High Specific Capacitance and Energy Density. *Molecules* **2019**, *24*, 2503. [CrossRef] [PubMed]
15. Aziz, S.B.; Hamsan, M.H.; Karim, W.O.; Kadir, M.F.Z.; Brza, M.A.; Abdullah, O.G. High Proton Conducting Polymer Blend Electrolytes Based on Chitosan: Dextran with Constant Specific Capacitance and Energy Density. *Biomolecules* **2019**, *9*, 267. [CrossRef] [PubMed]

16. Aziz, S.B.; Hamsan, M.H.; Kadir, M.F.Z.; Karim, W.O.; Abdullah, R.M. Development of Polymer Blend Electrolyte Membranes Based on Chitosan: Dextran with High Ion Transport Properties for EDLC Application. *Int. J. Mol. Sci.* **2019**, *20*, 3369. [CrossRef] [PubMed]
17. Tamilselvi, P.; Hema, M. Conductivity studies of LiCF3SO3 doped PVA: PVdF blend polymer electrolyte. *Phys. B* **2014**, *437*, 53–57. [CrossRef]
18. Yusof, Y.M.; Shukur, M.F.; Illias, H.A.; Kadir, M.F.Z. Conductivity and electrical properties of corn starch–chitosan blend biopolymer electrolyte incorporated with ammonium iodide. *Phys. Scr.* **2014**, *89*, 35701–35711. [CrossRef]
19. Munar, A.; Andrio, A.; Iserte, R.; Compañ, V. Ionic conductivity and diffusion coefficients of lithium salt polymer electrolytes measured with dielectric spectroscopy. *J. Non-Cryst. Solids* **2011**, *357*, 3064–3069. [CrossRef]
20. Aziz, S.B.; Abidin, Z.H.Z. Ion-transport study in nanocomposite solid polymer electrolytes based on chitosan: Electrical and dielectric analysis. *J. Appl. Polym. Sci.* **2015**, *132*, 41774. [CrossRef]
21. Aziz, S.B.; Abidin, Z.H.Z. Electrical and morphological analysis of chitosan: AgTf solid electrolyte. *Mater. Chem. Phys.* **2014**, *144*, 280–286. [CrossRef]
22. Aziz, S.B.; Mamand, S.M. The Study of Dielectric Properties and Conductivity Relaxation of Ion Conducting Chitosan: NaTf Based Solid Electrolyte. *Int. J. Electrochem. Sci.* **2018**, *13*, 10274–10288. [CrossRef]
23. Aziz, S.B.; Brza, M.A.; Kadir, M.F.Z.; Hamsan, M.H.; Abidin, Z.H.Z.; Tahir, D.A.; Abdullah, O.G. Investigation on Degradation and Viscoelastic Relaxation of Li Ion in Chitosan Based Solid Electrolyte. *Int. J. Electrochem. Sci.* **2019**, *14*, 5521–5534. [CrossRef]
24. Aziz, S.B. Li^+ ion conduction mechanism in poly (ε-caprolactone)-based polymer electrolyte. *Iran Polym. J.* **2013**, *22*, 877. [CrossRef]
25. Go, R.L.; Poutot, G.; Delaunay, D.; Fulchiron, R.; Koscher, E. Study and modeling of heat transfer during the solidification of semi-crystalline polymers. *Int. J. Heat Mass Transf.* **2005**, *48*, 5417–5430.
26. Leo'n, S.; Navas, J.J.; Alema'n, C. PCSP: A computer program to predict and analyze the packing in crystalline polymers. *Polymer* **1999**, *40*, 7351–7358. [CrossRef]
27. Aziz, S.B.; Abidin, Z.H.Z.; Arof, A.K. Effect of silver nanoparticles on the DC conductivity in chitosan–silver triflate polymer electrolyte. *Phys. B* **2010**, *405*, 4429–4433. [CrossRef]
28. Aziz, S.B.; Abidin, Z.H.Z.; Kadir, M.F.Z. Innovative method to avoid the reduction of silver ions to silver nanoparticles in silver ion conducting based polymer electrolytes. *Phys. Scr.* **2015**, *90*, 35808. [CrossRef]
29. Lima, C.G.A.; de Oliveira, R.S.; Figueir'O, S.D.; Wehmann, C.F.; G'oes, J.C.; Sombra, A.S.B. DC conductivity and dielectric permittivity of collagen–chitosan films. *Mater. Chem. Phys.* **2006**, *99*, 284–288. [CrossRef]
30. Thakur, Y.; Zhang, B.; Dong, R.; Lu, W.; Iacob, C.; Runt, J.; Bernholc, J.; Zhang, Q.M. Generating High Dielectric Constant Blends from Lower Dielectric Constant Dipolar Polymers using Nanostructure Engineering. *Nano Energy* **2016**. [CrossRef]
31. He, Y.; Zhu, B.; Inoue, Y. Hydrogen bonds in polymer blends. *Prog. Polym. Sci.* **2004**, *29*, 1021–1051.
32. Machappa, T.; Prasad, M.V.N.A. AC conductivity and dielectric behavior of polyaniline/sodium metavenadate (PANI/NaVO3) composites. *Phys. B Condens. Matter* **2009**, *404*, 4168–4172. [CrossRef]
33. Raj, C.J.; Varma, K.B.R. Synthesis and electrical properties of the $(PVA)_{0.7}(KI)_{0.3}{\cdot}\times H_2SO_4 (0 \leq x \leq 5)$ polymer electrolytes and their performance in a primary Zn/MnO_2 battery. *Electrochim. Acta* **2010**, *56*, 649–656.
34. Selvasekarapandian, S.; Baskaran, R.; Hema, M. Complex AC impedance, transference number and vibrational spectroscopy studies of proton conducting PVAc–NH_4SCN polymer electrolytes. *Phys. B Condens. Matter* **2005**, *357*, 412–419. [CrossRef]
35. Malathi, J.; Kumaravadivel, M.; Brahmanandhan, G.M.; Hema, M.; Baskaran, R.; Selvasekarapandian, S. Structural, thermal and electrical properties of PVA–LiCF3SO3 polymer electrolyte. *J. Non-Cryst. Solids* **2010**, *356*, 2277–2281. [CrossRef]
36. Wan, Y.; Creber, K.A.M.; Peppley, B.; Bui, V.T. Ionic conductivity of chitosan membranes. *Polymer* **2003**, *44*, 1057–1065. [CrossRef]
37. Hu, P.; Chai, J.; Duan, Y.; Liu, Z.; Cui, G.; Chen, L. Progress in nitrile-based polymer electrolytes for high performance lithium batteries. *J. Mater. Chem. A* **2016**, *4*, 10070. [CrossRef]
38. Carvalho, L.M.; Gueágan, P.; Cheradame, H.; Gomes, A.S. Variation of the mesh size of PEO-based networks filled with TFSILi: From an Arrhenius to WLF type conductivity behavior. *Europ. Polym. J.* **2000**, *36*, 401–409. [CrossRef]

39. Othman, L.; Chew, K.W.; Osman, Z. Impedance spectroscopy studies of poly(methyl methacrylate)-lithium salts polymer electrolyte systems. *Ionics* **2007**, *13*, 337–342. [CrossRef]
40. Mao, G.; Perea, R.F.; Howells, W.S.; Price, D.L.; Saboungi, M.-L. Relaxation in polymer electrolytes on the nanosecond timescale. *Nature* **2000**, *405*, 163–165. [CrossRef]
41. Rault, J. Origin of the Vogel-Fulcher-Tammann law in glass-forming materials: The α-β bifurcation. *J. Non-Cryst. Solids* **2000**, *271*, 177–217. [CrossRef]
42. Sakurai, K.; Maegawa, T.; Takahashi, T. Glass transition temperature of chitosan and miscibility of chitosan/poly(N-vinyl pyrrolidone) blends. *Polymer* **2000**, *41*, 7051–7056. [CrossRef]
43. Golodnitsky, D.; Strauss, E.; Peled, E.; Greenbaum, S. Review—On Order and Disorder in Polymer Electrolytes. *J. Electrochem. Soc.* **2015**, *162*, A2551–A2566. [CrossRef]
44. Long, L.; Wang, S.; Xiao, M.; Meng, Y. Polymer electrolytes for lithium polymer batteries. *J. Mater. Chem. A* **2016**, *4*, 10038–10069. [CrossRef]
45. Munshi, M.Z.A. Handbook of Solid State Batteries and Capacitors. *World Sci.* **1996**. [CrossRef]
46. Kumar, J.S.; Subrahmanyam, A.R.; Reddy, M.J.; Rao, U.V.S. Preparation and study of properties of polymer electrolyte system (PEO + $NaClO_3$). *Mater. Lett.* **2006**, *60*, 3346–3349. [CrossRef]
47. Reddy, M.J.; Sreekanth, T.; Rao, U.V.S. Study of the plasticizer effect on a (PEO + $NaYF_4$) polymer electrolyte and its use in an electrochemical cell. *Solid State Ion.* **1999**, *126*, 55–63. [CrossRef]
48. Aziz, S.B.; Hazrin, Z.; Abidin, Z. Electrical conduction mechanism in solid polymer electrolytes: New concepts to arrhenius equation. *J. Soft Matter* **2013**. [CrossRef]
49. Sreekanth, T.; Reddy, M.J.; Ramalingaiah, S.; Rao, U.V.S. Ion-conducting polymer electrolyte based on poly ethylene oxide complexed with $NaNO_3$ salt-application as an electrochemical cell. *J. Power Sources* **1999**, *79*, 105–110. [CrossRef]
50. Aziz, S.; Abdullah, R.; Rasheed, M.; Ahmed, H. Role of ion dissociation on DC conductivity and silver nanoparticle formation in PVA: AgNt based polymer electrolytes: Deep insights to ion transport mechanism. *Polymers* **2017**, *9*, 338. [CrossRef]
51. Awadhia, A.; Agrawal, S.L. Structural, thermal and electrical characterizations of PVA: DMSO: NH_4SCN gel electrolytes. *Solid State Ion.* **2007**, *178*, 951–958. [CrossRef]
52. Ramya, C.S.; Selvasekarapandian, S.; Hirankumar, G.; Savitha, T.; Angelo, P.C. Investigation on dielectric relaxations of PVP–NH_4SCN polymer electrolyte. *J. Non-Cryst. Solids* **2008**, *354*, 1494–1502. [CrossRef]
53. Aziz, S.B. Role of dielectric constant on ion transport: Reformulated Arrhenius equation. *Adv. Mater. Sci. Eng.* **2016**. [CrossRef]
54. Polu, A.R.; Kumar, R. Preparation and characterization of PEG–$Mg(CH_3COO)_2$–CeO_2 composite polymer electrolytes for battery application. *Bull. Mater. Sci.* **2014**, *37*, 309–314. [CrossRef]
55. Aziz, S.B. Electrical and Dielectric Properties of Solid and Nanocomposite Polymer Electrolytes Based on Chitosan. Ph.D. Thesis, University of Malaya, Kuala Lumpur, Malaysia, May 2012.
56. Natesan, B.; Karan, N.K.; Katiyar, R.S. Ion relaxation dynamics and nearly constant loss behavior in polymer electrolyte. *Phys. Rev. E Stat. Nonlin. Soft Matter Phys.* **2006**, *74*, 42801. [CrossRef]
57. Natesan, B.; Karan, N.K.; Rivera, M.B.; Aliev, F.M.; Katiyar, R.S. Segmental relaxation and ion transport in polymer electrolyte films by dielectric spectroscopy. *J. Non-Cryst. Solids* **2006**, *352*, 5205–5209. [CrossRef]
58. Aziz, S.B.; Abdullah, R.M.; Kadir, M.F.Z.; Ahmed, H.M. Non suitability of silver ion conducting polymer electrolytes based on chitosan mediated by barium titanate ($BaTiO_3$) for electrochemical device applications. *Electrochim. Acta* **2019**, *296*, 494–507. [CrossRef]
59. Hamsan, M.H.; Shukur, M.F.; Aziz, S.B.; Kadir, M.F.Z. Dextran from Leuconostoc mesenteroides-doped ammonium salt-based green polymer electrolyte. *Bull. Mater. Sci.* **2019**, *42*, 57. [CrossRef]
60. Baskaran, R.; Selvasekarapandian, S.; Hirankumar, G.; Bhuvaneswari, M.S. Vibrational, ac impedance and dielectric spectroscopic studies of poly(vinylacetate)–N,N–dimethylformamide–$LiClO_4$ polymer gel electrolytes. *J. Power Sources* **2004**, *134*, 235–240. [CrossRef]
61. Aziz, S.B. Study of electrical percolation phenomenon from the dielectric and electric modulus analysis. *Bull. Mater. Sci.* **2015**, *38*, 1597. [CrossRef]
62. Aziz, S.B.; Woo, T.J.; Kadir, M.F.Z.; Ahmed, H.M. A conceptual review on polymer electrolytes and ion transport models. *J. Sci.* **2018**, *3*, 1–17. [CrossRef]

63. Aziz, S.B.; Brza, M.A.; Mohamed, P.A.; Kadir, M.F.Z.; Hamsan, M.H.; Abdulwahid, R.T.; Woo, H.J. Increase of metallic silver nanoparticles in Chitosan: AgNt based polymer electrolytes incorporated with alumina filler. *Results Phys.* **2019**, *13*, 102326. [CrossRef]
64. Das-Gupta, D.K. Molecular processes in polymer electrets. *J. Electrost.* **2001**, *51*, 159–166. [CrossRef]
65. Aziz, S.B.; Rasheed, M.A.; Abidin, Z.H.Z. Optical and electrical characteristics of silver ion conducting nanocomposite solid polymer electrolytes based on chitosan. *J. Elec. Mater.* **2017**, *46*, 6119. [CrossRef]
66. Suthanthiraraj, S.A.; Sheeba, D.J.; Paul, B.J. Impact of ethylene carbonate on ion transport characteristics of PVdF–$AgCF_3SO_3$ polymer electrolyte system. *Bull. Mater. Sci.* **2009**, *44*, 1534–1539. [CrossRef]
67. Aziz, S.B. Morphological and optical characteristics of chitosan (1 − x): Cuox ($4 \leq x \leq 12$) based polymer nano-composites: Optical dielectric loss as an alternative method for tauc's model. *Nanomaterials* **2017**, *7*, 444. [CrossRef] [PubMed]
68. Aziz, S.B.; Abdullah, R.M. Crystalline and amorphous phase identification from the tanδ relaxation peaks and impedance plots in polymer blend electrolytes based on [CS: AgNt]x: PEO (x − 1) ($10 \leq x \leq 50$). *Electrochim. Acta* **2018**, *285*, 30–46. [CrossRef]
69. Patsidis, A.; Psarras, G.C. Dielectric behaviour and functionality of polymer matrix–ceramic $BaTiO_3$ composites. *Express Polym. Lett.* **2008**, *2*, 718–726. [CrossRef]
70. Aziz, S.B.; Abidin, Z.H.Z.; Arof, A.K. Influence of silver ion reduction on electrical modulus parameters of solid polymer electrolyte based on chitosan-silver triflate electrolyte membrane. *Express Polym. Lett.* **2010**, *4*, 300–310. [CrossRef]
71. Aziz, S.B.; Karim, W.O.; Qadir, K.; Zafar, Q. Proton ion conducting solid polymer electrolytes based on chitosan incorporated with various amounts of barium titanate ($BaTiO_3$). *Int. J. Electrochem. Sci.* **2018**, *13*, 6112–6125. [CrossRef]
72. Mohomed, K.; Gerasimov, T.G.; Moussy, F.; Harmon, J.P. A broad spectrum analysis of the dielectric properties of poly(2-hydroxyethyl methacrylate). *Polymer* **2005**, *46*, 3847–3855. [CrossRef]
73. Jayathilaka, P.A.R.D.; Dissanayake, M.A.K.L.; Albinsson, I.; Mellander, B.-E. Dielectric relaxation, ionic conductivity and thermal studies of the gel polymer electrolyte system PAN/EC/PC/LiTFSI. *Solid State Ion.* **2003**, *156*, 179–195. [CrossRef]
74. Koops, C.G. On the dispersion of resistivity and dielectric constant of some semiconductors at audio frequencies. *Phys. Rev.* **1951**, *83*, 121–124. [CrossRef]
75. Louati, B.; Hlel, F.; Guidara, K. Ac electrical properties and dielectric relaxation of the new mixed crystal $(Na_{0.8}Ag_{0.2})_2PbP_2O_7$. *J. Alloy. Compd.* **2009**, *486*, 299–303. [CrossRef]
76. Idris, N.H.; Senin, H.B.; Arof, A.K. Dielectric spectra of LiTFSI-doped chitosan/PEO blends. *Ionics* **2007**, *13*, 213–217. [CrossRef]
77. Saroj, A.L.; Singh, R.K. Thermal, dielectric and conductivity studies on PVA/Ionic liquid [EMIM][$EtSO_4$] based polymer electrolytes. *J. Phys. Chem. Solids* **2012**, *73*, 162–168. [CrossRef]
78. Tiong, T.S.; Buraidah, M.H.; Teoet, L.P. Conductivity studies of poly(ethylene oxide)(PEO)/poly(vinyl alcohol) (PVA) blend gel polymer electrolytes for dye-sensitized solar cells. *Ionics* **2016**, *22*, 2133–2142. [CrossRef]
79. Chandra, A.; Agrawal, R.C.; Mahipal, Y.K. Ion transport property studies onPEO–PVP blended solid polymer electrolyte membranes. *J. Phys. D Appl. Phys.* **2009**, *42*, 135107. [CrossRef]
80. Agrawal, R.C.; Chandra, A.; Bhatt, A.; Mahipal, Y.K. Investigations on ion transport properties of and battery discharge characteristic studies on hot-pressed Ag^+-ion-conducting nano-composite polymer electrolytes: (1 − x) [90 PEO: 10 $AgNO_3$]: $xSiO_2$. *New J. Phys.* **2008**, *10*, 43023. [CrossRef]
81. Winie, T.; Ramesh, S.; Arof, A.K. Studies on the structure and transport properties of hexanoyl chitosan-based polymer electrolytes. *Phys. B* **2009**, *404*, 4308–4311. [CrossRef]
82. Majid, S.R.; Arof, A.K. Proton-conducting polymer electrolyte films based on chitosan acetate complexed with NH_4NO_3 salt. *Phys. B* **2005**, *355*, 78–82. [CrossRef]
83. Arya, A.; Sharma, A.L. Optimization of salt concentration and explanation of two peak percolation in blend solid polymer nanocomposite films. *J. Solid State Electrochem.* **2018**, *22*, 2725. [CrossRef]
84. Patla, S.K.; Ray, R.; Karmakar, S.; Das, S.; Tarafdar, S. Nanofiller-Induced Ionic Conductivity Enhancement and Relaxation Property Analysis of the Blend Polymer Electrolyte Using Non Debye Electric Field Relaxation Function. *J. Phys. Chem. C* **2019**, *123*, 5188–5197. [CrossRef]
85. Alexander, M.; Jeffrey, H.; Mativetsky, M. Supramolecular Approaches to Nanoscale Morphological Control in Organic Solar Cells. *Int. J. Mol. Sci.* **2015**, *16*, 13381–13406.

86. Wei, D. Dye Sensitized Solar Cells. *Int. J. Mol. Sci.* **2010**, *11*, 1103–1113. [CrossRef]
87. Wu, G.; Hsieh, L.; Chien, H. Enhanced Solar Cell Conversion Efficiency Using Birefringent Liquid Crystal Polymer Homeotropic Films from Reactive Mesogens. *Int. J. Mol. Sci.* **2013**, *14*, 21319–21327. [CrossRef]

International Journal of
Molecular Sciences

MDPI

Article

Improving Wettability: Deposition of TiO_2 Nanoparticles on the O_2 Plasma Activated Polypropylene Membrane

Babak Jaleh [1,*], Ehsan Sabzi Etivand [1], Bahareh Feizi Mohazzab [1], Mahmoud Nasrollahzadeh [2] and Rajender S. Varma [3,*]

1 Department of Physics, Faculty of Science, University of Bu-Ali Sina, Hamedan 65174, Iran
2 Department of Chemistry, Faculty of Science, University of Qom, Qom 3716146611, Iran
3 Regional Centre of Advanced Technologies and Materials, Department of Physical Chemistry, Faculty of Science, Palacky University, Šlechtitelů 27, 783 71 Olomouc, Czech Republic
* Correspondence: jaleh@basu.ac.ir (B.J.); varma.rajender@epa.gov (R.S.V.); Tel.: +98-91-2211-4707 (B.J.); +1-(513)-487-2701 (R.S.V.); Fax: +98-813-838-1470 (B.J.); +1-(513)-569-7677 (R.S.V.)

Received: 30 May 2019; Accepted: 2 July 2019; Published: 5 July 2019

Abstract: Radio frequency plasma is one of the means to modify the polymer surface namely in the activation of polypropylene membranes (PPM) with O_2 plasma. Activated membranes were deposited with TiO_2 nanoparticles by the dip coating method and the bare sample and modified sample (PPM5-TiO_2) were irradiated by UV lamps for 20–120 min. Characterization techniques such as X-ray diffraction (XRD), Attenuated total reflection technique- Fourier transform infrared spectroscopy (ATR-FTIR), Thermogravimetric analysis (TGA), X-ray photoelectron spectroscopy (XPS), Scanning electron microscope (SEM) and water contact angle (WCA) measurements were applied to study the alteration of ensuing membrane surface properties which shows the nanoparticles on the sample surface including the presence of Ti on PPM. The WCA decreased from 135° (PPM) to 90° (PPM5-TiO_2) and after UV irradiation, the WCA of PPM5-TiO_2 diminished from 90° to 40°.

Keywords: PP membrane; O_2 plasma; TiO_2 nanoparticles; UV treatment; hydrophilicity

---

1. Introduction

Polypropylene membranes (PPM), due to good porosity, high void volume, and high thermal stability, have wide-ranging applications. The low energy surface and hydrophobicity of the membrane often leads to membrane fouling [1,2]. To address this dilemma, membrane surface treatments have been applied aimed at altering the surface wettability and chemical properties. A wide range of methods have been deployed for altering the surface properties, such as plasma [3–9], UV irradiation [3], ion irradiation [4] and chemical coating [5,6]. These treatments have been used to attain the following goals: produce special functional groups at the surface for interactions with other functional groups, amend surface energy, change hydrophobicity or hydrophilicity, and alteration of surface morphology [7]. The PPM has been used as a bioinspired substrate for separation applications [8]. Among the present methods for the modification of the surface, plasma treatment is a rather familiar approach [9], which improves the adhesion properties, biocompatibility and wettability [10]. PPM is poor in hydrophobicity and biocompatibility due to the lack of functional groups, which restricts its biomedical usage and possible application in aqueous solution separation and hence the surface modifications induced alterations in bio-compatibility and hydrophilicity. Presently, adsorption and permeation properties of porous membranes can be altered by the deposition of a layer on to their active surface. For example, a hydrophilic layer on the porous membrane can reduce protein binding and enhance flux [2] or alternatively changes can be affected via the addition of nanoparticles such as Al_2O_3 [11], ZnO [12],

Fe_3O_4 [13] and TiO_2 [14–21]. Among these nanoparticles, TiO_2 has photo-catalytic and desirable hydrophilicity properties [22]. Studies have been conducted to deposit TiO_2 nanoparticles on flat polymeric polyethersulfone ultrafiltration (UF) membranes, to decrease the fouling problem [19] with some promising results on the effect of TiO_2 nanoparticles on UF membranes [23]. Some studies have investigated altering the surface wettability of PPM by means of plasma; compared to the present work, the plasma conditions are different [24,25].

In this work, TiO_2 nanoparticles were deposited on PP by dip-coating to improve the surface hydrophilicity. At first, the PP membranes were activated by plasma in the range of 1 to 5 min. The high energy species such as electrons, atoms, and radicals in RF plasma interact with the PP surface, leading to the modifications of the surface functionality and the morphology for deposition of TiO_2. The surface morphology of the PP membrane and PP-TiO_2 membrane was investigated by SEM, to analyze the distribution of nanoparticles. Moreover, the effect of short UV treatment time on the wettability of the surface of the sample was examined by deployment of characterization techniques to study the ensuing final products.

Wettability has a vital role in the use of polymeric materials in industry and medical science. Oxygen plasma treatment is a common method aimed at fabricated materials in many research fields. Oxygen as a reactant gas that contributes to the fabrication of desirable materials by the surface reaction. During oxygen plasma treatment, the formation of oxygen functional groups ensues. Creating polar groups, oxygen-containing functional groups, has many benefits, especially in changing the wettability of polymers and creating space for bonding nanoparticles [26].

The wetting property plays an important role in the interface of a liquid and a solid surface, especially in polymer applications. External stimuli such as light illumination, temperature, solvents and others, can change surface wetting behavior by changing the morphology of stimuli-sensitive materials. One of the most important ways to change surface wettability is light illumination. There is a variety of responses to light illumination in different materials, but the semiconductor has the same photo-responsive mechanism that has been studied in detail [27]. In brief, since a semiconductor does not have a too large band gap energy, if a photon or light has an energy equal or greater than the band gap energy, the electron in valance band (VB) can absorb energy and jump to the conduction band (CB), resulting in the generation of holes (electron deficiencies) [28]. The charge carriers could recombine by vanishing absorbed energy in the form of light (photon generation) or heat (lattice vibration). Thus, these charge carriers are responsible for carrying out photo-oxidation or photo-reduction reactions [27].

Polymers containing active groups, i.e., oxygen-containing groups (nanoparticles), can be directly absorbed by a chemical reaction between nanoparticles and a polymer surface. In contrast, in inert polymers, nanoparticles are defused and trapped into polymer chains' free volume [29]. Introducing oxygen-containing groups, e.g., chemical and plasma methods, at the surface of inert polymers contributes to the deposition of nanoparticles [25]. Having a large number of hydrophilic hydroxyl groups, metal oxides show good adherent properties for deposition at the surface of the hydrophilic membrane [30]. It is envisaged that deposition of metal oxides on the membranes with superior chemical stability, such as TiO_2 or ZrO_2, would lead to an improvement in their separation performances and to extend their applications in diverse fields.

2. Results and Discussion

2.1. Scanning Electron Microscope (SEM)

To assess the influence of plasma exposure time on the deposition, SEM analysis was performed. The PPM deposited was prepared with different plasma treatment times under the same deposition condition, as mentioned in Section 3.1. The SEM images of pure and deposited PP without plasma treatment are shown in Figure 1 wherein the PP membrane shows a porous surface and lamellar structure, and the deposition of TiO_2 on the surface of the inactivated PP membrane is not uniform. Besides, the TiO_2 nanoparticles were aggregated in some regions. The SEM images of S1, S2 and

PPM5/TiO_2 are depicted in Figure 2. As can be observed, the amount of TiO_2 nanoparticles on the activated samples surface increased by increasing plasma treatment time; comparing deposition degree confirmed that nanoparticles were almost uniformly deposited on the surface of PPM5/TiO_2. In view of the almost uniform deposition degree, the other analyses were performed for PPM5/TiO_2 alone.

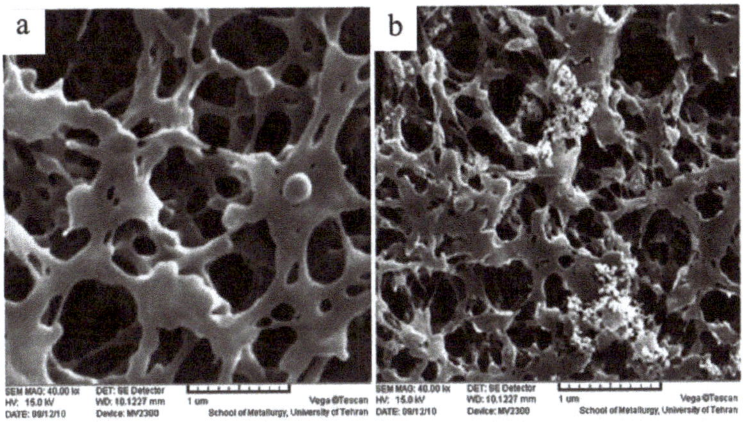

Figure 1. Scanning electron microscope (SEM) images of inactivated PPM (**a**) PPM/TiO_2 (**b**).

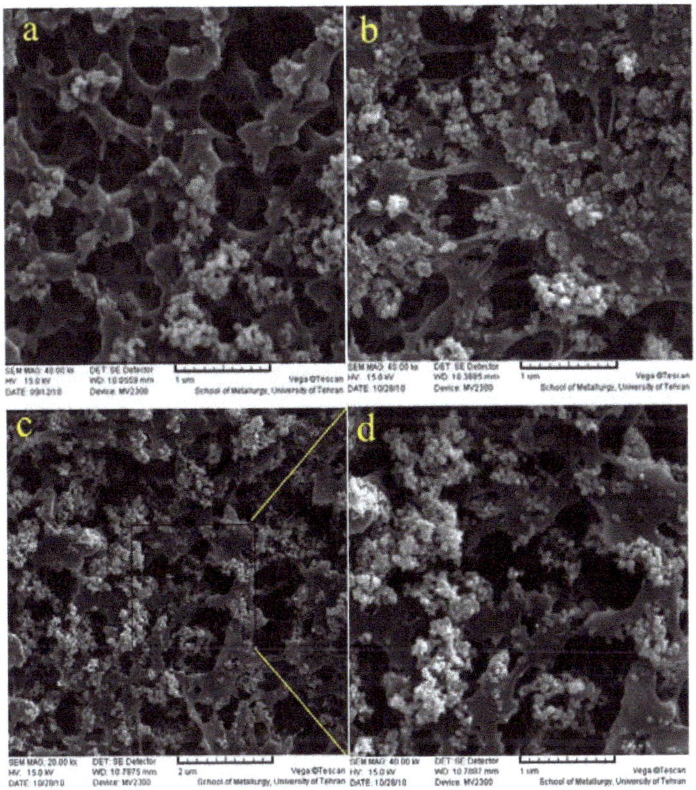

Figure 2. The SEM image of (**a**). PPM1/TiO_2; (**b**). PPM3/TiO_2 (**c**,**d**): PPM5/TiO_2.

2.2. *X-Ray Diffraction (XRD)*

XRD spectra of TiO_2, PPM membrane and PPM5/TiO_2 are shown in Figure 3. XRD spectrum of PP shows distinct peaks around 2θ = 14°, 16.9°, 18.5° and 21.8 ° attributed to the crystallographic plans (110), (040), (130), and (041), respectively [31,32]. The pattern of crystalline TiO_2 nanoparticles shows two characteristic peaks at 2θ = 25.28° and 2θ = 27.4° that indicate the anatase (101) and rutile (110) crystal phases, respectively. The sharp peak positions were in complete agreement with documented reports in the literature [28,33]. The XRD pattern of PP-TiO_2 shows several peaks for the PP membrane and one peak at 2θ = 25.2° attributed to the presence of TiO_2 (JCPDS 04-0477). Comparison with the TiO_2 pattern, the intensity of this peak was weak because of the low amassed value of TiO_2 on the membrane surface.

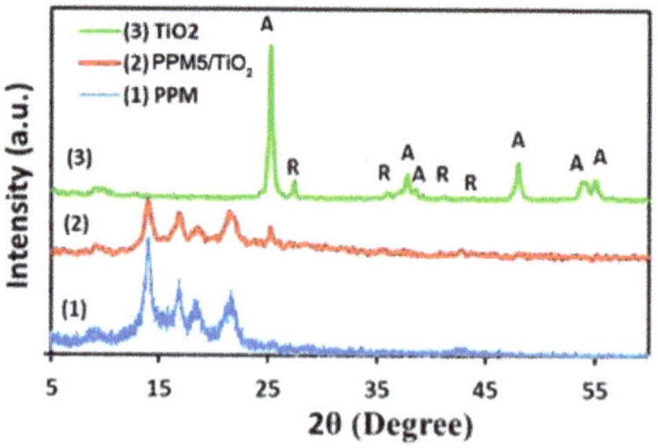

Figure 3. X-ray diffraction (XRD) patterns of (**1**): polypropylene membrane, (**2**): PPM5/TiO_2 and (**3**): TiO_2 (A: anatase, R: rutile).

2.3. *Thermal Gravimetric Analysis (TGA)*

Thermal stability of the PPM and PPM5/TiO_2. was studied by TGA analysis. Figure 4 illustrates the TGA thermograms of the PP membrane and PP membrane deposited with TiO_2 nanoparticles. As the curves show, both samples displayed similar behavior with a single mass loss zone. The onset degradation of PP membrane and deposited sample (PPM5/TiO_2) was observed around 107 °C and 213 °C, respectively. Increasing degradation temperature shows the presence of TiO_2 nanoparticles had a positive effect on thermal stability. To scrutinize the effect of deposition on thermal stability, three important points are listed in Table 1. According to the results, calculating the differential temperature between the samples during increasing temperature showed a fast decline, probably owing to the splitting of TiO_2 nanoparticles from the surface of the membrane [34].

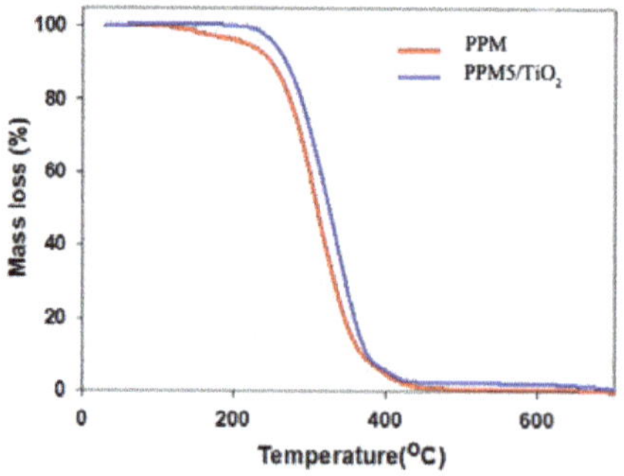

Figure 4. TGA thermograms of PP membrane and PPM5/TiO_2.

Table 1. Mass-loss temperature obtained from thermogravimetric analysis (TGA) thermogram of PPM and PPM5/TiO_2.

Samples	Mass Loss Temperature (±2 °C)		
	T_O	T_{50}	T_{90}
PP membrane	107	310	380
PPM5/TiO_2	213	358	394

The T_O, T_{50} and T_{90} performed temperature of onset degradation, 50% and 90% mass-loss, respectively.

2.4. Attenuated Total Reflection-Fourier Transform Infrared Spectroscopy (ATR-FTIR)

To study the influence of plasma and deposition of TiO_2 nanoparticles on the functional groups of the PP membrane, FTIR-ATR analysis was performed for fresh PPM, activated-PPM and PPM5/TiO_2. The PP spectrum was considered as a reference, and its peaks are shown in Figure 5a, and they are in good agreement with other reports [32,35,36]. As can be seen in Figure 5b, the peak at 1720 cm^{-1} (C = O stretching) is a consequence of the O_2 plasma treatment. The spectra of PPM5/TiO_2 (Figure 5c) shows an obviously declined intensity of the PP membrane peaks, which may be related to the immobilization of nanoparticles on the surface.

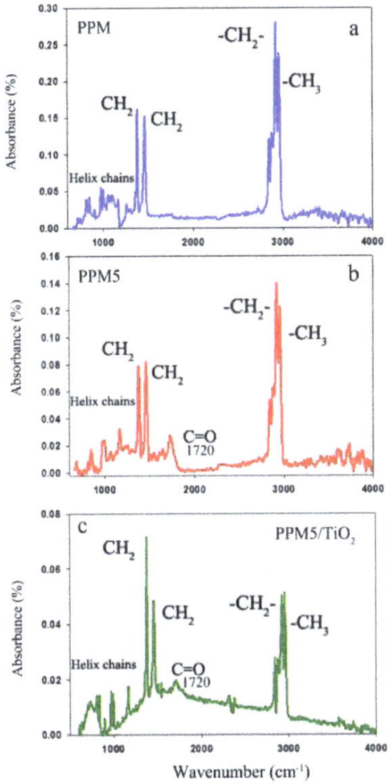

Figure 5. Attenuated total reflection- Fourier transform infrared spectroscopy (ATR-FTIR) spectrum of PPM (**a**), PPM5 (**b**) and PPM5/TiO_2 (**c**).

2.5. X-Ray Photoelectron Spectroscopy (XPS)

XPS spectroscopy, for the surface characterization of the polymer, was deployed to obtain compelling evidence for the presence of Ti on the surface of the samples. Figure 6 displays the surface chemical changes by precise XPS analysis. In comparison with the membrane [33], two obvious peaks appeared at 458.2 and 531.6 eV corresponding to Ti2p and O1s, respectively. The effect of O_2 plasma on the polypropylene membrane surface properties was investigated [36]. The XPS results led to the conclusion that oxygen containing functional groups were produced after the plasma treatment. Figure 6b shows the Ti (2p) high-resolution XPS spectrum with two bands in this region. The bands appeared at 463.8 and 458.2 eV binding energies corresponding to the Ti ($2p_{1/2}$) and Ti ($2p_{3/2}$) orbital of Ti atom, respectively.

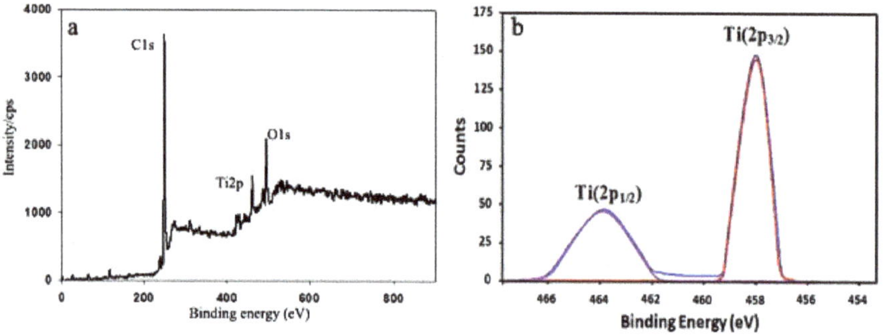

Figure 6. X-ray photoelectron spectroscopy (XPS) spectra of PPM5/TiO_2 (**a**), Resolved XPS spectra for Ti2p (**b**).

2.6. UV Treatment

In order to investigate the effect of UV treatment on the wettability of PP-TiO_2, a mercury lamp (TUV 30 W, Philips, Holland) with a wavelength of 254 nm was used; irradiation was performed under ambient conditions and in the range of 20–120 min duration.

When the surface became superhydrophilic, it absorbed more water molecules rather than impurities and organic matter and resisted adsorption. It has been widely reported that TiO_2 films will turn into superhydrophilic surfaces upon exposure to UV light, the phenomenon being termed photo-induced superhydrophilicity is initiated by the photo-generation of electrons and holes and their migration to the surface [28,37].

2.7. Wettability

The sessile drop method was employed to measure the water contact angles (WCAs). The mean value of all WCAs was accomplished by five measurements on the different position of the membrane. To study the influence of irradiation time on wettability, the WCAs were measured as irradiation time. Increasing wettability with a significant change has been reported previously due to the 5 min O_2 plasma treatment [36]. The increasing hydrophilicity of the activated PP membrane is possibly due to the presence of oxygen-containing functional groups, which has been corroborated by XPS results.

The increasing wettability (hydrophilic functional groups) of the PP membrane contributes to a good deposition of TiO_2 nanoparticles on the surface of the sample. The WCA measurements for PPM activated by O_2 plasma (5 min) and deposited with TiO_2 at different UV irradiation times are shown in Figure 7; WCAs decreased from 135° to 90° after plasma irradiation and 90° to 40° after UV irradiation. It has been widely reported that TiO_2 will turn into hydrophilic surfaces when exposed to UV light [37]. Compared to other reports [24], the plasma's step in the present work was more effective in changing the wettability of PPM.

When the TiO_2 surface was irradiated with UV light, an electron-hole pair was generated. The reaction could take place between the electron-hole pairs and absorbed H_2O and O_2 molecules. This mechanism led to an increase in wettability on the surface of TiO_2.

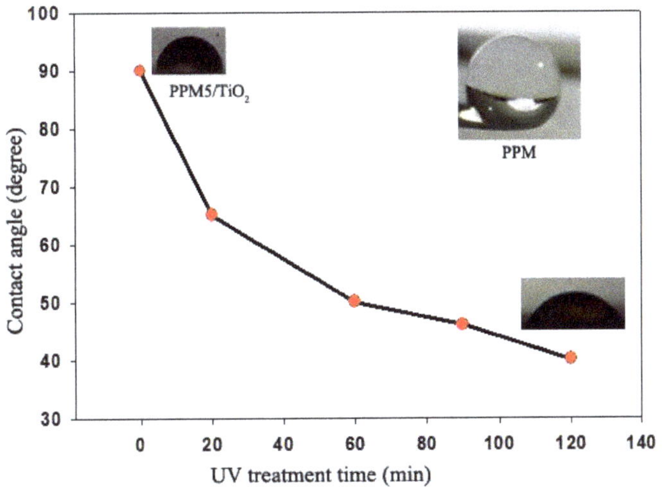

Figure 7. WCA (water contact angles) as a function of UV treatment time.

3. Materials and Methods

PP membranes with a diameter of 47 mm, a thickness of 190 μm and pore sizes of 0.22 μm were used for O_2 plasma treatment by radio frequency plasma (EMITECH KX1050, East Sussex, UK) TiO_2 powder (Degussa, Berlin, Germany, P-25) was used in these studies.

3.1. Preparation of Samples

Initially, to remove any chemical contamination from the PP membrane, it was washed with acetone, then dried at room temperature. Subsequently, the modification was carried out by means of 25 W O_2 plasma at 0.1 mbar pressure at different times, 1, 3 and 5 min. The TiO_2 suspensions were prepared using appropriate concentrations (0.05 wt%) of TiO_2 in pure ethanol (Merck, Kenilworth, NJ, USA). The TiO_2 suspension was homogenized via ultrasonication (40 °C for 20 min) and the suspension was deposited on the activated membrane by means of the dip-coating method three times with a 2 mm·s $^{-1}$ speed. The samples were dried between each deposition cycle. The specimens were named as PPMx/TiO_2, where x represents the plasma treatment time (x = 1, 3 and 5). Then the membranes were washed with distilled water, dried and finally illuminated by UV lamp (TUV 30 W, Philips, Amsterdam, Netherlands) for 20–120 min; the distance of being 20 cm from the lamp.

3.2. Membrane Characterization

X-ray diffraction (XRD, Unisantis xmd300, Georgesmarienhutte, Germany) measurements were recorded in 2θ in the range of 5–60°. The surface morphologies of PP and PPM5 deposited TiO_2 nanoparticles (PPM5/TiO_2) were examined by SEM (Philips, model XL30). Change of functional groups and influence nanoparticles deposition on the surface were studied using FT-IR-ATR (Bruker Alpha, Yokohama, Japan). Thermal gravimetric analysis (TGA) was accomplished by using a heating rate of 10 °C/min from 30 °C to 800 °C under air flow radiation. X-ray photoelectron spectroscopy (XPS) was recorded via an Al Kα X-ray source at 1486.6 eV. Finally, the alteration in wettability was examined using the WCA measuring system.

4. Conclusions

In continuation of our ongoing studies on PP membrane activated by means of O_2 plasma treatment, the present work dwells on the wettability as influenced by the deposition of TiO_2 nanoparticles on activated samples and UV irradiation. To overcome the hydrophobic properties of PPM, O_2 plasma treatment was utilized for various durations. Consequently, oxygen-containing functional groups appeared on the surface of PPM which facilitated the deposition of TiO_2 nanoparticles by the dip-coating method. The SEM depicts that most parts of TiO_2 were attached to the membrane surface as clusters. The XPS and XRD analyses confirmed the presence of TiO_2 nanoparticles at the surface of PPM. FTIR analysis affirmed the successful formation of oxygen-containing groups on the PP surface after a very short O_2 plasma treatment. The water contact angle (WCA) results led to the conclusion that significant improvement of the hydrophilicity of PPM5/TiO_2 was associated with the deposition of TiO_2. Deployment of UV treatment led to better effects in improving the hydrophilicity. Similarly, the flux and work efficiency of the membrane could be improved because of the enhanced hydrophilicity.

Author Contributions: Conceptualization, B.J.; methodology, B.J., E.S.E. and B.F.M.; resources, B.J.; writing—original draft, R.S.V., B.J. and M.N.; writing—review and editing, R.S.V., B.J. and M.N.

Funding: This research received no external funding.

Acknowledgments: The support of the Iranian Nano Council and Bu-Ali Sina University for this work are greatly appreciated.

Conflicts of Interest: The authors declare no conflict of interest.

References

1. Hu, M.X.; Yang, Q.; Xu, Z.K. Enhancing the hydrophilicity of polypropylene microporous membranes by the grafting of 2-hydroxyethyl methacrylate via a synergistic effect of photoinitiators. *J. Membr. Sci.* **2006**, *285*, 196–205. [CrossRef]
2. Yang, Q.; Xu, Z.K.; Dai, Z.W.; Wang, J.L.; Ulbricht, M. Surface modification of polypropylene microporous membranes with a novel glycopolymer. *Chem. Mater.* **2005**, *17*, 3050–3058. [CrossRef]
3. Liang, L.; Feng, X.; Peurrung, L.; Viswanathan, V. Temperature-sensitive membranes prepared by UV photopolymerization of N-isopropylacrylamide on a surface of porous hydrophilic polypropylene membranes. *J. Membr. Sci.* **1999**, *162*, 235–246. [CrossRef]
4. Qureshi, A.; Singh, D.; Singh, N.; Ataoglu, S.; Gulluoglu, A.N.; Tripathi, A.; Avasthi, D. Effect of irradiation by 140 Mev Ag11 + ions on the optical and electrical properties of polypropylene/TiO_2 composite. *Nucl. Instrum. Methods Phys. Res. Sect. B Beam Interact. Mater. At.* **2009**, *267*, 3456–3460. [CrossRef]
5. Li, C.; Yue, H.; Wang, Q.; Shi, M.; Zhang, H.; Li, X.; Dong, H.; Yang, S. A novel modified PP separator by grafting PAN for high-performance lithium–sulfur batteries. *J. Mater. Sci.* **2019**, *54*, 1566–1579. [CrossRef]
6. Song, Y.Z.; Zhang, Y.; Yuan, J.J.; Lin, C.E.; Yin, X.; Sun, C.C.; Zhu, B.; Zhu, L.P. Fast assemble of polyphenol derived coatings on polypropylene separator for high performance lithium-ion batteries. *J. Electroanal. Chem.* **2018**, *808*, 252–258. [CrossRef]
7. Tuominen, M.; Lahti, J.; Lavonen, J.; Penttinen, T.; Räsänen, J.P.; Kuusipalo, J. The influence of flame, corona and atmospheric plasma treatments on surface properties and digital print quality of extrusion coated paper. *J. Adhes. Sci. Technol.* **2010**, *24*, 471–492. [CrossRef]
8. Zhao, D.; Kim, J.F.; Ignacz, G.; Pogany, P.; Lee, Y.M.; Szekely, G. Bio-Inspired Robust Membranes Nanoengineered from Interpenetrating Polymer Networks of Polybenzimidazole/Polydopamine. *ACS Nano* **2019**, *13*, 125–133. [CrossRef]
9. Yu, H.Y.; Tang, Z.Q.; Huang, L.; Cheng, G.; Li, W.; Zhou, J.; Yan, M.G.; Gu, J.S.; Wei, X.W. Surface modification of polypropylene macroporous membrane to improve its antifouling characteristics in a submerged membrane-bioreactor: H_2O plasma treatment. *Water Res.* **2008**, *42*, 4341–4347. [CrossRef]
10. Fisher, E.R. A Review of Plasma-Surface Interactions During Processing of Polymeric Materials Measured Using the IRIS Technique. *Plasma Process. Polym.* **2004**, *1*, 13–27. [CrossRef]
11. Yan, L.; Li, Y.S.; Xiang, C.B.; Xianda, S. Effect of nano-sized Al_2O_3-particle addition on PVDF ultrafiltration membrane performance. *J. Membr. Sci.* **2006**, *276*, 162–167. [CrossRef]
12. Jian, P.; Yahui, H.; Yang, W.; Linlin, L. Preparation of polysulfone-Fe_3O_4 composite ultrafiltration membrane and its behavior in magnetic field. *J. Membr. Sci.* **2006**, *284*, 9–16. [CrossRef]
13. Chandramouleeswaran, S.; Mhaske, S.; Kathe, A.; Varadarajan, P.; Prasad, V.; Vigneshwaran, N. Functional behaviour of polypropylene/ZnO–soluble starch nanocomposites. *Nanotechnology* **2007**, *18*, 385702. [CrossRef]

14. Altan, M.; Yildirim, H. Mechanical and antibacterial properties of injection molded polypropylene/TiO_2 nano-composites: Effects of surface modification. *J. Mater. Sci. Technol.* **2012**, *28*, 686–692. [CrossRef]
15. Bottino, A.; Capannelli, G.; Comite, A. Preparation and characterization of novel porous PVDF-ZrO_2 composite membranes. *Desalination* **2002**, *146*, 35–40. [CrossRef]
16. Chin, S.S.; Chiang, K.; Fane, A.G. The stability of polymeric membranes in a TiO_2 photocatalysis process. *J. Membr. Sci.* **2006**, *275*, 202–211. [CrossRef]
17. Erdem, N.; Erdogan, U.H.; Cireli, A.A.; Onar, N. Structural and ultraviolet-protective properties of nano-TiO_2-doped polypropylene filaments. *J. Appl. Polym. Sci.* **2010**, *115*, 152–157. [CrossRef]
18. Luo, M.L.; Zhao, J.Q.; Tang, W.; Pu, C.S. Hydrophilic modification of poly (ether sulfone) ultrafiltration membrane surface by self-assembly of TiO_2 nanoparticles. *Appl. Surf. Sci.* **2005**, *249*, 76–84. [CrossRef]
19. Rahimpour, A.; Madaeni, S.; Taheri, A.; Mansourpanah, Y. Coupling TiO_2 nanoparticles with UV irradiation for modification of polyethersulfone ultrafiltration membranes. *J. Membr. Sci.* **2008**, *313*, 158–169. [CrossRef]
20. Velásquez, J.; Valencia, S.; Rios, L.; Restrepo, G.; Marín, J. Characterization and photocatalytic evaluation of polypropylene and polyethylene pellets coated with P25 TiO_2 using the controlled-temperature embedding method. *Chem. Eng. J.* **2012**, *203*, 398–405. [CrossRef]
21. Yang, S.; Gu, J.S.; Yu, H.Y.; Zhou, J.; Li, S.F.; Wu, X.M.; Wang, L. Polypropylene membrane surface modification by RAFT grafting polymerization and TiO_2 photocatalysts immobilization for phenol decomposition in a photocatalytic membrane reactor. *Sep. Purif. Technol.* **2011**, *83*, 157–165. [CrossRef]
22. Vatanpour, V.; Madaeni, S.S.; Khataee, A.R.; Salehi, E.; Zinadini, S.; Monfared, H.A. TiO_2 embedded mixed matrix PES nanocomposite membranes: Influence of different sizes and types of nanoparticles on antifouling and performance. *Desalination* **2012**, *292*, 19–29. [CrossRef]
23. Bae, T.H.; Tak, T.M. Effect of TiO_2 nanoparticles on fouling mitigation of ultrafiltration membranes for activated sludge filtration. *J. Membr. Sci.* **2005**, *249*, 1–8. [CrossRef]
24. Masaeli, E.; Morshed, M.; Tavanai, H. Study of the wettability properties of polypropylene nonwoven mats by low-pressure oxygen plasma treatment. *Surf. Interface Anal. Int. J. Devoted Dev. Appl. Tech. Anal. Surf. Interfaces Thin Film.* **2007**, *39*, 770–774. [CrossRef]
25. Szabová, R.; Černáková, L.; Wolfová, M.; Černák, M. Coating of TiO_2 nanoparticles on the plasma activated polypropylene fibers. *Acta Chim. Slovaca* **2009**, *2*, 70–76.
26. Chan, C.M.; Ko, T.M.; Hiraoka, H. Polymer surface modification by plasmas and photons. *Surf. Sci. Rep.* **1996**, *24*, 1–54. [CrossRef]
27. Kang, X.; Liu, S.; Dai, Z.; He, Y.; Song, X.; Tan, Z. Titanium Dioxide: From Engineering to Applications. *Catalysts* **2019**, *9*, 191. [CrossRef]
28. Jaleh, B.; Shahbazi, N. Surface properties of UV irradiated PC–TiO_2 nanocomposite film. *Appl. Surf. Sci.* **2014**, *313*, 251–258. [CrossRef]
29. Xu, Q.; Yang, J.; Dai, J.; Yang, Y.; Chen, X.; Wang, Y. Hydrophilization of porous polypropylene membranes by atomic layer deposition of TiO_2 for simultaneously improved permeability and selectivity. *J. Membr. Sci.* **2013**, *448*, 215–222. [CrossRef]
30. Chen, H.; Kong, L.; Wang, Y. Enhancing the hydrophilicity and water permeability of polypropylene membranes by nitric acid activation and metal oxide deposition. *J. Membr. Sci.* **2015**, *487*, 109–116. [CrossRef]
31. Wang, S.; Ajji, A.; Guo, S.; Xiong, C. Preparation of microporous polypropylene/titanium dioxide composite membranes with enhanced electrolyte uptake capability via melt extruding and stretching. *Polymers* **2017**, *9*, 110. [CrossRef] [PubMed]
32. Hernández-Aguirre, O.A.; Nunez-Pineda, A.; Tapia-Tapia, M.; Gomez Espinosa, R.M. Surface Modification of Polypropylene Membrane Using Biopolymers with Potential Applications for Metal Ion Removal. *J. Chem.* **2016**, *2016*. [CrossRef]
33. Feizi Mohazzab, B.; Jaleh, B.; Kakuee, O.; Fattah-alhosseini, A. Formation of titanium carbide on the titanium surface using laser ablation in n-heptane and investigating its corrosion resistance. *Appl. Surf. Sci.* **2019**, *478*, 623–635. [CrossRef]
34. Feizi Mohazzab, B.; Jaleh, B.; Nasrollahzadeh, M.; Issaabadi, Z. Journey on Greener Pathways via Synthesis of Pd/KB Polymeric Nanocomposite as a Recoverable Catalyst for the Ligand-Free Oxidative Hydroxylation of Phenylboronic Acid and Suzuki–Miyaura Coupling Reaction in Green Solvents. *Catal. Lett.* **2019**, *149*, 169–179. [CrossRef]

35. Fonouni, M.; Yegani, R.; Tavakkoli, A.; Mollazadeh, S. Investigating the Effect of Various Oxidizing Agents on the Surface Functionalization of Microporous Polypropylene Membranes. *J. Text. Polym.* **2016**, *4*, 92–100.
36. Jaleh, B.; Parvin, P.; Wanichapichart, P.; Saffar, A.P.; Reyhani, A. Induced super hydrophilicity due to surface modification of polypropylene membrane treated by O_2 plasma. *Appl. Surf. Sci.* **2010**, *257*, 1655–1659. [CrossRef]
37. Naghdi, S.; Jaleh, B.; Shahbazi, N. Reversible wettability conversion of electrodeposited graphene oxide/titania nanocomposite coating: Investigation of surface structures. *Appl. Surf. Sci.* **2016**, *368*, 409–416. [CrossRef]

International Journal of
Molecular Sciences

MDPI

Article

Poly(2,6-Dimethyl-1,4-Phenylene Oxide)-Based Hydroxide Exchange Separator Membranes for Zinc–Air Battery

Ali Abbasi [1,2], Soraya Hosseini [1,2], Anongnat Somwangthanaroj [1], Ahmad Azmin Mohamad [3] and Soorathep Kheawhom [1,2,*]

1 Department of Chemical Engineering, Faculty of Engineering, Chulalongkorn University, Bangkok 10330, Thailand
2 Computational Process Engineering Research Laboratory, Chulalongkorn University, Bangkok 10330, Thailand
3 School of Materials and Mineral Resources Engineering, Universiti of Sains Malaysia, Nibong Tebal, Pulau Pinang 14300, Malaysia
* Correspondence: soorathep.k@chula.ac.th; Tel.: +66-81-490-5280

Received: 26 June 2019; Accepted: 25 July 2019; Published: 26 July 2019

Abstract: Rechargeable zinc–air batteries are deemed as the most feasible alternative to replace lithium–ion batteries in various applications. Among battery components, separators play a crucial role in the commercial realization of rechargeable zinc–air batteries, especially from the viewpoint of preventing zincate ($Zn(OH)_4{}^{2-}$) ion crossover from the zinc anode to the air cathode. In this study, a new hydroxide exchange membrane for zinc–air batteries was synthesized using poly (2,6-dimethyl-1,4-phenylene oxide) (PPO) as the base polymer. PPO was quaternized using three tertiary amines, including trimethylamine (TMA), 1-methylpyrolidine (MPY), and 1-methylimidazole (MIM), and casted into separator films. The successful synthesis process was confirmed by proton nuclear magnetic resonance and Fourier-transform infrared spectroscopy, while their thermal stability was examined using thermogravimetric analysis. Besides, their water/electrolyte absorption capacity and dimensional change, induced by the electrolyte uptake, were studied. Ionic conductivity of PPO–TMA, PPO–MPY, and PPO–MIM was determined using electrochemical impedance spectroscopy to be 17.37, 16.25, and 0.29 mS/cm, respectively. Zincate crossover evaluation tests revealed very low zincate diffusion coefficient of 1.13×10^{-8}, and 0.28×10^{-8} cm^2/min for PPO–TMA, and PPO–MPY, respectively. Moreover, galvanostatic discharge performance of the primary batteries assembled using PPO–TMA and PPO–MPY as initial battery tests showed a high specific discharge capacity and specific power of ~800 mAh/g_{Zn} and 1000 mWh/g_{Zn}, respectively. Low zincate crossover and high discharge capacity of these separator membranes makes them potential materials to be used in zinc–air batteries.

Keywords: zinc–air battery; separator; hydroxide exchange membrane; anion-exchange membrane; ionic channel; polyphenylene oxide

1. Introduction

Recently, metal–air batteries have attracted high interest of researchers and industry as post lithium–ion technology. Among all metal–air batteries, aqueous zinc–air battery is a relatively established technology (known to the scientific community since the late nineteenth century) with high potential to be used in future energy requirements [1–3]. This type of battery has a very high theoretical energy density of 1086 Wh·kg^{-1} (including oxygen), five times higher than that of existing lithium–ion batteries [4]. Moreover, their production cost is estimated to be very low (~10 \$·kW^{-1}·h^{-1})

in comparison to lithium–ion batteries. Technically and economically, the zinc–air battery is deemed to be the most feasible alternative for lithium–ion batteries in various applications. Despite the early start and great potential, the development of zinc–air batteries has been limited by issues related to separator, electrolyte, metal electrode, and air catalyst [5].

Despite being an essential part of a zinc–air battery, the separator has not received its deserved attention compared to other components of the battery. The primary function of the separator is to prevent physical contact between the anode and cathode, while providing high ionic conductivity and selectivity to facilitate hydroxide ion transport with the aim of completing the battery circuit during its operation. An ideal separator must be chemically stable in contact with highly alkaline environment and electrochemically stable within a wide working potential window. Also it must be able to prevent zincate ($Zn(OH)_4{}^{2-}$) crossover and short-circuit due to the formation of zinc dendrites during battery charging. Besides, high electrical resistance and ionic conductivity is highly desirable [6].

Practically, the separator for zinc–air batteries is usually a porous polypropylene membrane such as Celgard® with the porosity of 10–20 μm. The porosity of these membranes results in their high ionic conductivity, which is crucial for a separator material. Nonetheless, due to the high porosity, beside hydroxide ions, zincate ions, which are produced through the oxidation of zinc during the battery discharge, also pass through the membrane from the anode to the cathode [7]. At the cathode side, because of asymmetric water evaporation-induced electrolyte deficiency, zincate ions are precipitated as zinc oxide (ZnO) on the catalyst surface, forming a resistive layer to ion/electron conduction which in turn, leads to higher polarization of air electrode and capacity loss of zinc–air batteries [8].

Even though there have been several attempts to develop a separator membrane with desired characteristics to be used in zinc–air batteries, more efforts are required in this field. Dewi et al. used a polyelectrolyte containing a sulfonium cation as a zinc–air battery separator. Even though they incorrectly assumed zinc ions in the electrolyte to be in the form of a cation (Zn^{2+}), the separator was highly effective in preventing zinc ions crossover to the cathode side, leading to the capacity increase by more than six times compared to polypropylene-based Celgard® separator [7]. More recently, electrospun polyetherimide (PEI) nanofibers impregnated with polyvinyl alcohol (PVA) has been used as a highly efficient separator for rechargeable zinc–air battery. In the prepared permselective membrane, PEI, which is known for its high chemical resistance in the alkaline environment, provides good mechanical stability. On the other hand, the pores of the electrospun nanofibers filled with PVA electrolyte offers high ionic conductivity. In such a design, zincate crossover decreased dramatically due to the bulky size of the zincate ions, making it difficult to pass through the separator compared to hydroxide ions [8]. In another study, similar impregnated nanofibrous mat concept with electrospun PVA/polyacrylic acid (PAA) nanofibers along with the impregnation solution of Nafion bearing pendant sulfonate groups was used. In this case, the nanofibers acted as the ionic conductive pathways for hydroxide ions, and the impregnation solution played the role of an anion repelling component, preventing bulky zincate ions transport from the anode to the cathode side [9].

Basically, the ionic conductivity of polymeric materials is determined by two factors: the ionic mobility and the ion exchange capacity (IEC). To increase hydroxide conductivity, improving IEC seems to be an easier way than improving hydroxide mobility. However, increasing IEC always results in excessive water uptake, which is a negative side effect, leading to severe swelling or even dissolution at higher temperatures. Moreover, increasing IEC usually leads to an increased zincate crossover. Therefore, a better and more efficient approach to expanding the hydroxide conductivity and selectivity is to enhance hydroxide mobility while keeping the IEC at a medium level. Such a growth in the hydroxide conductivity could be realized by reforming the ionic channels in polymer structure through hydrophilic/hydrophobic microphase separation. In this approach, a cation which is usually quaternary ammonium ($-NR_3{}^+$), is covalently bound to a hydrophobic polymer backbone and the anion (hydroxide) is dissociated in aqueous phase [10]. Several polymeric materials have been used to develop this type of separator membrane such as poly(arylene ether)s, poly-(phenylene)s, poly(ether imide)s, poly(styrene)s, poly(olefin)s, and poly(phenylene oxide)s [11].

Recently, there has been high interest in poly (phenylene oxide) (PPO) as the polymer backbone. PPO exhibits low-cost; commercial availability; high thermal, mechanical, and chemical stabilities; and facile postfunctionalization [12]. Li et al. designed and synthesized comb-shaped quaternized copolymer using PPO as the base polymer. The synthesis was carried out using Menshutkin reaction with N,N-dimethyl-1-hexadecylamine (DMHDA), and subsequent hydroxide exchange. In this process, DMHDA was attached to the benzylic position of PPO through covalent bonding from nitrogen atom, forming a quaternary ammonium group directly attached to the polymeric backbone with a pendant long hydrocarbon chain. The polymer exhibited phase-separated morphology with enhanced ionic conductivity and alkaline stability, making it a potential material as hydroxide exchange membrane for alkaline fuel cells [13]. In another study, Yang et al. prepared various quaternized PPO-based anion-exchange membranes using a series of saturated heterocyclic compounds including 1-methylpyrrolidine (MPy), 1-ethylpyrrolidine (EPy), 1-butylpyrrolidine (BPy), 1-methylpiperidine (MPrD), 1-ethylpiperidine (EPrD), and N-methylmorpholine (NMM). Also, the comparison of their physicochemical characteristics with TMA-quaternized PPO as the benchmark polymer was performed. The quaternized membranes showed different ionic conductivity and alkaline stability, depending on the quaternization agent used. For example, the polymer containing 1-methylpyrolidine exhibited ionic conductivity of 27 $mS{\cdot}cm^{-1}$ at 80 °C and excellent alkaline stability, keeping 87% of its original conductivity after being soaked in 1 M KOH at 80 °C for 500 h [14].

Although this strategy has already been used in fuel cell technology, its application in zinc–air batteries, where an alkaline stable hydroxide exchange membrane is required as well, is very rare. The main focus of this study is to synthesize PPO-based microphase separated hydroxide exchange membrane through the attachment of three various quaternary ammonium molecules to the main polymeric backbone with the aim of developing ionic channels for selective transfer of hydroxide ions. Even though the main aim of developing new separators with acceptable ionic conductivity and low zincate crossover is for overcoming challenges associated with rechargeable zinc–air batteries, this phase of our study includes separators characterization and their application in a primary zinc–air battery as an indication of their applicability in secondary batteries. The separators are characterized using physicochemical and electrochemical characterization tools to investigate their properties, including their ionic conductivity and zincate diffusion coefficient. Then, they are implemented in a primary zinc–air battery to study their influence on battery performance. The ionic conductivity, along with suppressed zincate crossover, makes this new separator an excellent candidate to be used in primary and secondary zinc–air batteries. In the next phase, the application of the separators in rechargeable zinc–air batteries will be investigated.

2. Results and Discussion

2.1. Separator Membrane Synthesis

Schematic view of the membrane synthesis process is illustrated in Figure 1. In the first stage, a mixture of N-bromosuccinimide (NBS) and benzoyl peroxide (BPO) is used for the bromination reaction. NBS acts as the bromination agent, and BPO is the initiator of the reaction. Controlling the temperature and reaction condition is very crucial to ensure that the bromination reaction takes place dominantly in the benzylic position and aromatic hydrogen atoms are not being involved in the reaction. Quaternization using three amine molecules including trimethylamine (TMA), 1-methylpyrrolidine (MPY), and 1-methylimidazole (MIM) was carried out at room temperature in 48 h using an excessive amount of the amine molecules to make sure of complete reaction. Subsequently, the quaternized polymer was casted to fabricate uniform and flexible films with the thickness of 40 to 60 μm. In the final stage, the films were soaked in potassium hydroxide (KOH), 7 M solution for 72 h for completing hydroxide exchange of bromine ions in the polymer structure. To make sure of the ion exchange process, every 24 h, the hydroxylation solution was changed to a fresh solution [15]. Three PPO-based separator membranes containing trimethylamine (PPO–TMA), 1-Methylpyrrolidine (PPO–MPY) and 1-methylimidazole (PPO–MIM) were synthesized.

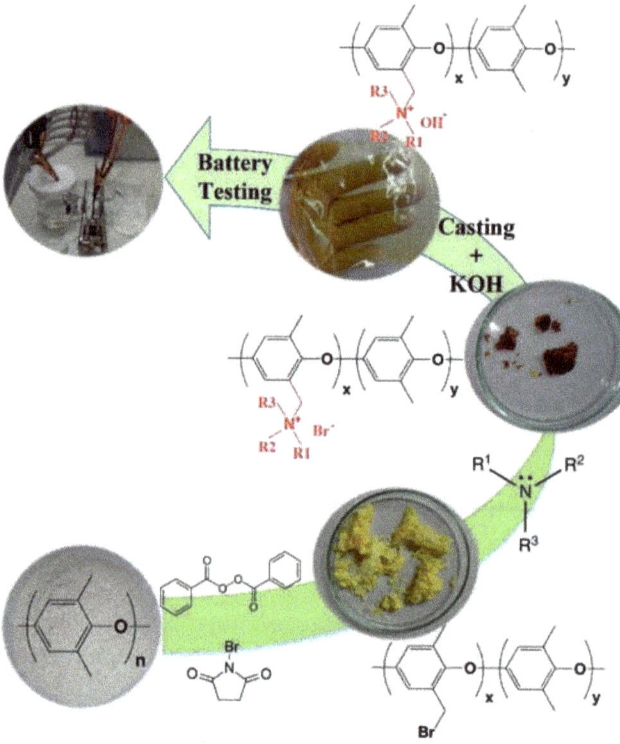

Figure 1. Schematic diagram of the membrane preparation process.

2.2. *Structural Changes*

2.2.1. Proton Nuclear Magnetic Resonance

Figure 2 shows the proton nuclear magnetic resonance (^{1}HNMR) spectrum of the brominated PPO. As previously reported in [16], pure PPO shows two characteristic peaks at around 2.1 and 6.5 ppm, assigned to the methyl (benzylic: $-CH_3$) and aryl protons, respectively. After bromination, a new peak appeared at 4.3 ppm, which is attributed to the brominated methyl protons ($-CH_2Br$). Furthermore, no peak was observed at 6 ppm (assigned to shifted aryl protons due to the bromination of neighboring aryl protons), showing that the bromination occurred dominantly at benzylic positions [17]. Considering the ratio of the integral area of CH_3 and CH_2, the bromination degree of the PPO was calculated to be 39%. Successful bromination reaction was confirmed using ^{1}HNMR spectrum [14].

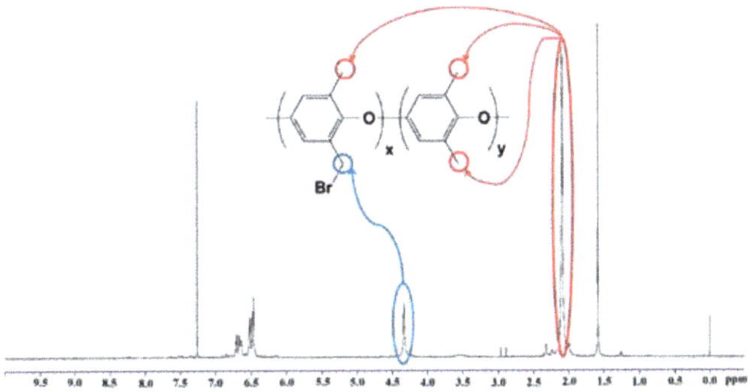

Figure 2. ^{1}HNMR spectra of BrPPO in $CDCl_3$. The blue line indicates the characteristic peak of brominated methyl protons ($-CH_2Br$). The red line indicates the characteristic peak of methyl (benzylic: $-CH_3$).

2.2.2. Fourier-Transform Infrared Spectroscopy

A successful synthesis of BrPPO and associated quaternized membranes was also confirmed by Fourier-transform infrared (FTIR) spectroscopy, as shown in Figure 3. Characteristic peaks of phenyl group of PPO appeared at about 1600 and 1470 cm^{-1}, corresponding to C=C stretching of the benzene ring and C–H stretching, respectively [14,16]. Compared to PPO, a new peak appeared at about 987 cm^{-1} for BrPPO, which was assigned to C-Br stretching [18]. After quaternization using TMA, MPY, and MIM, C-Br peak disappeared, approving complete quaternization reaction. Furthermore, for all quaternized samples, a broad peak appeared at 3200–3600 cm^{-1}, attributed to O–H stretching of the water molecules absorbed into the samples due to their increased hydrophilicity after quaternization. Moreover, for PPO–MIM, two strong absorption peaks appeared at about 750 and 1540 cm^{-1}, related to the presence of imidazolium cations in the sample [19]. FTIR results also show successful bromination of PPO and nucleophilic substitution of BrPPO with quaternization agents.

Furthermore, to have an initial evaluation of the separator membranes' chemical stability, PPO–TMA and PPO–MPY were soaked in KOH, 7 M solution for 150 h at 30 °C, and then, after washing and drying, they were analyzed using FTIR. As could be seen in Figure 3d,f, they showed similar characteristic peaks after soaking in the alkaline solution, and no significant change was observed in their spectrum, showing their stability in the solution used for zinc–air batteries.

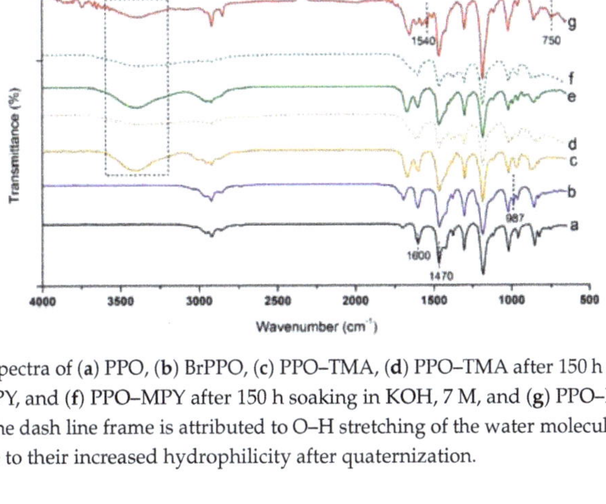

Figure 3. FTIR spectra of (**a**) PPO, (**b**) BrPPO, (**c**) PPO–TMA, (**d**) PPO–TMA after 150 h soaking in KOH, 7 M (**e**) PPO–MPY, and (**f**) PPO–MPY after 150 h soaking in KOH, 7 M, and (**g**) PPO–MIM. The broad peak shown in the dash line frame is attributed to O–H stretching of the water molecules absorbed into the samples due to their increased hydrophilicity after quaternization.

2.3. Thermal Properties

Figure 4 shows thermogravimetric analysis (TGA) curves for various separator membranes in their hydroxide form. Pristine PPO shows a single degradation step at 438 °C, as studied before [14]. The decomposition of BrPPO happened in two steps, with the first related to the degradation of brominated parts at 312 °C and the second for the PPO backbone decomposition at 438 °C [20]. Furthermore, all quaternized samples showed a three-step similar degradation behavior. The first step at approximately 90 to 150 °C was ascribed to the removal of moisture absorbed onto hydrophilic samples. The intensity of this peak for PPO–MIM was lower than those of PPO–TMA and PPO–MPY, showing its lower hydrophilicity, as confirmed by its lower water uptake (Table 1). The second peak at ~180–250 °C was attributed to the degradation of various quaternary ammonium groups, and the final degradation step was for PPO backbone at 405–420 °C [14]. Second degradation peak of PPO–MIM, related to the decomposition of 1-methylimidazolium quaternary ammonium group, occurred at a higher temperature compared to PPO–TMA and PPO–MPY. This could be attributed to the presence of an aromatic ring in its structure, leading to higher thermal stability [21]. TGA results revealed that all PPO-based separator membranes meet the thermal stability requirement for being used in zinc–air battery applications, which are usually operated at room temperature, or at the highest temperature below 80 °C [5].

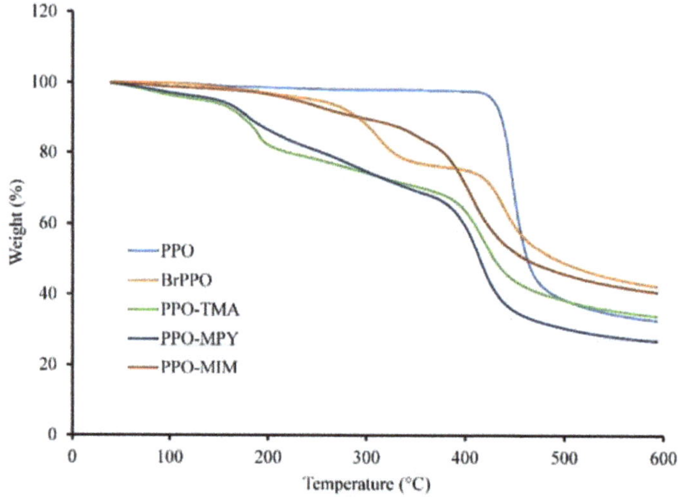

Figure 4. TGA curves of PPO, BrPPO, PPO–TMA, PPO–MPY, and PPO–MIM.

2.4. Electrochemical Characterization

2.4.1. Ionic Conductivity

Separator membranes need to absorb water/electrolyte to be able to conduct ions. Table 1 shows water and electrolyte uptake for various PPO-based separators along with their area and volume change induced by the uptake. As could be seen, PPO–TMA and PPO–MPy show high water uptake of 89 and 78%, respectively, showing their high hydrophilicity. However, the water uptake of PPO–MIM was as low as 13%, resulting from its low tendency to water absorption. This water uptake led to corresponding area and volume increase in the membranes, as shown in the table. Similar results have been obtained in previous studies for these three membranes [14].

Since in the zinc–air cell KOH, 7 M is used as the electrolyte, the uptake and dimensional change of the membranes were also measured in this electrolyte. All the membranes absorbed less electrolyte compared to water (~30% for PPO–TMA and PPO–MPY, and only 3% for PPO–MIM), leading to lower dimensional change. Also, very low electrolyte uptake of PPO–MIM was reflected in the ionic conductivity measurements, showing very low conductivity of 0.003 mS/cm determined using Nyquist plot of electrochemical impedance spectroscopy (EIS) (Figure 5). For PPO–TMA and PPO–MPy, the ionic conductivity was calculated to be 17.37 and 16.25 mS/cm. Due to deficient electrolyte uptake and low ionic conductivity of PPO–MIM, it was not included in the rest of the study.

Slightly higher ionic conductivities have been reported for the same separator membranes, which could be attributed to the higher measurement temperature and lower KOH solution concentration. In this study, the measurements were carried out in KOH, 7 M solution to mimic the real cell operation condition. As can be seen in Table 1, the separator membranes absorb much less electrolyte than they do in water, resulting in lower measured ionic conductivity.

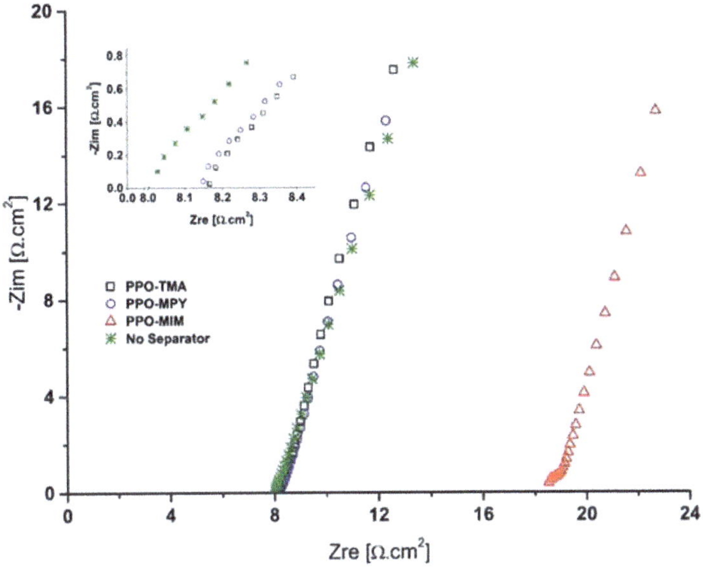

Figure 5. Nyquist plot of electrochemical impedance spectroscopy (EIS) for determining ionic conductivity of PPO-based separator membranes.

Table 1. Basic properties of PPO–TMA, PPO–MPy, and PPO–MIM separator membranes.

Sample	DI Water			KOH, 7 M			Thickness (μm)	Area (cm^2)	R_b (Ω) [2]	Ionic Conductivity (mS/cm)—σ	Zincate Diffusion Coefficient ($\times 10^{-8}$ cm^2/min)—D
	Uptake ΔW (wt%)	Area Change ΔA (%)	Volume Change ΔV (%)	Uptake ΔW (wt%)	Area Change ΔA (%)	Volume Change ΔV (%)					
PPO–TMA	89	65	119	31	11	39	50	1.766	0.1630	17.37	1.13
PPO–MPy	78	41	76	30	11	39	40	1.766	0.1394	16.25	0.28
PPO–MIM [1]	13	14	30	3	0	0	30	1.766	5.8014	0.29	N/A

[1] Due to very low electrolyte uptake of PPO–MIM and its very low ionic conductivity, this membrane was not included in the rest of the study. [2] R_b values were obtained by deducting the value obtained for the cell without using any separator (4.4606 Ω) from the resistance values measured for each sample.

2.4.2. Zincate Crossover

To realize electrochemically rechargeable Zn–air batteries, minimizing zincate crossover from the anode to the cathode is essential. In this regard, highly selective separator materials with acceptable hydroxide conductivity and limited zincate ion conductivity play an important role. In this study, a two-chamber diffusion cell was used to quantitatively investigate the zincate crossover through the developed separator membranes. The left chamber contained KOH, 7 M solution plus 0.5 M dissolved ZnO in the form of zincate ions and the right chamber contained only KOH, 7 M solution. The separator membrane was place between two chambers and the concentration gradient-driven crossover of zincate ions to the right chamber was measured as a function of time using inductively coupled plasma atomic emission spectroscopy (ICP–OES).

As could be seen in Figure 6a, the concentration of zinc ions in the right chamber increased with time for both PPO–TMA and PPO–MPY separator membranes. However, the slope of increase for PPO–MPY was lower than that of PPO–TMA, showing its lower zincate crossover during time. Furthermore, the diffusion coefficient of zincate ions through the separators was calculated using Equation 5. Besides, Figure 6b depicts the diffusion coefficient for two separator membranes. As can be seen, the diffusion coefficient for PPO–TMA and PPO–MPY was 1.13×10^{-8} and 0.28×10^{-8} cm^2/min, respectively, indicating lower zincate crossover for PPO–MPY compared to PPO–TMA. Nevertheless, in comparison with other separators previously studied, such as Celgard3501 (232.4×10^{-7} cm^2/min), PVA/PAA film (110.2×10^{-7} cm^2/min), Nafion bearing electrospun PVA/PAA mat (4.1×10^{-7} cm^2/min) [9] PEI nanofibers impregnated with PVA (5.0×10^{-6} cm^2/min) [8], and even Nafion (0.4×10^{-7} cm^2/min) [9]; PPO-based membranes have much lower zincate diffusion coefficient, making them highly desirable for zinc–air battery separators. This could be ascribed to the formation of ionic channels in the polymer structure through hydrophilic/hydrophobic microphase separation. As a result, the crossover of bulky zincate ions through the membrane separator was suppressed, but smaller hydroxide ions could pass through the separator.

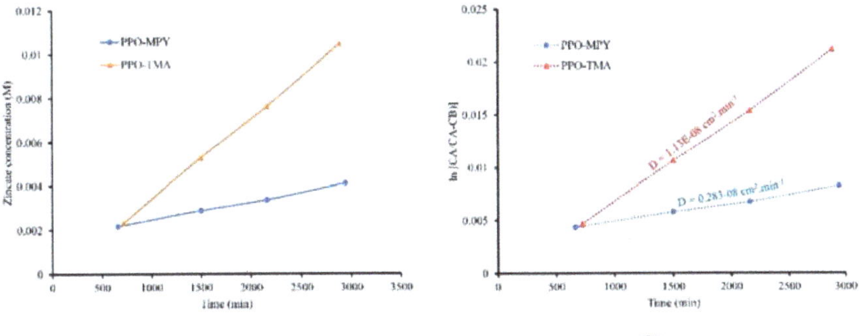

(a) (b)

Figure 6. Variation of the zincate concentration in the right chamber vs. time (**a**) and zincate diffusion calculation curves (**b**) for PPO–TMA and PPO–MPY separators.

2.4.3. Electrochemical Stability

The electrochemical stability window, which is defined as the width of voltage where no appreciable faradaic current flows, could be studied using cyclic voltammetry (CV). A wide electrochemical window is very crucial for the application of the membranes in batteries [22]. As could be seen in Figure 7, for PPO–TMA, there was no significant decomposition of separator components observed in a range of −1.5 to +1.5 V, providing a stability widow of 3 V. For PPO–MPy, the stability window was measured to be 4 V, which is even higher than that of PPO–TMA. These extensive electrochemical stability windows, wider than most of the previously studied separator materials [8,9,22,23], make PPO-based separator membranes a highly stable candidate to be used in battery applications, without any worry about the membranes themselves being involved in the electrochemical reaction during charge and discharge cycles.

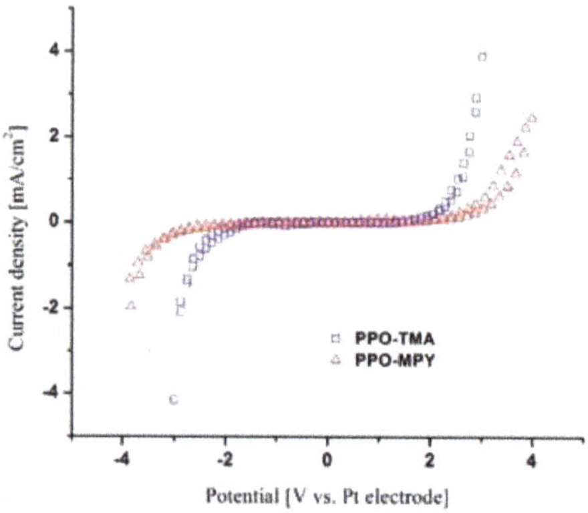

Figure 7. Cyclic voltammograms of PPO–TMA and PPO–MPY.

2.5. *Discharge Performance*

Zincate crossover affects the zinc–air battery performance in two main ways: (1) high crossover of zincate ions makes zinc ions inaccessible for the anode electrode during the charging process, negatively affecting battery performance. This is important for electrically rechargeable zinc–air batteries. (2) Due to electrolyte deficiency at the cathode side, zincate ions are precipitated as zinc oxide on the surface of catalyst particles, resulting in higher polarization of air electrode and capacity loss of zinc–air batteries because of formation of a resistive layer to ion/electron conduction. This will affect both primary and secondary zinc–air batteries [8]. Even though the main purpose of developing new separator material with low zincate crossover is to overcome the challenges associated with electrically rechargeable zinc–air batteries, in this study the PPO-based separators were used in a primary battery to have an indication of its applicability in secondary batteries, which will be the next stage of our work.

The prepared separator membranes were integrated into a homemade zinc–air battery with effective separator and cathode area of 1.77 cm^2. Figure 8 shows the polarization characteristics of pristine zinc plate in KOH, 7 M in a zinc–air cell using various separator membranes. The voltage and power of the cells show a strong dependency on the discharge current. Except for the initial sharp drop, all the cells showed a linear voltage decrease with an increase in the discharge current, revealing the dominance of ohmic losses on the cell performances [24]. Filter paper-based separator (FPS) was used as a benchmark separator. The intensity of ohmic losses for various separator membranes were in the order of FPS > PPO–TMA > PPO–MPY, revealing their ohmic resistance in the cell operation condition. The cells with FPS, PPO–TMA, and PPO–MPY showed maximum discharge current density of 89, 104, and 117 mA/cm^2 with a maximum power density of 55, 61, and 70 mW/cm^2, respectively. The comparison demonstrates lower ohmic loss for PPO-based membranes compared to FPS as a benchmark separator. It could be attributed to the formation of ionic channels in the structure of the membranes, facilitating hydroxide ions traveling through the separator.

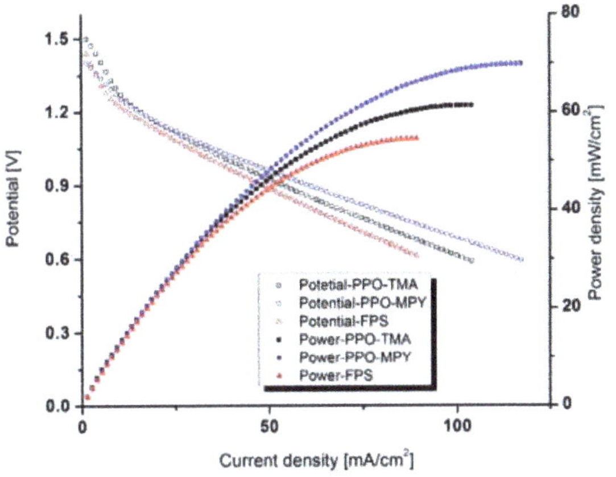

Figure 8. Polarization characteristics of the cells with various separators of PPO–MPY, PPO–TMA, and filter paper-based separator (FPS).

Galvanostatic discharge profiles of zinc–air cells assembled using PPO–TMA and PPO–MPY separator membranes at discharge current densities in the range of 2.5 to 15 mA/cm^2 are presented in Figure 9. For PPO–TMA (Figure 9a), the highest discharge capacity and power of 803 mAh/g_{Zn} and 932 mWh/g_{Zn} was obtained for discharge current of 15 mA/cm^2 at 0.9 V cut-off voltage. With decreasing the discharge current density, discharge time increased, and discharge capacity decreased to ~770 mAh/g_{Zn} for 2.5 mA/cm^2. This could be attributed to the higher hydrogen evolution and zinc corrosion reaction due to the elongated cell operation time [24]. However, because of having higher discharge voltage at lower discharge current densities (lower ohmic loss), the power increased to ~1015 mWh/g_{Zn} for 2.5 mA/cm^2 discharge current density. A similar trend was observed for the cell using PPO–MPY as a separator membrane, as could be seen in Figure 9b. The highest specific capacity of 795 mAh/g_{Zn} was achieved at a discharge current density of 15 mA/cm^2, decreasing to 772 mAh/g_{Zn} for 2.5 mA/cm^2. The power of the cell increased from 931 mWh/g_{Zn} for 15 mA/cm^2 to 996 mWh/g_{Zn} for 2.5 mA/cm^2 discharge current density. The cells assembled using both PPO-based separator membranes exhibited very high specific capacity and power [8].

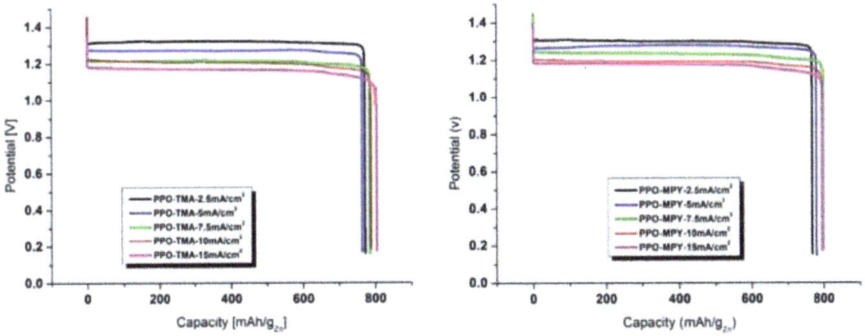

(a) (b)

Figure 9. Galvanostatic discharge profile of the cell with PPO–TMA (**a**) and PPO–MPY (**b**) as a separator membrane at discharge current densities of 2.5 to 15 mA/cm^2.

A comparison between the specific capacity of the cells assembled using PPO–TMA, PPO–MPY, and FPS at discharge current densities of 5 and 15 mA/cm^2 is shown in Figure 10. As could be seen, for 5 mA/cm^2, the capacity of the cells is in the same range of ~770 mAh/g_{Zn} for all three membranes. However, with increasing the discharge current density to 15 mA/cm^2, the capacity of the cells using PPO-based membranes increases to around 800 mAh/g_{Zn} due to shorter discharge time and so, lower hydrogen evolution reaction. Nonetheless, for FPS containing cell, the capacity decreased to ~730 mAh/g_{Zn}. It reveals the dominance of ohmic losses for this separator membrane in high discharge current densities. These results are in agreement with the polarization curve of PPO-based membranes and FPS, showing a lower ohmic loss for PPO-based separators.

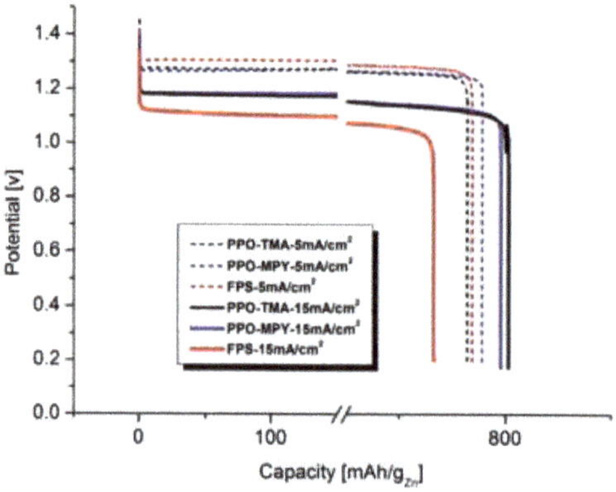

Figure 10. Galvanostatic discharge profile of the cells with PPO–TMA, PPO–MPY, and FPS as separator membrane at discharge current densities of 5 and 15 mA/cm^2.

3. Materials and Methods

3.1. Materials

Poly(2,6-dimethyl-1,4-phenylene oxide) powder (PPO), 1-methylpyrrolidine (≥98%), and manganese (IV) oxide (5 μm, 99.99%) were supplied by Sigma-Aldrich (St. Louis, MO, USA). *N*-bromosuccinimide (NBS) for synthesis, benzoyl peroxide (BPO, with 25% H_2O), and chlorobenzene were purchased from Merck Millipore. 1-methyl-2-pyrrolidinone (NMP), ethanol, methanol, and toluene all with Grade AR were purchased from QRëC (New Zealand). *N*,*N*-dimethylformamide (DMF) (LOBA Chemie, Grade AR, Mumbai, India), trimethylamine (TCI, ca. 25% in methanol), 1-methylimidazole (TCI, ≥99%), chloroform (BDH Chemicals, UK), and KOH plates (Kemaus, Australia) were used as received without further purification. Nickel (Ni) foam as cathode current collector with a purity of 99.97%, 100 pores per inch (PPI), and 1 mm thick was purchased from Qijing Trading Co., Ltd. (Wenzhou, China). Carbon black (Vulcan® XC-72, Cabot Corporation, Boston, MA, USA), BP-2000 (BLACK PEARLS® 2000, Cabot Corporation), and zinc plate (0.1 mm thick 99.99%, Shandong Yr Electronic Co., Ltd., Shandong, China) were used as received. Poly (styrene-co-butadiene) (Sigma-Aldrich, butadiene 4 wt%) was used to prepare binder for the cathode. Poly(vinyl acetate) (PVAc) from TOA Paint Public Co., Ltd. (Samut Prakarn, Thailand) and No. 4 Whatman filter paper (Sigma-Aldrich) were used to prepare the benchmark separator.

3.2. Separator Membrane Synthesis

PPO-based phase-separated hydroxide exchange separator membranes were synthesized using a process described previously [19]. The first stage of separator preparation was bromination of PPO at the benzylic position. A certain amount of PPO was dissolved in chlorobenzene at 50 °C under nitrogen atmosphere to prepare 5 wt/*v*% solution. Then, the temperature was increased to 80 °C, NBS (with PPO:NBS weight ratio of 1:0.8) and BPO (with NBS:BPO weight ratio of 1:0.05) were added to the mixture, and the reaction continued for 4 h under nitrogen atmosphere. The solution was then cooled down to room temperature and poured into an excessive amount of methanol for product precipitation. A pale-yellow fiber-like product was separated and then dissolved in chloroform for purification. The solution was precipitated again in an excessive amount of ethanol, filtered and dried in the vacuum oven at 60 °C overnight.

The next stage was to introduce quaternary ammonium functional group. Brominated PPO (BrPPO) was dissolved in NMP with the concentration of 7.5 wt/*v*%, and then an excessive amount of the quaternization agent (trimethylamine, 1-methylpyrrolidine, and 1-methylimidazole) was added to the solution. The reaction continued for 48 h in room temperature, and then the product was precipitated using toluene, washed a few times and then dried in a vacuum oven at 65 °C for 24 h [14].

In the next stage, a certain amount of quaternized BrPPO (Q-BrPPO) was dissolved in DMF at room temperature to prepare 30 wt/*v*% casting solution. The membrane was fabricated by casting the prepared solution onto a clean surface. Finally, the prepared cast films were soaked in degassed KOH, 7 M solution for 72 h to exchange the bromine ions in the polymer structure with hydroxide. Fresh hydroxide exchange solutions were used every 24 h to make sure of complete ion exchange [15].

To evaluate the performance of the synthesized separator membranes, a benchmark membrane using Whatman filter paper was also prepared. For this purpose, both sides of No.4 Whatman filter paper was coated with a 24 wt% PVAc solution and dried in an oven at 60 °C for 15 min.

3.3. Structural/Physicochemical Characterization

The ^{1}HNMR spectrum of BrPPO was obtained on a Bruker, Avance III HD 500 MHz using deuterated trichloromethane ($CDCl_3$) as a solvent to determine the bromination degree of PPO. A Perkin Elmer, Spectrum One Fourier-transform infrared (FTIR) spectrometer in the frequency range of 4000 to 600 cm^{-1} was carried out to study the chemical changes in the molecular structure during various synthesis stages. Thermogravimetric analysis was performed using SDT-Q600 TGA instrument in the temperature range of 30 to 600 °C at a heating rate of 10 °C min^{-1} to evaluate the thermal stability of the samples.

To measure water/electrolyte uptake capacity of the membranes, they were soaked in water/KOH, 7 M solution for 24 h and the weight differences before and after soaking were used for the measurements using Equation 1 [8].

$$\Delta W\ (\%) = [(W_{wet} - W_{dry})/W_{dry}] \times 100 \quad (1)$$

where W_{wet} and W_{dry} are the weights of the membranes after and before soaking in water/electrolyte, respectively. Similarly, dimensional changes of the membranes induced by water/electrolyte uptake were calculated by Equations 2 and 3 [8].

$$\Delta A\ \text{(area-based)}\ (\%) = [(A_{wet} - A_{dry})/A_{dry}] \times 100 \quad (2)$$

$$\Delta V\ \text{(volume-based)}\ (\%) = [(V_{wet} - V_{dry})/A_{dry}] \times 100 \quad (3)$$

3.4. Electrochemical Characterization

The ionic conductivity of the prepared membranes was measured using a potentiostat/galvanostat with impedance measurement unit (AMETEK, PAR VersaSTAT 3A) in the frequency range of 1 Hz–100 kHz with the excitation voltage of 10 mV$_{RMS}$ at room temperature. For this measurement, a diffusion cell with two chambers was used. EIS measurement with and without the membrane placed between two chambers containing KOH, 7 M was performed, and the difference in the bulk resistance (R_b) of the two measurements was used to calculate the ionic conductivity of the membranes using Equation 4 [25].

$$\sigma = l/R_b \cdot A \quad (4)$$

σ is the ionic conductivity (S/cm), R_b is the bulk resistance (Ω), and l and A are thickness (cm) and area (cm^2) of the membrane, respectively.

Cyclic voltammetry (CV) was carried out using an AMETEK, PAR VersaSTAT 3A potentiostat/galvanostat to evaluate electrochemical stability window of the membranes. Two-electrode configuration tests using 1 × 1 cm^2 platinum (Pt) working and counter electrodes were carried out for CV with a scan rate of 0.05 mV/s.

To study zincate ion ($Zn(OH)_4^{2-}$) crossover characteristics of the membranes, a kind of diffusion cell with two chambers was used [9]. The left chamber contained 50 mL of KOH, 7 M solution plus 0.5 M dissolved ZnO in the form of zincate ions while the right chamber contained only 50 mL of KOH, 7 M solution (Figure 11). The separator membrane was placed between two chambers. The chambers were stirred continuously to prevent concentration polarization. The concentration of zinc ions in the right chamber was measured in a predetermined time interval (12–14 h) using inductively coupled plasma optical emission spectroscopy (ICP-OES) to obtain time-dependent concentration variation graph. Moreover, the diffusion coefficient of zincate ions across the separator membranes was calculated from the experimental data using Equation 5 [9].

$$\ln(C_A/(C_A - C_B)) = (D \cdot A/V_B \cdot L) \cdot t \quad (5)$$

D is the diffusion coefficient of zincate ions across the separator membrane (cm^2/min), t is the time (min), V_B is the solution volume in the right chamber (deficiency chamber), A is the effective surface area (cm^2) of the separator, L is the thickness (cm) of the separator, and C_A and C_B are the concentration of zincate ions (mol/L) in the left and right chambers, respectively.

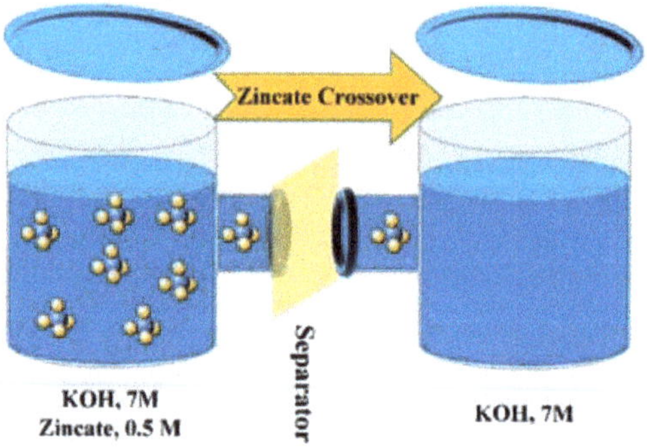

Figure 11. Schematic view of the diffusion cell used to measure zincate crossover.

3.5. Discharge Performance

A homemade zinc–air cell was used to evaluate the performance of the separator membranes. In this cell, the separator was in direct contact with the cathode, and the anode was a 1 × 1 cm^2 pure zinc plate immersed in 40 mL of KOH, 7 M solution (Figure 12). To prepare the cathode, a Ni-foam was used as the current collector and gas diffusion layer. One side of the foam was coated with a mixture of BP-2000 (30%) and PTFE (70%), dispersed in ethanol and pressed using a hot-press at 350 °C for 15 min (air diffusion side). For preparing the catalyst side of the foam, a mixture of MnO_2 (1.2 g: 30%), BP-2000 (1.4 g: 35%) and VXC-72 (1.4 g: 35%) was used. The mixture was stirred in 35 mL toluene for 2 h, and then 5 mL of 7.5 wt% poly(styrene-co-butadiene) solution in toluene (as a binder) was added and stirred for another 2 h. The final mixture was coated onto the Ni foam and pressed using a manual hot-press at 150 °C for 10 min. The size of the circular cathode used in the battery tests was 15 mm in the diameter.

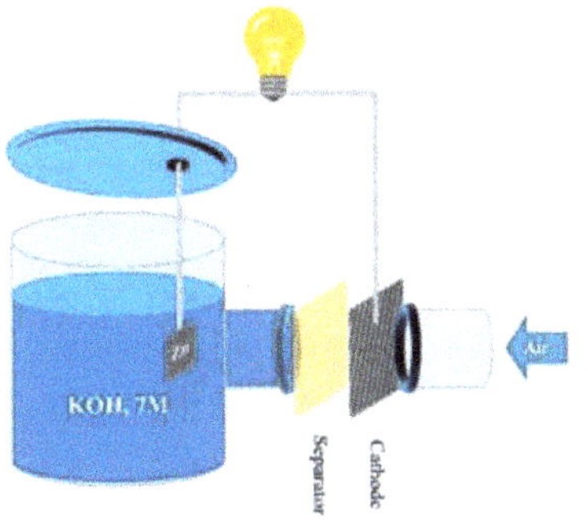

Figure 12. Schematic view of the discharge zinc–air cell.

Discharge performance was measured using a Battery Testing System (NEWARE, Shenzhen, China) at room temperature. The cell was discharged at a constant discharge current in the range of 2.5 to 15 mA/cm^2. For all experiments, the cut-off voltage was 0.9 V.

4. Conclusions

Three PPO-based hydroxide exchange separator membranes, containing TMA, MPY, and MIM as quaternization agents, were developed. PPO–TMA and PPO–MPY exhibited excellent characteristics, required for rechargeable zinc–air batteries. They offered a good ionic conductivity of ~0.17 mS/cm along with very low zincate diffusion coefficient of 1.13×10^{-8} and 0.28×10^{-8} cm^2/min for PPO–TMA and PPO–MPY, respectively. Besides, their excellent chemical and thermal stability, and wide electrochemical stability window of higher than 3 V make them a suitable candidate separator for the batteries. Polarization characteristics of the batteries, using these membranes, showed improved discharge current density, high discharge capacity, and high discharge power. The results concluded that PPO–TMA and PPO–MPY significantly enhanced the performances of the batteries. Also, they represent a promising candidate separator for rechargeable zinc–air batteries.

Author Contributions: Conceptualization, A.A. and S.K.; methodology, A.A., A.S., A.A.M. and S.K.; investigation: A.A. and S.H.; formal analysis, A.A.; writing—original draft preparation, A.A.; writing—review and editing, S.K.; supervision, S.K.; funding acquisition, S.K.; project administration, S.K.

Funding: The Thailand Research Fund (RSA6180008) and Rachadapisek Sompote Fund, Chulalongkorn University are acknowledged.

Acknowledgments: A.A. and S.H. thank the support from Rachadapisek Sompote Fund for Postdoctoral Fellowship, Chulalongkorn University.

Conflicts of Interest: The authors declare no conflicts of interest.

Abbreviations

+FPS	Filter paper-bases separator
MIM	1-Methylimidazolium
MPY	1-Methylpyrolinine
TMA	Trimethylamine
PPO	Poly (2,6-dimethyl-1,4-phenylene oxide)
PPO–MIM	Quaternized PPO using 1-methylimidazolium
PPO–MPY	Quaternized PPO using 1-methylpyrolinine
PPO–TMA	Quaternized PPO using trimethylamine

References

1. Hosseini, S.; Han, S.J.; Arponwichanop, A.; Yonezawa, T.; Kheawhom, S. Ethanol as an electrolyte additive for alkaline zinc-air flow batteries. *Sci. Rep.* **2018**, *8*, 11273. [CrossRef] [PubMed]
2. Lao-atiman, W.; Bumroongsil, K.; Arpornwichanop, A.; Bumroongsakulsawat, P.; Olaru, S.; Kheawhom, S. Model-Based Analysis of an Integrated Zinc-Air Flow Battery/Zinc Electrolyzer System. *Front. Energy Res.* **2019**, *7*, 15. [CrossRef]
3. Liu, S.; Han, W.; Cui, B.; Liu, X.; Sun, H.; Zhang, J.; Lefler, M.; Licht, S. Rechargeable Zinc Air Batteries and Highly Improved Performance through Potassium Hydroxide Addition to the Molten Carbonate Eutectic Electrolyte. *J. Electrochem. Soc.* **2018**, *165*, A149–A154. [CrossRef]
4. Fang, G.; Zhou, J.; Pan, A.; Liang, S. Recent Advances in Aqueous Zinc-Ion Batteries. *Acs Energy Lett.* **2018**, *3*, 2480–2501. [CrossRef]
5. Li, Y.; Dai, H. Recent advances in zinc-air batteries. *Chem. Soc. Rev.* **2014**, *43*, 5257–5275. [CrossRef]
6. Fu, J. *Material Design and Engineering for Polymer Electrolyte Membrane Zinc-Air Batteries*; UWSpace: Ontario, Canada, 2018.
7. Dewi, E.L.; Oyaizu, K.; Nishide, H.; Tsuchida, E. Cationic polysulfonium membrane as separator in zinc–air cell. *J. Power Sources* **2003**, *115*, 149–152. [CrossRef]
8. Lee, H.-J.; Lim, J.-M.; Kim, H.-W.; Jeong, S.-H.; Eom, S.-W.; Hong, Y.T.; Lee, S.-Y. Electrospun polyetherimide nanofiber mat-reinforced, permselective polyvinyl alcohol composite separator membranes: A membrane-driven step closer toward rechargeable zinc–air batteries. *J. Membr. Sci.* **2016**, *499*, 526–537. [CrossRef]
9. Kim, H.-W.; Lim, J.-M.; Lee, H.-J.; Eom, S.-W.; Hong, Y.T.; Lee, S.-Y. Artificially engineered, bicontinuous anion-conducting/-repelling polymeric phases as a selective ion transport channel for rechargeable zinc-air battery separator membranes. *J. Mater. Chem. A* **2016**, *4*, 3711–3720. [CrossRef]
10. Pan, J.; Chen, C.; Li, Y.; Wang, L.; Tan, L.; Li, G.; Tang, X.; Xiao, L.; Lu, J.; Zhuang, L. Constructing ionic highway in alkaline polymer electrolytes. *Energy Environ. Sci.* **2014**, *7*, 354–360. [CrossRef]
11. Zhu, L.; Pan, J.; Wang, Y.; Han, J.; Zhuang, L.; Hickner, M.A. Multication Side Chain Anion Exchange Membranes. *Macromolecules* **2016**, *49*, 815–824. [CrossRef]
12. Zhou, J.; Zuo, P.; Liu, Y.; Yang, Z.; Xu, T. Ion exchange membranes from poly(2,6-dimethyl-1,4-phenylene oxide) and related applications. *Sci. China Chem.* **2018**, *61*, 1062–1087. [CrossRef]
13. Li, N.; Yan, T.; Li, Z.; Thurn-Albrecht, T.; Binder, W.H. Comb-shaped polymers to enhance hydroxide transport in anion exchange membranes. *Energy Environ. Sci.* **2012**, *5*, 7888–7892. [CrossRef]
14. Yang, Y.; Xu, Y.; Ye, N.; Zhang, D.; Yang, J.; He, R. Alkali Resistant Anion Exchange Membranes Based on Saturated Heterocyclic Quaternary Ammonium Cations Functionalized Poly(2,6-dimethyl-1,4-phenylene oxide)s. *J. Electrochem. Soc.* **2018**, *165*, F350–F356. [CrossRef]
15. Varcoe, J.R.; Atanassov, P.; Dekel, D.R.; Herring, A.M.; Hickner, M.A.; Kohl, P.A.; Kucernak, A.R.; Mustain, W.E.; Nijmeijer, K.; Scott, K.; et al. Anion-exchange membranes in electrochemical energy systems. *Energy Environ. Sci.* **2014**, *7*, 3135–3191. [CrossRef]
16. Msomi, P.F.; Nonjola, P.; Ndungu, P.G.; Ramonjta, J. Quaternized poly(2.6 dimethyl-1.4 phenylene oxide)/polysulfone blend composite membrane doped with ZnO-nanoparticles for alkaline fuel cells. *J. Appl. Polym. Sci.* **2018**, *135*, 45959. [CrossRef]
17. Tongwen, X.; Weihua, Y. Fundamental studies of a new series of anion exchange membranes: Membrane preparation and characterization. *J. Membr. Sci.* **2001**, *190*, 159–166. [CrossRef]
18. Zhu, L.; Zimudzi, T.J.; Wang, Y.; Yu, X.; Pan, J.; Han, J.; Kushner, D.I.; Zhuang, L.; Hickner, M.A. Mechanically Robust Anion Exchange Membranes via Long Hydrophilic Cross-Linkers. *Macromolecules* **2017**, *50*, 2329–2337. [CrossRef]
19. Yang, J.; Liu, C.; Hao, Y.; He, X.; He, R. Preparation and investigation of various imidazolium-functionalized poly(2,6-dimethyl-1,4-phenylene oxide) anion exchange membranes. *Electrochim. Acta* **2016**, *207*, 112–119. [CrossRef]
20. Dang, H.-S.; Jannasch, P. A comparative study of anion-exchange membranes tethered with different hetero-cycloaliphatic quaternary ammonium hydroxides. *J. Mater. Chem. A* **2017**, *5*, 21965–21978. [CrossRef]

21. Si, Z.; Sun, Z.; Gu, F.; Qiu, L.; Yan, F. Alkaline stable imidazolium-based ionomers containing poly(arylene ether sulfone) side chains for alkaline anion exchange membranes. *J. Mater. Chem. A* **2014**, *2*, 4413–4421. [CrossRef]
22. Yang, C.-C.; Lin, S.-J. Alkaline composite PEO–PVA–glass-fibre-mat polymer electrolyte for Zn–air battery. *J. Power Sources* **2002**, *112*, 497–503. [CrossRef]
23. Vassal, N.; Salmon, E.; Fauvarque, J.F. Electrochemical properties of an alkaline solid polymer electrolyte based on P(ECH-co-EO). *Electrochim. Acta* **2000**, *45*, 1527–1532. [CrossRef]
24. Hosseini, S.; Lao-atiman, W.; Han, S.J.; Arpornwichanop, A.; Yonezawa, T.; Kheawhom, S. Discharge Performance of Zinc-Air Flow Batteries Under the Effects of Sodium Dodecyl Sulfate and Pluronic F. *Sci. Rep.* **2018**, *8*, 14909. [CrossRef] [PubMed]
25. Wu, G.M.; Lin, S.J.; You, J.H.; Yang, C.C. Study of high-anionic conducting sulfonated microporous membranes for zinc-air electrochemical cells. *Mater. Chem. Phys.* **2008**, *112*, 798–804. [CrossRef]

International Journal of
Molecular Sciences

MDPI

Article

Highly Conductive and Water-Swelling Resistant Anion Exchange Membrane for Alkaline Fuel Cells

Qianqian Ge [1], Xiang Zhu [1] and Zhengjin Yang [2,*]

[1] Polymer Composites Group, School of Chemistry & Chemical Engineering, Anhui University, Hefei 230601, China

[2] CAS Key Laboratory of Soft Matter Chemistry, Collaborative Innovation Center of Chemistry for Energy Materials, School of Chemistry and Materials Science, University of Science and Technology of China, Hefei 230026, China

* Correspondence: yangzj09@ustc.edu.cn; Tel.: +86-551-6360-1587

Received: 12 June 2019; Accepted: 12 July 2019; Published: 15 July 2019

Abstract: To ameliorate the trade-off effect between ionic conductivity and water swelling of anion exchange membranes (AEMs), a crosslinked, hyperbranched membrane (C-HBM) combining the advantages of densely functionalization architecture and crosslinking structure was fabricated by the quaternization of the hyperbranched poly(4-vinylbenzyl chloride) (HB-PVBC) with a multiamine oligomer poly(*N*,*N*-Dimethylbenzylamine). The membrane displayed well-developed microphase separation morphology, as confirmed by small angle X-ray scattering (SAXS) and transmission electron microscopy (TEM). Moreover, the corresponding high ionic conductivity, strongly depressed water swelling, high thermal stability, and acceptable alkaline stability were achieved. Of special note is the much higher ratio of hydroxide conductivity to water swelling (33.0) than that of most published side-chain type, block, and densely functionalized AEMs, implying its higher potential for application in fuel cells.

Keywords: hyperbranched polymer; crosslinking; alkaline fuel cells; ionic conductivity; water swelling

1. Introduction

Polymer electrolyte membrane fuel cells recently appeared as one of the most promising energy-conversion devices owing to their simplified operation, higher power density, and easier maintenance over conventional fuel cells with liquid solution as electrolyte [1,2]. The commonly used solid polymer electrolyte is a perfluorinated sulfonic acid-based membrane known as Nafion, which shows high proton conductivity, excellent mechanical properties, and good chemical stability [3,4]. Despite the extraordinary performance of proton exchange membrane fuel cells (PEMFCs) assembled with Nafion, the strong acidic conditions restrict the utilization of highly stable catalysts, for example, platinum or platinum-containing metal alloys [5,6]. To improve these deficiencies while offering a nice alternative to acidic systems, alkaline fuel cells (AFCs) operating at high pH, permit the usage of non-platinum catalysts [5,7,8] such as silver, cobalt, or nickel and, therefore, has received considerable attention over the past few decades. Additionally, AFCs also exhibit distinct advantages over PEMFCs in terms of faster oxygen reduction kinetics and lower crossover of fuels owing to the opposite direction of electroosmotic drag [5,9,10]. However, exploring the alkaline electrolyte, that is, anion exchange membranes (AEMs), which is the key component acting as a separator between oxidant and fuel chambers and a conductor of hydroxide ions [11–13], with high conductivity, lower swelling, improved mechanical, and chemical stability compared with Nafion, is an important technical challenge.

High conductivity, which reflects the transporting efficiency of anions, is considered as the fundamental performance indicator for AEMs. However, the intrinsic lower mobility of OH^- and the less-developed microphase-separated morphology of aromatic polymers compared with perfluorinated structures naturally lead to lower ionic conductivity in AEMs compared with well-known proton exchange membranes (PEMs). To explore alternative AEMs with high ionic conductivity, a significant amount of research [14,15] has been carried out. Among which, increasing ion exchange capacity (IEC) values of ionomers is the most common and convenient strategy. However, high ionic conductivity obtained in this way is always at the expense of severe water uptake and a concomitant decline in the mechanical strength, especially at elevated temperatures [16,17]. Consequently, IEC values are typically restricted to be at a moderate level. Another reasonable route is incorporating functional groups with stronger basicity such as quaternary guanidinium groups [18] and quaternary 1,8-Diazabicyclo[5.4.0]undec-7-ene [19], aiming to enhance the dissociation ability and the simultaneous increase in ionic conductivity. However, limited success in increasing conductivity was achieved in this way.

Notably, the subsequent investigation indicates that regulating the configuration of polymer backbones and the arrangement of ionic functional groups are effective strategies for improving ionic conductivity [5,20,21]. With respect to the configuration of polymer backbones, block copolymers are found to have well-defined hydrophilic–hydrophobic phase separated morphology [3,17,22–24]. Additionally, it is verified that positioning the functional groups on side chains rather than backbones can largely enhance their mobility to aggregate into ionic clusters and promote the microphase separation [9,21,25,26]. Moreover, decreasing the distance between the functional groups, that is, densely functionalization based on the block or side-chain type architecture, can further promote microphase separation and enhance the ionic conductivity [1,16,23,27–29]. However, most traditional densely functionalization strategies involved preparing multi-cation precursors via multi-steps, complicated synthesis, and purification procedures, which adds to the production cost and complexity. Our previous work reported the synthesis of hyperbranched oligomer poly-4-vinylbenzyl chloride (HB-PVBC) via one-step atom transfer radical polymerization (ATRP) reaction [30], which was a defining moment in hyperbranched AEM synthesis. The quaternization of HB-PVBC is expected to yield AEMs with ionic functional groups densely arranged. This specific structure is anticipated to provide a readily and effective strategy for preparing densely functionalized AEM with clear microphase separated morphology and high conductivity.

Though great progress has been acquired for increasing ionic conductivity, maintaining depressed water swelling at the same time remains challenging, especially at elevated temperatures [7]. To get over this dilemma between ionic conductivity and water swelling, significant research attention has been rendered and it has been approved that crosslinking is an effective strategy to obtain mechanically robust AEMs [2,4,29–33]. For instance, the crosslinking membrane CBQAPPO-3 exhibits the highest hydroxide conductivity of 33 mS cm^{-1} at room temperature and, simultaneously, benefitting from the crosslinking structure, much lower water uptake of 47.0% and linear swelling ratio (LSR) of 12.3% in comparison with that of the uncrosslinked counterpart BQAPPO-0.23 (water uptake = 120.4% and LSR = 24.1%) were acquired, leading to improved mechanical strength from 6.55 MPa to around 23 MPa [29,34]. More recently, cross-linking was also employed to toughen AEMs. As expected, the crosslinked membrane exhibits the significant improvement in water uptake (less than 7%), LSR (8%–5%, much smaller than uncrosslinked membranes) and tensile strength (higher than 18 MPa) [33]. The results showed that the crosslinking structure provides an efficient and convenient route for preparing membranes with much lower water swelling and the corresponding improved mechanical property.

Thus, we wonder whether the combination of high cation density and crosslinking could lead to AEMs with high ionic conductivity and depressed water swelling. To verify this hypothesis, we designed and fabricated crosslinked, hyperbranched AEMs (defined as C-HBM) by crosslinking the hyperbranched oligomer HB-PVBC with a multiamine poly (*N*,*N*-Dimethylbenzylamine), aiming to combine the advantages of the densely functionalization architecture and crosslinking. The properties of the membrane such as microphase separated morphology, water uptake, LSR, ionic conductivity, alkaline stability, and thermal properties were thoroughly investigated.

2. Results and Discussion

2.1. Synthesis and Characterization

For the synthesis of *N,N*-Dimethylbenzylamine (Scheme 1a), 4-vinylbenzyl chloride was first dropped into an excessive amount of dimethyl amine aqueous solution to alleviate the formation of quaternary ammonium salt by-product. The mixture is then extracted and vacuum distilled to produce purified *N,N*-Dimethylbenzylamine. The purity and chemical structure is identified by proton nuclear magnetic resonance (^{1}H NMR) analysis, as shown in Figure 1a. The expected chemical shifts and intensities for *N,N*-Dimethylbenzylamine were observed accordingly. The appearance of the peak at 2.2 ppm arising from methyl groups along with the shift of benzyl methylene groups from 4.3 ppm to 3.4 ppm indicates that the benzylic chloride groups has been successfully aminated. ^{1}H NMR (400 MHz, $CDCl_3$) δ 7.37 (d, *J* = 8.1 Hz, 2H), 7.26 (d, *J* = 8.1 Hz, 2H), 6.71 (dd, *J* = 17.6, 10.9 Hz, 1H), 5.73 (dd, *J* = 17.6, 0.9 Hz, 1H), 5.22 (dd, *J* = 10.9, 0.9 Hz, 1H), 3.41 (s, 2H), 2.23 (s, 6H).

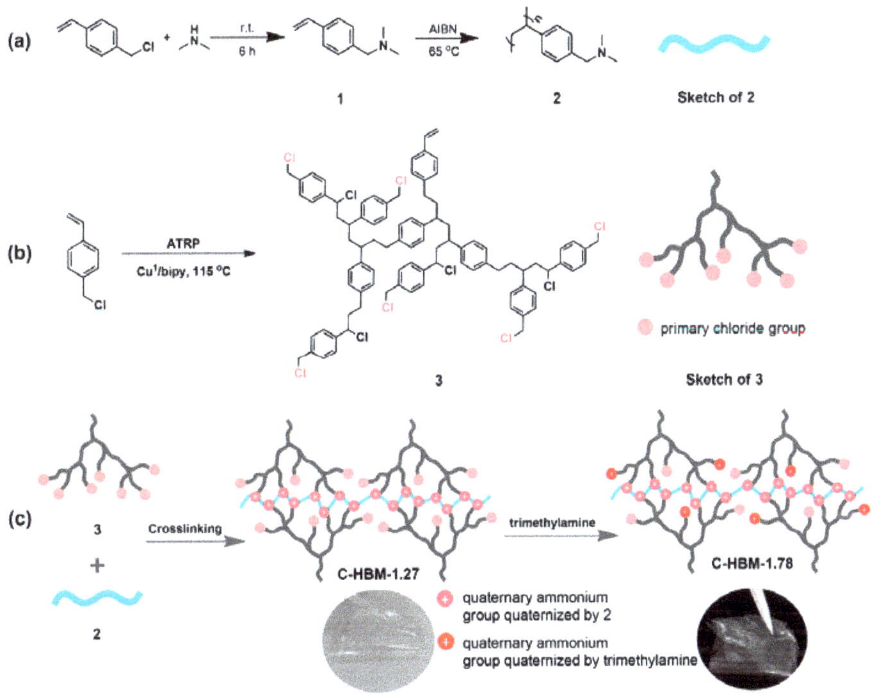

Scheme 1. Schematic synthesis of the quaternization reagent poly(*N,N*-Dimethylbenzylamine), the hyperbranched oligomer poly-4-vinylbenzyl chloride (HB-PVBC) and the crosslinked, hyperbranched membrane C-HBM-1.78 (ion exchange capacity (IEC)=1.78 g^{-1}). (**a**) Synthesis of the monomer 1 (*N,N*-Dimethylbenzylamine) through the amination of 4-vinylbenzyl chloride with dimethyl amine, and polyamine 2 (poly(*N,N*-Dimethylbenzylamine)) via radical polymerization; (**b**) synthesis of the hyperbranched oligomer HB-PVBC via atom transfer radical polymerization (ATRP) technique; and (**c**) the quaternization of HB-PVBC with poly(*N,N*-Dimethylbenzylamine) to yield densely functionalized ionomer C-HBM-1.27; further quaternization of C-HBM-1.27 with trimethylamine leads to membrane C-HBM-1.78 with a higher IEC value. AIBN, 2,2′-azodiisobutyronitrile.

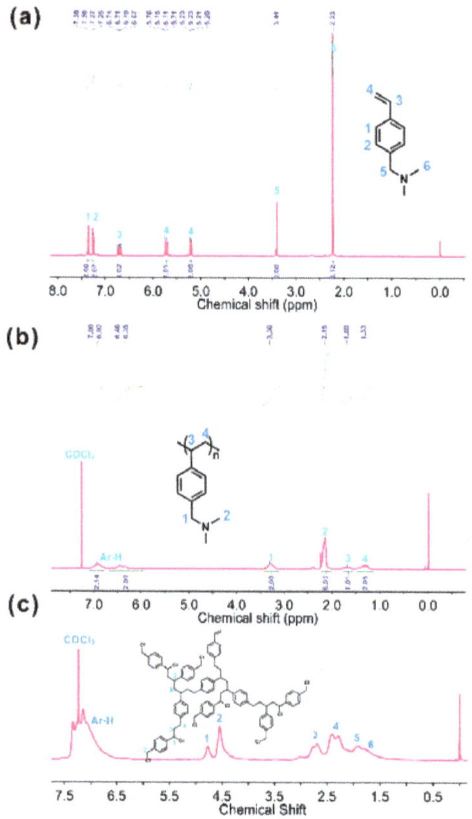

Figure 1. Proton nuclear magnetic resonance (^{1}H NMR) spectra of (**a**) the monomer *N*,*N*-Dimethylbenzylamine, (**b**) polyamine poly(*N*,*N*-Dimethylbenzylamine), and (**c**) the hyperbranched oligomer HB-PVBC.

The purified *N*,*N*-Dimethylbenzylamine was subjected to radical polymerization initiated by 2,2′-azodiisobutyronitrile (AIBN) to produce the polyamine poly(*N*,*N*-Dimethylbenzylamine) (Scheme 1a). Gel permeation chromatography (GPC) and ^{1}H NMR techniques were used to monitor the structure. As shown in Figure 1b, the disappearance of vinyl groups at 5.2 ppm, 5.7 ppm, and 6.7 ppm within *N*,*N*-Dimethylbenzylamine in combination with the appearance of new peaks at 1.0–2.5 ppm assigned to methylene and methylidyne protons demonstrate the successful polymerization of *N*,*N*-Dimethylbenzylamine. ^{1}H NMR (400 MHz, $CDCl_3$) δ 6.96 (d, *J* = 30.0 Hz, 2H), 6.40 (d, *J* = 45.8 Hz, 2H), 3.30 (s, 2H), 2.15 (s, 6H), 1.69 (s, 1H), 1.33 (s, 2H). The molecular weight and polydispersity of poly(*N*,*N*-Dimethylbenzylamine) determined by GPC are 1316 g mol^{-1} and 1.18, respectively (Figure 2a).

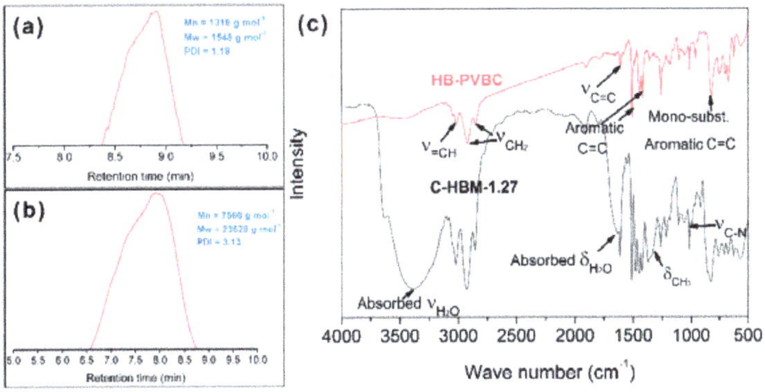

Figure 2. Gel permeation chromatography (GPC) traces of polyamine poly(*N,N*-Dimethylbenzylamine) (**a**) and the hyperbranched oligomer HB-PVBC (**b**); IR spectra of oligomer HB-PVBC and membrane C-HBM-1.27 (**c**). PDI, polydispersity.

The hyperbranched HB-PVBC was synthesized according to our previous procedure [30] (Scheme 1b) and its chemical structure was verified by ^{1}H NMR (Figure 1c). As estimated by GPC shown in Figure 2b, the molecular weight of HB-PVBC is 7560 g mol^{-1}. As shown in Scheme 1b, all the primary chloride groups are densely distributed surrounding the oligomer HB-PVBC. The quaternization of primary chloride groups with tertiary amine is expected to endow the resulting ionomer with densely distributed functional groups. Herein, the polyamine poly(*N,N*-Dimethylbenzylamine) is utilized as the quaternization agent to yield densely functionalized AEM, thereby promoting the aggregation of ionic domains to form interconnected hydrophilic channels. Figure 2c shows the reflectance Fourier transform infrared (FTIR) spectra of the oligomer HB-PVBC and membrane C HBM-1.27 (the titrated IEC value of 1.27 mmol g^{-1}). The peak at around 3400 cm^{-1} arising from the stretching vibration of absorbed water molecules of quaternary ammonium salt groups is detected, confirming the successful implementation of quaternization and crosslinking. The titrated IEC value of 1.27 mmol g^{-1} further confirmed the successful functionalization of HB-PVBC. The membrane C-HBM-1.27 at fully hydrated state displayed tensile strength (TS) of 13.37 MPa and elongation at break (Eb) of 3.31% tested on dynamic mechanical analyzer. Immersing the membrane in trimethylamine aqueous solution at room temperature for 24 h can further quaternize the residual primary chloride groups, leading to an increased IEC of 1.78 mmol g^{-1}. The yield membrane shows similar mechanical properties (TS = 12.37 MPa, Eb = 5.20%), and is denoted as C-HBM-1.78. Obviously, the mechanical property is inferior to most published cross-linking membranes [4,31,34], which is probably ascribed to the rigid structure of HB-PVBC and the multiaimne poly(*N,N*-Dimethylbenzylamine). As higher ionic content, that is, IEC value, generally leads to higher ionic conductivity, membrane C-HBM-1.78 is chosen for thorough investigation in the following section.

2.2. Membrane Morphology

Transmission electron microscopy (TEM) was carried out for membrane C-HBM-1.78 stained with iodine ions. As shown in Figure 3a, the dark areas represent the soft hydrophilic regions mainly composed of water and quaternary ammonium clusters, while the brighter ones correspond to hard hydrophobic matrix mainly composed of aromatic rings [8]. Well-developed microphase separation morphology with interconnected ionic channels is observed for C-HBM-1.78 throughout the view. Further observation regarding atomic force microscopy (AFM, Figure 3b) confirms the formation of well-defined microphase separated morphology.

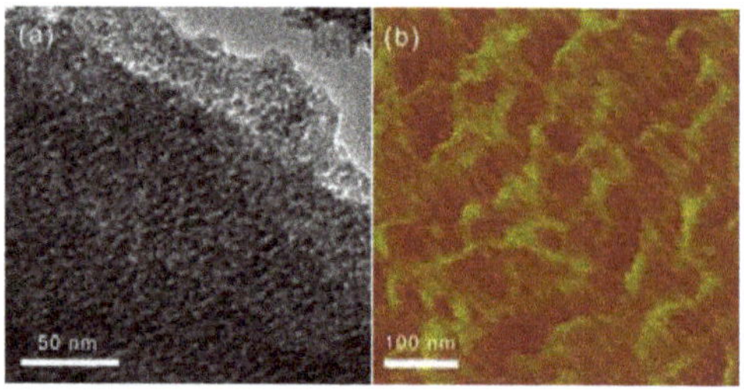

Figure 3. Microphase separated morphology images regarding (**a**) atomic force microscopy (AFM) and (**b**) transmission electron microscopy (TEM) of membrane C-HBM-1.78.

Quantitative information on the size of the ionic clusters was obtained from the small-angle X-ray scattering (SAXS) results (Figure 4). A distinct peak at 0.8 nm^{-1} (corresponding to an interdomain Bragg spacing of 7.9 nm) [23] emerged in the scattering profile, indicating the aggregation of ionic clusters in the densely functionalized AEM C-HBM-1.78. The image observed in TEM and AFM in combination with the SAXS profile demonstrated that the well-established ionic channels were formed within membrane C-HBM-1.78. This outstanding morphology is probably attributed to the specific structure of hyperbranched ionomer with densely distributed quaternary ammonium groups, which is believed to promote the self-assemble of quaternary ammonium groups to aggregate into ionic domains by narrowing the distance between quaternary ammonium groups during the membrane forming process. The formation of interconnected hydrophilic channels can thus be facilitated, which is beneficial for building effective ion conduction pathways, and guaranteeing high ion conductivity for membrane C-HBM-1.78. Despite that characteristic hydrophilic–hydrophobic phase separation is observed, it is apparent from the TEM and the SAXS data that no long-range, regular order of the ionic phase exists in membrane C-HBM-1.78 [35].

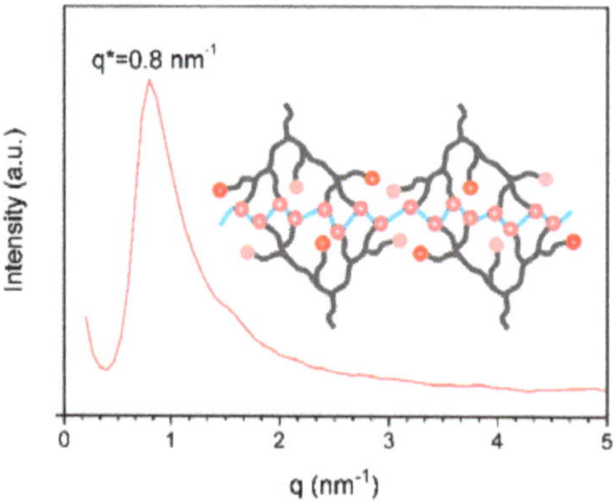

Figure 4. Small-angle X-ray scattering (SAXS) profile for membrane C-HBM-1.78 measured at room temperature under dry state (the insert is the schematic chemical structure of C-HBM-1.78).

2.3. Ionic Conductivity and Water-Swelling Resistance Property

For a practical fuel cell application, ionic conductivity and water uptake is of particular importance. An ideal AEM should have high ionic conductivity and depressed water adsorption. The ionic conductivity for membrane C-HBM-1.27 and C-HBM-1.78 as a function of temperature is displayed in Figure 5a. Owing to the high sensitivity of hydroxide conductivity in the atmosphere, the chloride conductivity is measured and plotted as well. As expected, both the hydroxide conductivity and chloride conductivity increases with increasing temperature as a result of the enhanced dissociation efficiency and bigger water retention capacity. Higher ion content, that is, IEC value, indeed leads to higher conductivity. Specifically, the chloride conductivity for C-HBM-1.27 at 30 °C is 25.6 mS cm^{-1}, while it rises to 39.7 mS cm^{-1} at 60 °C. For membrane C-HBM-1.78, higher chloride conductivity up to 34.2 mS cm^{-1} at 30 °C and 53.9 mS cm^{-1} at 60 °C is achieved.

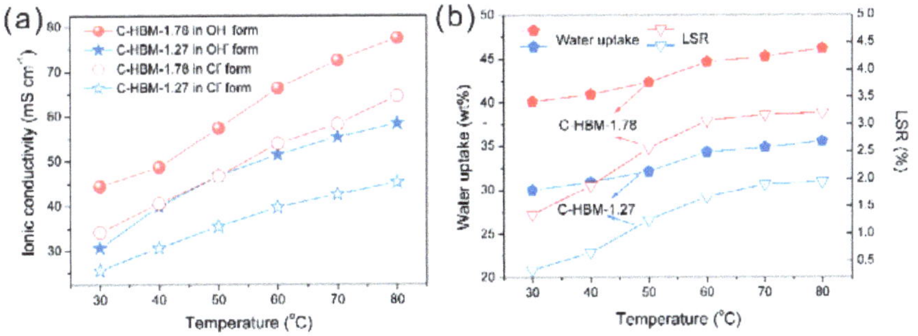

Figure 5. The ionic conductivity (**a**) and water uptake and linear swelling ratio (LSR) (**b**) as a function of temperature of membrane C-HBM-1.27 and C-HBM-1.78.

For better comparison, the hydroxide conductivity normalized on the basis of IEC (σ_{OH-}/IEC) is put forward, which indicates the efficiency of quaternary ammonium groups in the membrane for transporting hydroxide ions. As shown in Figure 6, membrane C-HBM-1.78 displayed higher σ_{OH-}/IEC of 25 than that of most published AEMs, including the side-chain-type [36–38], crosslinked [4,33,39], block [10,17,40], and densely functionalized AEMs [16,28,41], indicating its more efficient utilization of functional groups. This anion-conducting property should benefit from the well-developed microphase separated morphology along with interconnected hydrophilic channels, which can provide a smooth pathway for ion transportation. The high ionic conductivity is thus achieved, implying greater potential for applications in AFCs. However, as we all know, the high ion conductivity is strongly dependent on high water contents.

Introducing the cross-linking structure into AEMs is proven to be an effective method of restraining the water swelling. The water adsorption capacity and water swelling ratio are measured at the temperature range of 30–80 °C. As plotted in Figure 5b, both C-HBM-1.27 and C-HBM-1.78 showed increased water uptake with increasing temperature. Furthermore, higher water uptake is observed for membrane C-HBM-1.78 than that of C-HBM-1.27 owing to the higher ion content. For C-HBM-1.78, an acceptable water uptake of 46.1% and extremely low LSR of 3.2% at 80 °C was obtained. The excellent water-swelling resistance undoubtedly originated from the crosslinking structure and well established microphase separated morphology, which was confirmed by TEM, AFM, and SAXS. To highlight the outstanding advantages of the ionomer structure combining dense functionalization and crosslinking, the hydroxide conductivity and water-swelling resistance property at 30 °C of membrane C-HBM-1.78 are listed in Table 1 and compared with reported AEMs in Figure 6. Apparently, the σ_{OH-}/water uptake for our membrane Cr-M-1.78 of 1.1 is much higher than most AEMs with the side-chain-type (0.5–0.7) [36–38], crosslinking (0.1–0.4) [4,33,39], and block (0.2–0.6) [10,17,40] architectures, indicating its higher utilization efficiency of water molecules. An outstanding σ_{OH-}/LSR value of 33.0 for membrane C-HBM-1.78 is also obtained, which is around fifteen-fold higher than the side-chain-type (2.0) [36–38], crosslinking (0.8–2.9) [4,33,39], and block (0.5–2.2) [10,17,40] AEMs, and five-fold higher than the densely functionalized AEMs (0.9–6.4) [16,28,41]. In general, a great advantage of membrane C-HBM-1.78, including the higher conductivity in combination with largely depressed water swelling, is detected. This illustrates the effectiveness of combining dense functionalization and crosslinking to alleviate the trade-off effect between ionic conductivity and water swelling.

Table 1. Comparison regarding ion exchange capacity (IEC), hydroxide conductivity, water uptake, and linear swelling ratio (LSR) of C-HBM-1.78 with reported membrane samples.

Membrane Sample	Structure	Titrated IEC (mmol g^{-1})	σ_{OH-}	Water uptake (wt%)	LSR (%)	σ_{OH-}/IEC	σ_{OH-}/(100*Water Uptake)	σ_{OH-}/(100*LSR)	Reference
C-HBM-1.78 [a]	Crosslinking, Dense functionalization	1.78	44.5	40.1	1.4	25.0	1.1	33.0	This work
ImPES-1.0 [a]		1.83	42.7	83.9	21.7	23.3	0.5	2.0	[36]
OBuTMA-AAEPs-1.2(OH^-) [b]	Side chain type	1.87	30.0	62.0	14.7	16.0	0.5	2.0	[37]
gQAPPO [a]		1.78	27.0	40.0	13.2	15.2	0.7	2.0	[38]
CBQAPPO-4 [b]		2.37	43.0	110.0	15.0	18.1	0.4	2.9	[33]
J10-PPO [c]	Crosslinking	2.08	24.0	228.0	32.0	11.5	0.1	0.8	[4]
SIPN-95-2 [a]		1.75	35.5	213.1	37.5	20.3	0.2	1.0	[39]
ABA-QA-3 [a]		1.81	44.0	86.3	20.0	24.3	0.5	2.2	[10]
SEBS-CH_2-QA-1.5 [a]	Block	1.23	30.2	47.5	29.6	24.6	0.6	1.0	[17]
PAEK-QTPM-30 [c]		1.58	13.0	88.6	27.5	8.2	0.2	0.5	[40]
12-CQP-2 [b]		1.40	25.4	10.7	4.0	18.1	2.4	6.4	[16]
ImOH-HBPSf-40 [a]	Dense functionalization	2.16	31.0	143.9	36.3	14.4	0.2	0.9	[41]
QPAEN-0.4 [a]		1.78	47.3	34.2	11.2	26.6	1.4	4.2	[28]

[a] The σ_{OH-}, water uptake, and LSR were tested at 30 °C; [b] the σ_{OH-}, water uptake, and LSR were tested at 25 °C; [c] the σ_{OH-}, water uptake, and LSR were tested at 20 °C.

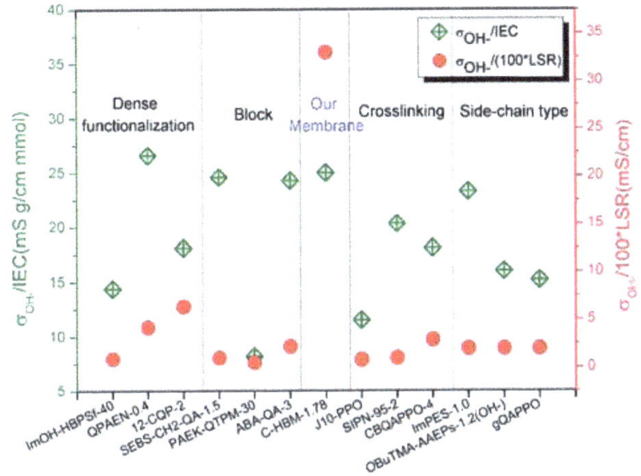

Figure 6. Comparison regarding effective hydroxide conductivity (σ_{OH^-}/IEC, σ_{OH^-}/LSR) of our membrane C-HBM-1.78 with reported densely functionalized, block, crosslinked, and side-chain-type membranes.

2.4. Alkaline Stability and Thermal Stability

Fuel cells usually operate under harsh basic conditions and elevated temperatures, which requires robust alkali-resistance and good thermally stable AEM. The alkaline stability of membrane C-HBM-1.78 is determined by examining the variation of IEC values and chloride conductivity after immersing membrane samples in a 1 mol L^{-1} NaOH aqueous solution at 60 °C at different exposure times. The membrane samples maintained their toughness, flexibility, and appearance after stability tests as long as 240 h, suggesting no decomposition of aromatic main chains. Figure 7 shows the decline in IEC values and chloride conductivity of membrane C-HBM-1.78 with the conditioning time. The chloride conductivity of the treated membrane C-HBM-1.78 at the 10th day remains at 22.5 mS cm^{-1} at 30 °C, which satisfies the basic requirement of the hydroxide conductivity (over 10 mS cm^{-1}) for fuel cell operation. The IEC values appeared as 1.28 mmol g^{-1}, 72% of the retention after 10 days treatment was achieved. It has already been demonstrated that the quaternary ammonium groups tend to disintegrate in alkaline solution because of the displacement of the ammonium group by OH^- via a direct nucleophilic substitution and Hofmann elimination when β-H atoms are present. In the present study, there is no β-H atom; the C-HBM-1.78 was thus degraded mainly by the nucleophilic substitution in which the hydroxide ions attack the α-carbon of ammonium cations. The outstanding alkali-resistance performance should benefit from the well-established hydrophilic–hydrophobic separation and crosslinking structure, where OH^- is confined into hydrophilic domains and the aromatic rings can be well protected by the additional hydrophobic structure.

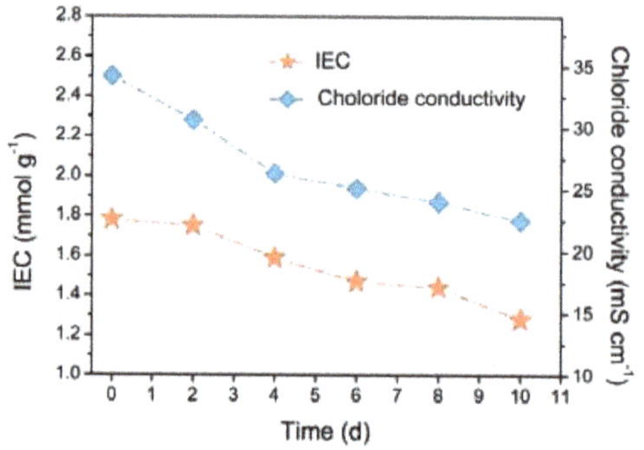

Figure 7. Time courses of chloride conductivity and IEC values of membrane C-HBM-1.78, in 1 mol L^{-1} NaOH aqueous solution at 60 °C. The membrane sample was converted into chloride form before conductivity measurement to avoid the influence of atmosphere CO_2.

The thermal stability of C-HBM-1.78 was investigated by thermogravimetic analysis (TGA) analysis. As shown in the TGA and differential thermal gravity (DTG) curve in Figure 8, a three-step degradation profile is observed for membrane C-HBM-1.78 from room temperature to 700 °C. The first weight loss occurred at around 218 °C, presumably corresponding to the degradation of the benzyltrimethylammonium chloride groups, and indicates the upper limit for practical application temperature of membrane C-HBM-1.78 in fuel cells. The weight loss between 218 and 700 °C is related to the degradation of the aromatic chains. In general, the thermal stability of C-HBM-1.78 can fulfill the practical application of fuel cells, which are typically operated at 60–80 °C.

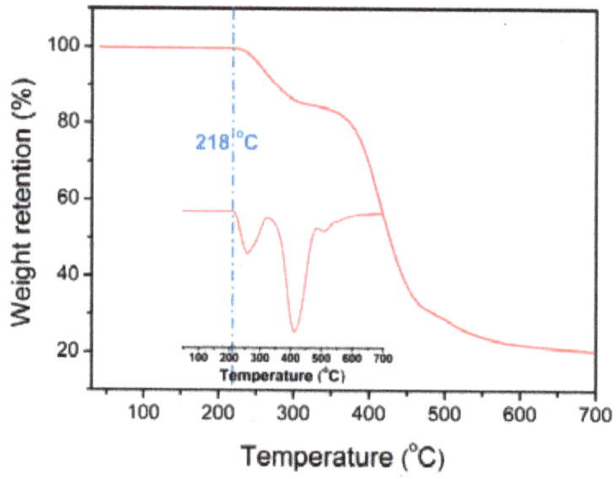

Figure 8. Thermogravimetic analysis (TGA) and differential thermal gravity (DTG, illustration) curves of membrane C-HBM-1.78.

3. Materials and Methods

3.1. Materials

The hyperbranched oligomer HB-PVBC was prepared and purified according to our previous procedure [34]. The chemical structure was examined by ^{1}H NMR, and the molecular weight (Mn = 7560 g mol^{-1}, Mw = 23629 g mol^{-1}) and polydispersity (PDI = 3.13) were determined by GPC. The monomer 4-vinylbenzyl chloride, trimethylamine aqueous solution (30%), and N-methyl pyrrolidone (NMP) were purchased from Energy Chemical (Shanghai, China) and used as received. The initiator AIBN was purchased from Energy Chemical (Shanghai, China) and recrystallized from ethanol. Cuprous chloride (CuCl) was purchased from Energy Chemical (Shanghai, China) and purified according to the reported procedure [42]. Dimethyl amine aqueous solution (33 wt%), anhydrous magnesium sulfate ($MgSO_4$), sodium chloride (NaCl), sodium sulfate (Na_2SO_4), and silver nitrate ($AgNO_3$) were purchased from Sinopharm Chemical Reagent Co. Ltd. (Shanghai, China). Potassium chromate (K_2CrO_4) was purchased from Alfa Aesar (Shanghai, P.R. China) and potassium iodide (KI) was purchased from Aladdin (Shanghai, P.R. China). Unless otherwise noted, all reagents are of analytical grade and were used as received. Distilled water was used throughout the experiment.

3.2. *Preparation of Monomer N,N-Dimethylbenzylamine (1) and Polyamine Poly(N,N-Dimethylbenzylamine) (2)*

The monomer *N,N*-Dimethylbenzylamine was synthesized by treating 4-vinylbenzyl chloride with dimethyl amine at room temperature according to a modified procedure [43]. Specifically, to a dimethyl amine solution (33 wt%, 50 mL, 0.25 mol) in a 250 mL, three-necked, round-bottomed flask equipped with a magnetic stir bar, 4-vinylbenzyl chloride (25 mL, 0.18 mol) was added dropwise via a dropping funnel over a period of 3 h under vigorous stirring. The reaction was allowed to proceed at room temperature for another 3 h. Then, the organic layer was separated by a separating funnel and dried over anhydrous $MgSO_4$, and the residue was vacuum distilled to obtain the monomer *N,N*-Dimethylbenzylamine from the mixture of *N,N*-Dimethylbenzylamine and poly(*N,N*-Dimethylbenzylamine) in 56% yield. ^1H NMR was used to characterize the structure.

Poly(*N,N*-Dimethylbenzylamine) was prepared via radical polymerization using AIBN as an initiator and NMP as the solvent. In a typical polymerization procedure, *N,N*-Dimethylbenzylamine (1.36 g, 8 mmol), NMP (0.6 mL), and AIBN (0.013 g, 0.08 mmol) were introduced into a 25 mL, round-bottomed flask equipped with a magnetic stir bar. The reaction mixture was filled nitrogen for 20 min to remove oxygen prior to the polymerization process, and subsequently placed into an oil bath at 65 °C. After 12 h, the reaction was stopped and the mixture was diluted with NMP and precipitated into water. This crude product was further purified for two times, and finally vacuum dried at 50 °C for 24 h to give the white product poly(*N,N*-Dimethylbenzylamine). ^1H NMR was utilized to prove its successful preparation and GPC was performed to characterize its molecular weight.

3.3. *Quaternization of HB-PVBC (3) with Polyamine Poly(N,N-Dimethylbenzylamine) (2)*

The densely functionalized AEM was fabricated via facile Menshutkin reaction of HB-PVBC and poly(*N,N*-Dimethylbenzylamine). Because of the particular structure of hyperbranched oligomer HB-PVBC with densely distributed benzylic chloride groups, the quaternization of HB-PVBC with polyamine poly(*N,N*-Dimethylbenzylamine) consequently yields a crosslinked AEM with densely arranged quaternary ammonium groups. In a typical procedure, to a solution of HB-PVBC (0.3 g) in NMP (6 mL), the solution of poly(*N,N*-Dimethylbenzylamine) (0.1 g) in NMP (2 mL) was slowly added under vigorous stirring. After stirring at room temperature for 10 min, the solution was directly cast on a flat, clean glass plate and then evaporated at 70 °C for 12 h, producing a transparent membrane. The membranes in OH^- form were obtained by immersing the membrane in NaOH aqueous solution (1 M) at room temperature for 24 h. The membrane sample was then washed thoroughly and immersed in distilled water for 48 h to remove residual NaOH before use.

3.4. *Characterization*

3.4.1. Nuclear Magnetic Resonance (NMR) and Fourier Transform Infrared (FTIR)

The chemical structure of *N,N*-Dimethylbenzylamine, poly(*N,N*-Dimethylbenzylamine) and HB-PVBC was determined by ^1H NMR (AVANCEII, 400MHz) with chloroform-d ($CDCl_3$) as solvents. FTIR was performed on membrane sample and hyperbranched oligomer HB-PVBC on a TENSOR27 FT-IR Spectrometer (Germany) under ambient conditions with a resolution of 5 cm^{-1} and a wide spectral range of 4000–500 cm^{-1} to characterize its structure.

3.4.2. Gel Permeation Chromatography (GPC)

The molecular weight and polydispersity (PDI = Mw/Mn) of poly(*N*,*N*-Dimethylbenzylamine) and HB-PVBC were determined by GPC on a PL 120 Plus (Agilent Technologies co., Ltd., China) equipped with a differential refractive index detector. The PL Gel Mixed Carbon 18 SEC columns connected in series were used to achieve the separation. Freshly prepared polymer samples in tetrahydrofuran (THF, HPLC grade) were passed through a 0.45 μm polytetrafluoroethylene syringe filter prior to injection. HPLC grade THF containing 0.03 wt% LiCl was used as the eluent at a flow rate of 1.0 mL min^{-1}. The detecting system was calibrated with a 2 mg mL^{-1} polystyrene (Mw = 110K).

3.4.3. Membrane Morphology Characterization

The microphase separated morphology of membrane C-HBM was examined by transmission electron microscopy (TEM), atomic force microscopy (AFM), and small angle X-ray scattering (SAXS). The membrane sample for TEM was prepared as follows: membranes were stained by soaking in a 1 mol L^{-1} potassium iodide (KI) aqueous solution at room temperature for 72 h, then washed with distilled water many times to remove the absorbed KI, and dried under vacuum at 40 °C. The stained membrane sample was then sectioned to yield slices with a thickness of 60–100 nm using a LEICA UC7FC7 ultramicrotome and coated on a copper grid. The electron micrograph was taken on a JEM-2100 transmission electron microscope operated at an accelerating voltage of 200 kV.

AFM observations in tapping mode were performed on a membrane sample in dry state with a veeco diInnova scanning probe microscope (SPM), using micro fabricated cantilevers with a force constant of approximately 20 N m^{-1}.

SAXS measurement was carried out on the SAXSess mc2 X-ray scattering system (Anton Paar). SAXS measurement was performed with Cu Kα radiation operating at 2 kW (40 kV and 50 mA). The distance between the sample and detector was approximately 260 mm and the wavelength of X-rays was 1.542 Å. The exposure time was 30 min for the sample.

3.4.4. Ion Exchange Capacity (IEC)

IEC was measured by Mohr's method. The membrane sample was firstly soaked in NaCl aqueous solution (1 M) for 24 h, then washed with distilled water to remove the absorbed NaCl and dried to a constant weight and weighed as W_{dry}. Finally, the membrane was immersed in Na_2SO_4 aqueous solution (0.5 M) for 24 h to exchange Cl^- from the membrane with SO_4^{2-}. The released Cl^- ions were then titrated with $AgNO_3$ aqueous solution (0.05 M) using K_2CrO_4 as a colorimetric indicator. The IEC value can thus be calculated from the amount of $AgNO_3$ consumed in the titration process and the mass of the dry membrane in Cl^- form, as shown in Equation (1).

$$\mathrm{IEC}(mmol/g) = V_{AgNO_3} \times C_{AgNO_3}/W_{dry} \quad (1)$$

3.4.5. Hydroxide Conductivity and Chloride Conductivity

The OH^- conductivity and chloride conductivity of membrane C-HBM were measured using a four-point probe technique on an Autolab PGSTAT 302N (Eco Chemie, Switzerland) equipped with a Teflon cell. During the measurement, the membrane sample was set into the Teflon cell, in which it was in contact with two current-collecting electrodes and two potential-sensing electrodes. Then, the cell was completely immersed in distilled water, and the impedance spectrum in galvanostatic mode and with an ac current amplitude of 0.1 mA over frequencies ranging from 1 MHz to 100 Hz was collected. Bode plots were used to determine the frequency region over which the magnitude of the impedance was constant. The ionic resistance of membrane was then obtained from a Nyquist plot and the ionic conductivity was calculated according to Equation (2):

$$\sigma = L/RWd \quad (2)$$

where L is the distance between two potential-sensing electrodes (here, 1 cm); R is the absolute ohmic resistance of the membrane sample; and W (here, 1 cm) and d are the width and thickness of the membrane, respectively.

3.4.6. Water Uptake and Linear Swelling Ratio (LSR)

Membrane samples in Cl^- form were immersed in distilled water at given temperatures for 24 h, then removed, and the membrane surfaces were quickly wiped dry with tissue paper. Water uptake and LSR of the membranes were calculated from the mass and length of wet and dry samples as follows:

$$WU(wt\%) = \left(W_{wet} - W_{dry}\right)/W_{dry} \times 100\% \quad (3)$$

$$LSR(\%) = \left(L_{wet} - L_{dry}\right)/L_{dry} \times 100\% \quad (4)$$

where W_{wet} and L_{wet} are the weight and length of the wet membrane, and W_{dry} and L_{dry} are those of the dry membrane after dried at 60 °C in a vacuum oven for 24 h, respectively.

3.4.7. Dynamic Mechanical Analyzer

The mechanical property of C-HBM at fully hydrated state was tested on a Q800 dynamic mechanical analyzer (TA Instruments) at a stretch rate of 0.5 N min^{-1}.

3.4.8. Thermogravimetic Analysis (TGA)

The thermal behavior of the membrane C-HBM was examined on a Perkin-Elmer Pyris-1 analyzer (USA) from 30 to 700 °C at a heating rate of 10 °C min^{-1} under a nitrogen atmosphere.

3.4.9. Alkaline Stability

To assess the alkali-resistance property of C-HBM, membrane samples were immersed in 1 mol L^{-1} NaOH aqueous solutions at 60 °C for 10 days. They were then taken out and thoroughly washed with distilled water prior to the measurement of IEC values and ionic conductivity. The chloride conductivity was measured by exchanging OH^- with Cl^- thoroughly to avoid the influence of atmosphere CO_2.

4. Conclusions

In conclusion, to alleviate the trade-off effect between ionic conductivity and water swelling, a crosslinking hyperbranched AEM was designed and fabricated, aiming to combine both advantages of dense functionalization and crosslinking. As expected, a much higher ratio of hydroxide conductivity to water swelling than that of the common side-chain-type, block, and single dense functionalization and single crosslinking AEMs was observed. Additionally, outstanding alkaline stability and thermal stability were also obtained for membrane C-HBM-1.78. The combination of excellent hydroxide conductivity and outstanding water-swelling resistance makes membrane C-HBM-1.78 attractive as AEM materials for fuel cell applications. Furthermore, this strategy proposed herein opens up new possibilities for overcoming the trade-off effect between ionic conductivity and water swelling for ion exchange membranes.

Author Contributions: Q.G. designed the study, conducted most experiments, and prepared the manuscript; Z.Y. supervised the study and revised the manuscript; X.Z. prepared the oligomer HB-PVBC and TEM sample, and titrated the IEC values.

Funding: This research was funded by the National Natural Science Foundation of China (No. 21808001), the Research Foundation from college of Chemistry & Chemical Engineering of Anhui University (No. J04100301), and the Start-up Grant from Anhui University (No. J01006153).

Acknowledgments: We acknowledge Abhishek Narayan Mondal for modifying the manuscript syntax.

Conflicts of Interest: The authors declare no conflict of interest.

Abbreviations

PEMs	Proton exchange membranes
AEMs	Anion exchange membranes
AFCs	Alkaline fuel cells
PEMFCs	Proton exchange membrane fuel cells
NMP	N-methyl pyrrolidone
HB-PVBC	Hyperbranched poly(4-vinylbenzyl chloride)
AIBN	2,2′-azodiisobutyronitrile
ATRP	Atom transfer radical polymerization
IEC	Ion exchange capacity
LSR	Linear swelling ratio
GPC	Gel permeation chromatography
NMR	Nuclear magnetic resonance
FTIR	Fourier transform infrared
TEM	Transmission electron microscopy
AFM	Atomic force microscopy
SAXS	Small angle X-ray scattering
DSC	Differential scanning calorimetry
TGA	Thermogravimetic analysis

References

1. Wang, J.; Gu, S.; Xiong, R.; Zhang, B.; Xu, B.; Yan, Y. Structure–Property Relationships in Hydroxide-Exchange Membranes with Cation Strings and High Ion-Exchange Capacity. *Chem. Sus. Chem.* **2015**, *8*, 4229–4234. [CrossRef] [PubMed]
2. Han, J.; Zhu, L.; Pan, J.; Zimudzi, T.J.; Wang, Y.; Peng, Y.; Hickner, M.A.; Zhuang, L. Elastic long-chain multication cross-linked anion exchange membranes. *Macromolecules* **2017**, *50*, 3323–3332. [CrossRef]
3. Tanaka, M.; Fukasawa, K.; Nishino, E.; Yamaguchi, S.; Yamada, K.; Tanaka, H.; Bae, B.; Miyatake, K.; Watanabe, M. Anion conductive block poly(arylene ether)s: Synthesis, properties, and application in alkaline fuel cells. *J. Am. Chem. Soc.* **2011**, *133*, 10646–10654. [CrossRef] [PubMed]
4. Zhu, L.; Zimudzi, T.J.; Wang, Y.; Yu, X.; Pan, J.; Han, J.; Kushner, D.I.; Zhuang, L.; Hickner, M.A. Mechanically robust anion exchange membranes via long hydrophilic cross-linkers. *Macromolecules* **2017**, *50*, 2329–2337. [CrossRef]
5. He, G.; Li, Z.; Zhao, J.; Wang, S.; Wu, H.; Guiver, M.D.; Jiang, Z. Nanostructured Ion-Exchange Membranes for Fuel Cells: Recent Advances and Perspectives. *Adv. Mater.* **2015**, *27*, 5280–5295. [CrossRef] [PubMed]
6. Li, L.; Lin, C.X.; Wang, X.Q.; Yang, Q.; Zhang, Q.G.; Zhu, A.M.; Liu, Q.L. Highly conductive anion exchange membranes with long flexible multication spacer. *J. Membr. Sci.* **2018**, *553*, 209–217. [CrossRef]
7. Li, N.; Yan, T.; Li, Z.; Thurn-Albrecht, T.; Binder, W.H. Comb-shaped polymers to enhance hydroxide transport in anion exchange membranes. *Energy Environ. Sci.* **2012**, *5*, 7888–7892. [CrossRef]
8. Akiyama, R.; Yokota, N.; Otsuji, K.; Miyatake, K. Structurally Well-Defined Anion Conductive Aromatic Copolymers: Effect of the Side-Chain Length. *Macromolecules* **2018**, *51*, 3394–3404. [CrossRef]
9. Dang, H.-S.; Weiber, E.A.; Jannasch, P. Poly (phenylene oxide) functionalized with quaternary ammonium groups via flexible alkyl spacers for high-performance anion exchange membranes. *J. Mater. Chem. A* **2015**, *3*, 5280–5284. [CrossRef]
10. Lin, C.X.; Wang, X.Q.; Li, L.; Liu, F.H.; Zhang, Q.G.; Zhu, A.M.; Liu, Q.L.; Lin, C.X.; Wang, X.Q.; Li, L. Triblock copolymer anion exchange membranes bearing alkyl-tethered cycloaliphatic quaternary ammonium-head-groups for fuel cells. *J. Power Sources* **2017**, *365*, 282–292. [CrossRef]
11. Clark, T.J.; Robertson, N.J.; Kostalik, H.A., IV; Lobkovsky, E.B.; Mutolo, P.F.; Abruna, H.D.; Coates, G.W. A Ring-Opening Meta thesis Polymerization Route to Alkaline Anion Exchange Membranes: Development of Hydroxide-Conducting Thin Films from an Ammonium-Functionalized Monomer. *J. Am. Chem. Soc.* **2009**, *131*, 12888–12889. [CrossRef] [PubMed]
12. Merle, G.; Wessling, M.; Nijmeijer, K. Anion exchange membranes for alkaline fuel cells: A review. *J. Membr. Sci.* **2011**, *377*, 1–35. [CrossRef]

13. Smitha, B.; Sridhar, S.; Khan, A.A. Solid polymer electrolyte membranes for fuel cell applications-a review. *J. Membr. Sci.* **2005**, *259*, 10–26. [CrossRef]
14. Pan, Z.F.; An, L.; Zhao, T.S.; Tang, Z.K. Advances and challenges in alkaline anion exchange membrane fuel cells. *Prog. Energy Combust. Sci.* **2018**, *66*, 141–175. [CrossRef]
15. Gottesfeld, S.; Dekel, D.R.; Page, M.; Bae, C.; Yan, Y.; Zelenay, P.; Kim, Y.S. Anion exchange membrane fuel cells: Current status and remaining challenges. *J. Power Sources* **2018**, *375*, 170–184. [CrossRef]
16. Hossain, M.M.; Hou, J.; Liang, W.; Ge, Q.; Xian, L.; Mondal, A.N.; Xu, T. Anion exchange membranes with clusters of alkyl ammonium group for mitigating water swelling but not ionic conductivity. *J. Membr. Sci.* **2017**, *550*, 101–109. [CrossRef]
17. Lin, C.X.; Wang, X.Q.; Hu, E.N.; Yang, Q.; Zhang, Q.G.; Zhu, A.M.; Liu, Q.L. Quaternized triblock polymer anion exchange membranes with enhanced alkaline stability. *J. Membr. Sci.* **2017**, *541*, 358–366.
18. Liu, L.; Li, Q.; Dai, J.; Wang, H.; Jin, B.; Bai, R. A facile strategy for the synthesis of guanidinium-functionalized polymer as alkaline anion exchange membrane with improved alkaline stability. *J. Membr. Sci.* **2014**, *453*, 52–60. [CrossRef]
19. He, Y.; Pan, J.; Wu, L.; Ge, L.; Xu, T. Facile preparation of 1, 8-Diazabicyclo [5.4.0] undec-7-ene based high performance anion exchange membranes for diffusion dialysis applications. *J. Membr. Sci.* **2015**, *491*, 45–52. [CrossRef]
20. Gao, X.; Lu, F.; Liu, Y.; Sun, N.; Zheng, L. The facile construction of an anion exchange membrane with 3D interconnected ionic nano-channels. *Chem. Commun.* **2017**, *53*, 767–770. [CrossRef]
21. Zhu, L.; Pan, J.; Christensen, C.M.; Lin, B.; Hickner, M.A. Functionalization of poly (2, 6-dimethyl-1, 4-phenylene oxide) s with hindered fluorene side chains for anion exchange membranes. *Macromolecules* **2016**, *49*, 3300–3309. [CrossRef]
22. Lai, A.N.; Wang, L.S.; Lin, C.X.; Zhuo, Y.Z.; Zhang, Q.G.; Zhu, A.M.; Liu, Q.L. Phenolphthalein-based poly (arylene ether sulfone nitrile) s multiblock copolymers as anion exchange membranes for alkaline fuel cells. *ACS Appl. Mater. Interfaces* **2015**, *7*, 8284–8292. [CrossRef] [PubMed]
23. Lin, C.X.; Wu, H.Y.; Li, L.; Wang, X.Q.; Zhang, Q.G.; Zhu, A.M.; Liu, Q.L. Anion Conductive Triblock Copolymer Membranes with Flexible Multication Side Chain. *ACS Appl. Mater. Interfaces* **2018**, *10*, 18327–18337. [CrossRef] [PubMed]
24. Kim, E.; Lee, S.; Woo, S.; Park, S.H.; Yim, S.D.; Shin, D.; Bae, B. Synthesis and characterization of anion exchange multi-block copolymer membranes with a fluorine moiety as alkaline membrane fuel cells. *J. Power Sources* **2017**, *359*, 568–576. [CrossRef]
25. Li, X.; Nie, G.; Tao, J.; Wu, W.; Wang, L.; Liao, S. Assessing the influence of side-chain and main-chain aromatic benzyltrimethyl ammonium on anion exchange membranes. *ACS Appl. Mater. Interfaces* **2014**, *6*, 7585–7595. [CrossRef] [PubMed]
26. Lin, C.X.; Huang, X.L.; Guo, D.; Zhang, Q.G.; Zhu, A.M.; Ye, M.L.; Liu, Q.L. Side-chain-type anion exchange membranes bearing pendant quaternary ammonium groups via flexible spacers for fuel cells. *J. Mater. Chem. A* **2016**, *4*, 13938–13948. [CrossRef]
27. He, Y.; Zhang, J.; Liang, X.; Shehzad, M.A.; Ge, X.; Zhu, Y.; Hu, M.; Yang, Z.; Wu, L.; Xu, T. Achieving high anion conductivity by densely grafting of ionic strings. *J. Membr. Sci.* **2018**, *559*, 35–41. [CrossRef]
28. Hu, E.N.; Lin, C.X.; Liu, F.H.; Wang, X.Q.; Zhang, Q.G.; Zhu, A.M.; Liu, Q.L. Poly(arylene ether nitrile) anion exchange membranes with dense flexible ionic side chain for fuel cells. *J. Membr. Sci.* **2018**, *550*, 254–265. [CrossRef]
29. He, Y.; Pan, J.; Wu, L.; Zhu, Y.; Ge, X.; Ran, J.; Yang, Z.; Xu, T. A novel methodology to synthesize highly conductive anion exchange membranes. *Sci. Rep.* **2015**, *5*, 13417. [CrossRef]
30. Ge, Q.; Liu, Y.; Yang, Z.; Wu, B.; Hu, M.; Liu, X.; Hou, J.; Xu, T. Hyper-branched anion exchange membranes with high conductivity and chemical stability. *Chem. Commun.* **2016**, *52*, 10141. [CrossRef]
31. Wu, L.; Pan, Q.; Varcoe, J.R.; Zhou, D.; Ran, J.; Yang, Z.; Xu, T. Thermal crosslinking of an alkaline anion exchange membrane bearing unsaturated side chains. *J. Membr. Sci.* **2015**, *490*, 1–8. [CrossRef]
32. Xu, W.; Zhao, Y.; Yuan, Z.; Li, X.; Zhang, H.; Vankelecom, I.F. Highly stable anion exchange membranes with internal cross-linking networks. *Adv. Funct. Mater.* **2015**, *25*, 2583–2589. [CrossRef]
33. Hao, J.; Gao, X.; Jiang, Y.; Zhang, H.; Luo, J.; Shao, Z.; Yi, B. Crosslinked high-performance anion exchange membranes based on poly (styrene-b-(ethylene-co-butylene)-b-styrene). *J. Membr. Sci.* **2018**, *551*, 66–75. [CrossRef]

34. He, Y.; Liang, W.; Pan, J.; Yuan, Z.; Ge, X.; Yang, Z.; Jin, R.; Xu, T. A mechanically robust anion exchange membrane with high hydroxide conductivity. *J. Membr. Sci.* **2016**, *504*, 47–54. [CrossRef]
35. Mohanty, A.D.; Chang, Y.R.; Yu, S.K.; Bae, C. Stable Elastomeric Anion Exchange Membranes Based on Quaternary Ammonium-Tethered Polystyrene-b-poly(ethylene-co-butylene)-b-polystyrene Triblock Copolymers. *Macromolecules* **2015**, *48*, 7085–7095. [CrossRef]
36. Zhuo, Y.Z.; Lai, A.L.; Zhang, Q.G.; Zhu, A.M.; Ye, M.L.; Liu, Q.L. Enhancement of hydroxide conductivity by grafting flexible pendant imidazolium groups into poly (arylene ether sulfone) as anion exchange membranes. *J. Mater. Chem. A* **2015**, *3*, 18105–18114. [CrossRef]
37. Zhang, Z.H.; Wu, L.; Varcoe, J.; Li, C.R.; Ong, A.L.; Poyntonb, S.; Xu, T.W. Aromatic polyelectrolytes via polyacylation of pre-quaternized monomers for alkaline fuel cells. *J. Mater. Chem. A* **2013**, *1*, 2595–2601. [CrossRef]
38. Yang, Z.J.; Zhou, J.H.; Wang, S.W.; Hou, J.Q.; Wu, L.; Xu, T.W. A strategy to construct alkali-stable anion. exchange membranes bearing ammonium groups via flexible spacers. *J. Mater. Chem. A* **2015**, *3*, 15015–15019. [CrossRef]
39. Pan, J.; Zhu, L.; Han, J.J.; Hickner, M.A. Mechanically Tough and Chemically Stable Anion Exchange. Membranes from Rigid-Flexible Semi-Interpenetrating Networks. *Chem. Mater.* **2015**, *27*, 6689–6698. [CrossRef]
40. Shen, K.Z.; Zhang, Z.P.; Zhang, H.B.; Pang, J.H.; Jiang, Z.H. Poly(arylene ether ketone) carrying hyperquaternized pendants: Preparation, stability and conductivity. *J. Power Sources* **2015**, *287*, 439–447. [CrossRef]
41. Li, S.S.; Gan, R.J.; Li, L.; Li, L.D.; Zhang, F.X.; He, G.H. Highly branched side chain grafting for enhanced conductivity and robustness of anion exchange membranes. *Ionics* **2018**, *24*, 189–199. [CrossRef]
42. Opsteen, J.A.; Van Hest, J.C.M. Modular synthesis of ABC type block copolymers by "click" chemistry. *J. Polym. Sci. Part A Polym. Chem.* **2010**, *45*, 2913–2924. [CrossRef]
43. Liu, Y.; Qi, P.; Wang, Y.; Zheng, C.; Liang, W.; Xu, T. In-situ crosslinking of anion exchange membrane bearing unsaturated moieties for electrodialysis. *Sep. Purif. Technol.* **2015**, *156*, 226–233. [CrossRef]

International Journal of
Molecular Sciences

MDPI

Article

Correlations of Ion Composition and Power Efficiency in a Reverse Electrodialysis Heat Engine

Fabao Luo [1,2,*], Yang Wang [3], Maolin Sha [1] and Yanxin Wei [1]

1 School of Chemistry and Chemical Engineering, Hefei Normal University, Hefei 230061, China; franksha@aliyun.com (M.S.); yxwei73@mail.ustc.edu.cn (Y.W.)
2 Anhui Province Key Laboratory of Environment-friendly Polymer Materials, Anhui University, Hefei 230601, China
3 CAS Key Laboratory of Soft Matter Chemistry, Collaborative Innovation Center of Chemistry for Energy Materials, School of Chemistry and Materials Science, University of Science and Technology of China, Hefei 230026, China; youngw@mail.ustc.edu.cn
* Correspondence: fbluo@mail.ustc.edu.cn

Received: 30 September 2019; Accepted: 18 November 2019; Published: 22 November 2019

Abstract: The main objective of this study is to explore the influence of ion composition on the trans-membrane potential across the ion exchange membrane (IEM), and thus offers a reference for the deep insight of "reverse electrodialysis heat engine" running in the composite systems. In comparison to the natural system (river water | seawater), the performance of the reverse electrodialysis (RED) stack was examined using $NaHCO_3$, Na_2CO_3, and NH_4Cl as the supporting electrolyte in the corresponding compartment. The effect of flow rates and the concentration ratio in the high salt concentration compartment (HCC)/low salt concentration compartment (LCC) on energy generation was investigated in terms of the open-circuit voltage (OCV) and power density per membrane area. It was found that the new system (0.49 M NaCl + 0.01 M $NaHCO_3$|0.01 M $NaHCO_3$) output a relatively stable power density (0.174 $W{\cdot}m^{-2}$), with the open-circuit voltage 2.95 V under the low flow rate of 0.22 cm/s. Meanwhile, the simulated natural system (0.5 M NaCl|0.01 M NaCl) output the power density 0.168 $W{\cdot}m^{-2}$, with the open-circuit voltage 2.86 V under the low flow rate of 0.22 cm/s. The findings in this work further confirm the excellent potential of RED for the recovery of salinity gradient energy (SGP) that is reserved in artificially-induced systems (wastewaters).

Keywords: salinity gradient power; reverse electrodialysis; concentration difference; electrolyte composition

1. Introduction

With the exhaustion of conventional fossil fuels and the excessive emission of greenhouse gas (CO_2), the demands on renewable energy has grown in the past decades. Salinity gradient energy (SGP) was recognized as one kind of blue energy which was reserved in seawater and river water. It was estimated that 2.4–2.6 TW energy was available by discharging rivers into oceans [1,2], based on Gibbs free energy of mixing. Using reverse electrodialysis (RED) as an energy conversion strategy, it is possible to recover SGP in the natural environment through an economically competitive and environmentally friendly manner. It has attracted growing attention because of its inherent advantages, such as clean and pollution-free, and simple installation [3–5]. RED uses ion exchange membranes as the separators, and allows the perm-selective transportation for cations and anions [6–8]. When the solution with different concentration is introduced into the corresponding compartment, the ions move across the correlative membranes, and the ion flux is transferred into the electron flux on the electrode. Generally, redox ion pairs (i.e., Fe^{3+}/Fe^{2+}) were used as the supporting electrolyte in the anode and cathode, or salt/base/acid supporting electrolyte to create water electrolysis circumstance [9,10]. The overall potential could be raised according to the requirements, by repeatedly assembling the membrane pairs (one membrane pair: 1 pc. cation exchange membrane + 1 pc. anion exchange membrane).

The natural system of river water|seawater was the dominant case which has been comprehensively investigated because of its abundant deposits [2,11,12]. The investigations on the optimization of ion exchange membranes or the membrane stack assembly were also the hotspots for the improvement of power density and the SGP recovery efficiency [5,11–15].

Some works reported a novel "reverse electrodialysis heat engine" technique that used ammonium bicarbonate (NH_4HCO_3) as the supporting electrolyte, and the waste heat as the "driven-force" to push the reaction $NH_4HCO_3 \leftrightarrow NH_3+CO_2+H_2O$ forwards. In the discharging circle, NH_3 and CO_2 gas dissolved in the concentrated solution, and the free ions (NH_4^+, HCO_3^-, CO_3^{2-}) diffuse across the ion exchange membrane (IEM) following the classical RED principle. Consequently, the decomposition of NH_4HCO_3 is the charging circle which recovers the diffused NH_4HCO_3 into the NH_3 and CO_2 gas. Recently, it was found the power density of the reverse electrodialysis heat engine system was slightly lower than the natural system on the basis of identical molar concentrations [16,17]. Meanwhile, the solution pH flow cell for converting waste carbon dioxide into electricity was also investigated [18].

Here in this work, we investigated the ion composition on the influence of energy recovery in the discharging circle of the heat engine. By using $NaHCO_3$, Na_2CO_3, NH_4Cl, and the composited solutions of NaCl + $NaHCO_3$, NaCl + Na_2CO_3, NaCl + NH_4Cl as the supporting electrolyte, we investigated the changes in the trans-membrane potential and power recovery in respect to the ion composition. Then, we provide deep insights into the reliability of the heat engine in complex compositions.

2. Results and Discussion

2.1. The Influence of Ion Species on Trans-Membrane Voltage

Lots of theoretical models and experimental studies have focused on improving the RED performance by means of salt concentration difference between adjacent ion exchange membrane, according to the Nernst theory [19]. Here in this work, RED was operated by introducing high electromotive force and changing the ion composition in both the high salt concentration compartment (HCC) and low salt concentration compartment (LCC). The system used in the experiments are listed in Table 1.

Table 1. The ion composition in the high salt concentration compartment (HCC) and low salt concentration compartment (LCC) in the experiments.

Salt Type	HCC	LCC
Case No.1	0.5 M NaCl	0.01 M NaCl
Case No.2	0.5 M NaCl	0.01 M $NaHCO_3$
Case No.3	0.49 M NaCl + 0.01 M $NaHCO_3$	0.01 M $NaHCO_3$
Case No.4	0.49 M NaCl +0.005 M Na_2CO_3	0.005 M Na_2CO_3
Case No.5	0.49 M NaCl + 0.01 M NH_4Cl	0.01 M NH_4Cl

The trans-membrane voltage for the anion exchange membrane (AEM) and cation exchange membrane (CEM) was investigated and given in Figure 1. It was found that the trans-membrane voltage across the CEMs and AEMs was around 85 mV in the simulated seawater|river water system (0.5 M NaCl|0.01 M NaCl), which is close to the theoretical Nernst potential. The finding here proves the ion exchange membranes used in the experiment have the desired transport number, and perform perm-selectivity toward the counter-ions well (i.e., CEM allows the transition of cations (counter-ion) and block the anions (co-ion), vice versa for AEM). When the electrolyte in the solution was changed to Na_2CO_3, $NaHCO_3$, or NH_4Cl, the changes in transmembrane voltage drop was noticed. However, the changes in AEMs and CEMs are absolutely different. For system Case No.2, the trans-membrane voltage on AEM increases to ca. 93 mV. It was further increased to ca. 97 mV on AEM in the 0.49 M NaCl + 0.01 M $NaHCO_3$|0.01 M $NaHCO_3$ system, without significant changes in CEM. However, the trans-membrane voltage on AEM decreased to 90 and 80 mV when changing the system to 0.49 M NaCl + 0.005 M Na_2CO_3|0.005 M Na_2CO_3 and 0.49 M NaCl + 0.01 M NH_4Cl|0.01 M NH_4Cl. The trans-membrane voltage changes in CEM was only found in the 0.49 M NaCl + 0.01 M NH_4Cl|0.01 M NH_4Cl system (Case, No.5) system, which is about 78 mV. The changes were mainly attributed to ion species transporting across the membrane matrix, and could be calculated according to the Nernst equation for the multi-component case.

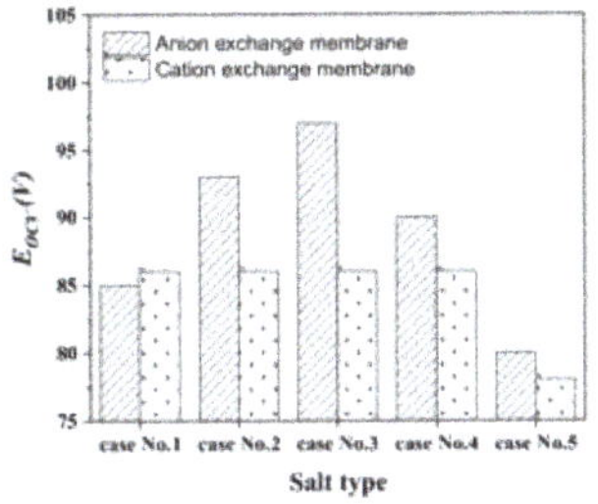

Figure 1. The transmembrane voltages on the AEM and CEM for Case No. 1–5.

The total stack electric resistance was investigated prior to the practical RED operation to survey the fundamental information of the RED stack. The electric resistance was recorded using the electric load under the OCV mode, by maintaining the flow rate at 135 mL·min^{-1} for all the five cases (to avoid the hydrodynamic influence). It is found in Figure 2 that Case No. 3 has the lowest total stack electric resistance (ca. 18 Ω) in the five operations. Case No. 5 has the highest total stack electric resistance (ca. 20.6 Ω), which may be owing to the introduction of the NH_4^+ ion. The NH_4^+ ion has corresponding different physicochemical characters (bare ion radius (0.148 nm), hydrated radius (0.331 nm), hydration free energy (29.5 kJ/mol-ion)), in comparison to the Na^+ ion (bare ion radius (0.117 nm), hydrated radius (0.358 nm), hydration free energy (365 kJ/mol-ion)) [20]. They perform absolute different properties when transporting across IEMs.

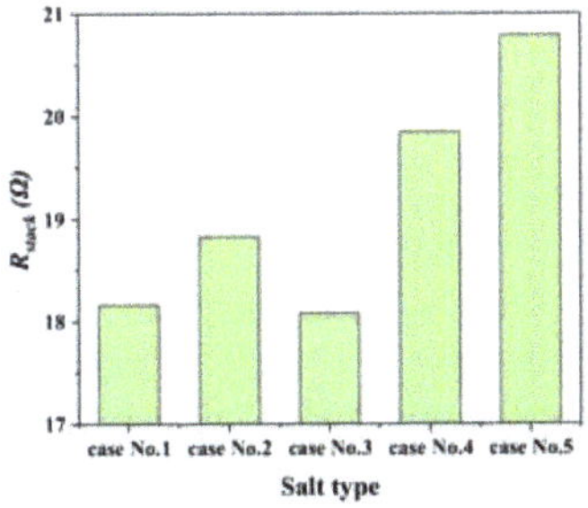

Figure 2. The total stack resistance tested in the open-circuit voltage (OCV) mode, for Case No. 1–5.

2.2. Polarization Curves of RED

The polarization curves for experiments Case No.1 and Case No.3 were investigated for the understanding of the influence of ion composition on power density. When the external load was connected with RED stack, the voltage output U could be calculated as the difference between the electromotive force E_{OCV} and the voltage drop across the internal resistance R_{stack} ($U = E_{OCV} - I\,R_{stack}$). The changes in voltage output (black line) and power density (blue line) in the function of the electric current was plotted as the two polarization curves, and are shown in Figure 3. It was found that OCV for both operations decreases with the loading current. The maximum E_{OCV} (OCV) are 2.95 and 3.06 V, while perform the same short-circuit current (0.155 A) for both cases. The maximum power density was 0.180 W $\cdot m^{-2}$ at the electric current of 0.083 A for Case No.1 (Figure 3a), and 0.176 $W \cdot m^{-2}$ at the electric current of 0.072 A for Case No.3 (Figure 3b). The current when RED reached the maximum outputting power density is lower for Case No.3 in comparison to Case No.1. This may be attributed to the difference in physicochemical properties for species HCO_3^- and Cl^-, and then reflected as the thermodynamic difference when mixing in the RED stack.

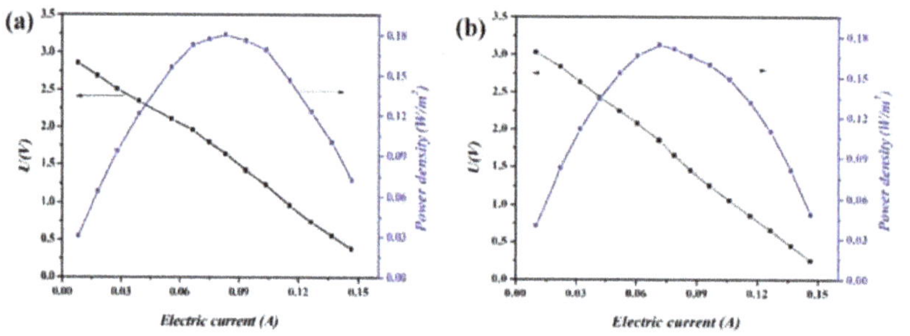

Figure 3. The changes in voltage output (black line) and power density (blue line) in the function of the electric current. (**a**) HCC: 0.5 $mol \cdot L^{-1}$ NaCl; LCC: 0.01 $mol \cdot L^{-1}$ NaCl. (**b**) HCC: 0.49 $mol \cdot L^{-1}$ NaCl + 0.01 $mol \cdot L^{-1}$ $NaHCO_3$; LCC: 0.01 $mol \cdot L^{-1}$ $NaHCO_3$. The flow rate ratio for HCC and LCC was kept at 180 $mL \cdot min^{-1}$.

2.3. The Discharging Property of RED

By investigating the changes in internal resistance (R_{stack}), open-circuit voltage (E_{OCV}), and power density (P_{gross}), the RED was investigated. This section presents the effects of flow rate and LCC electrolyte composition on the SGP recovery. The flow rates were changed from 45 to 225 $mL \cdot min^{-1}$, with respect to the changes in boundary layer resistance (R_{BL}).

To investigate the effect of ion composition in HCC on SGP recovery, 0.5 M NaCl (Case No. 1) in HCC was changed to 0.49 M NaCl + 0.01 M $NaHCO_3$ (Case No. 2). Figure 4a shows the effects of flow rate on OCV for Case No.1 and Case No. 2. It was found that the OCV increased from 2.26 to 3.14 V for Case No. 1 when the flow rate was increased from 45 to 225 $mL \cdot min^{-1}$. A similar trend was also found for Case No. 2 (increase from 2.35 to 3.14 V when increasing flow rate from 45 to 225 $mL \cdot min^{-1}$). This phenomenon could be appropriately explained according to the Nernst equation in the multi-component system (Equation 4). The species type and their corresponding activity in the solution, as well as the interaction with the functional group in the membrane matrix co-induced the transmembrane voltage changes for the different electrolyte. Otherwise, the increment on flow rate decreased the boundary layer thickness, and thus mitigated the concentration polarization in the boundary layer, which effectively improves the practical concentration difference on the two sides of the ion exchange membrane. The increment of transmembrane voltage drop by increasing the solution flowing rate could then be explained accordingly. For the power density, it increased from 0.108 to 0.162 $W \cdot m^{-2}$ and 0.125 to 0.176 $W \cdot m^{-2}$ for Case No. 1 and Case No. 2, respectively, by increasing the flow rate from 45 to 225 $mL \cdot min^{-1}$. The increment is significant just by doubling the flow rate to 90 $mL \cdot min^{-1}$, in comparison to the operation at the rate of 135, 180, and 225 $mL \cdot min^{-1}$. The powder density difference when using 0.5 M NaCl|0.01 M $NaHCO_3$ and 0.49 M NaCl + 0.01 M $NaHCO_3$|0.01 M $NaHCO_3$ as the supporting electrolyte in HCC|LCC was mainly attributed to the transition difference in the membrane phase and the concentration distribution in the boundary layer, which was discussed above.

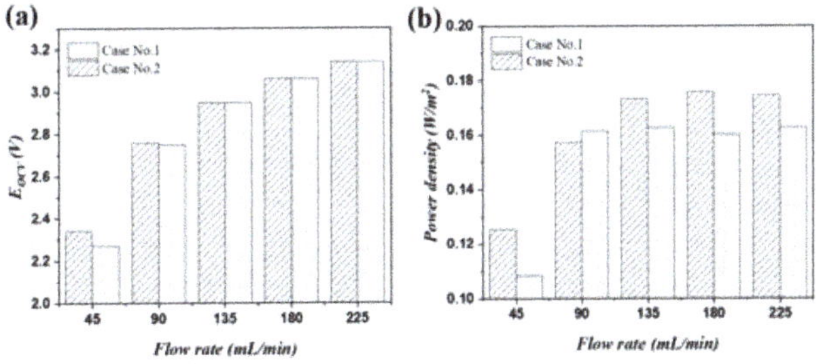

Figure 4. The role of NaCl, $NaHCO_3$ as the electrolyte in the LCC and their influence on reverse electrodialysis (RED) transmembrane voltage and power density. (**a**) Open-circuit voltage, (**b**) power density.

The concentration difference on the two sides of the membrane is one critical aspect which determines the performance of the RED stack. Therefore, we investigated the process by changing the electrolyte concentration in both the LCC and HCC, and the results are given in Figure 5. Figure 5 demonstrates the changes in E_{OCV} and the power density of the RED stack under different flow rates. The legends in the figure for five cases are illustrated in Table 2. It is interesting to find OCV increase with the flowing rate for every set of experiments. For example, when using 0.02 M $NaHCO_3$ as the electrolyte in the LCC, OCV gradually increases from 2.16 to 2.78 V when increasing the flow rate from 45 to 225 mL·min^{-1}. The finding here further confirms the critical role of the concentration polarization effect on the RED, which is not only a general case in natural systems (seawater|river water), but also in complex artificial complex systems. By maintaining all operations under the same flowing rate, it was found that the OCV decreases with the increment of $NaHCO_3$ concentration in the LCC. For example, it decreases from 2.44 to 2.16 V by increasing the $NaHCO_3$ concentration from 0.005 M $NaHCO_3$ to 0.02 M $NaHCO_3$ (flow rate of 45 mL·min^{-1}).

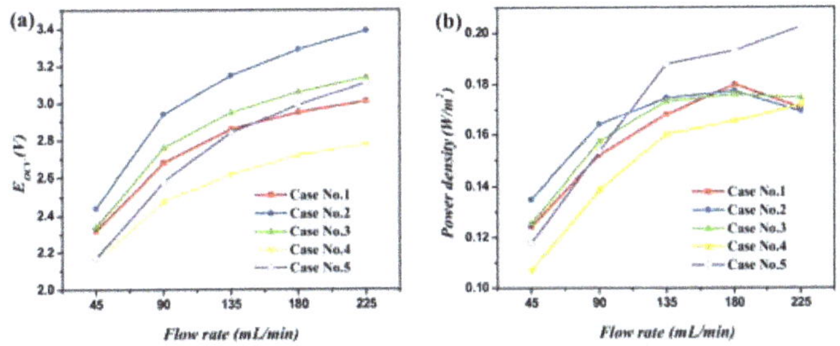

Figure 5. Influence of the concentration ratio of HCC and LCC on RED performance under different flow rates. (**a**) Open-circuit voltage, (**b**) power density. Note. The HCC salt solutions for Case No.1–5 were 0.5 M NaCl, 0.495 M NaCl + 0.005 M $NaHCO_3$, 0.49 M NaCl + 0.01 M $NaHCO_3$, 0.48 M NaCl + 0.02 M $NaHCO_3$, 0.99 M NaCl + 0.01 M $NaHCO_3$ solution, respectively.

Table 2. The ion composition in the HCC and LCC in the experiments.

Salt Type	HCC	LCC
Case No.1	0.5 M NaCl	0.01 M NaCl
Case No.2	0.495 M NaCl + 0.005 M $NaHCO_3$	0.005 M $NaHCO_3$
Case No.3	0.49 M NaCl + 0.01 M $NaHCO_3$	0.01 M $NaHCO_3$
Case No.4	0.48 M NaCl +0.02 M $NaHCO_3$	0.02 M $NaHCO_3$
Case No.5	0.99 M NaCl + 0.01 M $NaHCO_3$	0.01 M $NaHCO_3$

The power density increases with the flowing rate for every case of the experiments. For example, when using 0.02 M $NaHCO_3$ as the supporting electrolyte in the LCC, the power density gradually increases from 0.107 to 0.160 $W{\cdot}m^{-2}$ by increasing the flow rate from 45 to 135 $mL{\cdot}min^{-1}$. Additionally, the power density was relatively stable when the flow rate reached 135 $mL{\cdot}min^{-1}$. By maintaining all the operations under the same flowing rate, it was found that using 0.49 M NaCl + 0.01 M $NaHCO_3$|0.01 M $NaHCO_3$ as the supporting electrolyte in HCC|LCC (Case No.3) obtained a relatively high and stable power density. The highest power density (0.202 $W{\cdot}m^{-2}$) was obtained for the system of 0.99 M NaCl + 0.01 M $NaHCO_3$|0.01 M $NaHCO_3$ (Case No.5). The powder density was mainly determined by two factors, i.e., the OCV and stack resistance. The difference of power density among the five cases was mainly attributed to the ion transition difference in the membrane phase and the concentration distribution in the boundary layer, which was discussed above. When the concentration of $NaHCO_3$ was below 0.01 M in LCC, the main influence of power density was the membrane stack internal resistance. In contrast, when the concentration of $NaHCO_3$ was over 0.01 M in LCC, the main influence of power density was OCV. Therefore, the RED stack with the feed system of 0.49 M NaCl +0.01 M $NaHCO_3$|0.01 M $NaHCO_3$ was chosen for further experiments, with high energy utilized efficiency as well as a higher open-circuit voltage.

The RED performance was further investigated by changing the electrolyte in the LCC to Na_2CO_3 and NH_4Cl, and HCC to the composites of NaCl + Na_2CO_3 and NaCl + NH_4Cl, respectively. Figure 6a gives the changes in open-circuit voltages on the total RED stack by changing the flow rate. It was found that OCV on RED by using Na_2CO_3 as the LCC electrolyte was slightly lower than that using $NaHCO_3$ under operating conditions, except for the same flow rate at 45 mL/min. The OCV was further decreased by replacing the electrolyte in LCC to NH_4Cl. The highest OCV was 3.14, 3.09, and 2.93 V for the $NaHCO_3$, Na_2CO_3, and NH_4Cl electrolyte, respectively. The same trend was also found in the power density when changing the flow rates. The power density increases with an increase in the flow rate for all the three electrolytes (the electrolyte composition in LCC). The power density of three investigated electrolytes follows the order: $NaHCO_3$ > Na_2CO_3 > NH_4Cl (see Figure 6b). The highest power densities were 0.175, 0.175, and 0.143 W/m^2 for the $NaHCO_3$, Na_2CO_3, and NH_4Cl systems at the 225 mL/min flowing rate, respectively.

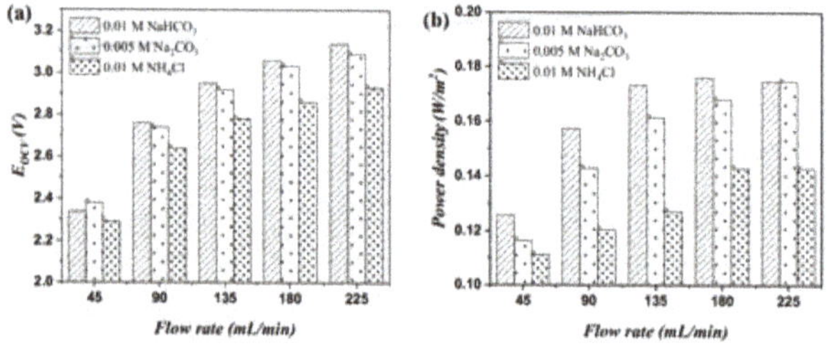

Figure 6. The changes in OCV and power density under different flow rates. (**a**) Open-circuit voltage, (**b**) power density. The legends represent electrolyte kind.

3. Materials and Methods

3.1. Materials

The cation exchange membranes and anion exchange membranes used in the experiments were CJ-MC-3 and CJ-MA-2, respectively (Hefei ChemJoy Polymers Co., Ltd., Hefei, China). The main properties of the ion exchange membranes are listed in Table 3. Before the experiments, the cation and anion exchange membranes were immersed in a 0.5 $mol{\cdot}L^{-1}$ NaCl solution for 24 h to change them into corresponding Na^+ and Cl^- form. The reagents used in the study, including NaCl, NH_4Cl, $NaHCO_3$, Na_2CO_3, $K_3[Fe(CN)_6]$, and $K_4[Fe(CN)_6]$, were all analytical grade and purchased from Sinopharm Chemical Reagent Co., Ltd., Shanghai, China. Deionized water was used throughout the experiments.

Table 3. The main characteristics of the membranes used in the experiments.

Properties	CJ-MC-3	CJ-MA-2
Thickness/mm	0.220	0.185
Ion exchange capacity/mmol/g	1.50	1.25
Water Uptake/%	35	32
Resistance/$\Omega \cdot cm^2$	3.0	1.5
Transfer number/%	98	99
Break stress/MPa	>3.5	>3.5

3.2. *Transmembrane Voltage Test*

The transmembrane voltage test was at room temperature using a self-designed experimental setup [21]. The set-up was composed of one membrane and two compartments: HC and LC compartments. The two compartments were separated by the test membrane using a quadrate clip. The two work electrodes were brought close to testing membrane surfaces to record the potential drop across the testing membranes which was recorded by a digital multimeter (VICTOR, VC890C+, VICTOR® YITENSENTM). The effective area of the membranes was 3.8 cm^2.

3.3. *RED Experimental Setup*

The experimental setup of RED was designed and installed in our lab. A schematic diagram of the RED principle is illustrated in Figure 7. The RED setup mainly contains (1) a cathode plate and an anode plate, which was made of titanium coated with ruthenium and iridium with the same effective area and acted as the electron conductor; (2) several cell pairs of cation and anion exchange membranes which were alternately arranged. Twenty repeated cell pairs were used with a total effective area of 20 $\times$ 2 $\times$ 189 cm^2 (9 $\times$ 21 cm); (3) silica gel spacers with the thickness of 0.75 mm were adopted to separate the anion and cation exchange membrane. In the RED process, three flow streams, i.e., electrode compartment, high salt concentration compartment (HC), and low salt concentration compartment (LC) were established. Electrode rinse solution was circulated with supporting electrolyte using the peristaltic pump (BT600L, Baoding Lead fluid Technology Co., Ltd., Baoding, China) at a flow rate of 90 $mL \cdot min^{-1}$. The same flow rate was maintained in HC and LC with two peristaltic pumps (BT600L, Lead fluid, China). The flow stream was under "feed-and-bleed" mode and circulated through the HC and LC with various velocities (45–225 $mL \cdot min^{-1}$).

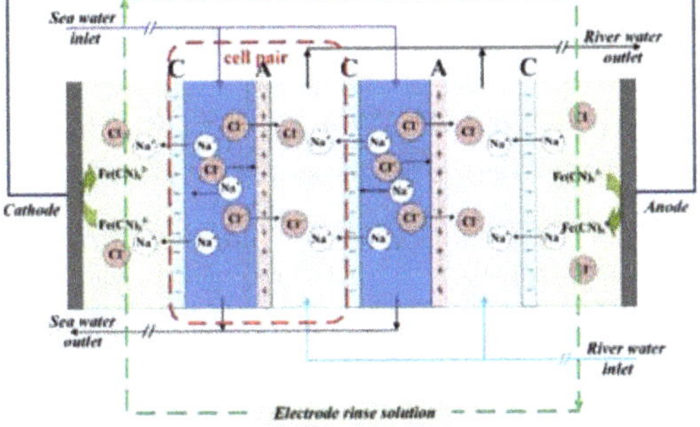

Figure 7. A schematic diagram of the principle of reverse electrodialysis, RED.

3.4. RED Performance Tests

A programmable DC electronic load (FT6300A, Shenzhen Faith Technology Co., Ltd., Shenzhen, China) was connected between the anode and cathode to measure the internal resistance, open-circuit voltage (OCV), and the electric current of the RED membrane stack. The stepwise current was set with a scanning rate of 10 mA/s, from 0 A to the maximum current (when the voltage of the membrane stack became reversed). The response curves of voltage vs. electric current and power density vs. electric current were recorded by a computer for each set condition. The electrochemical performance of the RED membrane stack was obtained by evaluating the open-circuit voltage, maximum power density, and maximum current. The open-circuit voltage was determined from the vertical axis intercept of the polarization curves, and the maximum current was obtained from the horizontal axis intercept in the polarization curves.

3.5. Determination of the RED System

The line flow velocity (V) can be defined as the mean fluid velocity inside a single spacer-filled channel. It can be estimated by Equation (1) [10,22].

$$V = \frac{Q}{N \cdot \delta \cdot b \cdot \varepsilon_{sp}}, \quad (1)$$

where Q is the volumetric flow rate ($mL \cdot min^{-1}$) in HC or LC inlet, δ is the spacer thickness (0.075 cm), b is the compartment width (9 cm for the small stack), and ε_{sp} is the spacer porosity (75% for the woven spacer used in this study). For simplicity, the flow rate in the experiment is expressed as the volumetric flow rate.

The power of the RED stack was calculated using Equation (2)

$$P = UI \quad (2)$$

where P is the power of the membrane stack (W), U is the voltage of the membrane stack (V), I is the scanned current (A).

To test the performance of RED, LC and HC were fed with solutions of 0.01 and 0.5 $mol \cdot L^{-1}$ NaCl, respectively. Electrode rinse solution consisted of 0.05 M $K_3[Fe(CN)_6]$, 0.05 M $K_4[Fe(CN)_6] \cdot 3H_2O$ and 0.25M NaCl, and was circulated in the electrode compartment at the flow rate of 90 $mL \cdot min^{-1}$.

To investigate the influence of ion composition on the RED performance, the HC was also fed with 0.5 $mol \cdot L^{-1}$ NaCl and 0.49 $mol \cdot L^{-1}$ NaCl + 0.01 $mol \cdot L^{-1}$ $NaHCO_3$, while the LC was 0.01 $mol \cdot L^{-1}$ $NaHCO_3$. Otherwise, the LC solution concentration was changed (0.005 to 0.02 M $NaHCO_3$), while fixing the HC solution as 0.495 $mol \cdot L^{-1}$ NaCl + 0.0051 $mol \cdot L^{-1}$ $NaHCO_3$ and 0.48 $mol \cdot L^{-1}$ NaCl + 0.02 M $NaHCO_3$, to test the influence of LC solution. The HC was also fed with 0.99 $mol \cdot L^{-1}$ NaCl + 0.01 $mol \cdot L^{-1}$ $NaHCO_3$, while the LC was 0.01 $mol \cdot L^{-1}$ $NaHCO_3$, to test the influence of HC solution concentration.

4. Theory

The theoretical Nernst potential over one cell ion exchange membrane for sodium chloride single medium system was calculated using the Nernst equation in Equation (3) [23,24].

$$E_{cell} = \alpha_{CEM}\frac{RT}{F}\ln\left(\frac{\gamma_c^{Na^+} \cdot C_c^{Na^+}}{\gamma_d^{Na^+} \cdot C_d^{Na^+}}\right) + \alpha_{AEM}\frac{RT}{F}\ln\left(\frac{\gamma_c^{Cl^-} \cdot C_c^{Cl^-}}{\gamma_d^{Cl^-} \cdot C_d^{Cl^-}}\right). \quad (3)$$

For sodium chloride and sodium bicarbonate double medium system, considering the two kinds anions of Cl^- and $HCO_3{}^-$ migration through anion exchange membrane, as well as referring to the multi-valent ions' climbing description in a RED process, the multi-ionic expression of the Nernst equation was revised as below.

$$E_{\text{cell}} = \alpha_{CEM}\frac{RT}{F}\ln(\frac{\gamma_c^{Na^+}\cdot C_c^{Na^+}}{\gamma_d^{Na^+}\cdot C_d^{Na^+}}) + \alpha_{AEM}\frac{RT}{F}[\ln(\frac{\gamma_c^{Cl^-}\cdot C_c^{Cl^-}}{\gamma_d^{Cl^-}\cdot C_d^{Cl^-}}) + \ln(\frac{\gamma_c^{HCO_3^-}\cdot C_c^{HCO_3^-}}{\gamma_d^{HCO_3^-}\cdot C_d^{HCO_3^-}})]. \quad (4)$$

Then, the theoretical open circuit potential of the whole membrane stack was simplified into Equation (5).

$$E_{OCV} = N\frac{(\alpha_{CEM} + \alpha_{AEM})RT}{zF}\ln(\frac{a_c}{a_d}). \quad (5)$$

The power produced is determined by the electrochemical potential drop across the membrane stack (E_{OCV}), the stack resistance, and external load resistance (R_{load}) resulting in Equation (6) [12,25].

$$P = I^2R_{load} = \frac{E_{OCV}{}^2}{(R_{stack} + R_{load})^2}R_{load}. \quad (6)$$

The maximum power output of the RED system is obtained when R_{load} equals the resistance of the stack (R_{stack}). Thus, the maximum power output can be simplified into Equation (7).

$$P_{max} = \frac{E_{OCV}{}^2}{4R_{stack}}. \quad (7)$$

Consequently, the gross power density (power output per unit membrane area, P_{gross}) was calculated from P_{max}, which is shown in Equation (8) [10].

$$P_{gross} = \frac{P_{max}}{2AN} = \frac{E_{OCV}{}^2}{8ANR_{stack}}, \quad (8)$$

where P_{gross} is the maximum gross power density ($W{\cdot}m^{-2}$), P_{max} is maximum power output (W), A is effective area of a single ion exchange membrane (m^2).

The total electric resistance of the RED stack includes the parts of electrodes, electrolytes, membranes, and diffusion boundary layers on the membrane-solution interface. Simplified models neglect the resistance of diffusion boundary layers and combine its contribution with membrane resistance and express the overall resistance (Ω) as Equation (9) [10].

$$R_{stack} = \frac{N}{A}(R_a + R_c + \frac{d_c}{\kappa_c} + \frac{d_d}{\kappa_d}) + R_{el}, \quad (9)$$

where A is a single effective membrane area (cm^2); R_a is the area resistance ($\Omega{\cdot}cm^2$) of anion exchange membrane; R_c is the area resistance ($\Omega{\cdot}cm^2$) of cation exchange membrane; R_{el} is the resistance (Ω) of electrodes; d_c is the thickness (cm) of HC; d_d is the thickness of LC; κ_c is the specific conductivity ($mS{\cdot}cm^{-1}$) of the concentrated solution; and κ_d is the specific conductivity ($mS{\cdot}cm^{-1}$) of the diluted solution.

The electrode resistance was negligible when the repeated membrane pairs were more substantial than 20 pcs. Then the entire RED stack resistance was expressed as Equation (10) [26,27].

$$R_{stack} = \frac{N}{A}(R_{ohmic} + R_{\Delta C} + R_{DL}), \quad (10)$$

where R_{ohmic} is the membrane resistance ascribed to the ionic transport through the membranes, which is equal to the one cell resistance discussed above. $R_{\Delta C}$ is the resistance ascribed to the reduced electromotive forces as a consequence of the change in the concentration of the bulk solution. It considers the change of the solution concentration from the inlet to the outlet of the solution compartment with the spatial difference of membrane potential (y-axis). R_{BL} is the boundary layer resistance due to concentration polarization which is induced by diffusive boundary layer near the membranes at lower flow rates (*x*-axis). Both $R_{\Delta C}$ and R_{BL} are kinds of non-ohmic resistance.

The net power density (P_{net}, in W·cm^{-2}) that generates in a RED stack is the difference between the gross power density and the power consumed on solution pumping (pressure drop over the inlet and outlet channel). Thus, the net power density can be expressed as Equation (11) [10].

$$P_{net} = P_{gross} - P_{pump} = \frac{E_{OCV}^2}{8ANR_{stack}} - \frac{\Delta p_c Q_c + \Delta p_d Q_d}{NA}, \tag{11}$$

where Δp_c and Δp_d are the pressure drops along HC and LC, respectively. Q_c and Q_d are the volumetric flow rates of HC and LC, respectively.

5. Conclusions

This work investigated the influence of ion composition on the trans-membrane potential across the ion exchange membrane (IEM), for a better understanding of "reverse electrodialysis heat engine" when running in the complex ion composition. The artificially prepared solutions $NaHCO_3$, Na_2CO_3, and NH_4Cl circulate inside the RED stack, as the alternative of the conventional NaCl systems. The evaluative criteria of the RED performance, i.e., open-circuit voltage, gross power density, and short-circuit current were introduced for the evaluation of the process. It was found that the new system (0.49 M NaCl + 0.01 M $NaHCO_3$|0.01 M $NaHCO_3$) output a relatively stable power density (0.174 W·m^{-2}), with the open-circuit voltage 2.95 V under the low flow rate of 0.22 cm/s. Meanwhile, the simulated natural system (0.5 M NaCl|0.01 M NaCl) output the power density 0.168 W·m^{-2}, with the open-circuit voltage 2.86 V under the low flow rate of 0.22 cm/s. The work will advance the understanding of reverse electrodialysis heat engine process, as well as the performance of RED when running in complex systems (wastewater).

Author Contributions: F.L. designed the study, conducted experiments, and prepared the manuscript; M.S. revised the manuscript; Y.W. (Yang Wang) and Y.W. (Yanxin Wei) carried out formal analysis.

Funding: This project was supported by the Anhui Provincial Outstanding Young Talent Support Program (gxyq2017051), the Anhui Provincial Natural Science Foundation (1808085QB36, 1808085MB35), Hefei Normal University College Scientific Research Project (2017QN15) and Foundation for Outstanding Talents in Higher Education of Anhui(gxbjZD32).

Conflicts of Interest: The authors declare no conflict of interest.

Abbreviations

A	the effective area of a single membrane (m^2)
a_c	the activity of the concentrated salt solution (mol·L^{-1})
a_d	the activity of the diluted salt solution (mol·L^{-1})
AEM	anion exchange membrane
b	compartment width (cm)
C_c	the concentration of HC (mol·L^{-1})
C_d	the concentration of LC (mol·L^{-1})
CEM	cation exchange membrane
d_c	the thickness of HCC (cm)
d_d	the thickness of LCC (cm)
F	Faraday constant (96485 C·mol^{-1})
HCC	high salt concentration compartment
I	current (A)
LCC	low salt concentration compartment
N	number of cell pairs for RED

OCV	open circuit voltage (V)
P	output power (W)
P_{gross}	maximum gross power density ($W \cdot m^{-2}$)
P_{max}	maximum output power (W)
P_{net}	net power density ($W \cdot m^{-2}$)
P_{pump}	power density consumed on pumping ($W \cdot m^{-2}$)
Q	volumetric flow rate ($mL \cdot min^{-1}$)
Q_c	the volumetric flow rate of HC ($L \cdot s^{-1}$)
Q_d	volumetric flow rate of LC ($L\ s^{-1}$)
R	gas constant (8.314 $J \cdot mol^{-1} \cdot K^{-1}$)
R_a	the area resistance of anion exchange membrane ($\Omega \cdot m^2$)
R_{BL}	boundary layer resistance (Ω)
R_c	the area resistance of cation exchange membrane ($\Omega \cdot m^2$)
R_{el}	the resistance of electrodes (Ω)
R_{load}	load resistance (Ω)
R_{ohmic}	ohmic resistance (Ω)
R_{stack}	stack resistance (Ω)
$R_{\triangle C}$	concentration difference resistance (Ω)
RED	reverse electrodialysis
SGP	salinity gradient power
T	absolute temperature (K)
U	voltage output
V	line flow velocity ($cm \cdot min^{-1}$)
z	electrochemical valence
α_{AEM}	permselectivity of the anion exchange membrane
α_{CEM}	permselectivity of the cation exchange membrane
δ	spacer thickness (cm)
ε_{sp}	spacer porosity
κ_c	the specific conductivity of the concentrated solution ($mS \cdot cm^{-1}$)
κ_d	the specific conductivity of the diluted solution ($mS \cdot cm^{-1}$)
Δp_c	the pressure drops along HC (KPa)
Δp_d	the pressure drops along LC (KPa)

References

1. Ortiz-Imedio, R.; Gomez-Coma, L.; Fallanza, M.; Ortiz, A.; Ibañez, R.; Ortiz, I. Comparative performance of Salinity Gradient Power-Reverse Electrodialysis under different operating conditions. *Desalination* **2019**, *457*, 8–21. [CrossRef]
2. Post, J.W.; Hamelers, H.V.M.; Buisman, C.J.N. Energy recovery from controlled mixing salt and fresh water with a reverse electrodialysis system. *Environ. Sci. Technol.* **2008**, *42*, 5785–5790. [CrossRef]
3. Logan, B.E.; Elimelech, M. Membrane-based processes for sustainable power generation using water. *Nature* **2012**, *488*, 313–319. [CrossRef]
4. Ran, J.; Wu, L.; He, Y.B.; Yang, Z.J.; Wang, Y.M.; Jiang, C.X.; Ge, L.; Bakangura, E.; Xu, T.W. Ion exchange membranes: New developments and applications. *J. Membr. Sci.* **2017**, *522*, 267–291. [CrossRef]
5. Mei, Y.; Tang, C.Y.Y. Recent developments and future perspectives of reverse electrodialysis technology: A review. *Desalination* **2018**, *425*, 156–174. [CrossRef]
6. Długołęcki, P.; Nymeijer, K.; Metz, S.; Wessling, M. Current status of ion exchange membranes for power generation from salinity gradients. *J. Membr. Sci.* **2008**, *319*, 214–222. [CrossRef]
7. Vermaas, D.A.; Bajracharya, S.; Sales, B.B.; Saakes, M.; Hamelers, B.; Nijmeijer, K. Clean energy generation using capacitive electrodes in reverse electrodialysis. *Energ. Environ. Sci.* **2013**, *6*, 643–651. [CrossRef]
8. Varcoe, J.R.; Atanassov, P.; Dekel, D.R.; Herring, A.M.; Hickner, M.A.; Kohl, P.A.; Kucernak, A.R.; Mustain, W.E.; Nijmeijer, K.; Scott, K.; et al. Anion-exchange membranes in electrochemical energy systems. *Ener.g Environ. Sci.* **2014**, *7*, 3135–3191. [CrossRef]

9. Veerman, J.; Saakes, M.; Metz, S.J.; Harmsen, G.J. Reverse electrodialysis: Evaluation of suitable electrode systems. *J. Appl. Electrochem.* **2010**, *40*, 1461–1474. [CrossRef]
10. Luo, F.B.; Wang, Y.M.; Jiang, C.X.; Wu, B.; Feng, H.Y.; Xu, T.W. A power free electrodialysis (PFED) for desalination. *Desalination* **2017**, *404*, 138–146. [CrossRef]
11. Tufa, R.A.; Pawlowski, S.; Veerman, J.; Bouzek, K.; Fontananova, E.; di Profio, G.; Velizarov, S.; Crespo, J.G.; Nijmeijer, K.; Curcio, E. Progress and prospects in reverse electrodialysis for salinity gradient energy conversion and storage. *Appl. Energ.* **2018**, *225*, 290–331. [CrossRef]
12. Hong, J.G.; Zhang, B.P.; Glabman, S.; Uzal, N.; Dou, X.M.; Zhang, H.G.; Wei, X.Z.; Chen, Y.S. Potential ion exchange membranes and system performance in reverse electrodialysis for power generation: A review. *J. Membr. Sci.* **2015**, *486*, 71–88. [CrossRef]
13. Ciofalo, M.; La Cerva, M.; Di Liberto, M.; Gurreri, L.; Cipollina, A.; Micale, G. Optimization of net power density in Reverse Electrodialysis. *Energy* **2019**, *181*, 576–588. [CrossRef]
14. Gao, H.P.; Zhang, B.P.; Tong, X.; Chen, Y.S. Monovalent-anion selective and antifouling polyelectrolytes multilayer anion exchange membrane for reverse electrodialysis. *J. Membr. Sci.* **2018**, *567*, 68–75. [CrossRef]
15. Pawlowski, S.; Crespo, J.G.; Velizarov, S. Profiled Ion Exchange Membranes: A Comprehensible Review. *Int. J. Mol. Sci.* **2019**, *20*, 165. [CrossRef]
16. Zhu, X.P.; He, W.H.; Logan, B.E. Influence of solution concentration and salt types on the performance of reverse electrodialysis cells. *J. Membr. Sci.* **2015**, *494*, 154–160. [CrossRef]
17. Vermaas, D.A.; Veerman, J.; Saakes, M.; Nijmeijer, K. Influence of multivalent ions on renewable energy generation in reverse electrodialysis. *Energ. Environ. Sci.* **2014**, *7*, 1434–1445. [CrossRef]
18. Kim, T.; Logan, B.E.; Gorski, C.A. A pH-Gradient Flow Cell for Converting Waste CO2 into Electricity. *Environ. Sci. Tech. Let.* **2017**, *4*, 49–53. [CrossRef]
19. Geise, G.M.; Curtis, A.J.; Hatzell, M.C.; Hickner, M.A.; Logan, B.E. Salt Concentration Differences Alter Membrane Resistance in Reverse Electrodialysis Stacks. *Environ. Sci. Tech. Let.* **2014**, *1*, 36–39. [CrossRef]
20. Jiang, C.X.; Wang, Q.Y.; Li, Y.; Wang, Y.M.; Xu, T.W. Water electro-transport with hydrated cations in electrodialysis. *Desalination* **2015**, *365*, 204–212. [CrossRef]
21. Jiang, C.X.; Zhang, D.Y.; Muhammad, A.S.; Hossain, M.M.; Ge, Z.J.; He, Y.B.; Feng, H.Y.; Xu, T.W. Fouling deposition as an effective approach for preparing monovalent selective membranes. *J. Membr. Sci.* **2019**, *580*, 327–335. [CrossRef]
22. Tedesco, M.; Brauns, E.; Cipollina, A.; Micale, G.; Modica, P.; Russo, G.; Helsen, J. Reverse Electrodialysis with saline waters and concentrated brines: A laboratory investigation towards technology scale-up. *J. Membr. Sci.* **2015**. [CrossRef]
23. Chen, X.; Jiang, C.X.; Shehzad, M.A.; Wang, Y.M.; Feng, H.Y.; Yang, Z.J.; Xu, T.W. Water-Dissociation-Assisted Electrolysis for Hydrogen Production in a Salinity Power Cell. *ACS Sustain. Chem. Eng.* **2019**, *7*, 13023–13030. [CrossRef]
24. Tedesco, M.; Hamelers, H.V.M.; Biesheuvel, P.M. Nernst-Planck transport theory for (reverse) electrodialysis: I. Effect of co-ion transport through the membranes. *J. Membrane. Sci.* **2016**, *510*, 370–381. [CrossRef]
25. Güler, E. *Anion Exchange Membrane Design for Reverse Electrodialysis*; Universiteit Twente: Enschede, The Netherlands, 2014.
26. Pawlowski, S.; Sistat, P.; Crespo, J.G.; Velizarov, S. Mass transfer in reverse electrodialysis: Flow entrance effects and diffusion boundary layer thickness. *J. Membr. Sci.* **2014**, *471*, 72–83. [CrossRef]
27. Vermaas, D.A.; Saakes, M.; Nijmeijer, K. Enhanced mixing in the diffusive boundary layer for energy generation in reverse electrodialysis. *J. Membr. Sci.* **2014**, *453*, 312–319. [CrossRef]

International Journal of
Molecular Sciences

MDPI

Article

Partial Fluxes of Phosphoric Acid Anions through Anion-Exchange Membranes in the Course of NaH_2PO_4 Solution Electrodialysis

Olesya Rybalkina, Kseniya Tsygurina, Ekaterina Melnikova, Semyon Mareev, Ilya Moroz, Victor Nikonenko *and Natalia Pismenskaya

Kuban State University, 149 Stavropolskaya st., 350040 Krasnodar, Russia
* Correspondence: v_nikonenko@mail.ru; Tel.: +7-918-41-45-816

Received: 2 July 2019; Accepted: 19 July 2019; Published: 23 July 2019

Abstract: Electrodialysis (ED) with ion-exchange membranes is a promising method for the extraction of phosphates from municipal and other wastewater in order to obtain cheap mineral fertilizers. Phosphorus is transported through an anion-exchange membrane (AEM) by anions of phosphoric acid. However, which phosphoric acid anions carry the phosphorus in the membrane and the boundary solution, that is, the mechanism of phosphorus transport, is not yet clear. Some authors report an unexpectedly low current efficiency of this process and high energy consumption. In this paper, we report the partial currents of $H_2PO_4^-$, HPO_4^{2-}, and PO_4^{3-} through Neosepta AMX and Fujifilm AEM Type X membranes, as well as the partial currents of $H_2PO_4^-$ and H^+ ions through a depleted diffusion layer of a 0.02 M NaH_2PO_4 feed solution measured as functions of the applied potential difference across the membrane under study. It was shown that the fraction of the current transported by anions through AEMs depend on the total current density/potential difference. This was due to the fact that the pH of the internal solution in the membrane increases with the growing current due to the increasing concentration polarization (a lower electrolyte concentration at the membrane surface leads to higher pH shift in the membrane). The HPO_4^{2-} ions contributed to the charge transfer even when a low current passed through the membrane; with an increasing current, the contribution of the HPO_4^{2-} ions grew, and when the current was about 2.5 i_{lim}^{Lev} (i_{lim}^{Lev} was the theoretical limiting current density), the PO_4^{3-} ions started to carry the charge through the membrane. However, in the feed solution, the pH was 4.6 and only $H_2PO_4^-$ ions were present. When $H_2PO_4^-$ ions entered the membrane, a part of them transformed into doubly and triply charged anions; the H^+ ions were released in this transformation and returned to the depleted diffusion layer. Thus, the phosphorus total flux, j_P (equal to the sum of the fluxes of all phosphorus-bearing species) was limited by the $H_2PO_4^-$ transport from the bulk of feed solution to the membrane surface. The value of j_P was close to i_{lim}^{Lev}/F (F is the Faraday constant). A slight excess of j_P over i_{lim}^{Lev}/F was observed, which is due to the electroconvection and exaltation effects. The visualization showed that electroconvection in the studied systems was essentially weaker than in systems with strong electrolytes, such as NaCl.

Keywords: ion-exchange membrane; Fujifilm; Neosepta; phosphate transport; limiting current density; voltammetry

---

1. Introduction

Ampholytes are substances which have chemical structures and electrical charges that depend on the pH of the medium due to their participation in protonation–deprotonation reactions. Ampholytes comprise a large number of substances, including nutrients or valuable components of food. Among them there are peptides, amino acids, anthocyanins, orthophosphoric, tartaric, citric acid anions, etc. Electrophoresis and electrodialysis with ion-exchange membranes (IEMs) are used increasingly to extract these substances from wastewater [1–4], products of biomass processing [5], as well as liquid wastes in the food industry [6–11].

The attractiveness of these methods is conditioned by the possibility of using not only the difference in particle mobility, but also the ability to change the sign and magnitude of their electric charge when adjusting the pH. The peculiarity of electrodialysis processes is that the composition of the ampholyte-containing solution changes in space and time not only quantitatively, as in the case of strong electrolytes, but also qualitatively. Indeed, the electrodialysis of strong electrolytes (for example, NaCl) is accompanied only by an increase or decrease in their concentration [12]. In the case of ampholytes, the transformation of one form into another takes place not only in the solution, which is located in the intermembrane space [13,14], but also inside an IEM [15,16]. In over-limiting current regimes, generation of H^+ and OH^- ions in protonation–deprotonation reactions with the participation of fixed groups at the membrane/depleted solution boundary [17,18] can significantly affect this transformation. These reactions can cause the so-called barrier effect [19,20]. It lies in the fact that a change in the pH of an ampholyte-containing solution at the membrane surface facing the desalination compartment entails the transformation of ampholyte species. Depending on the value of pH change, the ampholyte ions, which migrate from the solution bulk towards the membrane as counterions, can turn into molecular (zwitterionic for amino acid) form or into a co-ion whose charge sign is opposite to the charge sign of the ampholyte particles in the bulk solution. The molecular (zwitterionic) form cannot be transported through the IEM at the same rate as the counterions; the coions are ejected by the electric field from the near-membrane region into the bulk solution. If the generation of H^+ and OH^- ions occurs at both membranes forming the desalination channel, the ampholyte cations, which are formed at the anion-exchange membrane (AEM) are delivered by the electric field to the cation-exchange membrane (CEM), where they are transformed into anions and return back to the AEM, where they change the charge sign again. This phenomenon is called the circulation effect [21]. Both effects are used for the purification of amino acids or carboxylic acids from mineral impurities [22], as well as for the separation of inorganic ampholytes, such as sulfates and phosphates [23].

The transformation of an ampholyte from one form to another is also possible when it enters or leaves the membrane. This is due to the fact that the internal solution of AEM is more alkaline than the external solution; the pH of the internal solution is one or two units higher than in that of the external solution [16,24–26]. The reason is that the H^+ ions are pushed out from an AEM as coions. Similarly, due to the Donnan exclusion of OH^- ions from a CEM, the internal solution of this membrane has a pH value 1–2 units lower than in the external solution. As a result of this pH shift, the effects of facilitated diffusion [27,28] and facilitated electromigration [21,29] occur inside IEMs. The essence of these effects lies in the fact that getting into the acidic (or alkaline) medium inside the IEM, an amino acid zwitterion acquires a charge opposite to the charge of the membrane's fixed sites. After such a transformation, it easily passes through the IEM as a counterion. In our recent work [30], a similar mechanism was described, which explains a relatively high transport of ammonium ions (coions) through an AEM. In this case, the positively charged ammonium ion enters the alkaline AEM medium and transforms into a molecular form, which is not affected by the Donnan exclusion.

Thus, the transport of ampholytes in systems with ion-exchange membranes is coupled with chemical reactions of protonation–deprotonation. The influence of these reactions on the behavior of membrane systems in conditions of applied electric current has been described in theoretical works [31,32]. The dependence of concentration profiles of ampholyte species as well as their transport numbers in the membrane and adjacent diffusion layers on the applied current density was calculated. However, experimental verification of the model predications was not carried out.

As for the transport of phosphoric acid anions through AEM, this subject is of considerable interest not only for theory but also for practice. Isolation, purification, and concentration of these anions from municipal wastewater [33–35], animal waste [2,36], and sludge generated after its biological treatment [37,38] not only reduces the anthropogenic impact on the environment, but also allows for obtaining cheap fertilizers with simultaneous production of electricity [38]. It is important that the current efficiency in the recovery of phosphorus from solutions containing $H_2PO_4^-$ ions is significantly lower as compared to similar processes in the case of nitrates, chlorides, and other ions that do not undergo protonation–deprotonation reactions [2,34,39–41]. With an increasing potential drop, the phosphorus recovery efficiency from the desalination compartment first grows rapidly, but then remains unchanged over a wide range of voltages [23]; the recovery efficiency largely depends on the pH of the treated solution [39]. Some researchers explain the low efficiency of phosphorus recovery by steric hindrances that arise when transporting large, highly hydrated phosphoric acid anions [2,34].

Paltrinieri et al. [42], using the material balance equations, came to the conclusion that electric current can be transported through AEMs by doubly charged HPO_4^{2-} anions, while the feed solution contains only an NaH_2PO_4 solution, where only the singly charged $H_2PO_4^-$ anions are present. Note that, as a rule, when choosing a current mode, researchers are guided by the limiting current, i_{lim}, which is found from the intersection of the tangents to the initial part and the inclined plateau of current–voltage characteristic (CVC) curves [34,40]. Another way to find i_{lim} is through the use of the Cowan–Brown method [43] for CVC processing [44].

In this work, we report the experimental CVC and partial current densities of all orthophosphoric acid anions, namely, the $H_2PO_4^-$, HPO_4^{2-}, and PO_4^{3-} ions, through Neosepta AMX and Fujifilm AEM Type X membranes in the case of a 0.02 M NaH_2PO_4 feed solution. As well, we found the fluxes of phosphorus-bearing species through the membranes under study and the partial currents of $H_2PO_4^-$ and H^+ ions in the depleted diffusion layer adjacent to the membrane. We compared all these characteristics with the results of a simulation made using the model developed earlier [31,32]. We showed that the phosphorus-bearing species through an AEM in the ED process was less than one would expect if judging by the conventional treatment of CVC. Accordingly, we focused on the determination of limiting current density using the tangent intersection method and the Cowan–Brown method.

2. Results and Discussion

2.1. Total and Partial Current–Voltage Characteristics

Figures 1 and 2 show the experimental and calculated total and partial current-voltage characteristics (CVCs) of an AMX membrane in NaCl (Figure 1) and NaH_2PO_4 (Figure 2) solutions. The calculations are made using a mathematical model developed earlier [31,32] and briefly described below.

In the case of NaCl (Figure 1), the shapes of the total and partial CVCs were close to those described in the literature [45,46] for strong electrolytes, which do not participate in the proton-exchange reactions. The experimental limiting current, which is determined by the tangent intersection method (Figure 1), was close to i_{lim}^{Lev}. Note that the shape of the CVCs in the range from 1.5 to 2.5 mA·cm^{-2} was rather particular. There was a region where the value of $\Delta\varphi'$ decreased when the current density increased. The differential resistance of the membrane system within this current range was negative: $R_{dif} = d(\Delta\varphi)/di < 0$. This "anomaly" is known for the AMX and some other IEMs [47], it is due to the early electroconvective vortex formation at electrically and geometrically heterogeneous surfaces. With growing current density in the indicated range of currents, the increasing electroconvection (occurring as electroosmosis of the first kind [48]) makes the depleted solution resistance lower. Approaching the limiting current density was manifested by a fast increase in the differential resistance with increasing i. Thus, to take into account the singularity $R_{dif} < 0$ when determining the limiting current density, we drew the tangents to close to the linear segments of the CVCs, just before and after the occurrence of the limiting state, as shown in Figure 1.

The partial current density of OH^- ions became noticeable at a reduced potential drop corresponding to the limiting current, then it slowly increased with increasing potential drop in the range corresponding to the CVC plateau (section II) and it sharply increased when $\Delta\varphi'$ exceeded 0.8 V, i.e., in the over-limiting regime (section III).

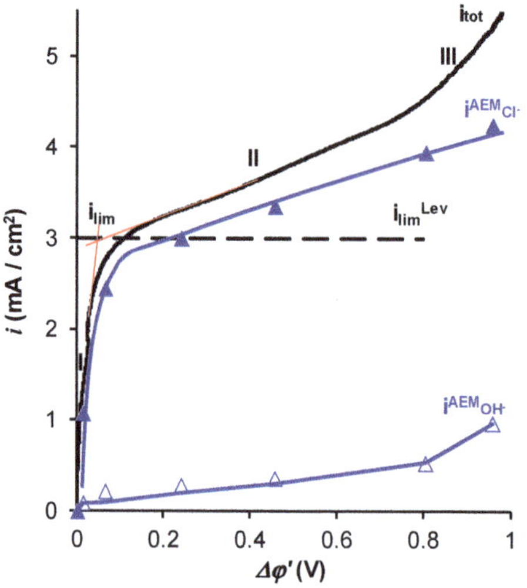

Figure 1. Total and partial current–voltage characteristics of the AMX membrane in a 0.02 M NaCl solution; i_{tot} is the total current density, $i^{AEM}_{Cl^-}$ and $i^{AEM}_{OH^-}$ are the current densities of the Cl^- and OH^- ions through the membrane, respectively. The dashed lines show the tangents to the CVC used to determine the experimental limiting current density.

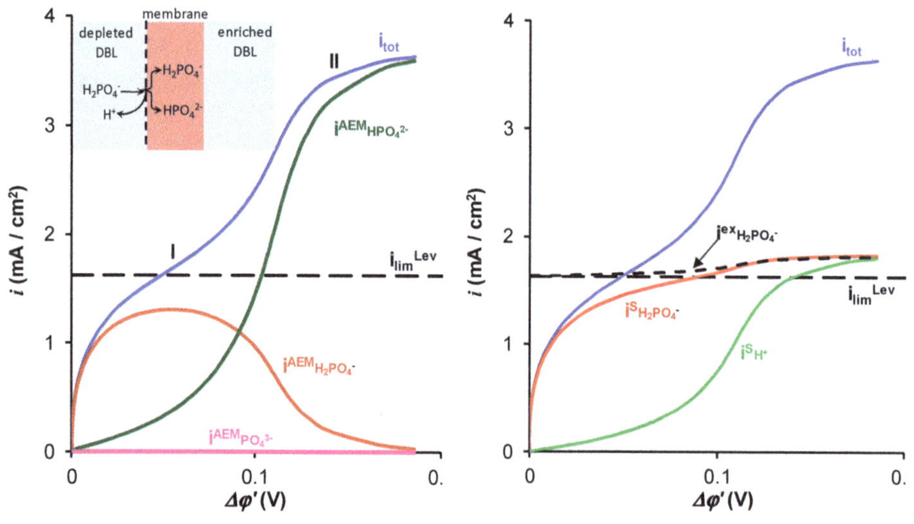

Figure 2. Theoretical total current density (i_{tot}) and partial currents of $H_2PO_4^-$ ($i^{AEM}_{H_2PO_4^-}$) and HPO_4^{2-} ($i^{AEM}_{HPO_4^{2-}}$) ions in an AMX membrane (**a**) as well as the partial currents of $H_2PO_4^-$ ($i^{s}_{H_2PO_4^-}$) and H^+ ($i^{s}_{H^+}$) ions in the depleted solution at the membrane surface (**b**) as functions of the corrected potential drop. Solid lines were calculated using the model [31,32]. Dashed lines show the limiting current $i^{Lev}_{\lim}$ calculated using the Leveque equation, Equation (6), and the exaltation current, $i^{ex}_{H_2PO_4^-}$, calculated using Equation (1). "I" and "II" show the first and second inclined plateaus, respectively.

The shape of $i^{AEM}_{OH^-}$ versus $\Delta\varphi'$ dependence shown in Figure 1 is fully consistent with well-established ideas about the development of water splitting with the catalytic participation of fixed-membrane groups: the generation of H^+ and OH^- ions at the membrane/solution interface begins when the concentration of NaCl in the boundary solution at the membrane surface reaches values of about 5×10^{-5} M [49], and the H^+ (OH^-) ions in the depleted solution start to compete with the Na^+ and Cl^- ions. This state occurs when the current density is close to the limiting current density at a potential drop across the depleted diffusion layer of about 0.25 V [49]. The rate of water splitting increases with an increase in the potential drop [50]. The partial current of Cl^- ions, $i_{Cl^-}^{AEM}$, continues to grow after reaching the value $i_{\lim}^{Lev}$ due to the effect of exaltation of the limiting current [51,52] and the development of electroconvection [49,53].

In the case of the NaH_2PO_4 solution, the model developed in References [31,32] predicts the presence of two inclined plateaus on the total CVC (Figure 2). In the region of the first plateau (about 0.05 V of the reduced potential drop), the partial current density of singly charged ions $H_2PO_4^-$ in the membrane reached a maximum value that was close to the limiting current calculated by the Leveque equation. The limiting current calculated by the Leveque equation is the limiting value of the partial current carried by $H_2PO_4^-$ ions in solution, where this current is limited by diffusion through the depleted diffusion layer. When $\Delta\varphi'$ is close to 0.05 V, the concentration of $H_2PO_4^-$ ions at the AEM surface reach a value which is much lower than the concentration value in the bulk solution, hence, its diffusion flux density attains a maximum [31,32].

Thus, the appearance of the first plateau was due to the saturation of the NaH_2PO_4 salt diffusion from the solution to the membrane surface (Figure 2b).

The decrease in NaH_2PO_4 concentration in the solution at the AEM surface led to a stronger Donnan exclusion of protons from the membrane. As a result, the pH of the AEM internal solution increased and a higher part of the singly charged phosphate $H_2PO_4^-$ ions transformed into doubly charged HPO_4^{2-} ions when crossing the membrane interface: $H_2PO_4^- \rightarrow HPO_4^{2-} + H^+$ (insert in Figure 2a). The protons released into the solution at the depleted membrane interface were involved in the charge transfer in the depleted DBL forming partial current density $i^s{}_{H+}$ (Figure 2b). When the fluxes of PO_4^{3-} and OH^- ions in the membrane were negligible, $i_{H+}^s = i_{HPO_4^{2-}}^{AEM}$ (Figure 2a). These protons appearing in solution and the doubly charged HPO_4^{2-} anions in the membrane caused the rise of current density above the first limiting current density, $i_{\lim}^1$. The second plateau (and limiting current, $i_{\lim}^2$) was observed when the membrane was completely converted into the HPO_4^{2-} form. In this state, the flux of protons, released when the $H_2PO_4^-$ anions entered the membrane and transformed into HPO_4^{2-}, was saturated.

Experimental values of the partial current densities through the AMX and Fuji membranes were calculated using the value of the HPO_4^{2-} ion flux density, $j_{H_2PO_4^-}^s$, which was measured using the method described in Section 3.2.1. Namely, $j_{H_2PO_4^-}^s$ was found from the rate of the NaH_2PO_4 removal from the diluate in the batch mode ED, Equation (14). Then, using Equations (15)–(20), it was possible to calculate the partial current densities of the $H_2PO_4^-$, HPO_4^{2-}, PO_4^{3-} and OH^- ions in a membrane, by applying these equations in pairs in a proper range of pH. For example, when the pH of the membrane internal solution was between 5 and 10, Equations (15) and (18) allow calculation of $i_{H_2PO_4^-}^{AEM}$ and $i_{HPO_4^{2-}}^{AEM}$. If the calculated value of $i_{H_2PO_4^-}^{AEM}$ is less than zero, it insinuates that the pH in the membrane is higher than 10, and the PO_4^{3-} anions are involved in current transfer instead of $H_2PO_4^-$. Then, Equations (19) and (20) must be used. As Figure 3 shows, the contribution of the PO_4^{3-} ions into the charge transfer in the membranes becomes significant only at relatively high current densities (exceeding approximately $3i_{\lim}{}^{Lev}$). Thus, we did not reach the conditions where the OH^- ions pass across the membrane because of the limitations of the measuring device.

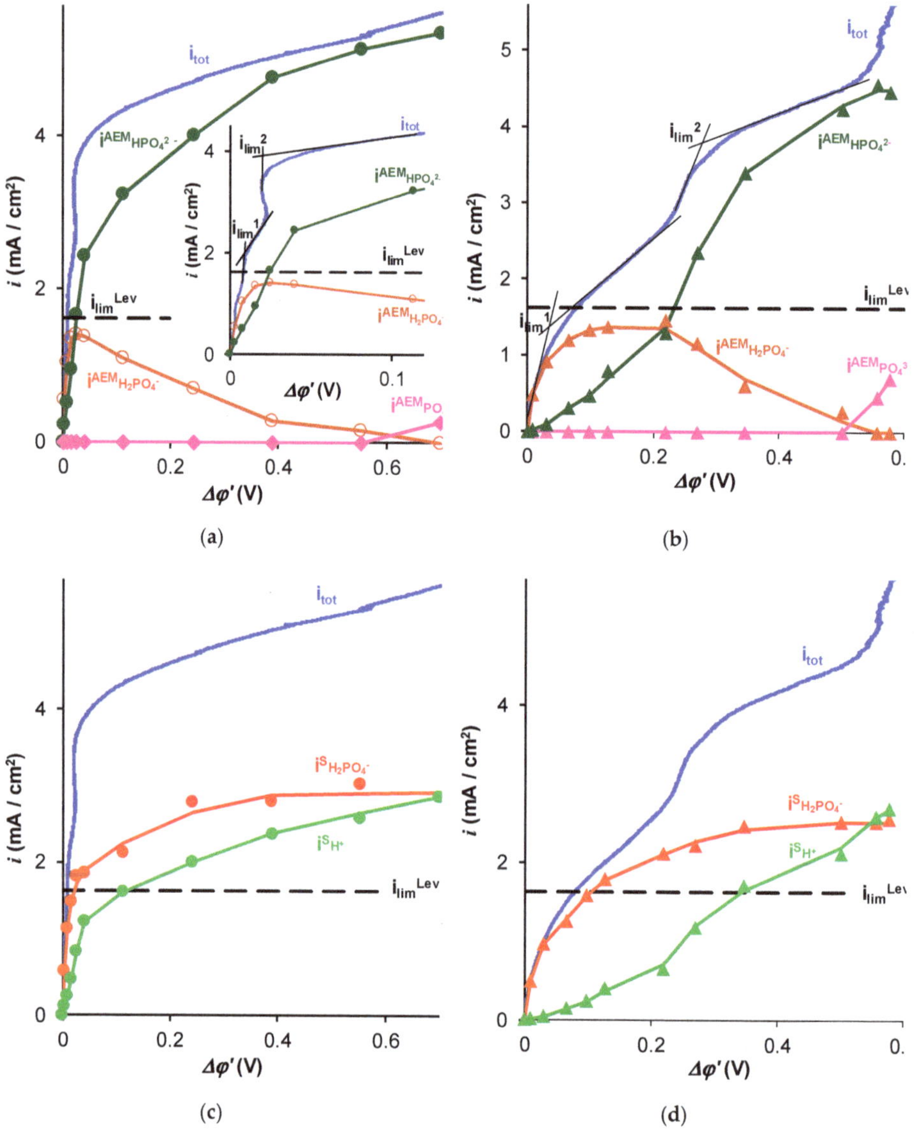

Figure 3. Experimental total current density (i_{tot}) and partial currents of $H_2PO_4^-$ ($i^{AEM}_{H_2PO_4^-}$), HPO_4^{2-} ($i^{AEM}_{HPO_4^{2-}}$), and PO_4^{3-} ($i^{AEM}_{PO_4^{3-}}$) ions in AMX (**a**) and AEM Fuji Type X (**b**) membranes, as well as partial currents of $H_2PO_4^-$ ($i^{s}_{H_2PO_4^-}$) and H^+ ($i^{s}_{H^+}$) in depleted solution near AMX (**c**) and AEM Fuji Type X (**d**) membranes as functions of the corrected potential drop. The curves connecting the markers are the fitting lines. The data were obtained in a 0.02 M NaH_2PO_4 solution. The dashed lines show the limiting current i^{Lev}_{lim} calculated using Equation (6).

When HPO_4^{2-} and PO_4^{3-} ions cross the membrane boundary with the enriched solution and appear in aqueous solution, they capture H^+ ions from water, which leads to an increase in pH of the enriched solution. It can be also interpreted as the generation of OH^- ions at the membrane/enriched solution interface. Thus, in membrane systems with NaH_2PO_4 solution, the process of generation of H^+ and OH^- ions is separated in space: the H^+ ions appear at the depleted solution/membrane interface and the OH^- ions at the membrane/enriched solution interface. The contribution of the OH^- transport in the charge transfer in the membrane becomes essential only at sufficiently high current densities/potential drops, when the pH of the internal membrane solution is greater than 13 (see Section 3.2.1.). Therefore, when pH < 13, the transport of OH^- ions across the membrane is negligible. In the experiments in this study, the situation where the transport of OH^- ions was measurable was not achieved because of the limitations of the measuring device.

2.1.1. Contributions of Electroconvection and Exaltation Effects.

The comparison of the results of the experiment (Figures 3 and 4a) and simulation using the model [31,32] (Figures 2 and 4a) for the values of i_{tot} and i_i as functions of $\Delta\varphi'$ showed that they were in relatively good agreement. The difference lies in the fact that for a given reduced potential drop (up to $\Delta\varphi'$ = 0.2 V, which is the maximum value for which the computations according to the model are possible), the experimental values of i_{tot} and $i^{AEM}_{H_2PO_4^-}$ were slightly higher than that predicted by the model. The experimental phosphorus flux density, $j_P = i^S_{H_2PO_4^-}/F$, through the membranes under study (Figure 4), determined by Equation (13), also exceeded the values calculated using this model [32].

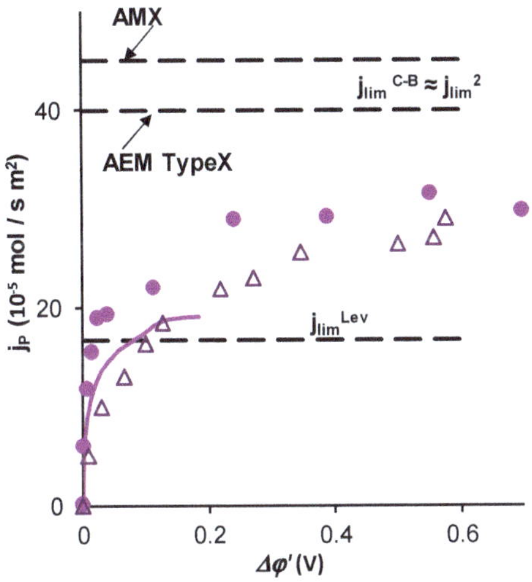

Figure 4. Experimental (markers) and calculated using the model [31,32] (solid line) total phosphorus flux density through the AMX (circles) and AEM Type-X (triangles) membrane vs. the reduced potential drop. Here $j_{lim}^{Lev} = i_{lim}^{Lev}/F$, where i_{lim}^{Lev} is calculated using Equation (6); $j_{lim}^{C\text{-}B} = i_{lim}^{C\text{-}B}/F$, where $i_{lim}^{C\text{-}B}$ is found from the CVC curves by the Cowan–Brown method [43].

The increase in the flux density of $H_2PO_4^-$ ions from the bulk solution to the AEM surface ($j^S_{H_2PO_4^-}$) over the value corresponding to the limiting current density i_{lim}^{Lev} can be caused only by two effects: (1) electroconvection, which reduces the effective thickness of the diffusion layer and (2) exaltation of the $H_2PO_4^-$ current. As shown in References [49,54], the contribution of gravitational convection in such systems is negligible. Figure 2b shows the calculated partial current density of $H_2PO_4^-$ ions, $\left(i^S_{H_2PO_4^-}\right)_{electrodif}$, which would occur if the ion transport was only due to electro-diffusion, taking into account the effect of exaltation [51], while the electroconvection was not taken into account. The calculation is carried out using the following equation [32]:

$$\left(i^{S}_{H_2PO_4^-}\right)_{electrodif} = i^{Lev}_{\lim} + i^{ex} = i^{Lev}_{\lim} + \frac{D_{H_2PO_4^-}}{D_{H^+}} i_{H^+} \quad (1)$$

where $i^{Lev}_{\lim}$ is the limiting value of the $H_2PO_4^-$ ion current density (calculated by Equation (6)) achieved by electro-diffusion in the absence of exaltation, and the second term, i^{ex}, describes the increase in the current density owing to the exaltation effect. This effect is due to the protons released from the $H_2PO_4^-$ ions when they enter the membrane. The H^+ ions carry a positive charge, which creates an additional electrostatic field attracting the $H_2PO_4^-$ anions from the bulk solution to the depleted membrane surface [51].

However, as can be seen from Figure 2b, the exaltation effect is too weak to provide the experimentally observed $H_2PO_4^-$ ions' flux in the diffusion layer near the membrane surface (equal to the total flux of phosphorus atoms through the membrane). Consequently, the main cause of growth should be electroconvection, which, as it known from References [48,49,53,55], significantly increases the mass transfer in dilute solutions of strong electrolytes. Our experiments show that in the systems under study, in the case of NaCl solution, the first large vortex structures (about 50 µm in size) are visualized at the outlet of the channel under study, when $i/i_{\lim}{}^{Lev} = 2$. With an increasing current, the vortex structures gradually spread along the entire length of the channel, and the dimensions of the vortices increase. At $i/i_{\lim}{}^{Lev} = 6$ (Figure 5a), the vortex structures occupy almost half of the 900 µm intermembrane space.

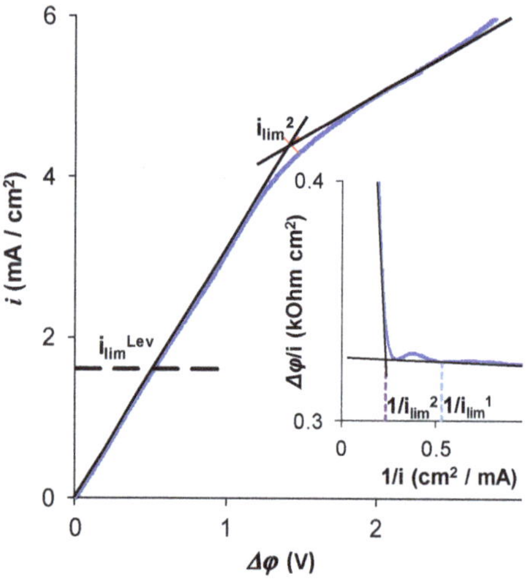

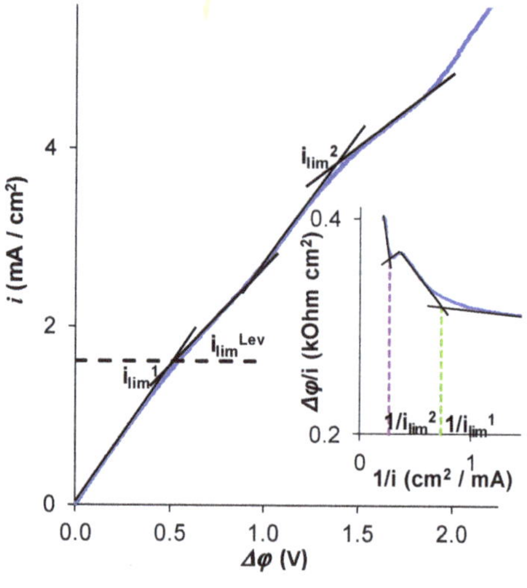

Figure 5. Current–voltage characteristics presented in the usual coordinates and the coordinates proposed by Cowan and Brown [43] (the insertion) for an AMX (**a**) and AEM Type X (**b**) membranes. $i_{\lim}{}^1$ and $i_{\lim}{}^2$ were the first and the second limiting currents determined by the tangent intersection method. The dashed lines in the insertions show the positions of the $1/i_{\lim}{}^1$ and $1/i_{\lim}{}^2$ points on the Cowan–Brown curves. The experiments were carried out in a 0.02 M NaH_2PO_4 solution.

The scenario of the electroconvective structures' development in the NaH_2PO_4 solution was the same as in the case of the NaCl solution. The difference was that the vortices of the same diameter emerged at essentially higher current densities, e.g., those with a diameter of 50 μm only formed when $i/i_{lim}^{Lev} = 6$ (Figure 6b).

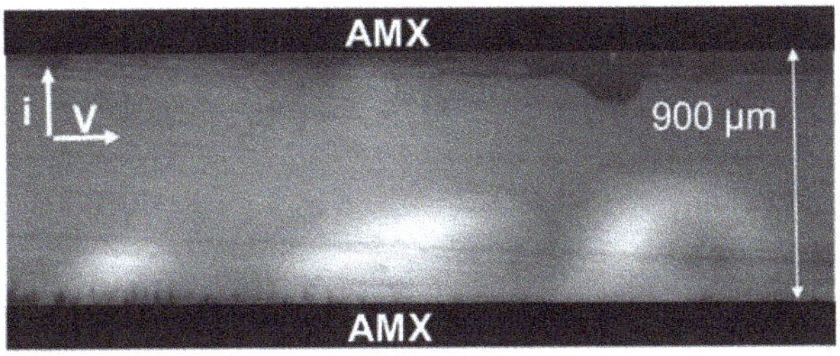

(a)

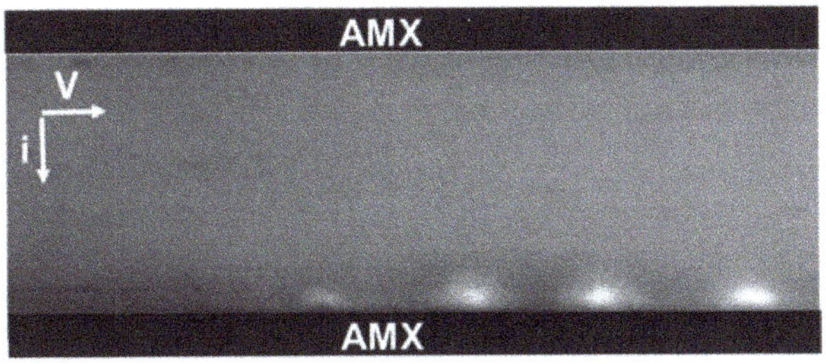

(b)

Figure 6. Visualization of the vortex structures at the AMX membrane surface in the desalination channel of the electrodialysis cell, where a 0.01 M NaCl solution (**a**) or a 0.01 M NaH_2PO_4 (**b**) was pumped. The intermembrane distance was h = 0.9 mm, the channel length was L = 3 mm, the linear flow velocity of the solution was V = 1 mm/s; $i/i_{lim}^{Lev} = 6$.

The reason for the weaker electroconvection in the case of the NaH_2PO_4 solution, apparently, was the high rate of H^+ ion generation at the same value of the i/i_{lim} ratio. Getting into the space charge region at the depleted membrane surface, the H^+ ions caused a decrease in its density [56]. In the case of strong electrolyte solutions, the generation of H^+ (and OH^-) ions occurred, as noted above, after reaching a certain threshold concentration of salt ions near the membrane surface. After overcoming this threshold (which requires a certain minimum potential drop), the H^+ and OH^- ions were generated as a result of "water splitting" occurring as proton-exchange reactions with catalytic participation of fixed groups at the membrane/solution interface [57–60]. In the case of the NaH_2PO_4 solution, the generation of H^+ ions occurred in the non-threshold mode as a result of the dissociation of a part of $H_2PO_4^-$ ions when they enter the membrane bulk. This process takes place as soon as an external electric field was applied to the membrane.

2.1.2. Rate of Phosphorus Removal from Feed Solution and Its Theoretical Assessment.

The rate of phosphorus removal from the feed solution is controlled by the flux density of the $H_2PO_4^-$ ions from the feed bulk solution to the anion-exchange membrane surface. Since the pH of the feed solution was close to 4.5, there were no other phosphorus-bearing species except for the $H_2PO_4^-$ ions. As mentioned above, the electro-diffusion flux of the $H_2PO_4^-$ ions was limited by the first limiting current density, i_{lim}^1 (which was very close to the theoretical value i_{lim}^{Lev}); exaltation and electroconvection can enhance this transport.

It should be noted that the charge transfer by doubly and triply charged anions of orthophosphoric acid through the membrane, as well as the transport of OH^- ions, are parasitic processes if the purpose of electrodialysis is the removal of phosphorus from the feed solution. Therefore, the phosphorus flux density, j_P, which is transported through the AEM by all the phosphorus-bearing particles, Equation (13), turns out to be significantly less than what one would expect if the formula $j = i_{lim}/F$ (where j is the effective transport; i_{lim} is the limiting current density through the membrane) is used by the analogy with strong electrolytes. In the practice of electrodialysis, the value of i_{lim} was found using CVC and applying the tangent intersection method [12], or the Cowan–Brown method [43], as explained above. In the last case, the integral resistance of the membrane system (equal to $\Delta\varphi/i$) was presented as a function of the inverse current density, $1/i$ (Figure 4).

As can be seen in Figure 5, the experimental determination of $i_{\text{lim}}{}^1$ was difficult if the CVC with the total potential drop, $\Delta\varphi$, was used, both when the tangent intersection method or the Cowan–Brown method were applied. However, the second plateau and the second limiting current density, $i_{\text{lim}}{}^2$, which corresponds to the state where the membrane is completely transformed into the form of doubly charged $HPO_4{}^{2-}$ ions, were essentially better pronounced on the CVC. When the Cowan–Brown coordinates (Figure 5a,b, insertions) were used, the region of the curve related to $i_{\text{lim}}{}^1$ was very close to a straight line. The limiting current density i_{lim}^{C-B}, usually found as the intersection point to two nearly linear regions of the $\Delta\varphi/i$ versus $1/i$ curve, was then rather close to the second limiting current, $i_{\text{lim}}{}^2$, since the singularity of the curve related to $i_{\text{lim}}{}^1$ in these coordinates was quasi invisible, especially if a more rough scale (as usual) was used. Thus, when the CVC is recorded with insufficient accuracy, confusion may occur: the second limiting current density can be taken as the current density, which determines the rate of phosphorus removal.

The first limiting current is difficult to see, apparently, because the approach to saturation in the $H_2PO_4{}^-$ ion diffusion transport in solution is accompanied by a rapid increase in the contribution to charge transfer of H^+ ions, released when $H_2PO_4{}^-$ enters the membrane. As shown by mathematical modeling [32], the plateau of the first limiting current was the less noticeable the smaller the difference in values of the dissociation constants related to the first (K_{a1}) and the second (K_{a2}) steps of the ampholyte dissociation. In the case of phosphoric acid, the difference between $pK_{a1} = 2.12$ and $pK_{a2} = 7.21$ was nevertheless rather large. This difference was essentially lower for a number of other weak acids, such as tartaric acid ($pK_{a1} = 2.98$ and $pK_{a2} = 4.34$ [61,62]). Thus, one can expect that the detection of the first limiting current would be even more difficult than in the case of phosphoric acid.

The practical question of how to detect the first limiting current density is quite important. Let us note that the plateau related to $i_{\text{lim}}{}^1$ was easily detectable when using the i versus $\Delta\varphi'$ coordinates (Figure 3a,b). The use of the i versus $\Delta\varphi$ coordinates (Figure 5a,b) makes the detection of $i_{\text{lim}}{}^1$ essentially more uncertain; the second limiting current may be taken for the first one. Nevertheless, if the maximum rate of phosphorus removal, $j_P{}^{\text{max}}$, is evaluated by $i_{\text{lim}}{}^2$, the obtained value will be approximately 1.5 times higher than the experimentally measured value of $j_P{}^{\text{max}}$ in this study. Apparently, it is particularly the overestimation of the limiting current, which is responsible for the unexpectedly high energy consumption and low current efficiency of the ED recovery of phosphorus and other ampholyte species, reported by several authors [10,34,39,41].

3. Materials and Methods

3.1. Membrane and Solutions

The Neosepta AMX anion-exchange membrane was manufactured (Astom, Japan) using a previously described method [63–65]. This membrane contained quaternary ammonium bases and a small amount of secondary and tertiary amines [66]. It had an undulated surface: there were "hills" and "valleys" located in staggered order. This order was caused by the geometry of PVC reinforcing fabric [67]. The average difference between the highest and the lowest points on the surface of the swollen membrane was about 30 μm [67]. The characteristic size of the geometric inhomogeneity along the surface (280 μm) was close in magnitude to the thickness of the diffusion layer adjacent to the membrane in the compartments of the electrodialysis cell used in the study. The AMX membrane thickness in a 0.02 M NaCl solution and the ion exchange capacity of swollen membrane were equal to 125 ± 5 μm and 1.23 ± 0.05 mmol·g^{-1} wet, respectively.

The basis of the homogeneous AEM Type X membrane (Fujifilm, The Netherlands) was a three-dimensional structure (substrate) of inert polyolefin fibers [68], which were obtained by electroforming [69] methods. The aerogel formed by the fibers was pressed to a predetermined thickness. The space among the fibers was then filled with aliphatic polyacrylamide [70,71] and functionalized with quaternary ammonium bases [72]. The AEM Type X thickness in a 0.02 M NaCl solution and the ion exchange capacity of swollen membrane were equal to 120 ± 5 μm and 1.50 ± 0.05 mmol·g^{-1} wet, respectively.

All membrane samples underwent a standard salt pretreatment [73] and then were equilibrated with a 0.02 M salt solution before experiments. The solutions of sodium chloride (NaCl) and monosodium phosphate (NaH_2PO_4) were prepared from a crystalline salt (analytical grade) provided by OJSC Vekton; the 0.10 M NaOH solution was prepared from a titrant (manufactured by Uralkhiminvest, Russia). NaOH was used to maintain a constant pH value of the solution circulating through the compartments of the measuring cell. Distilled water, with an electrical conductivity of 0.8 μS·cm^{-1}, pH = 6.0 ± 0.2, and temperature of 25 °C, was used to prepare the solutions. Table 1 shows several characteristics of the electrolyte solutions used in the experiments.

Table 1. Several characteristics of the electrolyte solutions and membrane systems under study. The data are related to the temperature, 25 °C.

Electrolyte, pH	Diffusion Coefficients at Infinite Dilution, D_i, 10^{-5} cm^{2}s^{-1}			Transport Numbers at Infinite Dilution		Theoretical Limiting Current Density in 0.02 M Solution *, mA cm^{-2}	Diffusion Layer Thickness *, μm
	Cation	Anion	Electrolyte	Cation	Anion		
NaCl pH = 5.70 ± 0.05	1.334	2.032 [61]	1.61 [75]	0.396	0.604	3.07	256
NaH_2PO_4 pH = 4.60 ± 0.05	[61,74]	0.959 [61]	1.11 [74]	0.581	0.419	1.62	225

* Calculated using the Leveque equation (Equation (6)) as described in the Material and Methods section.

Figure 7 shows the distribution of the phosphoric acid species (in mole fractions) as a function of pH. It is calculated using the equations for the equilibrium of protonation–deprotonation reactions on the first, second, and third steps:

$$H_3PO_4 + H_2O \Leftrightarrow H_2PO_4^- + H_3O^+ \tag{2}$$

$$H_2PO_4^- + H_2O \Leftrightarrow HPO_4^{2-} + H_3O^+ \tag{3}$$

$$HPO_4^{2-} + H_2O \Leftrightarrow PO_4^{3-} + H_3O^+ \tag{4}$$

Negative logarithms of the equilibrium constants of these reactions at 25 °C are equal to [61] 2.12 (pK_{a1}), 7.21 (pK_{a2}), and 12.34 (pK_{a3}).

3.2. Methods

The measurements of the electrochemical characteristics of AEMs were carried out at a temperature of 25 ± 1 °C using a flow-through four-compartment electrodialysis laboratory cell connected to an Autolab PGSTAT-100 electrochemical complex. The setup and the cell are described in detail in Reference [30,49]. A schematic design of the setup is shown in Figure 8. The intermembrane distance in the desalination compartment (14), *h*, was equal to 6.6 mm; the linear flow velocity of the electrolyte solution through each chamber, *V*, was 0.4 cm·s^{-1}; the polarizable area of the membrane was 2 × 2 cm^2; the tips of Luggin's capillaries (5) used to record the potential drop across the membrane under study were located at a distance of approximately 0.8 mm from its surfaces. The plexiglass frames separating the membranes in the electrodialysis cell (Figure 7) were equipped with special guides in the shape of a comb, which provided laminar regime of the solution flow in the cell compartments.

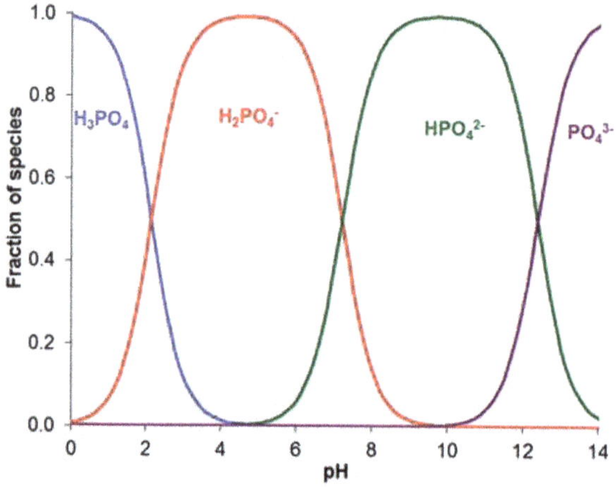

Figure 7. Distribution of the phosphoric acid species (in mole fractions) in a solution as a function of pH.

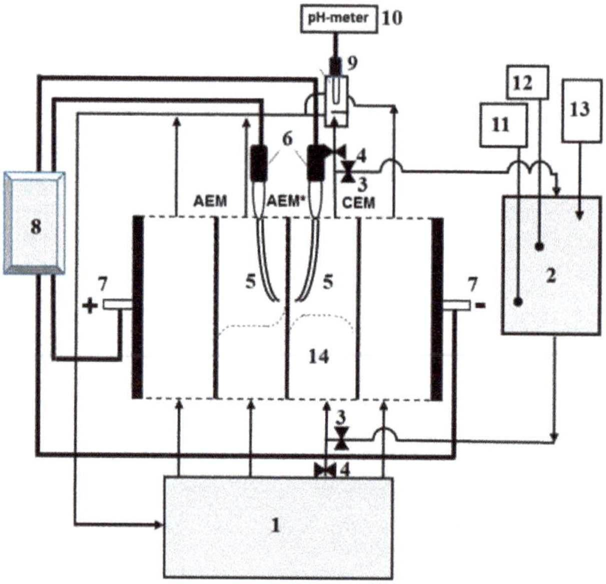

Figure 8. Schematic design of the setup used for determining the mass transfer and electrochemical characteristics of the cation-exchange membrane (CEM) and anion-exchange membrane (AEM) forming the desalination compartment (14). The setup included: an intermediate feed tank (1); an additional tank (2) for maintaining a constant pH; valves (3, 4); Luggin capillaries (5) connected with measuring Ag/AgCl electrodes (6); platinum polarizing electrodes (7); an electrochemical complex (an Autolab PGSTAT-100) (8); a flow cell (9) with an immersed combined electrode for pH measurement; a pH meter (10) connected to a computer; a combined electrode for pH measurement (11) connected to a pH meter; a conductivity cell (12) connected to a conductometer; a device (13) for maintaining a constant pH in the solution circulating through tank (2); AEM* is the anion-exchange membranes under study; CEM and AEM are the auxiliary membranes. The dotted lines schematically show the counterion concentration profiles in the cell compartments.

The current–voltage characteristics (CVC) were measured in the galvanodynamic mode at a current sweep rate of 0.02 $\mu A \cdot s^{-1}$. Measuring Ag/AgCl electrodes EVL-1M3.1 (Gomel, Belorussia) with a working area of several tenths of cm^2 immersed in saturated KCl solution were used. The volume of the feed solution in tank (1) at the beginning of the experiment was 5 dm^3. This solution was fed to all of the compartments of the cell and then returned to the same tank. Due to the relatively large volume of this solution and the fact that the diluate and concentrate were mixed before returning to tank (1), the deviation of the species' concentrations in the tank from their initial values during one run of the experiment did not exceed 1%.

The total potential drop, $\Delta\varphi$, measured using Luggin capillaries (5) depends, along with the potential drop across the polarized diffusion layers, on the resistance of the membrane and solution. The latter is a function of the distance between the membrane and capillaries (5) [76]. This distance is difficult to find and reproduce when replacing one membrane with another one. To exclude this difficulty, the corrected potential drop $\Delta\varphi'$ [77] is used instead of $\Delta\varphi$:

$$\Delta\phi' = \Delta\phi - iR_{ef} \tag{5}$$

where the effective resistance of the membrane system R_{ef} (Ohm cm^2) includes the ohmic resistance of the space (membrane + solution) among the measuring capillaries, as well as the diffusion resistance of the interphase boundaries of depleted and enriched diffusion layers [49]. The value of R_{ef} is found from the initial part of the CVC by extrapolation $i \rightarrow 0$ in the coordinates i versus $d(\Delta\varphi)\ di$.

The limiting current density, $i_{\lim}^{Lev}$, was calculated using the Leveque equation obtained in the framework of the convective–diffusion model [78]. A 1:1 electrolyte can be expressed as:

$$i_{\lim}^{Lev} = 1.47\left[\frac{FDC}{h(T_i - t_i)}\left(\frac{h^2V}{LD}\right)^{1/3} - 0.2\right] \tag{6}$$

where L is the desalination channel length; C (mole·dm^{-3}) is the electrolyte concentration of the feed solution at the entrance to the desalination channel; D is the electrolyte diffusion coefficient at infinite dilution (Table 1); t_i is the transport number of salt counter-ion in solution at infinite dilution (Table 1); T_i is the effective transport number of salt counter-ion in the membrane. The latter is defined as the fraction of the electric current transferred by ions "i" without restrictions on the concentration or pressure gradients; for the investigated membranes in the given solutions, the values of T_i for the salt coions (Na^+) are taken to be zero.

According to the definition above:

$$T_i = \frac{z_i j_i F}{i} = \frac{i_i}{i} \tag{7}$$

where z_i and j_i are the ion charge number and the flux density, F is the Faraday constant; $i_i = z_i j_i F$ is the partial current density of ion i.

Since j_i in general cases change with the coordinate normal to the membrane surface, T_i is a function of this coordinate. In particular, if the solution pH is close to 4.5, the main phosphorus-bearing species is $H_2PO_4^-$. In the membrane, pH is 1–3 units higher [16,24,26], therefore, together with $H_2PO_4^-$, the presence of doubly charged HPO_4^{2-}, and even triply charged PO_4^{3-}, is possible.

The average of the cell length thickness of the diffusion layer, δ, is estimated by combining the Leveque and the Peers equations [32]

$$\delta = 0.68h\left(\frac{LD}{h^2V}\right)^{1/3} \tag{8}$$

The values of $i_{lim}{}^{Lev}$ and δ for each electrolyte are summarized in Table 1.

3.2.1. Effective Transport Numbers, T_i

The values of the effective transport numbers, T_i, and partial currents of counter-ions are obtained using the same cell as when measuring the total CVC. For measuring T_i, desalination compartment (14) was fed with a solution from the additional tank (2). Before an experiment was run, the circuit consisting of the desalination channel, tank (2), and connecting tubes were filled with a 0.035 M solution of the electrolyte under study (NaCl or NaH_2PO_4). The other circuits of the cell were fed with a 0.02 M solution of NaCl or NaH_2PO_4 from tank (1). The initial electrolyte concentration in tank (2) was 0.035 M; it decreased during one experimental run to 0.01 M. In this concentration range, the coion (Na^+) transferred through the anion-exchange membrane can be neglected [79]. As well, in this concentration range, the concentration of carbonic acid dissociation products formed due to the continuous dissolution of atmospheric carbon dioxide in the working solution is negligible [80].

The initial volume of the solution in the desalination circuit (including tank (2), the desalination compartment and connecting tubes) was 0.100 ± 0.002 dm^3; the solution flows among the membranes at a velocity of 0.40 ± 0.01 cm·s^{-1}. The volume, concentration, and flow velocity of the solution were selected in a way to ensure quasi-stationary conditions for desalination during an experimental run. In separate experiments, it was found that to meet these conditions, the rate of electrical conductivity decreased in the desalting circuit (in the solution circulating through tank (2)) should not exceed 1% per minute [80].

The experiment was carried out at a constant corrected potential drop $\Delta\varphi'$. At equal time intervals (10 min), the electrical conductivity (κ), pH, and temperature of the solution in tank (2) (Figure 8) were recorded using a combined electrode for pH measurement (11) connected to the pH meter Expert 001 and the submersible conductometric cell (12) connected to the conductometer Expert 002 (LLC "Ekoniks expert", Russia). In all cases presented in this paper, the solution leaving the desalination compartment of the cell (compartment (14) in Figure 8) was acidified. To maintain a constant pH of the desalination stream close to 5.7 ± 0.05 (NaCl) or to 4.6 ± 0.05 (NaH_2PO_4), a 0.1 M NaOH solution was continuously added into tank (2) using a micro capillary (13). The electrolyte concentration in tank (2) was periodically determined by measuring the solution conductivity in this tank and using an equation, which connects the concentration of NaCl or NaH_2PO_4 solution with its conductivity.

The concentration of the solution in the desalination circuit changes with time due to the transfer of electrolyte ions through the membranes forming the desalination compartment to the neighboring compartments, as well as due to the addition of a titrant into this tank. The mass balance for a salt ion 1 (e.g., a cation) in the desalination circuit is described by Equation (9), under the assumption that the difference in concentrations in the different elements of this circuit (tank (2), the cell and connecting tubes) is negligible [30]:

$$\overline{V}\frac{dC}{dt} = -\frac{i(T_1^{CEM} - T_1^{AEM})S\,n}{z_1 F} + C_T\frac{d\overline{V_T}}{dt} \quad (9)$$

where i is the current density; T_1^{CEM} and T_1^{AEM} are the effective transport numbers of ion 1 in the cation-exchange membrane and anion-exchange membrane, respectively, which form the desalination compartment; C is the current salt concentration (NaCl or NaH_2PO_4) in tank (2); $\overline{V}$ is the volume of the solution in the desalination circuit; n is the number of desalination compartments ($n = 1$); C_T and $\overline{V}_T$ are the concentration and volume of the titrant, which is added to tank (2) to compensate for changes in pH caused by the H^+ (OH^-) generation at the membrane/solution interfaces; S is the area of active (polarizable) membrane surface. As it was mentioned above, the solution after passing the desalination compartment (14) was acidified, hence, a NaOH solution was added to the desalination stream.

The first term on the right-hand side of Equation (9) describes the decrease in the salt concentration in the tank (2) caused by the flow of electric current across the AEM and CEM; the second term describes the addition of a titrant in tank (2).

Consider the case of NaCl solution and a CEM under study. If the transfer of cations (Na^+) through the auxiliary AEM is neglected ($T_1^{AEM} = 0$), it follows from Equation (9) that:

$$j_{Na^+}^{CEM} = \frac{i_{Na^+}^{CEM}}{F} = \frac{iT_{Na^+}^{CEM}}{F} \approx -\frac{\overline{V}}{S}\frac{dC}{dt} + \frac{C_T}{S}\frac{d\overline{V_T}}{dt} \quad (10)$$

In the case where an AEM is under study, Equation (9) should be written for the Cl^- ions. Like as above, we assumed that the auxiliary membrane (CEM) was not permeable for anions, i.e., $T_{Cl^-}^{CEM} = 0$. Since the Cl^- ions were not present in the titrant added to tank (2), Equation (9) reads:

$$j_{Cl^-}^{AEM} = \frac{i_{Cl^-}^{AEM}}{F} = \frac{iT_{Cl^-}^{AEM}}{F} \approx -\frac{\overline{V}}{S}\frac{dC}{dt} \quad (11)$$

When the transport numbers of the salt cation in the CEM (Na^+) and the salt anion (Cl^-) in the AEM were determined, the transport numbers of the H^+ ion in the CEM and OH^- ion in the AEM were found according to Equation (12):

$$T_{H^+}^{CEM} = 1 - T_{Na^+}^{CEM},\ T_{OH^-}^{AEM} = 1 - T_{Cl^-}^{AEM} \quad (12)$$

In the case of the NaH_2PO_4 solution desalination, the calculation of the partial currents of sodium ions and protons in CEM was carried out according to Equations (10) and (12).

The determination of the partial currents of $H_2PO_4^-$, HPO_4^{2-}, PO_4^{3-}, and OH^- ions through an AEM was carried out using the material balance equations presented below. The following assumptions were made for their deduction.

1. The total phosphorus flux through an AEM, j_P^{AEM}, is equal to the sum of fluxes of all phosphorus-bearing ions entering the AEM from the diluate compartment. Since only the $H_2PO_4^-$ ions were present at pH = 4.6 in this compartment, we have, $j_{H_2PO_4^-}^s$:

$$j_{H_2PO_4^-}^s = j_P^{AEM} = j_{H_2PO_4^-}^{AEM} + j_{HPO_4^{2-}}^{AEM} + j_{PO_4^{3-}}^{AEM} = \frac{i_{H_2PO_4^-}^{AEM}}{z_{H_2PO_4^-}F} + \frac{i_{HPO_4^{2-}}^{AEM}}{z_{HPO_4^{2-}}F} + \frac{i_{PO_4^{3-}}^{AEM}}{z_{PO_4^{3-}}F} \quad (13)$$

where superscripts "s" and "AEM" relate to the solution at the diluate side of the membrane and the AEM, respectively.

The $j_{H_2PO_4^-}^s$ value can be determined experimentally by the rate of NaH_2PO_4 concentration decrease in the desalination stream (where the pH was kept constant), in accordance with an equation which is similar to Equation (11):

$$j_{H_2PO_4^-}^s = -\frac{\overline{V}}{S}\frac{dC_{NaH_2PO_4}}{dt} \quad (14)$$

2. From the distribution of the phosphoric acid species as a function of pH (Figure 8), it follows that only two sorts of phosphorous-containing species can be present in an AEM: either HPO_4^{2-} and HPO_4^{2-} (when the pH of the internal solution is in the range from 5 to 10) or HPO_4^{2-} and PO_4^{3-} (pH from 10 to 13). The concentrations of the remaining species in any of the three pH ranges presented above are so small that they can be neglected. The third possible pair, PO_4^{3-} and OH^- can coexist at pH > 13. At pH = 13, the molar fraction of PO_4^{3-} in the membrane is close to 0.8 and that of HPO_4^{2-}, 0.2, while the concentration of OH^- ions is 0.1 M, which is about 5% of the exchange capacity, hence, the total concentration of phosphate species. However, taking into account the high mobility of OH^- ions, it can be assumed that these ions will compete with the PO_4^{3-} ions, whose content is dominant at this pH value.

Since the H^+ ions being coions are excluded from the anion-exchange membrane, the pH of the membrane internal solution is 1 to 2 pH units higher than the pH of the external solution [16]. The coion exclusion increases with the dilatation of the external solution contacting the membrane, as it follows from the well-known Donnan equation [81]. With an increasing current density, the solution contacting the membrane surface at its diluate side becomes increasing diluted and the concentration of the H^+ ions in the near-surface membrane layer decreases, so that the pH of the internal solution in this layer increases. Therefore, the composition of the membrane near-surface layer on the diluate side varies with the current density: at relatively low current density (low concentration polarization), the pH in the membrane internal solution is relatively low, and this layer contains mainly ions $H_2PO_4^-$ and HPO_4^{2-}. With increasing current density, the concentration polarization increases, the external boundary solution becomes more diluted, the pH in the membrane internal solution becomes higher, and the membrane near-surface layer is enriched firstly with the PO_4^{3-} ions, and then, with the OH^- ions.

In the pH range from 5 to 10 (relatively low current densities), where only $H_2PO_4^-$ and HPO_4^{2-} are present in the membrane, the total current density, i, is carried exclusively by these ions:

$$i^{AEM}_{H_2PO_4^-} + i^{AEM}_{HPO_4^{2-}} = i \tag{15}$$

and, according to Equation (13):

$$j^s_{H_2PO_4^-} = \frac{i^{AEM}_{H_2PO_4^-}}{F} + \frac{i^{AEM}_{HPO_4^{2-}}}{2F} \tag{16}$$

where coefficient "2" in the denominator of the second term on the right-hand side stands for $z_{HPO_4^{2-}}$. Then, the partial current densities of singly and doubly charged phosphorus-bearing anions can be found from the two following equations:

$$i^{AEM}_{H_2PO_4^-} = 2Fj^s_{H_2PO_4^-} - i \tag{17}$$

$$i^{AEM}_{HPO_4^{2-}} = 2\left(i - Fj^s_{H_2PO_4^-}\right) \tag{18}$$

Thus, we can calculate the partial current densities of the $H_2PO_4^-$ and HPO_4^{2-} species, if we measure the current density, i, and the rate of the diluate desalination; the latter allows for the calculation of $j^s_{H_2PO_4^-}$, Equation (14).

In the range of pH from 10 to 13.5 (high current densities), where $i^{AEM}_{HPO_4^{2-}} + i^{AEM}_{PO_4^{3-}} = i$, in a similar way, we find:

$$i^{AOM}_{HPO_4^{2-}} = 2\left(3Fj^s_{H_2PO_4^-} - i\right) \tag{19}$$

$$i^{AOM}_{PO_4^{3-}} = 3\left(i - 2Fj^s_{H_2PO_4^-}\right) \tag{20}$$

The coefficients 2 and 3 in Equations (17)–(20) stands for the charge numbers $z_{HPO_4^{2-}}$ and $z_{PO_4^{3-}}$, respectively.

At even higher current densities, where all doubly charged orthophosphoric acid anions transform into triple-charged ones ($T_{HPO_4^{2-}} = i_{HPO_4^{2-}}/i = 0$)and the pH of the internal solution exceeds 13.5, the current is transported through the membrane by PO_4^{3-} and OH^- ions. The partial currents of these ions are:

$$i^{AEM}_{PO_4^{3-}} = 3Fj^s_{H_2PO_4^-} \tag{21}$$

$$i^{AEM}_{OH^-} = i - i^{AEM}_{PO_4^{3-}} \tag{22}$$

In accordance with Equations (2)–(4) and (13), at the AEM/solution interface, the partial flux of protons entering the depleted diffusion layer is:

$$j^s_{H^+} = j^{AEM}_{HPO_4^{2-}} + 2j^{AEM}_{PO^{3-}} \tag{23}$$

Taking into account that $i_k = j_k z_k F$, the partial current density of H^+ ions in the depleted solution near the membrane surface is:

$$i^s_{H^+} = \frac{i^{AEM}_{HPO_4{}^{2-}}}{2} + \frac{2i^{AEM}_{PO_4{}^{3-}}}{3} \tag{24}$$

3.2.2. Visualization of Electroconvective Vortices.

Visualization of vortex flows near an AMX membrane surface facing the desalination compartment was carried out using a technique similar to that described in Reference [82].

Two poly(methyl methacrylate) frames clung close to the ion-exchange membrane and formed a channel, with a 0.9 mm width and a 3 mm length. A 0.01 M NaCl or 0.01 M NaH_2PO_4 solution entered the desalination compartment through the holes in the clamping plates, in which the electrode chambers were located. The linear flow rate of the solution was 0.001 cm·s^{-1}. The distance from the investigated membranes to the measuring electrodes was 800 μm. To visualize vortexes, 10 μM rhodamine 6G was added to the solution. Rhodamine 6G is able to fluoresce in a narrow range of wavelengths and it is a large cation. The chlorine or hydrogen phosphate ions were the anions in the system under study. A SOPTOP CX40M optical microscope (China) with a 5× objective and a digital eyepiece camera were used to record these vortexes. The resolution of the digital optical system allows for the recording of the appearance of vortices with a diameter of 20 microns or greater. Digital video recording was carried out simultaneously while measuring the potential difference over the membrane, and the current density through the membrane was set as a function of time. The limiting current density was estimated from the current–voltage characteristic of the membrane using the tangent intersection method.

4. Theory

The calculation of the CVC, as well as the transport numbers, partial fluxes, and partial currents of ions in the membrane system was carried out using a stationary 1D model described in detail in References [31,32]. A three-layer system under direct current conditions was considered. It consisted of an anion-exchange membrane (AMX) and two adjacent diffusion boundary layers (DBLs). Migration and diffusion transport of neutral and negatively charged ampholyte species, as well as the Na^+, H^+, and OH^- ions in all three layers, is described using the Nernst–Planck equation under the local electroneutrality condition. The model assumes the independence of the current density and the sum of the fluxes of all phosphorus-bearing species from the coordinate. As well, local chemical equilibria were assumed among the ampholyte species involved in protonation–deprotonation reactions in the AEM and DBLs. In addition, the condition of the ion-exchange equilibrium (expressed by the Donnan equations) at the membrane/solution boundaries was applied. The ion diffusion coefficients in the solution (Table 1) were taken at the infinite dilution [83], because at the near-limiting current densities, the ion concentrations in the solution next to membrane surface were close to zero. The diffusion coefficients in the membrane were determined from the AMX membrane conductivity [84].

5. Conclusions

The transport of the singly charged phosphoric acid anions, $H_2PO_4{}^-$, across an AEM was accompanied by dissociation of a part of these anions when entering the membrane. The H^+ ions released in this dissociation returned to the depleted solution and participates in carrying electric charge. The generation of H^+ ions occurred without voltage threshold, which existed in the case of strong electrolytes. Depending on the current density, the $H_2PO_4{}^-$ anions could transform into $HPO_4{}^{2-}$ ions (at relatively low current densities, approximately at $i < 2.5\ i_{lim}{}^{Lev}$) or $PO_4{}^{3-}$ ions (at higher current densities). At relatively high current densities/voltages along with this transformation, water splitting with catalytic participation of membrane functional groups can occur.

The dissociation of the $H_2PO_4{}^-$ anion occurred because the pH of the membrane internal solution was higher than the pH of the external solution. The shift in pH was due to the Donnan exclusion of the H^+ ions, which were coions for the AEM. The degree of the H^+ exclusion (hence, the pH of the internal solution) depends on the electrolyte concentration in the boundary solution: the more dilute this solution, the higher the internal solution's pH. Since the concentration of the boundary solution decreased with increasing current density, the internal solution pH increased with increasing i, as a consequence, the equivalent fraction of doubly and triply charged anions in the membrane increased.

Each step of the $H_2PO_4^-$ anion dissociation had a response on the I–V curve. When the boundary electrolyte concentration became much lower than the bulk concentration, an inclined plateau of the first limiting electric current was observed. When $i > i_{lim}^1$, a more complete transformation of the $H_2PO_4^-$ anion into the HPO_4^{2-} ions in the membrane occurred, the H^+ ions liberated in this reaction gave rise to the electric current. The second plateau appeared when the membrane passed completely in the HPO_4^{2-} form. The next rise in the current density occurred due to the appearance of the PO_4^{3-} ions in the membrane and a new portion of the H^+ ions ejected into the depleted solution. When HPO_4^{2-} and PO_4^{3-} ions crossed the membrane boundary and into the enriched solution on the other side of the membrane, they captured H^+ ions from water, which led to an increase of pH in the enriched solution. It can be also interpreted as generation of OH^- ions at the membrane/enriched solution interface. Thus, in AEM/NaH_2PO_4 systems, the process of generation of H^+ and OH^- ions is separated in space: H^+ ions appear at the depleted solution/membrane interface and OH^- ions at the membrane/enriched solution interface.

Following from the above, phosphorus can be transferred across the membrane by singly, doubly or triply charged anions. However, the origin of all these species are the $H_2PO_4^-$ anions presented in the feed solution. Hence, the sum of the phosphorus-bearing species fluxes is equal to the flux of the $H_2PO_4^-$ anions from the bulk solution to the depleted membrane boundary. The $H_2PO_4^-$ ions are transported by electro-diffusion including the exaltation effect and by electroconvection. The Leveque equation gives the limiting current density not including the exaltation effect; the latter gives an increase in the phosphorus flux approximately equal to 10%, and electroconvection (in the conditions of our experiment) gives the contribution of about 60% to the overall flux. The occurrence of electroconvection was established via visualization of electroconvective vortices. It was shown that under the same value of i/i_{lim}^{Lev}, the size of electroconvective vortices in the case of 0.02 M NaH_2PO_4 solution was essentially lower than in the case of NaCl solution of the same concentration. Thus, what is important is that increasing current densities/voltages in ED systems does not lead to expected growth of the extraction degree of phosphorus. Increasing current density is spent particularly in the H^+ ions' transport. Electroconvection is less effective, as in the case of NaCl, since the H^+ ions reduce the space charge at the depleted solution/membrane interface. Hence, with increasing current density over i_{lim}^{Lev}, the current efficiency of the phosphorus recovery strongly decreases.

Another interesting remark is that the conventional application of the Cowan–Brown method to find the limiting current density, gives the second limiting current density. This current density is nearly two times higher than i_{lim}^1, which mainly determines the flux of phosphorus across the membrane.

Author Contributions: Conceptualization, N.P. and V.N.; Methodology, N.P.; Formal Analysis, S.M.; Investigation, O.R., K.T., E.M. and I.M.; Writing–Original Draft Preparation, N.P.; Writing–Review and Editing, V.N. and N.P.; Experimentation, O.R., K.T., E.M. and I.M.; Supervision, V.N. and N.P.; Project Administration, N.P.; Funding Acquisition, N.P. All authors approved the final article.

Funding: We are grateful to the Russian Science Foundation, grant No. 17-19-01486, for the financial support of this study. The authors thank the Core Facility "Environmental Analytical Center" of the Kuban State University (unique identifier RFMEFI59317X0008) for providing their equipment.

Conflicts of Interest: The authors declare no conflict of interest.

References

1. Paltrinieri, L.; Huerta, E.; Puts, T.; Van Baak, W.; Verver, A.B.; Sudhölter, E.J.R.; De Smet, L.C.P.M. Functionalized anion-exchange membranes facilitate electrodialysis of citrate and phosphate from model dairy wastewater. *Environ. Sci. Technol.* **2019**, *53*, 2396–2404. [CrossRef] [PubMed]

2. Shi, L.; Xie, S.; Hu, Z.; Wu, G.; Morrison, L.; Croot, P.; Hu, H.; Zhan, X. Nutrient recovery from pig manure digestate using electrodialysis reversal: Membrane fouling and feasibility of long-term operation. *J. Membr. Sci.* **2019**, *573*, 560–569. [CrossRef]
3. Tran, A.T.K.; Zhang, Y.; Lin, J.; Mondal, P.; Ye, W.; Meesschaert, B.; Pinoy, L.; Van Der Bruggen, B. Phosphate pre-concentration from municipal wastewater by selectrodialysis: Effect of competing components. *Sep. Purif. Technol.* **2015**, *141*, 38–47. [CrossRef]
4. Zhang, Y.; Desmidt, E.; Van Looveren, A.; Pinoy, L.; Meesschaert, B.; Van Der Bruggen, B. Phosphate separation and recovery from wastewater by novel electrodialysis. *Environ. Sci. Technol.* **2013**, *47*, 5888–5895. [CrossRef] [PubMed]
5. Handojo, L.; Wardani, A.K.; Regina, D.; Bella, C.; Kresnowati, M.T.A.P.; Wenten, I.G. Electro-membrane processes for organic acid recovery. *RSC Adv.* **2019**, *9*, 7854–7869. [CrossRef]
6. El Rayess, Y.; Mietton-Peuchot, M. Membrane technologies in wine industry: An overview. *Crit. Rev. Food Sci.* **2016**, *56*, 2005–2020. [CrossRef] [PubMed]
7. Martín, J.; Díaz-Montaña, E.J.; Asuero, A.G. Recovery of anthocyanins using membrane technologies: A review. *Crit. Rev. Anal. Chem.* **2018**, *48*, 143–175. [CrossRef] [PubMed]
8. Galier, S.; Roux-de Balmann, H. The electrophoretic membrane contactor: A mass transfer-based methodology applied to the separation of whey proteins. *Sep. Purif. Technol.* **2011**, *77*, 237–244. [CrossRef]
9. Durand, R.; Fraboulet, E.; Marette, A.; Bazinet, L. Simultaneous double cationic and anionic molecule separation from herring milt hydrolysate and impact on resulting fraction bioactivities. *Sep. Purif. Technol.* **2019**, *210*, 431–441. [CrossRef]
10. Chandra, A.; Tadimeti, J.G.D.; Chattopadhyay, S. Transport hindrances with electrodialytic recovery of citric acid from solution of strong electrolytes. *Chin. J. Chem. Eng.* **2018**, *26*, 278–292. [CrossRef]
11. Suwal, S.; Rozoy, É.; Manenda, M.; Doyen, A.; Bazinet, L. Comparative study of in situ and ex situ enzymatic hydrolysis of milk protein and separation of bioactive peptides in an electromembrane reactor. *ACS Sustain. Chem. Eng.* **2017**, *5*, 5330–5340. [CrossRef]
12. Strathmann, H. Electrodialysis, a mature technology with a multitude of new applications. *Desalination* **2010**, *264*, 268–288. [CrossRef]
13. Fidaleo, M.; Moresi, M. Concentration of trisodium citrate by electrodialysis. *J. Membr. Sci.* **2013**, *447*, 376–386. [CrossRef]
14. Myles, T.D.; Grew, K.N.; Peracchio, A.A.; Chiu, W.K.S. Transient ion exchange of anion exchange membranes exposed to carbon dioxide. *J. Power Sour.* **2015**, *296*, 225–236. [CrossRef]
15. Ramirez, P.; Alcaraz, A.; Mafe, S.; Pellicer, J. pH and supporting electrolyte concentration effects on the passive transport of cationic and anionic drugs through fixed charge membranes. *J. Membr. Sci.* **1999**, *161*, 143–155. [CrossRef]
16. Sarapulova, V.; Nevakshenova, E.; Pismenskaya, N.; Dammak, L.; Nikonenko, V. Unusual concentration dependence of ion-exchange membrane conductivity in ampholyte-containing solutions: Effect of ampholyte nature. *J. Membr. Sci.* **2015**, *479*, 28–38. [CrossRef]
17. Andersen, M.B.; Rogers, D.M.; Mai, J.; Schudel, B.; Hatch, A.V.; Rempe, S.B.; Mani, A. Spatiotemporal pH dynamics in concentration polarization near ion-selective membranes. *Langmuir* **2014**, *30*, 7902–7912. [CrossRef]
18. Femmer, R.; Mani, A.; Wessling, M. Ion transport through electrolyte/polyelectrolyte multi-layers. *Sci. Rep.* **2015**, *5*, 1–12. [CrossRef]
19. Zabolotskii, V.I.; Gnusin, N.P.; El'nikova, L.F.; Blednykh, V.M. Exhaustive purification of amino acids by removal of mineral impurities by electrodialysis with ion-exchange membranes. *J. App. Chem.-Ussr* **1986**, *59*, 127–131.
20. Shaposhnik, V.A.; Eliseeva, T.V. Barrier effect during the electrodialysis of ampholytes. *J. Membr. Sci.* **1999**, *161*, 223–229. [CrossRef]
21. Eliseeva, T.V.; Shaposhnik, V.A. Effects of circulation and facilitated electromigration of amino acids in electrodialysis with ion-exchange membranes. *Russ. J. Electrochem.* **2000**, *36*, 64–67. [CrossRef]
22. Dufton, G.; Mikhaylin, S.; Gaaloul, S.; Bazinet, L. How electrodialysis configuration influences acid whey deacidification and membrane scaling. *J. Dairy Sci.* **2018**, *101*, 7833–7850. [CrossRef] [PubMed]
23. Rotta, E.H.; Bitencourt, C.S.; Marder, L.; Bernardes, A.M. Phosphorus recovery from low phosphate-containing solution by electrodialysis. *J. Membr. Sci.* **2019**, *573*, 293–300. [CrossRef]

24. Franck-Lacaze, L.; Sistat, P.; Huguet, P.; Lapicque, F. Protonation and diffusion phenomena in poly(4-vinylpyridine)-based weak anion-exchange membranes. *J. Membr. Sci.* **2009**, *340*, 257–265. [CrossRef]
25. Koter, S.; Kultys, M. Modeling the electric transport of sulfuric and phosphoric acids through anion-exchange membranes. *Sep. Purif. Technol.* **2010**, *73*, 219–229. [CrossRef]
26. Kozmai, A.E.; Nikonenko, V.V.; Zyryanova, S.; Pismenskaya, N.D.; Dammak, L. A simple model for the response of an anion-exchange membrane to variation in concentration and pH of bathing solution. *J. Membr. Sci.* **2018**, *567*, 127–138. [CrossRef]
27. Vasil'eva, V.; Goleva, E.; Pismenskaya, N.; Kozmai, A.; Nikonenko, V. Effect of surface profiling of a cation-exchange membrane on the phenylalanine and NaCl separation performances in diffusion dialysis. *Sep. Purif. Technol.* **2019**, *210*, 48–59. [CrossRef]
28. Ueno, K.; Doi, T.; Nanzai, B.; Igawa, M. Selective transport of neutral amino acids across a double-membrane system comprising cation and anion exchange membranes. *J. Membr. Sci.* **2017**, *537*, 344–352. [CrossRef]
29. Ramírez, P.; Alcaraz, A.; Mafé, S. Modeling of amino acid electrodiffusion through fixed charge membranes. *J. Coll. Interface Sci.* **2001**, *242*, 164–173. [CrossRef]
30. Rybalkina, O.A.; Tsygurina, K.A.; Melnikova, E.D.; Pourcelly, G.; Nikonenko, V.V.; Pismenskaya, N.D. Catalytic effect of ammonia-containing species on water splitting during electrodialysis with ion-exchange membranes. *Electrochim. Acta* **2019**, *299*, 946–962. [CrossRef]
31. Belashova, E.D.; Pismenskaya, N.D.; Nikonenko, V.V.; Sistat, P.; Pourcelly, G. Current-voltage characteristic of anion-exchange membrane in monosodium phosphate solution. Modelling and experiment. *J. Membr. Sci.* **2017**, *542*, 177–185. [CrossRef]
32. Melnikova, E.D.; Pismenskaya, N.D.; Bazinet, L.; Mikhaylin, S.; Nikonenko, V.V. Effect of ampholyte nature on current-voltage characteristic of anion-exchange membrane. *Electrochim. Acta* **2018**, *285*, 185–191. [CrossRef]
33. Wang, X.; Zhang, X.; Wang, Y.; Du, Y.; Feng, H.; Xu, T. Simultaneous recovery of ammonium and phosphorus via the integration of electrodialysis with struvite reactor. *J. Membr. Sci.* **2015**, *490*, 65–71. [CrossRef]
34. Ward, A.J.; Arola, K.; Thompson Brewster, E.; Mehta, C.M.; Batstone, D.J. Nutrient recovery from wastewater through pilot scale electrodialysis. *Water Res.* **2018**, *135*, 57–65. [CrossRef] [PubMed]
35. Yang, Y.; Lin, L.; Tse, L.K.; Dong, H.; Yu, S.; Hoffmann, M.R. Membrane-separated electrochemical latrine wastewater treatment. *Environ. Sci. Water Res. Technol.* **2019**, *5*, 51–59. [CrossRef]
36. Shi, L.; Hu, Y.; Xie, S.; Wu, G.; Hu, Z.; Zhan, X. Recovery of nutrients and volatile fatty acids from pig manure hydrolysate using two-stage bipolar membrane electrodialysis. *Chem. Eng. J.* **2018**, *334*, 134–142. [CrossRef]
37. Ebbers, B.; Ottosen, L.M.; Jensen, P.E. Electrodialytic treatment of municipal wastewater and sludge for the removal of heavy metals and recovery of phosphorus. *Electrochim. Acta* **2015**, *181*, 90–99. [CrossRef]
38. Geng, Y.-K.; Wang, Y.; Pan, X.-R.; Sheng, G.-P. Electricity generation and in situ phosphate recovery from enhanced biological phosphorus removal sludge by electrodialysis membrane bioreactor. *Bioresour. Technol.* **2018**, *247*, 471–476. [CrossRef]
39. De Paepe, J.; Lindeboom, R.E.F.; Vanoppen, M.; De Paepe, K.; Demey, D.; Coessens, W.; Lamaze, S.; Verliefde, A.R.D.; Clauwaert, P.; Vlaeminck, S.E. Refinery and concentration of nutrients from urine with electrodialysis enabled by upstream precipitation and nitrification. *Water Res.* **2018**, *144*, 76–86. [CrossRef]
40. Gallya, C.R.; Benvenutia, T.; Trindade, C.M.; Rodrigues, M.A.S.; Zoppas-Ferreira, J.; Pérez-Herranz, V.; Bernardes, A.M. Electrodialysis for the tertiary treatment of municipal wastewater: Efficiency of ion removal and ageing of ion exchange membranes. *J. Environ. Chem. Eng.* **2018**, *3*, 5855–5869. [CrossRef]
41. Weinertova, K.; Honorato, R.S.; Stranska, E.; Nedela, D. Comparison of heterogeneous anion-exchanges for nitrate ion removal from mixed salt solution. *Chem. Pap.* **2018**, *72*, 469–478. [CrossRef]
42. Paltrinieri, L.; Poltorak, L.; Chu, L.; Puts, T.; van Baak, W.; Sudhölter, E.J.R.; de Smet, L.C.P.M. Hybrid polyelectrolyte-anion exchange membrane and its interaction with phosphate. *React. Funct. Polym.* **2018**, *133*, 126–135. [CrossRef]
43. Cowan, D.Q.; Brown, I.W. Effect of turbulence in limiting current in electrodialysis cell. *Ind. Eng. Chem.* **1959**, *51*, 1445–1449. [CrossRef]
44. Chandra, A.; Tadimeti, J.G.D.; Bhuvanesh, E.; Pathiwada, D.; Chattopadhyay, S. Switching selectivity of carboxylic acids and associated physic-chemical changes with pH during electrodialysis of ternary mixtures. *Sep. Purif. Technol.* **2018**, *193*, 327–344. [CrossRef]

45. Lopatkova, G.Y.; Volodina, E.I.; Pis'menskaya, N.D.; Fedotov, Y.A.; Cot, D.; Nikonenko, V.V. Effect of chemical modification of ion-exchange membrane MA-40 on its electrochemical characteristics. *Russ. J. Electrochem.* **2006**, *42*, 847–854. [CrossRef]
46. Zabolotskii, V.I.; Bugakov, V.V.; Sharafan, M.V.; Chermit, R.K. Transfer of electrolyte ions and water dissociation in anion-exchange membranes under intense current conditions. *Russ. J. Electrochem.* **2012**, *48*, 650–659. [CrossRef]
47. Pismenskaya, N.D.; Pokhidnia, E.V.; Pourcelly, G.; Nikonenko, V.V. Can the electrochemical performance of heterogeneous ion-exchange membranes be better than that of homogeneous membranes? *J. Membr. Sci.* **2018**, *566*, 54–68. [CrossRef]
48. Rubinstein, I.; Zaltzman, B. Equilibrium electro-osmotic instability in concentration polarization at a perfectly charge-selective interface. *Phys. Rev. Fluids* **2017**, *2*, 093702. [CrossRef]
49. Pismenskaya, N.D.; Nikonenko, V.V.; Belova, E.I.; Lopatkova, G.Y.; Sistat, P.; Pourcelly, G.; Larshe, K. Coupled convection of solution near the surface of ion exchange membranes in intensive current regimes. *Russ. J. Electrochem.* **2007**, *43*, 307–327. [CrossRef]
50. Kniaginicheva, E.; Pismenskaya, N.; Melnikov, S.; Belashova, E.; Sistat, P.; Cretin, M.; Nikonenko, V. Water splitting at an anion-exchange membrane as studied by impedance spectroscopy. *J. Membr. Sci.* **2015**, *496*, 78–83. [CrossRef]
51. Kharkats, Y.I. The mechanism of supralimiting currents at ion-exchange membrane electrolyte interfaces. *Sov. Electrochem.* **1985**, *21*, 917–920.
52. Urtenov, M.; Kirillova, E.; Seidova, N.; Nikonenko, V. Decoupling of the Nernst-Planck and Poisson equations. Application to a membrane system at overlimiting currents. *J. Phys. Chem. B* **2007**, *111*, 14208–14222. [CrossRef] [PubMed]
53. Nikonenko, V.V.; Mareev, S.A.; Pis'menskaya, N.D.; Uzdenova, A.M.; Kovalenko, A.V.; Urtenov, M.K.; Pourcelly, G. Effect of electroconvection and its use in intensifying the mass transfer in electrodialysis. *Russ. J. Electrochem.* **2017**, *53*, 1122–1144. [CrossRef]
54. Rybalkina, O.A.; Melnikova, E.D.; Pismenskiy, A.V. Influence of gravitational convection on current–voltage characteristics of an electromembrane stack in sodium dihydrogen phosphate solution. *Petrol. Chem.* **2018**, *58*, 114–120. [CrossRef]
55. Rubinstein, S.M.; Manukyan, G.; Staicu, A.; Rubinstein, I.; Zaltzman, B.; Lammertink, R.G.H.; Mugele, F.; Wessling, M. Direct observation of a nonequilibrium electro-osmotic instability. *Phys. Rev. Lett.* **2008**, *101*, 236101. [CrossRef]
56. Mishchuk, N.A. Concentration polarization of interface and non-linear electrokinetic phenomena. *Adv. Coll. Interface Sci.* **2010**, *160*, 16. [CrossRef]
57. Simons, R. Electric field effects on proton transfer between ionizable groups and water in ion exchange membranes. *Electrochim. Acta* **1984**, *29*, 151–158. [CrossRef]
58. Simons, R. Strong electric field effects on proton transfer between membrane-bound amines and water. *Nature* **1979**, *280*, 824–826. [CrossRef]
59. Zabolotskii, V.I.; Shel'deshov, N.V.; Gnusin, N.P. Dissociation of water molecules in systems with ion-exchange membranes. *Russ. Chem. Rev.* **1988**, *57*, 801–808. [CrossRef]
60. Slouka, Z.; Senapati, S.; Yan, Y.; Chang, H.-C. Charge inversion, water splitting, and vortex suppression due to DNA sorption on ion-selective membranes and their ion-current signatures. *Langmuir* **2012**, *29*, 8275–8283. [CrossRef]
61. Lide, D.R. *Handbook of Chemistry and Physics*; CRC Press: New York, NY, USA, 1995.
62. McMurry, J. *Fundamentals of Organic Chemistry*; Brooks/Cole; Cengage Learning: Belmont, NJ, USA, 2010.
63. Mizutani, Y. Structure of ion exchange membranes. *J. Membr. Sci.* **1990**, *49*, 121–144. [CrossRef]
64. Doi, S.; Yasukawa, M.; Kakihana, Y.; Higa, M. Alkali attack on anion exchange membranes with PVC backing and binder: Effect on performance and correlation between them. *J. Membr. Sci.* **2019**, *573*, 85–96. [CrossRef]
65. Doi, S.; Kinoshita, M.; Yasukawa, M.; Higa, M. Alkali attack on anion exchange membranes with PVC backing and binder: II prediction of electrical and mechanical performances from simple optical analyses. *Membranes* **2018**, *8*, 133. [CrossRef] [PubMed]
66. Güler, E.; van Baak, W.; Saakes, M.; Nijmeijer, K. Monovalent-ion-selective membranes for reverse electrodialysis. *J. Membr. Sci.* **2014**, *455*, 254–270. [CrossRef]

67. Mareev, S.A.; Butylskii, D.Y.; Pismenskaya, N.D.; Larchet, C.; Dammak, L.; Nikonenko, V.V. Geometric heterogeneity of homogeneous ion-exchange Neosepta® membranes. *J. Membr. Sci.* **2018**, *563*, 768–776. [CrossRef]
68. Ion Exchange Membranes for Water Purification. Brochure of Fujifilm Membrane Technology. Version number 1.0. 2018. Available online: https://www.fujifilmmembranes.com/water-membranes/ion-exchange-membranes (accessed on 1 July 2019).
69. Rabolt, J.F.; Lee, K.H.; Givens, S.R. Method of Solution Preparation of Polyolefin Class Polymers for Electrospinning Processing Included. U.S. Patent 8,083,983 B2, 27 December 2011.
70. Zhu, Y.; Ahmad, M.; Yang, L.; Misovich, M.; Yaroshchuk, A.; Bruening, M.L. Adsorption of polyelectrolyte multilayers imparts high monovalent/divalent cation selectivity to aliphatic polyamide cation-exchange membranes. *J. Membr. Sci.* **2017**, *537*, 177–185. [CrossRef]
71. Zhang, W.; Ma, J.; Wang, P.; Wang, Z.; Shi, F.; Liu, H. Investigations on the interfacial capacitance and the diffusion boundary layer thickness of ion exchange membrane using electrochemical impedance spectroscopy. *J. Membr. Sci.* **2016**, *502*, 37–47. [CrossRef]
72. Antheunis, H.; Hessing, J.; Van Berchum, B. Curable Compositions and Membranes. U.S. Patent 8,968,965 B2, 3 March 2015.
73. Berezina, N.P.; Timofeev, S.V.; Kononenko, N.A. Effect of conditioning techniques of perfluorinated sulphocationic membranes on their hydrophylic and electrotransport properties. *J. Membr. Sci.* **2002**, *209*, 509–518. [CrossRef]
74. Volkov, A.I.; Zharskii, I.M. *Comprehensive Chemical Handbook*; Modern School: Moscow, Russia, 2005.
75. Robinson, R.A.; Stokes, R.H. *Electrolyte Solutions*; Butterworths: London, UK, 1959.
76. Belova, E.I.; Lopatkova, G.Y.; Pismenskaya, N.D.; Nikonenko, V.V.; Larchet, C.; Pourcelly, G. Effect of anion-exchange membrane surface properties on mechanisms of overlimiting mass transfer. *J. Phys. Chem. B* **2006**, *110*, 13458–13469. [CrossRef]
77. Rösler, H.-W.; Maletzki, F.; Staude, E. A numerical study of the hydrodynamic stable concentration boundary layers in a membrane system under microgravitational conditions. *J. Membr. Sci.* **1992**, *72*, 171–179. [CrossRef]
78. Newman, J.S. *Electrochemical Systems*; Prentice Hall: Englewood Cliffs, NY, USA, 1973.
79. Sheldeshov, N.V.; Ganych, V.V.; Zabolotsky, V.I. Transport numbers of salt ions and water dissociation products in cation and anion-exchange membranes. *Sov. Electrochem.* **1991**, *23*, 11–15.
80. Laktionov, E.V.; Pismenskaya, N.D.; Nikonenko, V.V.; Zabolotsky, V.I. Method of electrodialysis stack testing with the feed solution concentration regulation. *Desalination* **2003**, *152*, 101–116. [CrossRef]
81. Helfferich, F.G. *Ion Exchange*; McGraw-Hill: New York, NY, USA, 1962; ASIN B0000CLGWI.
82. Kwak, R.; Guan, G.; Peng, W.K.; Han, J. Microscale electrodialysis: Concentration profiling and vortex visualization. *Desalination* **2013**, *308*, 138–146. [CrossRef]
83. Mareev, S.A.; Butylskii, D.Y.; Kovalenko, A.V.; Petukhova, A.V.; Pismenskaya, N.D.; Dammak, L.; Larchet, C.; Nikonenko, V.V. Accounting for the concentration dependence of electrolyte diffusion coefficient in the Sand and the Peers equations. *Electrochim. Acta* **2016**, *195*, 85–93. [CrossRef]
84. Zabolotsky, V.I.; Nikonenko, V.V. Effect of structural membrane inhomogeneity on transport properties. *J. Membr. Sci.* **1993**, *79*, 181–198. [CrossRef]

International Journal of
Molecular Sciences

MDPI

Article

Simple Preparation of $LaPO_4$:Ce, Tb Phosphors by an Ionic-Liquid-Driven Supported Liquid Membrane System

Jianguo Li [1,3], Hongying Dong [1,*], Fan Yang [2,3,*], Liangcheng Sun [3,4], Zhigang Zhao [2,3], Ruixi Bai [2,3] and Hao Zhang [2,3]

1 School of Chemical Engineering, Inner Mongolia University of Technology, Hohhot 010051, China
2 Key Laboratory of Design and Assembly of Functional Nanostructures, Fujian Provincial Key Laboratory of Nanomaterials, Fujian Institute of Research on the Structure of Matter, Chinese Academy of Sciences, Xiamen 361021, China
3 Xiamen Institute of Rare Earth Materials, Chinese Academy of Sciences, Xiamen 361021, China
4 Baotou Research Institute of Rare Earths, Baotou 014030, China
* Correspondence: donghongying@imut.edu.cn (H.D.); fanyang2013@fjirsm.ac.cn (F.Y.);
Tel.: +86-0471-6575722 (H.D.); +86-157-5073-3599 (F.Y.)

Received: 21 June 2019; Accepted: 3 July 2019; Published: 12 July 2019

Abstract: In this work, $LaPO_4$:Ce, Tb phosphors were prepared by firing a $LaPO_4$:Ce, Tb precipitate using an ionic-liquid-driven supported liquid membrane system. The entire system consisted of three parts: a mixed rare earth ion supply phase, a phosphate supply phase, and an ionic-liquid-driven supporting liquid membrane phase. This method showed the advantages of a high flux, high efficiency, and more controllable reaction process. The release rate of $PO_4{}^{3-}$ from the liquid film under different types of ionic liquid, the ratio of the rare earth ions in the precursor mixture, and the structure, morphology, and photoluminescence properties of $LaPO_4$:Ce, Tb were investigated by inductively coupled plasma-atomic emission spectroscopy, X-ray diffraction, Raman spectra, scanning electron microscopy, and photoluminescence emission spectra methods. The results showed that a pure phase of lanthanum orthophosphate with a monoclinic structure can be formed. Due to differences in the anions in the rare earth supply phase, the prepared phosphors showed micro-spherical (when using rare earth sulfate as the raw material) and nanoscale stone-shape (when using rare earth nitrate as the raw material) morphologies. Moreover, the phosphors prepared by this method had good luminescent properties, reaching a maximum emission intensity under 277 nm excitation with a predominant green emission at 543 nm which corresponded to the $^5D_4$-$^7F_5$ transition of Tb^{3+}.

Keywords: $LaPO_4$: Ce; Tb; ionic liquid; supported liquid membrane; photoluminescence

---

1. Introduction

Recently, rare-earth-ion-doped multicomponent compounds have attracted considerable attention due to their potential applications in the fields of electroluminescent devices, high-resolution displays, biological labels, and integrated optics [1–4]. Among these rare-earth-doped oxide phosphors, trivalent-cerium- and terbium-coactivated $LaPO_4$ is significant because of its low solubility in water, its high thermal stability, and its high-efficiency energy transfer between Ce^{3+} and Tb^{3+} [5–8]. Due to the 4f orbital properties of La^{3+}, lanthanide phosphate is transparent in the visible region and has been proven to be an ideal host structure for other lanthanide ions, resulting in luminescent materials in the UV-visible region [9–11]. In Ce^{3+} and Tb^{3+} co-doped $LaPO_4$, Ce^{3+} with optically allowed d–f transitions is an effective activator for Tb^{3+} emission [12,13]. $LaPO_4$:Ce^{3+}, Tb^{3+} powders have been widely used as the green component of three band emission type fluorescent lamps [14,15]. In addition,

$LaPO_4$:Ce^{3+}, Tb^{3+} phosphors have drawn continuous research attention in several other applications, including transparent fillers/markers, biomedical purposes, and plasma display panels [13,16–18].

Phosphor particles should be spherical in shape with no aggregation and their particle size should be in the micron range (<3 μm) with a narrow size distribution [16,19]. Spherical phosphor particles are more advantageous for the optical and geometric structure of the phosphor layer. The size of the phosphor affects the number of phosphor particles needed to produce the best coating for a particular application [20]. The shape, size distribution, and other microstructural characteristics of phosphors can be well controlled by different synthetic methods and reaction conditions. To date, several methods have been reported for the synthesis of phosphate phosphor materials, such as coprecipitation [21–23], solvothermal methods [24,25], electrospinning methods [26], solid-state methods [27], sol-gel processes [10,12], and spray pyrolysis [14]. Of these, coprecipitation is a common industrial synthetic method used to produce rare earth oxide powders and has the advantages of being feasible, low-cost, and environmentally friendly. Beyond that, fluorescent powders prepared by the coprecipitation method have uniform particle sizes, low agglomeration, and low phase impurities [21]. However, many factors such as reaction temperature, aging time, pH value, and solution concentration need to be controlled, which limits the development of coprecipitation methods. It is still challenging to simply prepare phosphors with favorable morphologies and excellent luminescent performance.

Ionic liquids are a kind of green solvent which includes a wide range of liquids, excellent thermal stability, a wide electrochemical window, and a low vapor pressure [28–32]. Recently, much attention has been paid to the study of ionic liquids in supported liquid membrane systems [33,34]. Ionic-liquid-driven supported liquid membrane systems have shown the advantages of high flux, high efficiency, strong durability, and environmental friendliness, and have made great progress in gas separation, organic separation, metal ion separation, and chemical reactions. Our team first committed to the use of an ionic-liquid-driven supported liquid membrane system to prepare $CePO_4$ inorganic nanomaterials. In doing so, we could easily control the morphologies (rod or sphere) of rare earth luminescent materials by adjusting the pH and the concentration of SO_4^{2-} [34]. Here, we used a facile ionic-liquid-driven supported liquid membrane method to prepare rare earth ion (Ce^{3+}, Tb^{3+}) co-doped $LaPO_4$ phosphors with different morphologies (spherical and stone-like shapes). The preparation procedure, the role of the ionic liquid supported liquid membrane, characterization of the crystal structure, and photoluminescent properties of the synthesized $LaPO_4$:Ce^{3+}, Tb^{3+} phosphors are reported in the following sections. This method has been proven to be easily controlled, simple, and mild, and the phosphors prepared by this method show good morphological and photoluminescent properties.

2. Results and Discussion

2.1. Ionic-Liquid-Driven HVHP Membrane Characterization

In this experiment, we use a microporous ionic-liquid-driven HVHP membrane as selective ion channels that can selectively transfer PO_4^{3-} from the PO_4^{3-} supply phase into the mixed rare earth ion supply phase to prepare phosphors. We referred to the methods of Krzysztof A. et al. [35] to characterize the ionic-liquid-driven HVHP membrane. A Raman study was performed to investigate the ionic liquid presence on the surface and inside the HVHP membrane. The spectra of the inner part of the ionic-liquid-driven HVHP membrane was recorded up to 40 μm below the surface. As shown in Figure 1, the Raman vibration modes of ionic liquids can be observed on the surface of and in the interior of their corresponding functional membranes, which proves ionic liquids' presence on the surface and inside the HVHP membranes. Figure 2 shows SEM micrographs of the cross-section of the untreated HVHP membrane, as well as SEM micrographs and the corresponding map microanalysis (B or S) of the cross-section of the resulting [C_4mim][BF_4]- and [C_4mim][Tf_2N]-driven HVHP membranes. The micrographs and map microanalysis show that the ionic liquid infiltrates the reticular surface of the membrane.

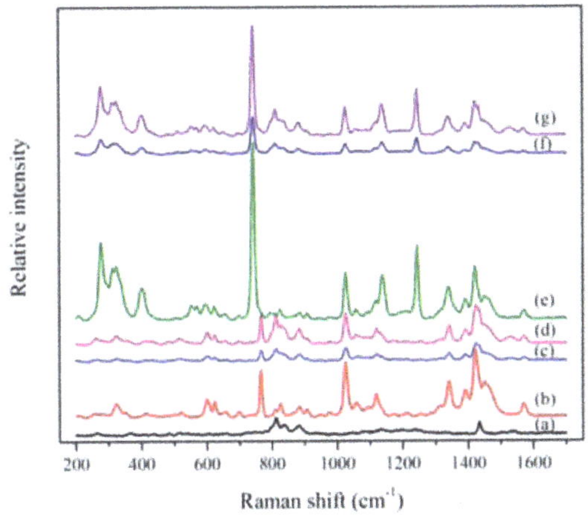

Figure 1. Comparison of Raman spectra of (a) untreated HVHP membrane, (b) [C_4mim][BF_4], (c) the surface of the [C_4mim][BF_4]-driven HVHP, (d) the internal part (40 μm below the surface) of the [C_4mim][BF_4]-driven HVHP, (e) [C_4mim][Tf_2N], (f) the surface of the [C_4mim][Tf_2N]-driven HVHP membrane and (g) the internal part (40 μm below the surface) of the [C_4mim][Tf_2N]-driven HVHP membrane.

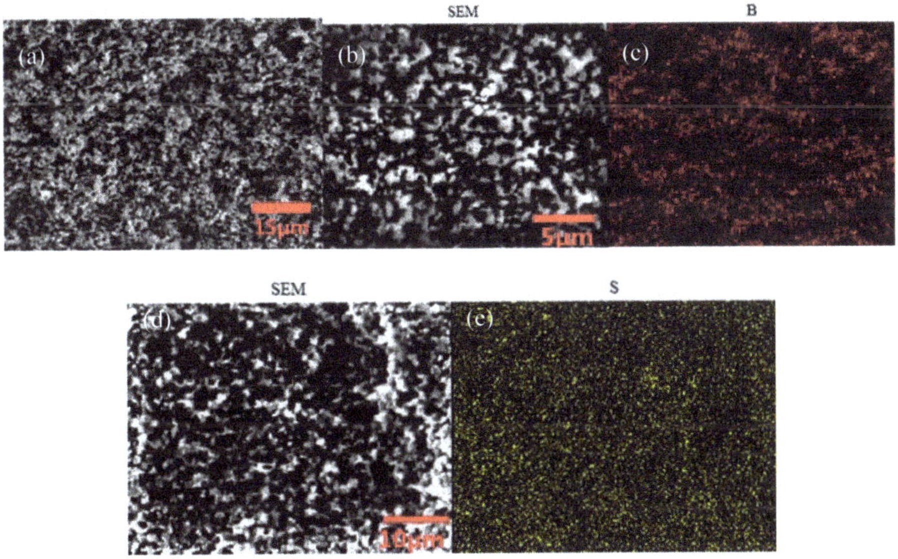

Figure 2. Plot of (**a**) micrographs of the cross-section of the untreated HVHP membrane, (**b**) micrographs of the cross-section of the [C_4mim][BF_4]-driven HVHP, (**c**) map microanalysis (B) of [C_4mim][BF_4]-driven HVHP, (**d**) micrographs of the cross-section of [C_4mim][Tf_2N] and (**e**) map microanalysis (S) of the [C_4mim][Tf_2N]-driven HVHP membrane.

2.2. Membrane Reaction Mechanism

We referred to the mechanism study of PanPan Zhao et al. [34] to propose a possible membrane reaction mechanism. The entire process can be divided into two parts: the liquid membrane transport stage and the precipitation reaction stage. The porous HVHP membrane is a good hydrophobic barrier which can effectively separate the two aqueous phases. After being immersed in an ionic liquid, the function of the HVHP film changes significantly. The microporous HVHP membrane containing an ionic liquid consists of selective ion channels which can selectively transfer PO_4^{3-} from the PO_4^{3-} supply phase into the mixed rare earth ion supply phase; the microporous HVHP membrane without ionic liquid cannot do this (Figure 3a). A precipitation reaction occurs upon PO_4^{3-} contacting the mixed rare earth ion supply phase on the other side of the HVHP membrane. Throughout the process, the cation (imidazolium) of the ionic liquid is responsible for the selective transfer of PO_4^{3-} from the PO_4^{3-} supply phase to the mixed rare earth ion supply phase. The anion is responsible for controlling the mixed rare earth ion supply phase and the release rate and ionic liquid hydrophobicity are correlated [34]. The hydrophilicity follows the order $[N(SO_2CF_3)_2]^- < [BF_4]^-$, so the release rate of PO_4^{3-} of the ionic-liquid-driven HVHP membrane is in this order (Figure 3a). The reaction appears to be a liquid-liquid extraction and occurs in the ionic liquid-film phase at the membrane interface. In addition, due to the thinness and high porosity of the porous HVHP membrane, the numerous ion transport channels are very short, meaning the precipitation reaction occurs quickly and efficiently. The experimental device and a schematic diagram of the reaction mechanisms are shown in Figure 4.

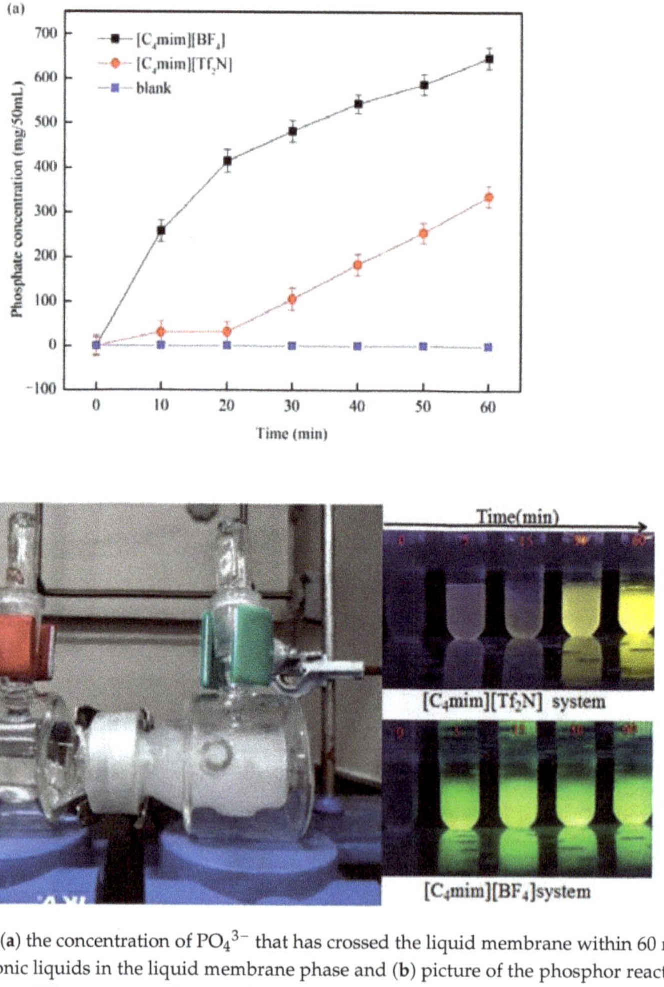

Figure 3. Plot of (**a**) the concentration of PO_4^{3-} that has crossed the liquid membrane within 60 min using different ionic liquids in the liquid membrane phase and (**b**) picture of the phosphor reaction process and images of the precursor solution color within 60 min when using different ionic liquids in the liquid membrane phase.

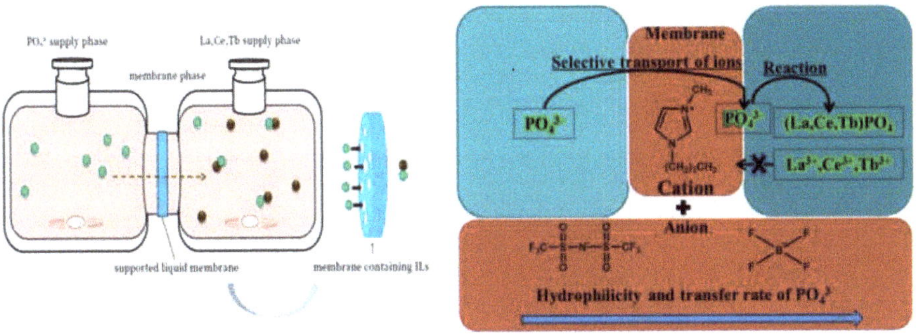

Figure 4. Ionic-liquid-driven supported liquid membrane system and schematic diagram of reaction mechanism.

2.3. *Transmittance of PO_4^{3-} under the Action of the Two Functional Membranes*

To investigate the transfer efficiency of PO_4^{3-} by the different functional membranes, we performed the following experiment devices: two glass units sandwiching blank-, [C_4mim][BF_4]- or [C_4mim][Tf_2N]-infiltrated membranes. The glass units were filled with 50 mL deionized water and 50 mL of the phosphoric acid solution (1 M). To ensure a homogeneous system, both solutions were stirred with a magnetic stirrer at 1000 rpm. Samples of 10 µL were taken from the deionized water phase every 10 min. Then, the samples were diluted and the transfer of PO_4^{3-} under the action of three kinds of membranes was measured with ICP. Figure 3a shows the changes of PO_4^{3-} concentration in 50 mL of a deionized water phase under the action of three different membranes within 60 min. The PO_4^{3-} concentration remained at 0 under the action of the blank-infiltrated membrane while the PO_4^{3-} concentration increased under the action of the ionic liquid functional membranes. This suggests that PO_4^{3-} cannot cross the blank-infiltrated membrane but can cross ionic liquid functional membranes. There is a clear difference between the [C_4mim][BF_4] functional membrane and the [C_4mim][Tf_2N] functional membrane, which indicates that the transfer efficiency of the [C_4mim][BF_4] functional membrane toward PO_4^{3-} is much greater than that of the [C_4mim][Tf_2N] functional membrane. Figure 3b shows a picture of the phosphor reaction process and images of the precursor solution color when using rare earth sulfates as the rare earth supply phase for the different ionic liquid systems at different times under 254 nm irradiation. As shown in Figure 3b, precipitation occurs only in the rare earth phases, which indicates that the rare earth ions cannot pass through the membrane channels but phosphate can. The sample solutions all emitted green fluorescence when under a 254 nm light source and the fluorescence brightness increased with increasing reaction time. This proves that the membrane transfer rate of PO_4^{3-} for the [C_4mim][BF_4] functional membrane is markedly faster than that for the [C_4mim][Tf_2N] functional membrane, which is consistent with the results of previous research. In the actual production application, we can choose an appropriate functional membrane with different ionic liquids to achieve the effect of controlling the rate of production.

2.4. *Study of the Proportion of $LaPO_4$:Ce^{3+}, Tb^{3+}*

To explore the proportion of product, we dissolved an appropriate amount of the precursor (preparation conditions: 1 mol phosphoric acid solution, rare earth sulfate solution, and [C_4mim][Tf_2N] functional membrane) in a moderate amount of hydrochloric acid at 60 °C for 30 min, diluted the sample to the right concentration, and then tested the rare earth ion concentration by ICP. Table 1 shows the molar ratio of the mixed solution of rare earth elements from the rare earth supply phase and the molar ratio of the precursor. A clear difference between the solution proportion and precursor proportion can be seen in Table 1 because the rare earth ions do not completely precipitate. Then, we obtained the molar ratio (a fitting degree of greater than 99%) commonly used in the production of phosphate (La:Ce:Tb = 55:30:15) by simply adjusting the molar ratio of the rare earth ions in solution.

Table 1. The molar ratio of a mixed solution of the rare earth elements and the molar ratio of the precursor.

Samples	La^{3+} Molal Percent	Ce^{3+} Molal Percent	Tb^{3+} Molal Percent
Initial solution	55.13	31.04	13.83
Initial precursors	53.40	34.99	11.61
After adjusting solution	55.70	26.20	18.09
After adjusting precursors	55.00	30.11	14.89

2.5. Structure and Morphologies of the $LaPO_4$:Ce^{3+}, Tb^{3+} Phosphors

In this study, we used an ionic-liquid-driven supported liquid membrane system to prepare phosphors. The whole system consisted of two glass units sandwiching a functional membrane ([C_4mim][BF_4] or [C_4mim][Tf_2N]). The glass units were filled with 50 mL of the rare earth mixture (rare earth sulfates or rare earth nitrates) and 50 mL of the phosphoric acid solution (1 M). The PO_4^{3-} crossed the functional membrane to react with the rare earth ions in this system. Finally, the phosphors were prepared by calcining the precursors. The powder samples prepared from different rare earth ion sources (rare earth nitrates and rare earth sulfates) in the [C_4mim][BF_4] functional membrane were labelled BN and BS, respectively, and powder samples prepared from different rare earth ion sources (rare earth nitrates and rare earth sulfates) in the [C_4mim][Tf_2N] functional membrane were labelled NN and NS, respectively.

X-ray diffraction patterns were employed to determine the phase purities and crystal structures of the phosphor products. Figure 5a shows the XRD patterns of the precursors prepared under different conditions (different rare earth solutions and different ionic liquids). The vertical bars show the standard hexagonal $LaPO_4$ peak positions (JCPDS No. 04-0635). Figure 5a shows that the diffraction peaks of all the precursors can be readily indexed to the hexagonal structure of $LaPO_4$ in the P6222 space group (JCPDS No 04-0635). Figure 5b shows the XRD patterns of the as-prepared $LaPO_4$:Ce, Tb phosphor samples prepared under different conditions (different rare earth solutions and different ionic liquids). The vertical bars show the standard monoclinic $LaPO_4$ peak positions (JCPDS No. 32-0493). From Figure 5b, it is obvious that peaks at 2θ = 19.04°, 21.74°, 27.08°, 28.88°, and 42.48° are present after annealing at 1000 °C, which may be attributed to the (011), (101), (200), (120), and (221) reflections of the monazite crystalline structure of lanthanum phosphate. A monoclinic phase (space group: P21/n) of pure $LaPO_4$ (JCPDS No. 32-0943) was obtained. By comparing the XRD pattern of the as-prepared precursors and $LaPO_4$:Ce, Tb phosphor samples, we found that after annealing at 1000 °C, the fluorescent powder XRD peaks were sharper, the crystallinity was better and the structure of $LaPO_4$:Ce, Tb had changed from a hexagonal to a monoclinic crystal phase. In addition, after the sample was calcined at 1000 °C, all the diffraction peaks shifted to the right compared with the standard diffraction peaks. This is because the radii of Ce^{3+} (~0.1034 nm) and Tb^{3+} (~0.0923 nm) are smaller than the radius of La^{3+} (~0.1061 nm) in the $LaPO_4$:Ce^{3+}, Tb^{3+} crystals, which leads to lattice contraction and a reduction of interplanar distance. Thus, based on the Bragg diffraction principle $2d\sin\theta = \lambda$, where the decrease of the d value increases the diffraction angle, the diffraction peak positions of the XRD patterns move towards larger angles [23,36].

The morphology, size, and microstructural details were investigated by scanning electron microscopy. Figure 6 shows the SEM micrographs of the precursors prepared under different conditions (different rare earth solutions and different ionic liquids). Notably, the morphologies of the precursors are similar in the different ionic liquid functional membranes but show different morphologies for the different rare earth sources. When the anion of the rare earth mixed solution was sulfate, the samples exhibited a spherical morphology with particle sizes in the range of 600–800 nm, and a rough surface which consisted of aggregates of smaller particles. When the anion of the rare earth mixed solution was nitrate, the samples exhibited a flower-like structure with a diameter of approximately 30 nm and a length of approximately 200 nm. According to our previous research, we believe the reason for the formation of this globular structure is due to the template effect of SO_4^{2-} [34]. Figure 7 shows SEM micrographs of the as-prepared $LaPO_4$:Ce, Tb phosphor samples prepared under different conditions (different rare earth solutions and different ionic liquids). Similarly to the precursors, the annealed samples had similar morphologies when prepared with different ionic liquid functional membranes but different morphologies when prepared with different rare earth sources. After sintering, the samples prepared with rare earth sulfates as the raw material maintained their spherical structure, but the aggregation between spheres was more severe than that in the precursor samples. The small particles on the spherical surface of the sintered samples were larger than those on the surface of the respective precursor. However, after sintering, the samples prepared with rare earth nitrates as the raw material changed from a nanowire flower-like structure to a stone-like structure. The particle sizes were in the range 30–300 nm. The results show that the crystal size of all the samples increased after calcining, and due to the templating effect of SO_4^{2-}, the samples with rare earth sulfates as the raw material continued to maintain their large micro-sized spherical morphology, while the shape of the samples prepared using rare earth nitrates as the raw material grew from a flower-like structure into a stone-like morphology.

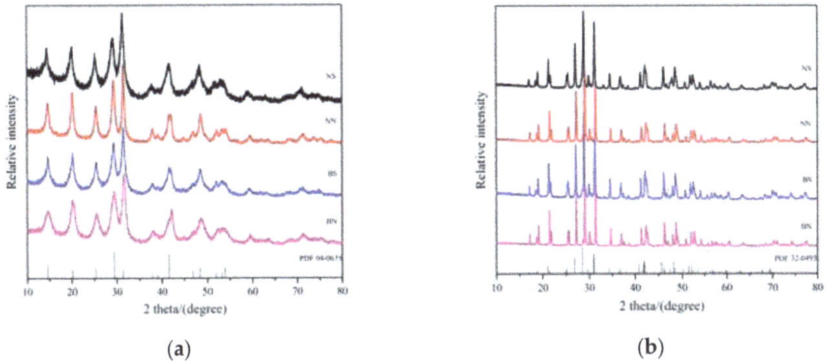

Figure 5. XRD patterns of the precursors (**a**) and calcined $LaPO_4:Ce^{3+}$, Tb^{3+} phosphors (**b**) prepared under different conditions.

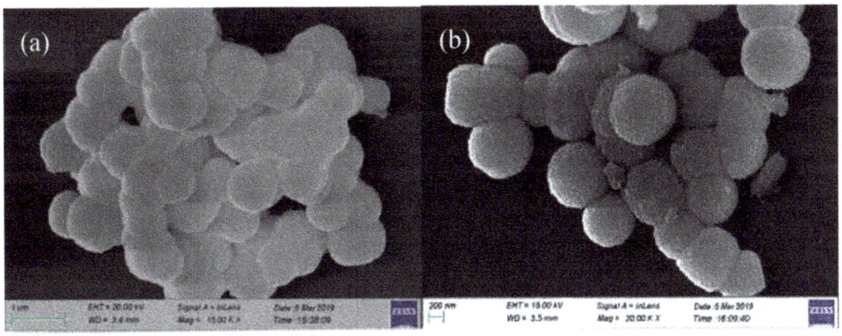

Figure 6. *Cont.*

Figure 6. SEM images of the precursors prepared under different conditions: (**a**) BS, (**b**) NS, (**c**) BN, and (**d**) NN. (EHT for extra high tension, WD for working distance)

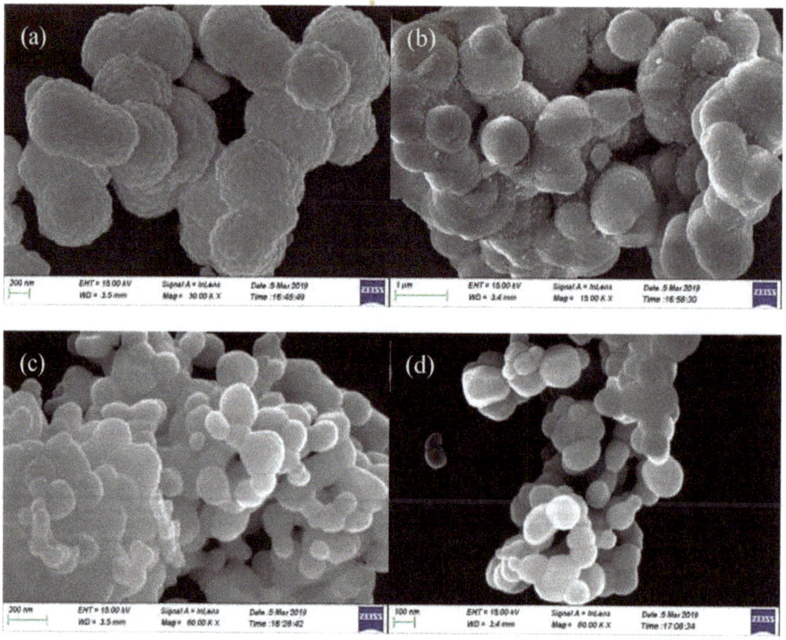

Figure 7. SEM images of calcined $LaPO_4$:Ce^{3+}, Tb^{3+} phosphors prepared under different conditions: (**a**) BS, (**b**) NS, (**c**) BN, and (**d**) NN. (EHT for extra high tension, WD for working distance)

2.6. Photoluminescent Properties of the $LaPO_4$:Ce^{3+}, Tb^{3+} Phosphors

Figure 8 shows the excitation spectra of all the calcined $LaPO_4$:Ce^{3+}, Tb^{3+} phosphors by monitoring the $^5D_4 \rightarrow {^7F_5}$ emission of Tb^{3+} (λ_{em} = 543 nm) at room temperature. As clearly shown in Figure 8, the obtained fluorescent materials absorb excitation energy in the range of 240–310 nm with a maximum excitation wavelength at 277 nm, which may be related to the f–d transitions of Ce^{3+}. In addition, several small peaks can be detected in the range of 310–400 nm, which could be caused by the f–f transitions of Tb^{3+} [37,38]. Because of the forbidden nature of these transitions, their oscillator strength is much weaker than that of the spin-allowed $4f^1$–$4f^05d^1$ Ce^{3+} transitions [22]. The excitation spectra consist of the strong excitation band of Ce^{3+} and the weak excitation bands of Tb^{3+}, revealing that Tb^{3+} are essentially excited by Ce^{3+}. In fact, several of the weak f–f excitation bands of Tb^{3+} are only present in the region of the Ce^{3+} emission. Thus, energy transfer from Ce^{3+} to Tb^{3+} occurs [21]. The emission spectra of all the calcined $LaPO_4$:Ce^{3+}, Tb^{3+} phosphors at an excitation of 277 nm are shown in Figure 9.

All the calcined $LaPO_4:Ce^{3+}$, Tb^{3+} phosphors show obvious photoluminescence in the spectral range of 450–650 nm, and the four emission peaks at 487, 543, 584, and 621 nm can be assigned to the $^5D_4$–$^7F_6$, $^5D_4$–$^7F_5$, $^5D_4$–$^7F_4$, and $^5D_4$–$^7F_3$ transitions, respectively, of Tb^{3+} [39,40]. Among these peaks, the green emission at 543 nm, which corresponds to the $^5D_4$–$^7F_5$ transition of Tb^{3+}, is the predominant peak. The spectral properties of the phosphors prepared by an ionic-liquid-driven supported liquid membrane system are essentially the same as those prepared by other synthetic methods, in which the improved ionic-liquid-driven supported liquid membrane system is a new and effective method to prepare $LaPO_4:Ce^{3+}$, Tb^{3+} phosphors.

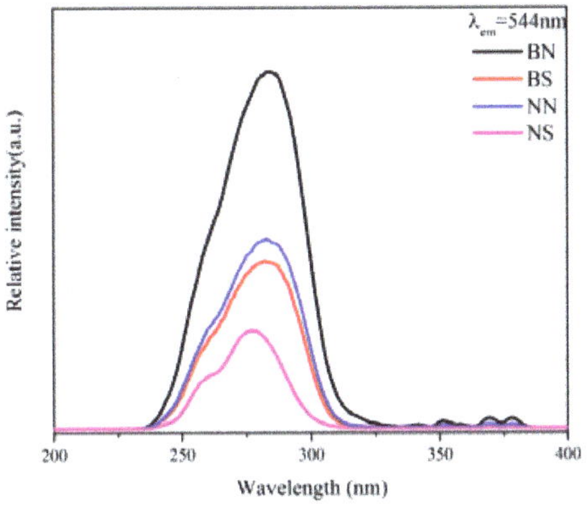

Figure 8. The excitation spectra of the calcined $LaPO_4:Ce^{3+}$, Tb^{3+} phosphors prepared under different conditions.

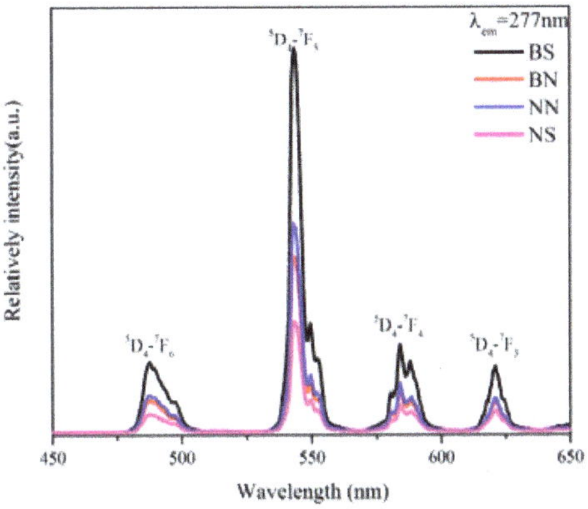

Figure 9. The emission spectra of the calcined $LaPO_4$: Ce^{3+}, Tb^{3+} phosphors prepared under different conditions.

3. Materials and Methods

3.1. Materials

The lanthanum sulfate hydrate, cerium sulfate hydrate, terbium sulfate hydrate, lanthanum nitrate hydrate, cerium nitrate hydrate, and terbium nitrate hydrate were provided by the Baotou Research Institute of Rare Earths, and the phosphoric acid solution was purchased from Aladdin (Shanghai, China). The chemicals used in the experiments were of analytical grade. The HVHP-04700 (pore size 0.45 µm and thickness 125 µm, ø: 5.5 cm, DUPAPORE®), a hydrophobic porous polyvinylidene fluoride film, was obtained from Millipore Corp. The ionic liquids (ILs) were selected from $[C_4mim][BF_4]$

and $[C_4mim][Tf_2N]$ produced by the Center for Green Chemistry and Catalysis, Lanzhou Institute of Chemical Physics, Chinese Academy of Sciences. Figure 10 shows the molecular structures of the ionic liquids used in this study.

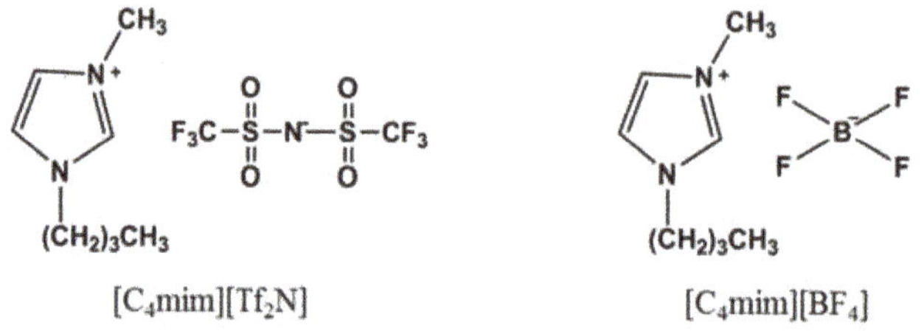

Figure 10. The molecular structures of the different ionic liquids used.

3.2. *Preparation of the Ionic-Liquid-Driven Supported Liquid Membrane, the La, Ce, and Tb Supply Phase, and the PO_4^{3-} Supply Phase*

To prepare the ionic-liquid-driven supported liquid membrane, the hydrophobic porous polyvinylidene fluoride film (HVHP-04700) was immersed in an ionic liquid (≥200 μL of either $[C_4mim][BF_4]$ or $[C_4mim][Tf_2N]$) for more than 2 h. For the La, Ce, and Tb supply phase, a lanthanum sulfate, cerium sulfate, and terbium sulfate mixed solution (or a lanthanum nitrate, cerium nitrate, and terbium nitrate mixed solution) was prepared in a calibrated volumetric flask by dissolving each compound in ultrapure water at a suitable La/Ce/Tb molar ratio. Each compound was placed in an ultrasound cleaner for 20 min to ensure complete dissolution. The PO_4^{3-} supply phase (1 M) was prepared in a calibrated volumetric flask by the dilution of concentrated phosphoric acid.

3.3. *$LaPO_4$:Ce^{3+}, Tb^{3+} Precursor Synthetic Process for the Membrane Reaction*

The experiment was carried out in a glass cell system with self-adjusting diffusion which consisted of two glass units sandwiching a functional membrane ($[C_4mim][BF_4]$ or $[C_4mim][Tf_2N]$). The glass units were filled with 50 mL of the rare earth mixture (rare earth sulfates or rare earth nitrates) at a certain ratio and 50 mL of the phosphoric acid solution (1 M). To ensure a homogeneous system, both solutions were stirred with a magnetic stirrer at 1000 rpm (Mini MR, IKA). After the complete reaction had occurred at room temperature, the white product was collected, centrifuged, and washed with ethanol more than 5 times, and then dried at 60 °C in a drying oven for 12 h to obtain the precursor. Under a set temperature (1000 °C), the $LaPO_4$:Ce, Tb precursor powder was calcined for 1 h under a reducing atmosphere to finally produced the green-emitting phosphors.

3.4. *Characterization Methods*

The concentration of P in the ultrapure water phase and the rare earth ion concentration of the $LaPO_4$:Ce^{3+}, Tb^{3+} samples were measured using a HORIBA-Jobin Yvon ULTIMA 2 series by inductively coupled plasma-atomic emission spectroscopy (ICP-AES). The Raman spectra were measured using the Horiba Jobin Yvon S.A.S. LabRAM Aramis. The structures and phase purities of the as-prepared $LaPO_4$:Ce^{3+}, Tb^{3+} samples were identified using X-ray diffraction analysis with a Bruker AXS D8 Advance Powder X-ray diffractometer (Cu Kα radiation, λ = 1.5418 Å). The morphologies, energy spectrum of membranes, and the as-prepared products were observed under a ZEISS SIGMA 500 field emission scanning electron microscope. The excitation and emission spectra were taken on an Edinburgh FLS980 spectrometer equipped with a 450 W ozone-free xenon arc lamp as the excitation source.

4. Conclusions

In summary, $LaPO_4$:Ce, Tb phosphors with monoclinic structures and good photoluminescence were successfully synthesized using a novel, controllable, and efficient ionic-liquid-driven supported liquid membrane system. The release rate of PO_4^{3-} from the liquid membrane with different ionic liquids was different. The phosphors prepared by this method exhibited micro-spherical (when using rare earth sulfates as the raw material) and nanoscale stone-shape (when using rare earth nitrates as the raw material) morphologies due to the influence of the different anions. These studies indicate that ionic-liquid-driven supported liquid membrane systems are a promising method for preparing $LaPO_4$:Ce^{3+}, Tb^{3+} phosphors.

Author Contributions: Formal analysis, J.L.; funding acquisition, F.Y.; investigation, J.L., Z.Z., R.B., and H.Z.; methodology, J.L.; project administration, F.Y.; resources, F.Y.; supervision, H.D., F.Y., and L.S.; writing—original draft, J.L.; writing—review and editing, H.D., F.Y., and L.S.

Funding: The project was sponsored by the Clean Nuclear Energy System Fuels and Materials Joint Innovation Key Laboratory of Fujian Province, The National Natural Science Foundation of China (51865044), the Key Laboratory of Baiyunobo Rare Earth Resource Researches and Comprehensive Utilization (2018Z2004), the Science and Technology Service Network Initiative (2017T3002), the Science and Technology Key R&D Programs of Jiangxi Province (20171ACH80013), and the Science and Technology Projects of Xiamen (3502Z20172031).

Acknowledgments: The authors are grateful to characterizations provided by the Public Technology Service Center, Xiamen Institute of Rare Earth Materials.

Conflicts of Interest: The authors declare no conflict of interest.

References

1. Eliseeva, S.V.; Bunzli, J.C. Lanthanide luminescence for functional materials and bio-sciences. *Chem. Soc. Rev.* **2010**, *39*, 189–227. [CrossRef] [PubMed]
2. Wang, G.; Peng, Q.; Li, Y. Lanthanide-Doped Nanocrystals: Synthesis, Optical-Magnetic Properties, and Applications. Accounts. *Chem. Res.* **2011**, *44*, 322–332. [CrossRef] [PubMed]
3. Richardson, F.S. Terbium(III) and Europium(III) Ions as luminescent probes and stains for biomolecular systems. *Chem. Rev.* **1982**, *82*, 541–552. [CrossRef]
4. Wang, N.; Zhang, S.; Zhang, X.; Wei, Y. Preparation of $LaPO_4$: Ce^{3+}, Tb^{3+} nanophosphors by mixed co-precipitation process and their photoluminescence properties. *Ceram. Int.* **2014**, *40*, 16253–16258. [CrossRef]
5. Niu, N.; Yang, P.; Wang, Y.; Wang, W.; He, F.; Gai, S.; Wang, D. $LaPO_4$:Eu^{3+}, $LaPO_4$:Ce^{3+}, and $LaPO_4$:Ce^{3+}, Tb^{3+} nanocrystals: Oleic acid assisted solvothermal synthesis, characterization, and luminescent properties. *J. Alloys Compd.* **2011**, *509*, 3096–3102. [CrossRef]
6. Yang, P.; Quan, Z.; Li, C.; Hou, Z.; Wang, W.; Lin, J. Solvothermal synthesis and luminescent properties of monodisperse $LaPO_4$:Ln (Ln=Eu^{3+}, Ce^{3+}, Tb^{3+}) particles. *J. Solid State Chem.* **2009**, *182*, 1045–1054. [CrossRef]
7. Hu, X.; Yan, S.; Ma, L.; Wan, G.; Hu, J. Preparation of $LaPO_4$: Ce, Tb phosphor with different morphologies and their fluorescence properties. *Powder Technol.* **2009**, *192*, 27–32. [CrossRef]
8. Yu, L.; Song, H.; Liu, Z.; Yang, L.; Lu, S.; Zheng, Z. Remarkable improvement of brightness for the green emissions in Ce^{3+} and Tb^{3+} co-activated $LaPO_4$ nanowires. *Solid State Commun.* **2005**, *134*, 753–757. [CrossRef]
9. Rao, R.P.; Devine, D.J. RE-Activated Lanthanide Phosphate Phosphors for PDP Applications. *J. Lumin.* **2000**, *87*, 1260–1263. [CrossRef]
10. Yu, M.; Lin, J.; Fu, J.; Han, Y.C. Sol–gel fabrication, patterning and photoluminescent properties of $LaPO_4$:Ce^{3+}, Tb^{3+} nanocrystalline thin films. *Chem. Phys. Lett.* **2003**, *371*, 178–183. [CrossRef]
11. Wang, Z.; Quan, Z.; Lin, J.; Fang, J. Polyol-Mediated Synthesis and Photoluminescent Properties of Ce^{3+} and/or Tb^{3+} Doped $LaPO_4$ Nanoparticles. *J. Nanosci. Nanotechnol.* **2005**, *5*, 1532–1536. [CrossRef] [PubMed]
12. Yu, M.; Wang, H.; Lin, C.K.; Li, G.Z.; Lin, J. Sol–gel synthesis and photoluminescence properties of spherical SiO_2@$LaPO_4$:Ce^{3+}/Tb^{3+}particles with a core-shell structure. *Nanotechnology* **2006**, *17*, 3245–3252. [CrossRef]

13. Dong, H.; Liu, Y.; Yang, P.; Wang, W.; Lin, J. Controlled synthesis and characterization of $LaPO_4$, $LaPO_4$:Ce^{3+} and $LaPO_4$:Ce^{3+}, Tb^{3+} by EDTA assisted hydrothermal method. *Solid State Sci.* **2010**, *12*, 1652–1660. [CrossRef]
14. Lenggoro, I.W.; Xia, B.; Mizushima, H.; Okuyama, K.; Kijima, N. Synthesis of $LaPO_4$:Ce, Tb phosphor particles by spray pyrolysis. *Mater. Lett.* **2001**, *50*, 92–96. [CrossRef]
15. Zhiqi, L.; Le, R.; Zhaowu, Z.; Dali, C.; Na, Z.; Minglai, L.; Meisheng, C.; Xiaowei, H. Synthesis of $LaPO_4$:Ce, Terbium by Co-Precipitation Method. *J. Rare Earths* **2006**, *24*, 137–140. [CrossRef]
16. Nunez, N.O.; Liviano, S.R.; Ocana, M. Citrate mediated synthesis of uniform monazite $LnPO_4$ (Ln = La, Ce) and Ln:$LaPO_4$ (Ln = Eu, Ce, Ce + Tb) spheres and their photoluminescence. *J. Colloid Interface Sci.* **2010**, *349*, 484–491. [CrossRef] [PubMed]
17. Song, W.-S.; Choi, H.-N.; Kim, Y.-S.; Yang, H. Formation of green-emitting $LaPO_4$:Ce, Tb nanophosphor layer and its application to highly transparent plasma displays. *J. Mater. Chem.* **2010**, *20*, 6929–6934. [CrossRef]
18. Song, W.-S.; Kim, Y.-S.; Yang, H. Construction of Highly Transparent Plasma Display Devices Using Hydrothermally Synthesized Green-Emitting $LaPO_4$:Ce, Tb Nanophosphors. *J. Electrochem. Soc.* **2011**, *158*, J137–J142. [CrossRef]
19. Duault, F.; Junker, M.; Grosseau, P.; Guilhot, B.; Iacconi, P.; Moine, B. Effect of different fluxes on the morphology of the $LaPO_4$:Ce, Tb phosphor. *Powder Technol.* **2005**, *154*, 132–137. [CrossRef]
20. Zhu, H.; Zhu, E.; Yang, H.; Wang, L.; Jin, D.; Yao, K. High-Brightness $LaPO_4$:Ce^{3+}, Tb^{3+} Nanophosphors: Reductive Hydrothermal Synthesis and Photoluminescent Properties. *J. Am. Ceram. Soc.* **2008**, *91*, 1682–1685. [CrossRef]
21. Yang, M.; You, H.; Liu, K.; Zheng, Y.; Guo, N.; Zhang, H. Low-temperature coprecipitation synthesis and luminescent properties of $LaPO_4$:Ln^{3+} (Ln^{3+} = Ce^{3+}, Tb^{3+}) nanowires and $LaPO_4$: Ce^{3+}, Tb^{3+}/$LaPO_4$ core/shell nanowires. *Inorg. Chem.* **2010**, *49*, 4996–5002. [CrossRef] [PubMed]
22. Ansari, A.A.; Khan, M.A.M. Structural and spectroscopic studies of $LaPO_4$:Ce/Tb@$LaPO_4$@SiO_2 nanorods: Synthesis and role of surface coating. *Vib. Spectrosc.* **2018**, *94*, 43–48. [CrossRef]
23. Dong, W.-L.; Zhang, X.-Y.; Shi, H.; Mi, X.-Y.; Wang, N.-L.; Han, K.-X. Synthesis and photoluminescence properties of $LaPO_4$:Ce^{3+}, Tb^{3+} nanophosphors by microwave-assisted co-precipitation method at low temperature. *Funct. Mater. Lett.* **2015**. [CrossRef]
24. Zhu, H.; Ou, G.; Gao, L. Hydrothermal synthesis of $LaPO_4$:Ce^{3+}, Tb^{3+}@$LaPO_4$ core/shell nanostructures with enhanced thermal stability. *Mater. Chem. Phys.* **2010**, *121*, 414–418. [CrossRef]
25. Fu, Z.; Bu, W. High efficiency green-luminescent $LaPO_4$:Ce, Tb hierarchical nanostructures: Synthesis, characterization, and luminescence properties. *Solid State Sci.* **2008**, *10*, 1062–1067. [CrossRef]
26. Hou, Z.; Wang, L.; Lian, H.; Chai, R.; Zhang, C.; Cheng, Z.; Lin, J. Preparation and luminescence properties of Ce^{3+} and/or Tb^{3+} doped $LaPO_4$ nanofibers and microbelts by electrospinning. *J. Solid State Chem.* **2009**, *182*, 698–708. [CrossRef]
27. Li, Y.; Chai, Q.; Liao, S.; Chen, Z.; He, Y.; Xia, Y.; Wu, W.; Li, B. Non-isothermal kinetics study with isoconversional procedure and DAEM: Thermal decomposition of $LaPO_4$:Ce,Tb·0.5H_2O. *Mater. Chem. Phys.* **2013**, *142*, 453–458. [CrossRef]
28. Zhang, Y.; Chen, J. Preparation of $REPO_4$ (RE = La-Gd) nanorods from an ionic liquid extraction system and luminescent properties of $CePO_4$:Tb^{3+}. *Rare Met.* **2016**. [CrossRef]
29. Bühler, G.; Stay, M.; Feldmann, C. Ionic liquid-based approach to doped nanoscale oxides: $LaPO_4$:RE (RE = Ce, Tb, Eu) and In_2O_3:Sn (ITO). *Green Chem.* **2007**, *9*, 924–926. [CrossRef]
30. Muntzeck, M.; Wilhelm, R. Influence of Ionic Liquids on an Iron(III) Catalyzed Three-Component Coupling/Hydroarylation/Dehydrogenation Tandem Reaction. *Int. J. Mol. Sci.* **2016**, *17*, 860. [CrossRef]
31. Park, J.; Jung, Y.; Kusumah, P.; Lee, J.; Kwon, K.; Lee, C.K. Application of ionic liquids in hydrometallurgy. *Int. J. Mol. Sci.* **2014**, *15*, 15320–15343. [CrossRef] [PubMed]
32. Lacrămă, A.M.; Putz, M.V.; Ostafe, V. A Spectral-SAR Model for the Anionic-Cationic Interaction in Ionic Liquids: Application to Vibrio fischeri Ecotoxicity. *Int. J. Mol. Sci.* **2007**, *8*, 842–863. [CrossRef]
33. Ramakul, P.; Mooncluen, U.; Yanachawakul, Y.; Leepipatpiboon, N. Mass transport modeling and analysis on the mutual separation of lanthanum(III) and cerium(IV) through a hollow fiber supported liquid membrane. *J. Ind. Eng. Chem.* **2012**, *18*, 1606–1611. [CrossRef]

34. Zhao, P.; Yang, F.; Zhao, Z.; Liao, Q.; Zhang, Y.; Chen, P.; Guo, W.; Bai, R. A simple preparation method for rare-earth phosphate nano materials using an ionic liquid-driven supported liquid membrane system. *J. Ind. Eng. Chem.* **2017**, *54*, 369–376. [CrossRef]
35. Bogdanowicz, K.; Sistat, P.; José, A.R.; Giamberini, M. Liquid crystalline polymeric wires for selective proton transport, part 2: Ion transport in solid-state. *Polymer* **2016**, *92*, 50–57. [CrossRef]
36. Shi, Y.; Wang, Y.; Wang, D.; Liu, B.; Li, Y.; Wei, L. Synthesis of Hexagonal Prism (La, Ce, Tb)PO_4 Phosphors by Precipitation Method. *Cryst. Growth Des.* **2012**, *12*, 1785–1791. [CrossRef]
37. Bühler, G.; Feldmann, C. Transparent luminescent layers via ionic liquid-based approach to $LaPO_4$:RE (RE = Ce, Tb, Eu) dispersions. *Appl. Phys. A* **2007**, *87*, 631–636. [CrossRef]
38. Pankratov, V.; Popov, A.I.; Chernov, S.A.; Zharkouskaya, A.; Feldmann, C. Mechanism for energy transfer processes between Ce^{3+} and Tb^{3+} in $LaPO_4$:Ce, Tb nanocrystals by time-resolved luminescence spectroscopy. *Phys. Status Solidi B* **2010**, *247*, 2252–2257. [CrossRef]
39. Pankratov, V.; Popov, A.I.; Kotlov, A.; Feldmann, C. Luminescence of nano- and macrosized $LaPO_4$:Ce, Tb excited by synchrotron radiation. *Opt. Mater.* **2011**, *33*, 1102–1105. [CrossRef]
40. Pankratov, V.; Popov, A.I.; Shirmane, L.; Kotlov, A.; Feldmann, C. $LaPO_4$:Ce, Tb and YVO_4:Eu nanophosphors: Luminescence studies in the vacuum ultraviolet spectral range. *J. Appl. Phys.* **2011**, *110*, 053522. [CrossRef]

International Journal of
Molecular Sciences

MDPI

Article

How Charge and Triple Size-Selective Membrane Separation of Peptides from Salmon Protein Hydrolysate Orientate Their Biological Response on Glucose Uptake

Loïc Henaux [1,2], Jacinthe Thibodeau [1,2], Geneviève Pilon [2,3], Tom Gill [4], André Marette [2,3] and Laurent Bazinet [1,2,*]

1 Department of Food Sciences and Laboratory of Food Processing and Electromembrane Processes (LTAPEM), Université Laval, Quebec, QC G1V 0A6, Canada; loic.henaux.1@ulaval.ca (L.H.); jacinthe.thibodeau.1@ulaval.ca (J.T.)

2 Institute of Nutrition and Functional Foods (INAF), Université Laval, Quebec, QC G1V 0A6, Canada; Genevieve.Pilon@criucpq.ulaval.ca (G.P.); Andre.Marette@criucpq.ulaval.ca (A.M.)

3 Québec Heart and Lung Institute, Université Laval, Department of Medicine, Quebec, QC G1V 4G5, Canada

4 Department of Process Engineering and Applied Science, Dalhousie University, P.O. Box 15,000, Halifax, NS B3H 4R2, Canada; Tom.Gill@Dal.Ca

* Correspondence: Laurent.Bazinet@fsaa.ulaval.ca; Tel.: +1-418-656-2131 (ext. 7445); Fax: +1-418-656-3353

Received: 22 March 2019; Accepted: 17 April 2019; Published: 20 April 2019

Abstract: The valorization of by-products from natural organic sources is an international priority to respond to environmental and economic challenges. In this context, electrodialysis with filtration membrane (EDFM), a green and ultra-selective process, was used to separate peptides from salmon frame protein hydrolysate. For the first time, the simultaneous separation of peptides by three ultrafiltration membranes of different molecular-weight exclusion limits (50, 20, and 5 kDa) stacked in an electrodialysis system, allowed for the generation of specific cationic and anionic fractions with different molecular weight profiles and bioactivity responses. Significant decreases in peptide recovery, yield, and molecular weight (MW) range were observed in the recovery compartments depending on whether peptides had to cross one, two, or three ultrafiltration membranes. Moreover, the Cationic Recovery Compartment 1 fraction demonstrated the highest increase (42%) in glucose uptake on L6 muscle cells. While, in the anionic configuration, both Anionic Recovery Compartment 2 and Anionic Recovery Compartment 3 fractions presented a glucose uptake response in basal condition similar to the insulin control. Furthermore, Cationic Recovery Compartment 3 was found to contain inhibitory peptides. Finally, LC-MS analyses of the bioassay-guided bioactive fractions allowed us to identify 11 peptides from salmon by-products that are potentially responsible for the glucose uptake improvement.

Keywords: electrodialysis with filtration membrane (EDFM); triple size-selective separation; glucose uptake; bioassay-guided validation; bioactive peptides

1. Introduction

Type 2 diabetes (T2D) is a complex multifactorial disorder resulting from insulin resistance in peripheral tissues, such as skeletal muscle, and pancreatic β-cell dysfunction [1]. This disease is growing at an alarming rate and is predicted to account for more than 350 million cases by 2030 [2]. Skeletal muscle is the major site of glucose uptake in the postprandial state and the development of insulin resistance in this tissue is considered a cornerstone in the pathogenesis of T2D. Interestingly, increased fish consumption has been suggested to protect against metabolic syndrome (MetS), type 2 diabetes and cardiovascular disease (CVD) in obese subjects [3,4]. Our group previously showed that fish protein is an important contributor to these beneficial effects. Indeed, Pilon et al. (2011) showed that salmon protein hydrolysate reduced inflammation in visceral adipose tissue and improved insulin sensitivity in an animal model of diet-induced obesity [5]. Furthermore, we recently reported that in a mouse model of obesity, a low molecular weight peptide (LMWP) fraction (<1 kDa) from the proteolytic digestion of salmon filleting waste improved glucose tolerance and lipid homeostasis [6]. Interestingly, other fish protein sources than salmon have been reported for their metabolic properties in animal models and also in humans [7–13]. Therefore, proteins from marine by-products definitively represent a high potential for the development of functional foods and nutraceutical products [7–13]. On the other hand, the value-added aspects of organic by-products have become a priority in order to respond to the sustainability, environmental, economic, and regulatory challenges [14,15].

Consequently, production, separation, and characterization of bioactive peptide (BP) by-products are important issues for the food and biopharmaceutical industries, and we are now aware that BPs' activity depends on specific molecular and chemical properties [16]. For example, the surface charge of peptides has been shown to be an important factor for the expression of their bioactivity, and they have to be selectively recovered to maximize their activity [17]. However, since enzymatic hydrolysis is used to liberate BPs from the protein matrix [18], generating complex peptide mixtures, a sustainable technique allowing for the selective purification and concentration of these BPs or peptide fractions from complex mixtures is needed. Also, it was demonstrated that BPs may have low molecular weight depending on the conditions of digestion and the types of proteolytic enzymes used [19,20]. Amongst the available technologies for peptide separation, pressure-driven processes such as ultrafiltration, nanofiltration [21–23], and chromatographic methods [24,25] are perhaps most frequently used. However, pressure-driven techniques sometimes fail to separate molecules of similar size and are susceptible to membrane fouling [26]. Chromatography is too costly, slow, not applicable for the fractionation of large sample volumes, and sometimes uses organic solvents [27]. Isoelectric focusing is perhaps a more biocompatible separation technology that is most often used on a laboratory scale and, more recently, on a larger scale, but as mentioned by Hashimoto et al. (2005, 2006) [28,29], the limiting volume (50 L), the degradation of agarose gels after prolonged time (8 h) of peptide fractionation, and high voltages (500–600 V) led researchers toward alternative methods for an optimal separation and purification. More recently, electrodialysis with filtration membrane (EDFM), a "green" and ultra-selective process, was developed for separation/concentration of bioactive ingredients. EDFM is based on the size exclusion capabilities of porous membranes with the charge selectivity of electrodialysis. In comparison with other common technologies used for biomolecule separation, EDFM has many unique advantages: it is environmentally-friendly, using no organic solvents or dangerous chemicals; it is highly selective for targeted molecules; it operates at low pressure and therefore reduces membrane fouling; it allows simultaneous molecular separation and concentration as well as preservation of the feed solution's commercial value. Recently, Roblet et al. (2016) used the EDFM process to fractionate a low molecular weight (<1 kDa) salmon protein hydrolysate and demonstrated that the basal glucose uptake as well as insulin-stimulated glucose uptake were enhanced by 40% and 31%, respectively, at pH 6 in the final feed compartment [30]. However, in that work, only one molecular weight cut-off (MWCO) of 20 kDa was used for the ultrafiltration (UF) membrane and consequently no discrimination according to the peptide size between recovery fractions and their glucose uptake response was possible.

Hence, in a context of eco-efficiency and to create value-added products, the objectives of the present study were (1) to simultaneously separate specific peptide fractions, according to their charges as well as MW, from a salmon protein hydrolysate by EDFM by stacking three UF membranes of different molecular weight exclusion limits (50, 20, and 5 kDa), (2) to characterize the peptide fractions obtained after separation in terms of molecular weight profiles and sequences, and (3) to measure in vitro the level of glucose uptake response of these fractions in the presence or absence of insulin stimulation in L6 skeletal muscle cells, following this charge and size separation.

2. Results and Discussion

2.1. Evolution of Peptide Concentration and Final Migration Rates

The patterns of peptide separation and concentration as a function of time in recovery compartments of both cationic and anionic configurations measured by the micro-BCA method are represented in Figure 1.

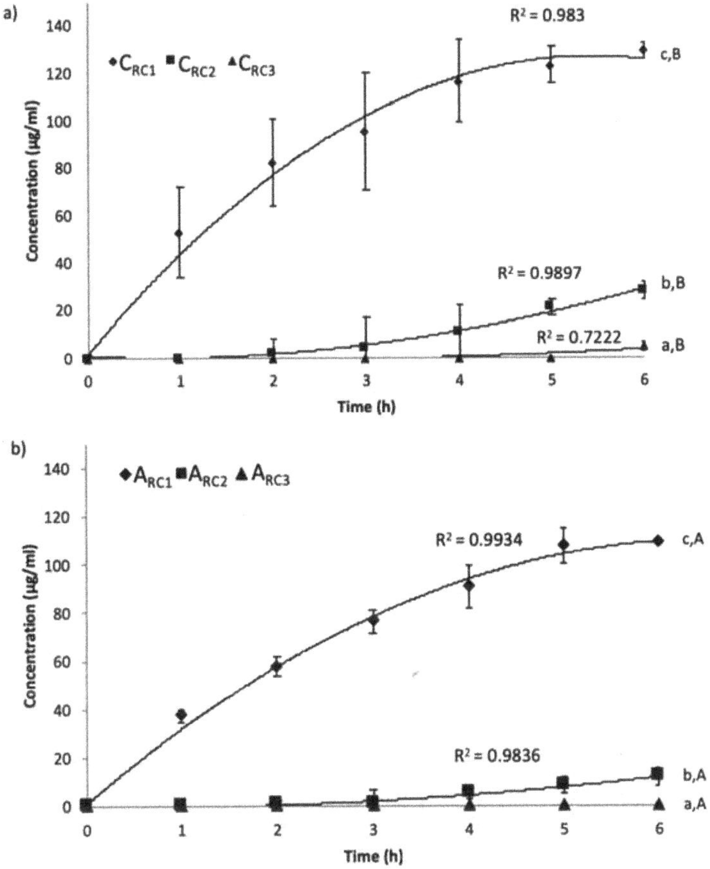

Figure 1. Evolution of peptide concentration in (**a**) cationic (Cationic Recovered Compartment 1, Cationic Recovered Compartment 2, Cationic Recovered Compartment 3, respectively, C_{RC1}, C_{RC2}, and C_{RC3}) and (**b**) anionic (Anionic Recovered Compartment 1, Anionic Recovered Compartment 2, Anionic Recovered Compartment 3, respectively A_{RC1}, A_{RC2}, and A_{RC3}) compartments during 6 h of the electrodialysis with ultrafiltration membrane (EDUF) process. Lowercase letters are used to compare the three recovered compartments of the same configuration where capital letters are used to compare the recovered compartments between anionic and cationic configurations. Values followed by different letters were statistically different.

As expected, significant differences ($p < 0.05$) were obtained concerning the peptide concentrations in the recovery compartments in the order of $C_{RC1} > C_{RC2} > C_{RC3}$ for the cationic configuration and $A_{RC1} > A_{RC2} > A_{RC3}$ for the anionic configuration. Indeed, final concentrations obtained for cationic fractions were 129.10 ± 3.57; 27.74 ± 3.61 and 4.78 ± 1.80µg/mL corresponding to peptide migration rates (Table 1) ranging between 0.13 and 3.9 g/m^2·h. At the same time, the final concentrations recovered in the anionic configuration were 108.91 ± 0.41 µg/mL and 11.51 ± 3.66 µg/mL, corresponding to peptide migration rates between 0.24 and 2.24 g/m^2·h (Table 1). No significant peptide migration ($p > 0.05$) was observed in the A_{RC3} compartment after 6 h of electrodialysis with ultrafiltration membrane (EDUF) treatment. However, in this particular sample, nitrogen content analysis by LECO (Table 1) showed the presence of very low concentrations of peptides (0.15%) (p/p) in the final powder, after lyophilisation of the recovery compartments.

Table 1. The relative energy consumption, peptide migration rate, and peptide concentration in cationic and anionic configuration compartments. Lowercase letters are used to compare the three recovered compartments of the same configuration, means with different lowercase letters are significantly different ($p < 0.05$). Whereas capital letters are used to compare the recovered compartments between anionic and cationic configuration, means with different capital letters are significantly different ($p < 0.05$).

EDUF Fractions	Peptide (%)	Peptide Migration Rate (g/m^2·h)	Relative Energy Consumption (Wh/g)
Unfractionated salmon protein hydrolysate (U_{SPH})	80.83 ± 2.14	—	—
EDUF Configuration a	—	—	512.56 ± 95.59 ^{b}
Cationic Final Feed Compartment (C_{FFC})	67.16 ± 1.72	—	—
Cationic Recovery Compartment 1 (C_{RC1})	9.20 ± 2.29	3.19 ± 0.14 ^{a,A}	—
Cationic Recovery Compartment 2 (C_{RC2})	1.79 ± 0.54	0.73 ± 0.06 ^{b,A}	—
Cationic Recovery Compartment 3 (C_{RC3})	0.33 ± 0.19	0.13 ± 0.06 ^{c,A}	—
EDUF Configuration b	—	—	849.71 ± 80.18 ^{a}
Anionic Final Feed Compartment (A_{FFC})	71.88 ± 0.94	—	—
Anionic Recovery Compartment 1 (A_{RC1})	6.67 ± 0.7	2.24 ± 0.21 ^{a,B}	—
Anionic Recovery Compartment 2 (A_{RC2})	0.37 ± 0.14	0.21 ± 0.06 ^{b,B}	—
Anionic Recovery Compartment 3 (A_{RC3})	0.15 ± 0.008	0.01 ± 0.04 ^{c,B}	—

Peptides have to migrate further from the inlet feed solution to reach compartment 2 and even further to reach compartment 3 as compared to compartment 1. So, only low molecular weight peptides and/or larger peptides with high charge density could reach compartments 2 and 3 and migrate through the UF membrane with a MWCO of 20 kDa (UF-20 kDa) and the UF membrane with a MWCO of 5 kDa (UF-5 kDa). This was confirmed by previous studies carried out on flaxseed cationic peptides [31] and snow crab anionic peptides [32] for a configuration composed of two UF membranes with different MWCOs. Differences observed between the cationic configuration and the anionic configuration could be due to the higher cationic peptide concentrations generated by the successive digestion with pepsin and trypsin/chymotrypsin. These results were in accordance with work by Udenigwe et al. (2012) [33] on flaxseed hydrolysate protein, where a higher peptide concentration was observed in the cationic compartment after the EDUF separation. Moreover, in the present study, for the C_{RC1} and A_{RC1} compartments, results showed a linear increase of the migration rate during the first four hours of EDUF treatment, and then a slowdown appeared in migration rates during the last two hours. These results could be due to an alteration of the membrane (UF membranes and ion-exchange membranes (IEMs)) integrity or to a membrane fouling. Indeed, the thickness and conductivity of each membrane of both configurations were determined before and after three repetitions. No differences were observed concerning the thickness whatever the membrane for both configurations. Nevertheless, a decrease of the conductivity of the UF membranes and the IEMs for both configurations could indicate an internal and irreversible fouling by peptides or free amino acids. Indeed, Suwal et al. (2015) have observed an irreversible fouling by free amino acids in internal nano-pores of IEM during EDUF separation [34].

2.2. Characterization of Peptide Profile by RP-UPLC-MS

Peptides of low molecular weight (MW) in the 301–500 Da range were the most prevalent in the unfractionated salmon protein hydrolysate (U_{SPH}) (46.6% of total abundance) (Figure 2a).

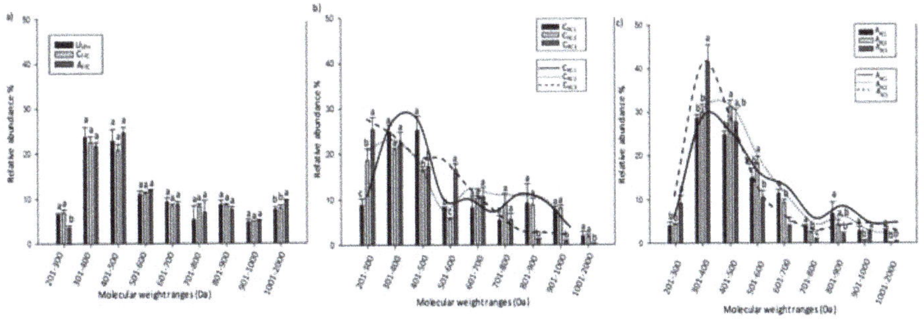

Figure 2. Profiles of peptide molecular weight in (**a**) U_{SPH}, C_{FFC}, and A_{FFC}, (**b**) cationic compartments (C_{RC1}, C_{RC2}, and C_{RC3}) and (**c**) anionic compartment (A_{RC1}, A_{RC2}, and A_{RC3}) generated after 6 h of the EDUF process. Means with different lowercase letters within a molecular weight range are significantly different ($p < 0.05$).

These results were in accordance with previous results obtained by Roblet et al. (2016) [30]. Indeed, similar relative abundances were found for molecular weight ranging from 301 to 500 Da. After 360 min of EDUF treatment, a significant decrease ($p < 0.005$) was observed for levels of peptides ranging from 201 to 300 Da in the Anionic Final Feed Compartment (A_{FFC}) and a significant increase ($p < 0.05$) for levels of peptides from 1001 to 2000 in the anionic collection compartments (A_{FFC}). No difference ($p < 0.05$) was observed between amounts of peptides accumulated in the Cationic Final Feed Compartment (C_{FFC}) and A_{FFC} compartments for the peptides from other size ranges (Figure 2a).

Figure 2b compares the abundance of the different peptides in terms of MW after 6 h of EDUF separation among the C_{RC1}, C_{RC2}, and C_{RC3} compartments. It appeared that the peptide abundances followed a normal distribution that was shifted toward the low molecular weight peptides (LMWPs) as the migration progressed. Indeed, maximal accumulation of peptides for the C_{RC1} (25.26 ± 3.08%) was observed for MWs ranging from 401 to 500 Da. The C_{RC2} maximal abundance of 20.96 ± 0.38% was observed for MWs ranging from 301 to 400 Da while for C_{RC3}, highest peptide accumulations (25.40 ± 2.69%) was observed for MWs ranging from 201 to 300 Da.

The peptide abundances obtained from the anionic configuration are shown in Figure 2c. Concerning anionic recovery compartments, the majority of peptides ranged in size from 301 to 500 Da. The results also demonstrated that A_{RC3} contained the highest peptide accumulations for MWs ranging from 201 to 300 Da and 301 to 400 Da (9.19 ± 1.82 and 41.64 ± 3.79% of the total accumulation, respectively) compared to A_{RC2} and A_{RC1}. Indeed, due to their highest charge and/or lower MWs, peptides ranging from 201 to 400 Da were more able to cross all UF membranes and reach the last compartment. Peptides with MWs between 401 to 500 Da and 501 to 600 Da were significantly higher in the A_{RC2} compared to the A_{RC1} and A_{RC3}, respectively. Finally, the level of high molecular weight peptides (HMWPs) (over 601 Da) was higher in the A_{RC1} (27.69 ± 1.85%) compared to the A_{RC2} and A_{RC3} (16.33 ± 1.58 and 9.40 ± 0.51%).

As expected, a decrease in the average size of peptides was observed as follows: $C_{FFC} > C_{RC1} > C_{RC2} > C_{RC3}$ for the cationic configuration and $A_{FFC} > A_{RC1} > A_{RC2} > A_{RC3}$ for the anionic configuration, which confirmed the high selectivity of the EDUF process.

2.3. Relative Energy Consumption

The relative amount of energy consumed is a measurement of the energy used for the migration of one gram of peptides and is reported in Table 1. The results were found to be 512 and 849 Wh/g for cationic and anionic configurations, respectively. The lowest energy consumption was observed for the cationic configuration due to its higher global migration rate. Indeed, as previously demonstrated by Koumfieg Noudou et al. (2016), the increase of the inlet peptide concentration resulted in a decrease of the relative energy consumption [35]. The relative energy consumption varied depending on the cell configuration, the voltage applied, and the peptide migration rate, as demonstrated in previous works, with values ranging from 3.53 to 631 Wh/g [32,35].

2.4. Glucose Uptake Experiments

The effects of salmon peptide (recovered, initial, and post treatment) fractions on in-vitro glucose uptake on L6 skeletal muscle cells using two different peptide concentrations (1 ng/mL and 1 μg/mL) were measured in basal and insulin-stimulated conditions. Results presented in Figure 3a show a significant enhancement ($p < 0.05$) of insulin-stimulated glucose uptake for the C_{FFC} at 1 ng/mL (29%) but not for the U_{SPH} and A_{FFC} (in absence or presence of insulin stimulation). However, glucose uptake was not affected by any of the fractions used (U_{SPH}, C_{FFC}, or A_{FFC}) at 1 μg/mL in the presence or absence of insulin. These results are in accordance with previous works of Roblet et al. (2016) [30], where a limited effect of the initial salmon protein hydrolysate was observed at 1 ng/mL and 1 μg/mL. While the final solution recovered in the feed compartment showed a significant enhancement of the glucose uptake at pH 6 [30].

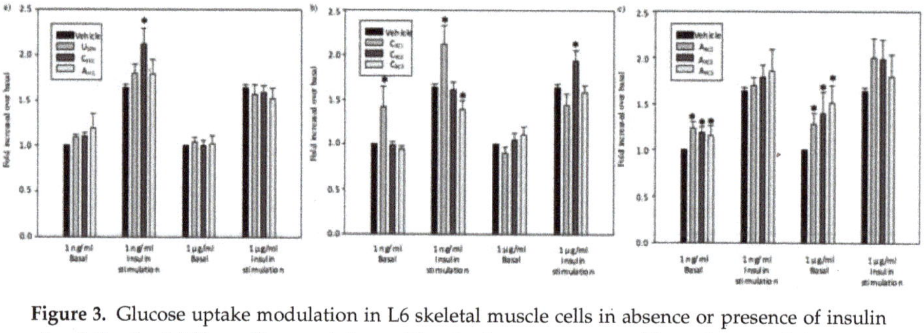

Figure 3. Glucose uptake modulation in L6 skeletal muscle cells in absence or presence of insulin stimulation by (**a**) U_{SPH}, C_{FFC}, and A_{FFC}, (**b**) cationic compartments (C_{RC1}, C_{RC2}, and C_{RC3}) and (**c**) anionic compartments (A_{RC1}, A_{RC2}, and A_{RC3}) generated after 6 h of the EDUF process. One asterisk indicate that mean values are significantly different ($p < 0.05$) than the mean value for the control.

As shown in Figure 3b, after 6 h of EDUF separation in the cationic configuration, only C_{RC1} showed a significant bioactivity ($p < 0.05$) at 1 ng/mL on both basal (42%) and insulin-stimulated glucose uptake (29%) but not at 1 μg/mL. Conversely, C_{RC2} significantly enhanced glucose uptake (18%) in the presence of insulin at 1 μg/mL but not at the lower concentration (Figure 3b). Interestingly, C_{RC3} was found to significantly decrease (15%) insulin-induced glucose uptake ($p < 0.05$) at 1 ng/mL but had no effect at the higher concentration or on basal glucose uptake. According to these results, it appeared that the C_{RC3} fraction has no anti-diabetic potential, suggesting that the C_{RC3} may contain some inhibitory peptides. The C_{FFC} fraction, obtained from the U_{SPH} at the end of the EDUF process, and thus depleted in cationic peptides, was found to have higher glucose uptake activity than U_{SPH} in the presence of insulin. In addition, C_{RC1} presented a similar or higher glucose uptake response than C_{FFC} independent of the condition. It is also important to note that, in the basal state, the C_{RC1} at 1 ng/mL was able to stimulate glucose uptake to the same extent as insulin alone. Additionally, at 1 μg/mL, since no significant increase in glucose uptake for both C_{FFC} and C_{RC1} was reported in the presence or absence of insulin stimulation, let us conclude that their glucose uptake response was not dose-dependent, suggesting that some neutral peptides in the fractions may have masked the bioactivity of the positive ones. Concerning the glucose uptake stimulation in the basal condition, these results were in accordance with previous works obtained by Roblet et al. (2016) [30]. Those authors demonstrated a significant enhancement of the glucose uptake in the absence of insulin stimulation for the cationic fraction at pH 6 as the C_{RC1} [30]. Difference appeared for the bioactivity in the presence of insulin stimulation. Indeed, in previous works, the glucose uptake was not affected by the cationic fraction while C_{RC1} and C_{RC2} showed a significant enhancement of the glucose uptake with insulin stimulation. These differences could be due to the EDUF configuration (three UF membranes with MWCOs of 50, 20, and 5 kDa vs. one UF membrane with a MWCO of 20 kDa) and separation parameters (duration: 6 h vs 1 h; electric field strength of 6 V/cm vs, 14 V/cm; initial peptide concentration of 0.7% vs. 2% [30]) which allowed for the recovery of a higher peptide concentration and a higher diversity of peptides. Moreover, amongst all cationic peptides separated in the different fractions, using Mass Profiler Professional software, MWs and retention time (Table 2) of seventeen peptides were found to be simultaneously and specifically present in all three bioactive fractions (C_{FFC}, C_{RC1}, and C_{RC2}). Thereafter using the Spectrum Mill MS Proteomics software and a specific protein Salmo salar database from NCBI [36], five peptides were identified; their sequences, net charges, and protein precursors are not shown here due to confidential issues (a patent application is in progress).

Table 2. Cationic and anionic peptides simultaneously present in each cationic (C_{FFC}, C_{RC1}, and C_{RC3}) and anionic (A_{RC1}, A_{RC2}, and A_{RC3}) bioactive fraction.

#	Retention Time (min)	Cationic Peptide's Molecular Weight (Da)	Frequency *	#	Retention Time (min)	Anionic Peptide's Molecular Weight (Da)	Frequency *
1	6.655	627.3711	8	1	9.878	416.2344	8
2	8.844	671.3281	8	2	10.172	456.2654	8
3	12.663	794.4654	9	3	14.991	531.2895	9
4	13.862	507.2681	9	4	15.240	409.1843	9
5	13.910	843.4589	9	5	15.271	503.2657	8
6	13.910	719.4219	9	6	15.750	502.2628	9
7	13.915	956.5451	9	7	16.536	444.2577	9
8	14.035	1085.6240	8	8	18.710	869.5485	9
9	14.141	801.4025	8	10	19.019	502.2713	8
10	16.378	805.4078	9	11	19.783	407.2053	9
11	16.491	372.2368	8	12	20.905	515.3020	8
12	18.962	473.3213	9	13	21.399	407.2056	9
13	18.963	643.4267	9	14	22.346	458.2737	9
14	21.274	634.3794	9	15	23.048	494.2369	9
15	25.261	409.2029	9	16	23.091	542.2369	9
16	27.316	434.2523	9	17	24.073	431.2728	9
17	30.407	1014.5737	9	18	24.089	592.2850	9
				19	26.698	829.3969	9
				20	26.855	458.2721	9
				21	29.435	660.3509	8

* For each configuration, nine fractions were compared (three compartments in triplicate). A frequency of nine means that the molecular mas was found in each compartment and for each repetition.

Concerning anionic fractions, all recovered fractions (A_{RC1}, A_{RC2}, and A_{RC3}) demonstrated a significant enhancement of the bioactivity ($p < 0.05$) for both concentrations (1 ng/mL and 1 µg/mL) at the basal level (Figure 3c) and a tendency (not statistically significant, $p > 0.05$) to be increased in insulin-stimulated conditions. Moreover, very interestingly, both A_{RC2} and A_{RC3} showed the same increase in glucose uptake ($p = 0.31$ and $p = 0.55$, respectively) compared with that of insulin, while A_{RC1} was not able to reach the same level of bioactivity ($p = 0.01$). That could be explained by the selectivity of the process leading to the concentration of bioactive peptides in the second and last compartments. Nevertheless, the A_{FFC} fraction that was depleted in anionic peptides did not show any improvement of glucose uptake. Roblet et al. (2016) [30] also observed a significant effect of anionic fractions recovered at both pH 3 and pH 9 on glucose uptake modulation in the absence of insulin stimulation, while a limited effect was observed for the anionic fraction obtained at pH 6 [30]. The greater effect obtained on glucose uptake modulation by the three A_{RC} (A_{RC1}, A_{RC2}, and A_{RC3}) fractions compared to the anionic fraction obtained in previous work could be explained, as for the cationic fractions, by differences concerning the EDFM configuration and separation parameters [32,37]. Using the same method as for cationic peptides, the MWs of twenty-one anionic peptides present in all bioactive anionic recovered fractions (A_{RC1}, A_{RC2}, and A_{RC3}) were identified (Table 2). Amongst these twenty-one peptides, six peptides were confirmed and characterized. As mentioned previously for cationic peptides, the characteristics and sequences of anionic peptides are not shown here due to an in-progress patent application.

Finally, anionic peptides increased glucose uptake in the absence of insulin stimulation, while cationic peptides increased it in the presence of insulin stimulation. In skeletal muscle cells, glucose uptake can be modulated by at least two different signaling pathways: IRS-1/PI3K/Akt (insulin dependent) and 5′-AMP-activated protein kinase (AMPK) (insulin independent) [38]. As previously explained, anionic peptides seem to stimulate the glucose uptake in the absence of insulin stimulation, while for cationic peptides, a better response was obtained in the insulin-stimulation condition; it is possible that the anionic and cationic peptides reported in Table 2 stimulate different pathways involved in glucose uptake. These two pathways are well known for their critical role for glucose transporter translocation to the muscle cell surface in the presence or absence of insulin. Moreover, these pathways were identified as therapeutic targets of anti-diabetic drugs such as metformin and thiazolidenediones, the activation of these pathways by EDUF-isolated salmon bioactive peptides could represent a therapeutic or preventive potential of T2D [39]. To verify this hypothesis, further investigation should be carried out to confirm if these SPH peptide fractions are potential activators of the IRS-1/PI3K/AKT and/or AMPK pathways.

3. Materials and Methods

3.1. Materials and Electrodialysis Cell

3.1.1. Electrodialysis Configurations

The electrodialysis cell used for the experiment was an MP type cell manufactured by ElectroCell Systems AB Company (Täby, Sweden). The cell had an effective surface area of 200 cm^2 and was composed by one anion-exchange membrane (AEM), one cation-exchange membranes (CEM), and three UF membranes with MWCO of 50, 20, and 5 kDa (UF-50 kDa, UF-20 kDa, UF-5 kDa, respectively) as illustrated in Figure 4. The electrodes used were a dimensionally-stable anode (DSA) and a 316 stainless steel cathode. The electrical potential for the electrodialysis with ultrafiltration membrane (EDUF instead of EDFM since the filtration membrane was a UF membrane) was supplied by a variable 0–100 V power source. One polypropylene spacer (0.74 mm) was stacked in each compartment to promote turbulence. Two different cell configurations allowing the separation of cationic or anionic charged peptides from salmon protein hydrolysate were tested in this study:

For both configurations, the cell was composed of five closed loops: three of them contained 1.5 L of a KCl solution (2 g/L) for the recovery compartments, one loop contained the feed compartment, and the last one contained the electrode rinsing solution (20 g/L Na_2SO_4, 3 L), and was split in half between the anode and the cathode compartments. The solutions were circulated using five centrifugal pumps, and the flow rates were set at 2 L/min using flow meters (the electrode rinsing solution was maintained at 4 L/min and split in half between the anode and the cathode compartments) (Blue-White Industries Ltd., Huntington Beach, CA, USA).

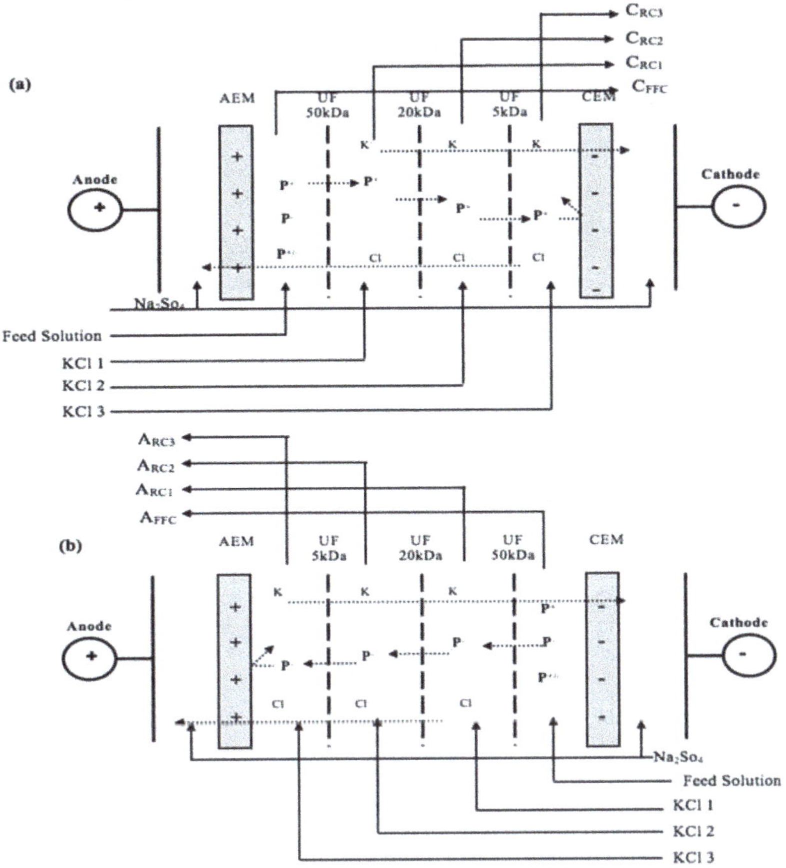

Figure 4. Layout showing EDUF membrane configurations, (**a**) cationic and (**b**) anionic, for the fractionation of the feed solution, which was an Unfractionated Salmon Protein Hydrolysate (U_{SPH}). AEM: anion-exchange membrane, CEM: cation exchange membrane, UF membrane: ultrafiltration membrane, P^+: cationic peptides; P^-: anionic peptides, $P^{+/-}$: neutral peptides, A_{RC}: anionic recovery compartments, and C_{RC}: cationic recovery compartments.

Cationic configuration—The first EDUF cell configuration, shown in Figure 4a, was for the separation of cationic peptides. The UF membranes were placed in the cell according to their exclusion limits starting from the anode side to allow the migration of cationic peptides on the basis of their size and charge. The compartment containing a KCl solution circulating between the UF-50 kDa and UF-20 kDa was named the cationic recovery compartment 1 (C_{RC1}). The cationic recovery compartment 2 (C_{RC2}) was located between the UF-20 kDa and UF-5 kDa, and the cationic recovery compartment 3 (C_{RC3}) between the UF-5 kDa and the CEM. The feed solution consisting of salmon protein hydrolysate (SPH, 1.5 L, 0.7% *w*/*v*) was circulated in the compartment between the UF-50 kDa and the AEM.

Anionic configuration—In this second configuration (Figure 4b), the UF membranes were arranged according to their MWCOs starting from the anode side to allow the migration of anionic peptides on the basis of their size and charge. The compartment containing a KCl solution circulating between the UF-50 kDa and UF-20 kDa membranes was called the "anionic recovery compartment 1" (A_{RC1}), anionic recovery compartment 2 (A_{RC2}) was located between the UF-20 kDa and UF-5 kDa membranes and finally the anionic recovery compartment 3 (A_{RC3}) between the UF-5 kDa and AEM. The feed solution (SPH) was circulated in the compartment between the UF-50 kDa and the CEM.

3.1.2. Electroseparation Protocol

The EDUF separations were performed according to the previous study of Roblet et al. (2016) [30]. Briefly, the EDUF separations were performed in batches for both cell configurations using a constant electrical field strength of 6 V/cm (corresponding to a current density varying between 0.005 and 0.008 A/cm^2 during the treatment), for 6 h, at a controlled temperature (~16 °C) [34]. The SPH was diluted with demineralized water to obtain a final protein concentration of 0.7% (*w*/*v*). Following the results obtained by Roblet et al. [30], the pH of the SPH and recovery (KCl) solutions were adjusted to pH 6 before each run with 0.1 N NaOH and/or 0.1 N HCl and maintained constant thereafter. For each treatment, 5 mL of SPH and recovery solutions were collected every hour for further analysis. The electrical conductivity of the feed solution and recovery solutions was maintained at a constant level by adding KCl, following the recommendations of Suwal et al. (2015) [34]. Three replicates of each condition were performed. Finally, a CIP (cleaning-in-place) was performed at the end of each replicate according to the following process: 10 min with an acid solution (HCL 0.1 N), 20 min with a basic solution (NaOH 0.1 N), and finally 10 min with an acid solution (HCL 0.1 N). Then, the system was rinsed with distilled water until reaching a pH of 6.

3.2. *Materials*

3.2.1. Hydrolysate Preparation

Salmon protein hydrolysate (SPH) was produced according to the procedure described previously by Jin, (2012) [40] and subsequently used by Chevrier et al. (2015) [6] and Roblet et al. (2016) [30]. Briefly, salmon frames were offered by Cooke Aquaculture. They were thawed, mechanically deboned, and homogenized in a 1.0 M NaOH solution. The proteins were isoelectrically precipitated at a pH of 4.5. Then, the proteins were first hydrolyzed with pepsin, and then by a mix of trypsin/chymotrypsin. Once hydrolysis was complete, the supernatant was filtered through a 5 µm pore size paper filter to remove insoluble molecules. Finally, the filtrate was ultrafiltered using a Prep/Scale Tangential Flow Filtration (TFF) 2.5 ft^2 cartridge with a molecular weight cut-off of 1 kDa (Millipore Corporation, Bedford, MA, USA). Permeates were collected, demineralized by conventional electrodialysis, and finally freeze-dried.

3.2.2. Chemicals

KCl was obtained from ACP Inc (Montreal, QC, Canada) and Na_2SO_4 from Laboratoire MAT (Québec city, QC, Canada). Formic acid, 1.0 M HCl, and 1.0 M NaOH solutions were from Fisher Scientific (Montreal, QC, Canada), trifluoroacetic acid was purchased from J.T. Baker (Phillipsburg, NJ, USA). NaCl, Acetonitrile optima® liquid chromatography-mass spectrometry (LC/MS), and water grade were from VWR international (Montréal, QC, Canada). Concerning the glucose uptake experiments, the alpha-Minimal Essential Medium (α-MEM), Fetal Bovine Serum (FBS), and trypsin (0,25% solution) were obtained from Invitrogen (Burlington, ON, Canada). The 2-déoxy-D-glucose (non-radioactive), $CaCl_2$, Hepes-Na, and MgSO4 were purchased from Sigma Aldrich (Oakville, ON, Canada). D-2-deoxy-[^{3}H] glucose was from Perkin Elmer (Woodbridge, ON, Canada) and Pierce® BCA Protein Assay Kit BCA was from Pierce Biotechnology (Rockford, IL, USA). Insulin was from CHUL's pharmacy (Québec, QC, Canada). Also, the L6 skeletal muscle cells line, derived from neonatal rat thigh skeletal muscle, were provided by Dr. A. Klip, Hospital for Sick Children (Toronto, ON, Canada).

3.2.3. Membranes

Three UF membranes made of polyether sulfone (PES) with molecular weight exclusion limits or molecular weight cut-off (MWCO) of 50, 20, and 5k Da were purchased from Synder filtration (Vacaville, CA, USA). Unlike in classical filtration processes where higher pressure is applied, previous papers published on EDMF demonstrated that the MWCO of UF membranes should be about ten times higher than the size of proteins or peptides to be successfully migrated due to steric hindrance from the hydration layer [41]. Indeed, in an electro-ultrafiltration module, Bargeman et al. (2002)

observed that the migration of α_{S2} casein f(183–207) was strongly reduced when a membrane with a MWCO of 20 kDa (six-times higher than the molecular weight of the peptide) was used due to the friction of peptides in the membrane pores [41]. This was also confirmed for EDUF by previous works by our team on peptides and chitosan oligomers [42]. While food-grade Neosepta CMX-SB cationic membranes and Neosepta AMX-SB anionic membranes were obtained from Tokuyama Soda Ltd. (Tokyo, Japan).

3.3. *Analyses*

3.3.1. pH

The pH of all solutions was measured and kept constant throughout the experiments using a pH-meter model SP20 (Thermo Orion, West Chester, PA, USA) equipped with a VWR Symphony epoxy gel combination pH electrode (VWR, Montreal, QC, Canada).

3.3.2. Relative Energy Consumption of the EDUF Process

The energy consumption during the EDUF process was calculated using Equation (1):

$$EC = \int_{t=0h}^{t=6h} I * U \, dt \tag{1}$$

where, EC is the energy (Wh), I the current intensity (A), and U the voltage (V). The relative energy consumption during EDUF treatment was then calculated by dividing the total energy by total grams of peptides obtained at the end of the treatment.

3.3.3. Peptide Concentration and Nitrogen Concentration Determination

To follow the peptide migration during the EDUF separation, the peptide concentrations in all the solutions were determined using micro bicinchoninic acid (μBCA) protein assay reagents (Pierce, Rockford, IL, USA) using bovine serum albumin (BSA) as the standard protein. The microplate was incubated with a mix of 150 μL of the sample and 150 μL of the working reagent, at 37 °C for 2 h. Then, the microplate was cooled to room temperature and the absorbance was read at 562 nm on a microplate reader (Thermomax, Molecular devices, Sunnyvale, CA, USA).

Nitrogen concentrations were analysed in final lyophilized fractions using a LECO Model 601-500 FP528 apparatus (LECO corporation, St. Joseph, MI, USA). Samples of 0.150 g were analyzed in duplicate. The protein content was determined using the protein factor of 6.25 (% Nitrogen × 6.25). The instrument was previously calibrated with ethylenediaminetetraacetic acid (EDTA).

3.3.4. Final Peptide Migration Rates

Final migration rates of peptides (MR) in recovery compartments were calculated using Equation (2):

$$MR = \frac{F \times L}{t \times S} \tag{2}$$

where, F is the concentration at t time in g/mL, L is the volume of the final solution in mL, t is the duration for reaching F concentration in one hour, and S is the total UF membrane area in m^2.

3.3.5. Reverse Phase Ultra Performance Liquide Chromatography (RP-UPLC) and Tandem Mass Spectrometry (MS/MS) Analyses

The RP-UPLC analyses were done according to the previous study from Durand et al. (2019) [43]. Briefly, a 1290 Infinity II UPLC (Agilent Technologies, Santa Clara, CA, USA) was used to separate samples before entering the samples in the mass spectrometer. The EDUF fractions were diluted to 0.5 mg/mL, then filtered through 0.22 μm PVDF filters into a glass vial. Then, 5 μL of each sample were loaded onto an Acquity UPLC CSH 130Å, 1.7 µm C18 column (2.1 mm i.d.× 150mm, Waters Corporation, Milford, MA, USA) at a flow rate of 400 µL/min and a temperature of 45 °C. A linear gradient from 2% to 25% over 50 min and ramping to 90% over 57 min were used. The gradient consisted of a solvent A, which was LC-MS grade water with 0.1% formic acid, and a solvent B, which was LC-MS grade ACN with 0.1% formic acid. Each sample was run in triplicate for statistical evaluation of technical reproducibility.

A hybrid ion mobility quadrupole TOF mass spectrometer (6560 high definition mass spectrometry (IM-Q-TOF), Agilent, Santa Clara, USA) was used to identify the composition of each EDUF fraction. All LC-MS/MS experiments were acquired using Q-TOF. Signals were recorded in positive mode at Extended Dynamic Range, 2 GHz, 3200 *m/z*, with a scan range between 100 to 2000 m/z. Nitrogen was used as the drying gas at 13.0 L/min and 150 °C, and as a nebulizer gas at 30 psig. The capillary voltage was set at 3500 V, the nozzle voltage at 300 V, and the fragmentor at 400 V. Data analyses were done using the Agilent Mass Hunter Software package (LC/MS Data Acquisition, Version B.07.00 and Qualitative Analysis for IM-MS, Version B.07.00 with BioConfirm Software, Agilent, Santa Clara, CA, USA). An additional search was done using the Spectrum Mill MS Proteomics Workbench Rev B.05.00.180. The Salmo salar protein database [36] was used to search for and identify potential peptides.

3.3.6. Glucose Uptake Experiments

Glucose uptake experiments were conducted as previously described by Roblet et al. (2016) [30]. L6 skeletal muscle cells were grown in an α-minimum essential medium (α-MEM) containing 2% (*v*/*v*) fetal bovine serum (FBS) in an atmosphere of 5% CO_2 at 37 °C [44]. Cells were plated at 600,000 cells/plate in 24-well plates to obtain about 25,000 cells/mL. The cells were incubated for 7 days to reach their complete differentiation to myotubes (7 days post-plating). L6 myotubes were deprived of FBS for 3 h, with a α-MEM containing 0% of FBS. Then, the cells were incubated for 75 min, with 10 µL of EDUF fractions at a concentration of 1 µg/mL and 1 ng/mL. Finally, insulin was added (10 µL at 1.10–5 M) for 45 min. Experiments were repeated nine times, and each repetition was run in triplicate. After experimental treatments, cells were rinsed once with 37 °C HEPES-buffered solution (20 mM HEPES, pH 7.4, 140 mM NaCl, 5 mM KCl, 2.5 mM$MgSO_4$, and 1 mM$CaCl_2$) and were subsequently incubated in HEPES-buffered solution containing 10 µM2-deoxyglucose and 0.3 µCi/mL2-deoxy-[^{3}H] glucose for 8 min. Then, the cells were rinsed three times with 0.9% NaCl solution at 4 °C and then frozen. The next day, the cells were disrupted by adding 500 µl of a 50 mM NaOH solution. The radioactivity was determined by scintillation counting. Protein concentrations were determined by the BCA method, and results of glucose uptake were expressed as relative value over the vehicle, which was the control.

3.3.7. Statistical Analyses

Peptide concentration, relative abundance, membrane conductivity and thickness, and glucose-transport array values between different peptide fractions were subjected to a one-way analysis of variance (Anova) using SAS software version 9.1 (SAS institute Inc., Cary, NC, USA) with a significant p values of 0.05 for acceptance. Duncan and Dunnett post-hoc tests were used.

The relative energy consumption was compared by student's t-test ($p < 0.05$ as probability level for acceptance).

4. Conclusions

The simultaneous separation of peptides by three UF membranes (50, 20, and 5 kDa MWCO) stacked in an electrodialysis system allowed for the generation of specific cationic and anionic fractions with different MW profiles and levels of glucose uptake response. As expected, significant decreases were observed concerning the peptide concentrations in the recovery compartments in the order of $C_{FFC} > C_{RC1} > C_{RC2} > C_{RC3}$ and $A_{FFC} > A_{RC1} > A_{RC2} > A_{RC3}$ for the cationic and anionic configurations, respectively. Moreover, the peptide profiles in terms of MWs followed the same tendency as the peptide concentrations with HMWPs concentrated in the feed compartment while LMWPs were able to cross the three UF membranes stacked in the electrodialysis and some reached the last compartment. For the first time, a triple size-separation by EDUF allowed for the concentration, in one step, of bioactive peptides in the C_{RC1} and inhibitor peptides in C_{CRC3}. Coupling the EDMF-based separation of peptides with bioassay-guided validation of their metabolic activity with LC-MS identification allowed for the identification of eleven potential antidiabetic peptides from a complex salmon frame protein hydrolysate containing more than 250 different peptides. Hence, a pre-separation by EDUF appears to be a new powerful tool and key step for accelerating peptide identification. Nevertheless, further mass spectrometry analysis is needed to identify and determine the distribution of each peptide in the fractions and if the bioactivity is linked to one or more peptides in those fractions. These peptides were recently synthesised and their bioactivity measurements during in-vitro tests, alone or in combination, are currently under way to confirm the anti-diabetic activity of these peptides.

Author Contributions: Conceptualization, L.H. and L.B.; methodology, L.H. and G.P.; software, L.H. and J.T.; validation, L.H., T.G., A.M. and L.B.; formal analysis, L.H.; investigation, L.H.; resources, L.H., G.P., T.G., A.M. and L.B.; data curation, L.H.; writing—original draft preparation, L.H.; writing—review and editing, L.H. and L.B.; visualization, L.H., J.T., G.P., T.G., A.M. and L.B.; supervision, L.B.; project administration, L.B.; funding acquisition, L.B.

Funding: This work was also supported by the "Conseil de recherches en sciences naturelles et en génie du Canada" (CRSNG), the Consortium de Recherche et Innovations en Bioprocédés Industriels au Québec (CRIBIQ grant number 2013-018-C10), and the Canadian Institutes for Health Research (CIHR grant number FH4-129922). Authors would also thanks Mitacs (FR12048) for their financial support by a Cluster grant program: Salmon peptides identification and purification and their insulin modulation.

Acknowledgments: The authors are thankful to Bruno Marcotte and Diane Gagnon for their technical assistance and Cooke Aquaculture for providing the salmon frames. Special gratitude goes to Shyam Suwal and Muhammad Javeed Akthar for the EDUF separation experiments.

Conflicts of Interest: The authors declare no conflict of interest. The funders had no role in the design of the study; in the collection, analyses, or interpretation of data; in the writing of the manuscript, or in the decision to publish the results.

References

1. Stumvoll, M.; Goldstein, B.J.; Van Haeften, T.W. Type 2 diabetes: Principles of pathogenesis and therapy. *Lancet* **2005**, *365*, 1333–1346. [CrossRef]
2. Wild, S.; Roglic, G.; Green, A.; Sicree, R.; King, H. Estimates for the year 2000 and projections for 2030. *Diabetes Care* **2004**, *27*, 1047–1053. [CrossRef]
3. Daviglus, M.L.; Stamler, J.; Orencia, A.J.; Dyer, A.R.; Liu, K.; Greenland, P.; Molly, M.K.; Morris, D.; Shekelle, R.B. Fish Consumption And The 30-Year Of Fatal Myocardial Infarction. *N. Engl. J. Med.* **1997**, *336*, 1046–1053. [PubMed]
4. Nkondjock, A.; Receveur, O. Fish-seafood consumption, obesity, and risk of type 2 diabetes: An ecological study. *Diabetes Metab.* **2003**, *29*, 635–642. [PubMed]
5. Pilon, G.; Ruzzin, J.; Rioux, L.E.; Lavigne, C.; White, P.J.; Froyland, L.; Jacques, H.; Bryl, P.; Beaulieu, L.; Marette, A. Differential effects of various fish proteins in altering body weight, adiposity, inflammatory status, and insulin sensitivity in high-fat-fed rats. *Metabolism* **2011**, *60*, 1122–1130. [CrossRef] [PubMed]
6. Chevrier, G.; Mitchell, P.L.; Rioux, L.E.; Hasan, F.; Jin, T.; Roblet, C.; Doyen, A.; Pilon, G.; St-Pierre, P.; Lavigne, C.; et al. Low-Molecular-Weight Peptides from Salmon Protein Prevent Obesity-Linked Glucose Intolerance, Inflammation, and Dyslipidemia in $LDLR^{-/-}$/ApoB100/100 Mice. *J. Nutr.* **2015**, *145*, 1415–1422. [CrossRef] [PubMed]
7. Lavigne, C.; Tremblay, F.; Asselin, G.; Jacques, H.; Marette, A. Prevention of skeletal muscle insulin resistance by dietary cod protein in high fat-fed rats. *Am. J. Physiol. Endocrinol. Metab.* **2001**, *281*, E62–E71. [CrossRef] [PubMed]
8. Lavigne, C.; Marette, A.; Jacques, H. Cod and soy proteins compared with casein improve glucose tolerance and insulin sensitivity in rats. *Am. J. Physiol. Endocrinol. Metab.* **2000**, *278*, E491–E500. [CrossRef] [PubMed]

9. Tremblay, F.; Lavigne, C.; Jacques, H.; Marette, A. Dietary Cod Protein Restores Insulin-Induced Activation of Phosphatidylinositol 3-Kinase/Akt and GLUT4 Translocation to the T-Tubules in Skeletal Muscle of High-Fat–Fed Obese Rats. *Diabetes Care* **2003**, *52*, 29–37. [CrossRef]
10. Ouellet, V.; Marois, J.; Weisnagel, S.J.; Jacques, H. Dietary cod protein improves insulin sensitivity in insulin-resistant men and women: A randomized controlled trial. *Diabetes Care* **2007**, *30*, 2816–2821. [CrossRef]
11. Ouellet, V.; Weisnagel, S.J.; Marois, J.; Bergeron, J.; Julien, P.; Gougeon, R.; Tchernof, A.; Holub, B.J.; Jacques, H. Dietary Cod Protein Reduces Plasma C-Reactive Protein in Insulin-Resistant Men and Women. *J. Nutr.* **2008**, *12*, 2386–2391. [CrossRef] [PubMed]
12. Wergedahl, H.; Liaset, B.; Gudbrandsen, O.A.; Lied, E.; Espe, M.; Muna, Z.; Mørk, S.; Berge, R.K. Fish protein hydrolysate reduces plasma total cholesterol, increases the proportion of HDL cholesterol, and lowers acyl-CoA:cholesterol acyltransferase activity in liver of Zucker rats. *J. Nutr.* **2004**, *134*, 1320–1327. [CrossRef] [PubMed]
13. Liaset, B.; Madsen, L.; Hao, Q.; Criales, G.; Mellgren, G.; Marschall, H.; Hallenborg, P.; Espe, M.; Frøyland, L.; Kristiansen, K. Fish protein hydrolysate elevates plasma bile acids and reduces visceral adipose tissue mass in rats. *Biochim. Biophys. Acta* **2009**, *1791*, 254–262. [CrossRef]
14. Lin, C.S.K.; Pfaltzgra, L.A.; Herrero-davila, L.; Mubofu, E.B.; Abderrahim, S.; Clark, J.H.; Koutinas, A.A.; Kopsahelis, N.; Stamatelatou, K.; Dickson, F.; et al. Food waste as a valuable resource for the production of chemicals, materials and fuels. *Energy Environ. Sci.* **2013**, *6*, 426–464. [CrossRef]
15. Koutinas, A.A.; Vlysidis, A.; Pleissner, D.; Kopsahelis, N.; Garcia, I.L.; Kookos, I.K.; Papanikolaou, S.; Kwanb, T.H.; Lin, C.S.K. Valorization of industrial waste and by-product streams via fermentation for the production of chemicals and biopolymers. *Green Chem.* **2014**, *43*, 2587–2627. [CrossRef]
16. Udenigwe, C.C.; Aluko, R.E. Food protein-derived bioactive peptides: Production, processing, and potential health benefits. *J. Food Sci.* **2012**, *71*, R11–R24. [CrossRef] [PubMed]
17. Jenssen, H.; Hamill, P.; Hancock, R.E.W. Peptide antimicrobial agents. *Clin. Microbiol. Rev.* **2006**, *19*, 491–511. [CrossRef]
18. Saadi, S.; Saari, N.; Anwar, F.; Abdul Hamid, A.; Ghazali, H.M. Recent advances in food biopeptides: Production, biological functionalities and therapeutic applications. *Biotechnol. Adv.* **2015**, *33*, 80–116. [CrossRef]
19. Yoshikawaa, M.; Fujitab, H.; Matobaa, N.; Takenakaa, Y.; Yamamotoa, T.; Yamauchia, R.; Tsurukia, H.; Takahatac, K. Bioactive peptides derived from food proteins preventing lifestyle-related diseases. *BioFactors* **2000**, *12*, 143–146. [CrossRef]
20. Korhonen, H.; Pihlanto, A. Food-derived Bioactive Peptides—Opportunities for Designing Future Foods. *Curr. Pharm. Des.* **2003**, *9*, 1297–1308. [CrossRef]
21. Deeslie, W.D.; Cheryan, M. Fractionation of Soy Protein Hydrolysates Using Ultrafiltration Membranes. J. Food Sci. **1992**, *57*, 411–413. [CrossRef]
22. Jeon, Y.J.; Byun, H.G.; . Kim, S.K. Improvement of functional properties of cod frame protein hydrolysates using ultrafiltration membranes. *Process Biochem.* **1999**, *35*, 471–478. [CrossRef]
23. Roblet, C.; Doyen, A.; Amiot, J.; Bazinet, L. Impact of pH on ultrafiltration membrane selectivity during electrodialysis with ultrafiltration membrane (EDUF) purification of soy peptides from a complex matrix. *J. Memb. Sci.* **2013**, *43*, 207–217. [CrossRef]
24. Ellegard, K.H.; Gammelgard-Larsen, C.; Sorensen, E.S.; Fedosov, S. Process scale chromatographic isolation, characterization and identification of tryptic bioactive casein phosphopeptides. *Int. Dairy J.* **1999**, *9*, 639–652. [CrossRef]
25. Recio, I.; Visser, S. Two ion-exchange chromatographic methods for the isolation of antibacterial peptides from lactoferrin. In situ enzymatic hydrolysis on an ion-exchange membrane. *J. Chromatogr. A* **1999**, *831*, 191–201. [CrossRef]
26. Brans, G.; Schroën, C.G.P.H.; Van Der Sman, R.G.M.; Boom, R.M. Membrane fractionation of milk: State of the art and challenges. *J. Memb. Sci.* **2004**, *243*, 263–272. [CrossRef]
27. Bargeman, G.; Houwing, J.; Recio, I.; Koops, G.-H.; Van Der Horst, C. Electro-membrane filtration for the selective isolation of bioactive peptides from an alpha(s2)-casein hydrolysate. *Biotechnol. Bioeng.* **2002**, *80*, 599–609. [CrossRef]

28. Hashimoto, K.; Sato, K.; Nakamura, Y.; Ohtsuki, K. Development of a large-scale (50 L) apparatus for ampholyte-free isoelectric focusing (autofocusing) of peptides in enzymatic hydrolysates of food proteins. *J. Agric. Food Chem.* **2005**, *53*, 3801–3806. [CrossRef]
29. Hashimoto, K.; Sato, K.; Nakamura, Y.; Ohtsuki, K. Development of continuous type apparatus for ampholyte-free isoelectric focusing (autofocusing) of peptides in protein hydrolysates. *J. Agric. Food Chem.* **2006**, *54*, 650–655. [CrossRef]
30. Roblet, C.; Akhtar, M.J.; Mikhaylin, S.; Pilon, G.; Gill, T.; Marette, A.; Bazinet, L. Enhancement of glucose uptake in muscular cell by peptide fractions separated by electrodialysis with filtration membrane from salmon frame protein hydrolysate. *J. Funct. Foods* **2016**, *22*, 337–346. [CrossRef]
31. Doyen, A.; Udenigwe, C.C.; Mitchell, P.L.; Marette, A.; Aluko, R.E.; Bazinet, L. Anti-diabetic and antihypertensive activities of two flaxseed protein hydrolysate fractions revealed following their simultaneous separation by electrodialysis with ultrafiltration membranes. *Food Chem.* **2014**, *145*, 66–76. [CrossRef] [PubMed]
32. Suwal, S.; Roblet, C.; Doyen, A.J.; Amiot, J.; Beaulieu, L.; Legault, J.; Bazinet, L. Electrodialytic separation of peptides from snow crab by-product hydrolysate: Effect of cell configuration on peptide selectivity and local electric field. *Sep. Purif. Technol.* **2014**, *127*, 29–38. [CrossRef]
33. Udenigwe, C.C.; Adebiyi, A.P.; Doyen, A.; Li, H.; Bazinet, L.; Aluko, R.E. Low molecular weight flaxseed protein-derived arginine-containing peptides reduced blood pressure of spontaneously hypertensive rats faster than amino acid form of arginine and native flaxseed protein. *Food Chem.* **2012**, *132*, 468–475. [CrossRef] [PubMed]
34. Suwal, S.; Roblet, C.; Amiot, J.; Bazinet, L. Presence of free amino acids in protein hydrolysate during electroseparation of peptides: Impact on system efficiency and membrane physicochemical properties. *Sep. Purif. Technol.* **2015**, *147*, 227–236. [CrossRef]
35. Koumfieg Noudou, V.Y.; Suwal, S.; Amiot, J.; Mikhaylin, S.; Beaulieu, L.; Bazinet, L. Simultaneous electroseparation of anionic and cationic peptides: Impact of feed peptide concentration on migration rate, selectivity and relative energy consumption. *Sep. Purif. Technol.* **2016**, *157*, 53–59. [CrossRef]
36. Available online: https://www.ncbi.nlm.nih.gov/protein/?term=salmo+salar (accessed on 11 July 2018).
37. Aider, M.; Brunet, S.; Bazinet, L. Effect of pH and cell configuration on the selective and specific electrodialytic separation of chitosan oligomers. *Sep. Purif. Technol.* **2008**, *63*, 612–619. [CrossRef]
38. Girón, DN.; Sevillano, N.; Salto, R.; Haidour, A.; Manzano, M.; Jiménez, M.L.; Rueda, R.; López-Pedrosa, J.M. Salacia oblonga extract increases glucose transporter 4-mediated glucose uptake in L6 rat myotubes: Role of mangiferin. *Clin. Nutr.* **2009**, *28*, 565–574. [CrossRef]
39. Zhou, G.; Myers, R.; Li, Y.; Chen, Y.; Shen, X.; Fenyk-melody, J.; Wu, M.; Ventre, J.; Doebber, T.; Fujii, NN.; et al. Role of AMP-activated protein kinase in mechanism of metformin action. *J. Clin Invest.* **2001**, *108*, 1167–1174. [CrossRef]
40. Jin, T. Separation and Purification of Antidiabetic Bioactive Peptide from Salmon and Cod Waste. Available online: https://dalspace.library.dal.ca/handle/10222/37804 (accessed on 19 April 2019).
41. Bargeman, G.; Koops, G.H.; Houwing, J.; Breebaart, I.; Van Der Horst, H.C.; Wessling, M. The development of electro-membrane filtration for the isolation of bioactive peptides: The effect of membrane selection and operating parameters on the transport rate. *Desalination* **2002**, *149*, 369–374. [CrossRef]
42. Aider, M.; Brunet, S.; Bazinet, L. Electroseparation of chitosan oligomers by electrodialysis with ultrafiltration membrane (EDUF) and impact on electrodialytic parameters. *J. Membr. Sci.* **2008**, *309*, 222–232. [CrossRef]
43. Durand, R.; Fraboulet, E.; Marette, A.; Bazinet, L. Simultaneous double cationic and anionic molecule separation from herring milt hydrolysate and impact on resulting fraction bioactivities. *Sep. Purif. Technol.* **2019**, *210*, 431–441. [CrossRef]
44. Tremblay, F.; Marette, A. Amino acid and insulin signaling via the mTOR/p70 S6 kinase pathway: A negative feedback mechanism leading to insulin resistance in skeletal muscle cells. *J. Biol. Chem.* **2001**, *276*, 38052–38060. [PubMed]

International Journal of
Molecular Sciences

MDPI

Article

Positive Impact of Pulsed Electric Field on Lactic Acid Removal, Demineralization and Membrane Scaling during Acid Whey Electrodialysis

Guillaume Dufton [1,2], Sergey Mikhaylin [1,2], Sami Gaaloul [3] and Laurent Bazinet [1,2,*]

1 Institute of Nutrition and Functional Foods (INAF), Dairy Research Center (STELA) and Department of Food Sciences, Université Laval, Québec, QC G1V 0A6, Canada; guillaume.dufton.1@ulaval.ca (G.D.); Sergey.Mikhaylin@fsaa.ulaval.ca (S.M.)
2 Laboratoire de Transformation Alimentaire et Procédés ÉlectroMembranaires (LTAPEM, Laboratory of Food Processing and ElectroMembrane Processes), Université Laval, Québec, QC G1V 0A6, Canada
3 Parmalat, Victoriaville, Québec, QC G6P 9V7, Canada; sami_gaaloul@parmalat.ca
* Correspondence: Laurent.Bazinet@fsaa.ulaval.ca; Tel.: +1-418-656-2131 (ext. 7445); Fax: +1-418-656-3353

Received: 17 January 2019; Accepted: 31 January 2019; Published: 13 February 2019

Abstract: The drying of acid whey is hindered by its high mineral and organic acid contents, and their removal is performed industrially through expensive and environmentally impacting serial processes. Previous works demonstrated the ability to remove these elements by electrodialysis alone but with a major concern—membrane scaling. In this study, two conditions of pulsed electric field (PEF) were tested and compared to conventional DC current condition to evaluate the potential of PEF to mitigate membrane scaling and to affect lactic acid and salt removals. The application of a PEF 25 s/25 s pulse/pause combination at an initial under-limiting current density allowed for decreasing the amount of scaling, the final system electrical resistance by 32%, and the relative energy consumption up to 33%. The use of pulsed current also enabled better lactic acid removal than the DC condition by 10% and 16% for PEF 50 s/10 s and 25 s/25 s, respectively. These results would be due to two mechanisms: (1) the mitigation of concentration polarization phenomenon and (2) the rinsing of the membranes during the pause periods. To the best of our knowledge, this was the first time that PEF current conditions were used on acid whey to both demineralize and deacidify it.

Keywords: acid whey; electrodialysis; pulsed electric field; demineralization; scaling; lactic acid removal

1. Introduction

Acid whey is the principal co-product of a wide variety of dairy products, such as fresh cheese, caseinate, and Greek yogurt. Its production is increasing every year to answer the actual popularity rise of aforementioned products, and its valorization represents a great deal for dairy industries [1,2]. However, most acid whey applications need a preemptive drying for volume, transport, and preservation enhancement, but acid whey processing is hindered by its high hygroscopic character resulting in major operational problems during its drying. Saffari & Langrish in 2014 and Chandrapala & Vasiljevic in 2017 explained the inability to produce good quality dried powders from acid whey as being due to its high lactic acid and mineral contents. Both lactic acid and mineral contents are responsible for a decrease in glass transition temperature and, consequently, to lower crystallization yields [3,4]. To allow acid whey subsequent valorization, the dairy industry used several processes, such as combinations of electrodialysis (ED), ion-exchange resins, and nanofiltration [5,6] to remove salts and lactic acid. Nevertheless, the cost and environmental impact of the above-mentioned processes are tremendous and mostly attributed to the use of ion-exchange resin [7]. Hence, in the recent years, a number of studies were published regarding acid whey processing by optimizing either nanofiltration [8,9] or ED processes [10,11], trying to provide better cost-effective and eco-efficient alternatives. So far, ED processes allowed for reaching suitable deacidification (44%) and demineralization (67%) rates that have been reported as sufficient for the acid whey to be spray-dried while obtaining powders of acceptable quality [9]. However, whatever the configuration or parameters used, these ED processes were subject to major scaling (mineral fouling) on membrane surfaces [11], making the transposition of ED alone, or without pretreatments, impossible at an industrial scale.

Membrane fouling and scaling during ED have been studied by several authors, and the major scaling agents reported and found in acid whey were calcium and magnesium [12–14]. To mitigate or even sometimes completely suppress scaling consisting of these minerals, a particularly effective method was recently reported in the literature: the use of pulsed electric fields (PEF). Cifuentes et al. (2011), Mikhaylin et al. (2014), and Andreeva et al. (2018) successfully demonstrated the favorable impact of using PEF on scaling mitigation during ED of model salt solutions [15–17]. Moreover, the application of PEF during ED was also reported to enhance ion migration thanks to a mitigation of the concentration polarization phenomenon [18–20]. In all previous studies of ED under PEF, demineralization efficiency and membrane scaling were the main issues. However, deacidification of real acid food solutions in conventional ED under PEF has never been reported in the literature.

In this context, the goal of the present study is to demonstrate the efficiency of applying PEF during ED for simultaneous deacidification and demineralization of acid whey. The specific objectives were, therefore, (1) to evaluate the impact of PEF on the deacidification and demineralization rates of acid whey, (2) to test the effect of different pulse/pause combinations on demineralization and deacidification rates, and (3) to ascertain the potential effect of PEF on membrane scaling mitigation.

2. Results and Discussion

2.1. *Whey and Recovery Solutions Analysis*

2.1.1. Lactate Migration

Regarding lactate migration, the application of the two pulse/pause combinations had a significant impact, in comparison with the DC current control condition, on the deacidification rates obtained ($P < 0.01$). Indeed, in the DC current condition, the lactate concentration decreased in the AWComp from 7.22 ± 0.14 g/L to 4.53 ± 0.19 g/L ($P < 0.001$), corresponding to a 37.2% ± 1.7% deacidification rate (Figure 1). In parallel, the corresponding acidification rate observed in the OAComp, with a recovery of 2.80 ± 0.10 g/L of lactate, was 38.8% ± 1.5% of the initial lactate concentration. For the PEF 50 s/10 s condition, the lactate concentration in the AWComp dropped from 7.07 ± 0.02 g/L to 4.13 ± 0.01 g/L ($P < 0.001$), corresponding to a deacidification rate of 41.6% ± 0.1%. In the OAComp, the final lactate concentration reached 3.01 ± 0.07 g/L which represented 42.6% ± 1.0% of the initial lactate concentration. Finally, for the PEF 25 s/25 s condition, lactate concentration in the AWComp decreased from 7.09 ± 0.08 g/L to 3.94 ± 0.11 g/L ($P < 0.001$) corresponding to a deacidification rate of 44.4% ± 1.1%. In parallel, the final lactate concentration in the OAComp reached 3.30 ± 0.07 g/L and an acidification rate of 46.6% ± 1.5% of the initial lactate concentration. The final deacidification rates obtained were in direct correlation with the final lactate concentration in the OAComp: DC current < PEF 50 s/10 s < PEF 25 s/25 s.

As expected, lactate migrated through the AEMs from the AWComp to the OAComp during the three effective hours of current application, equivalent to a number of charges transported of 10800 C. For all conditions, the lactate recovery was similar to the deacidification observed, meaning no lactate was trapped or involved in any fouling and/or scaling reaction [11]. The application of PEF was demonstrated to be efficient for the improvement of the deacidification rates: the 50 s/10 s pulse/pause condition allowed an increase of 10%, while the 25 s/25 s condition allowed an improvement of 16%. The effect of PEF on the migration rates has already been demonstrated on salt solutions by several authors [15,17–19] but, in the present study, it was demonstrated for the first time on lactate in acid whey. According to these authors, PEF has a favorable impact on the reduction of concentration polarization at the membrane vicinity and, thus, on the reduction of water splitting phenomenon. Indeed, by applying pulsed current, the pause lapses allow the ion's concentrations at the membrane's boundary layers to return to a value close to the bulk solution concentration. By doing so, when the current is applied again, the migration efficiency can momentarily exceed the maximum imposed by the limiting current density, thanks to a lower ohmic resistance and a better ion availability [20]. The higher migration rate obtained for the PEF 25 s/25 s condition confirmed the results reported previously by Cifuentes-Araya et al. (2011) [15] for the ratio $r = 1$ (10 s/10 s) on salt solutions. Pelletier et al. (2015) [21] also obtained their best deacidification conditions for pulse/pause combinations of ratio $r = 1$ (1 s/1 s and 2 s/2 s) on cranberry juice. Their experiments resulted in a 12 to 19% improvement of their migration rates in PEF modes in comparison to DC one, depending on the organic acid anion, for a longer run (4 effective hours) but at lower current density than in the present study.

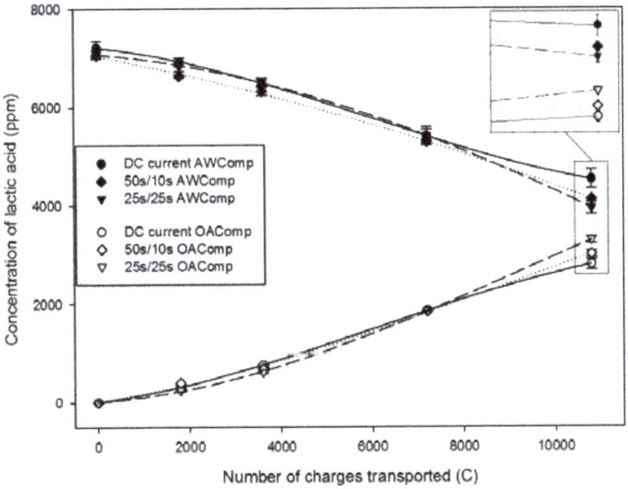

Figure 1. Evolution of lactic acid (in ppm) in the AWComp (black points) and OAComp (white points) for the DC (dots), pulsed electric field (PEF) 50 s/10 s (rhombuses), and PEF 25 s/25 s (inverted triangles) current conditions.

2.1.2. Protein Content

Before and after ED, the total protein content was determined to evaluate whether whey proteins were transferred or lost during the process. The results are shown in Table 1, and allowed us to conclude that the whole protein content was preserved.

Table 1. Total protein content in whey in g/L, before and after electrodialysis.

Time	DC	PEF 50 s/10 s	PEF 25 s/25 s
Before ED	9.70 ± 2.22a	7.56 ± 2.00a	7.35 ± 1.02a
After ED	7.67 ± 0.84a	6.35 ± 1.72a	6.84 ± 1.58a

Column marked with the same letter indicate no significant difference between the values ($P > 0.05$).

2.1.3. pH

The pH of whey, in the AWComp, significantly decreased for all conditions after the three effective hours of current application, or number of charges transported of 10800 C ($P < 0.01$). As shown in Figure 2a, the pH decreased from an averaged initial value of 4.58 ± 0.03 to final values of 4.03 ± 0.02 for DC current, 4.14 ± 0.08 for PEF 50 s/10 s, and 4.23 ± 0.01 for PEF 25 s/25 s. Due to higher variability during PEF 50 s/10 s condition, no significant difference was observed between this condition and both other (P = 0.74 and P = 0.11 with DC, and PEF 25 s/25 s current conditions, respectively). However, the pH variation between the beginning and the end of the treatment for DC and PEF 25 s/25 s current conditions was significantly different (P = 0.04). For all three conditions, the pH evolution in the AWComp presents similar profiles. They all present one inflection point at around 7200 C for the DC and PEF 50 s/10 s current conditions, and at around 6600 C for PEF 25 s/25 s. From the initial pH value, the application of either DC, PEF 50 s/10 s, or PEF 25 s/25 s current resulted, at the end of the process, in a pH decrease of 11%, 9%, and 7%, respectively. Regarding OAComp (Figure 2b), the pH evolution had a similar trend during the treatment for all conditions: an initial decrease to reach a minimum value followed by a subsequent increase ($P < 0.05$). However, the subsequent increase appeared at a different number of charges transported depending on the current condition. For the DC current condition, pH decreased from 6.84 ± 0.11 to a minimum of 5.86 ± 0.15 at 4200 C, and then increased to reach a final pH of 6.55 ± 0.34. For PEF 50 s/10 s and 25 s/25 s the pH reached respective minima of 5.82 ± 0.12 at 5400 C and 5.68 ± 0.03 at 6600 C, and then rose, to final respective pH values of 6.30 ± 0.15 and 5.95 ± 0.02 (Figure 2b). As for the pH evolution in AWComp, the values of the final pH in OAComp were statistically similar between the PEF 50 s/10 s and the two others current conditions (P = 0.39 and P = 0.18 with DC and PEF 25 s/25 s current conditions, respectively) while the PEF 25 s/25 s and DC current conditions were significantly different (P = 0.03).

pH decreases in the AWComp during acid whey deacidification were reported, in previous studies in DC current conditions [10,11], to be caused by the dissociation of lactic acid following lactate migration, and by protons generated from water splitting phenomenon appearing mainly, after a certain time, on the AEM diluate side, as observed by other authors on model salt solutions [22,23]. However, the results of the current study demonstrate that the application of an adjusted PEF ratio was efficient in the reduction of such pH variations. As already mentioned, PEF has been reported to have a favorable impact on the reduction of the water splitting phenomenon during salt solution electrodialysis. This effect is described by Sistat et al. (2015) [20] and Malek et al. (2013) [19] as the restoration of the ionic concentrations at the membrane boundary layers during the pause periods, hence limiting the concentration polarization. The higher final pH value in AWComp for the PEF 25 s/25 s current condition (Figure 2a), in comparison to the DC one, was probably caused by the lower production of protons in the whey through the diminution of water splitting phenomenon at the AEM's interface, in contact with the acid whey. Regarding OAComp (Figure 2b), the evolution of pH has been described in a previous study [11] for the DC current condition. The initial pH decrease in all conditions was likely caused by the migration of protons from the AWComp through the CEMs, followed by other slower anions, such as phosphates or organic acids, increasing the buffer capacity of the solution, hence, the pH stabilization after 4200 C. After 7200 C, the pH rose again due to the aforementioned water splitting phenomenon occurring after further demineralization of the AWComp. However, regarding the results of the PEF 25 s/25 s current condition, the decreasing phase lasted longer and remained constant while increasing at a drastically slower pace than in the DC current condition. This is further proof of the favorable impact of applying PEF on the water splitting phenomenon reduction.

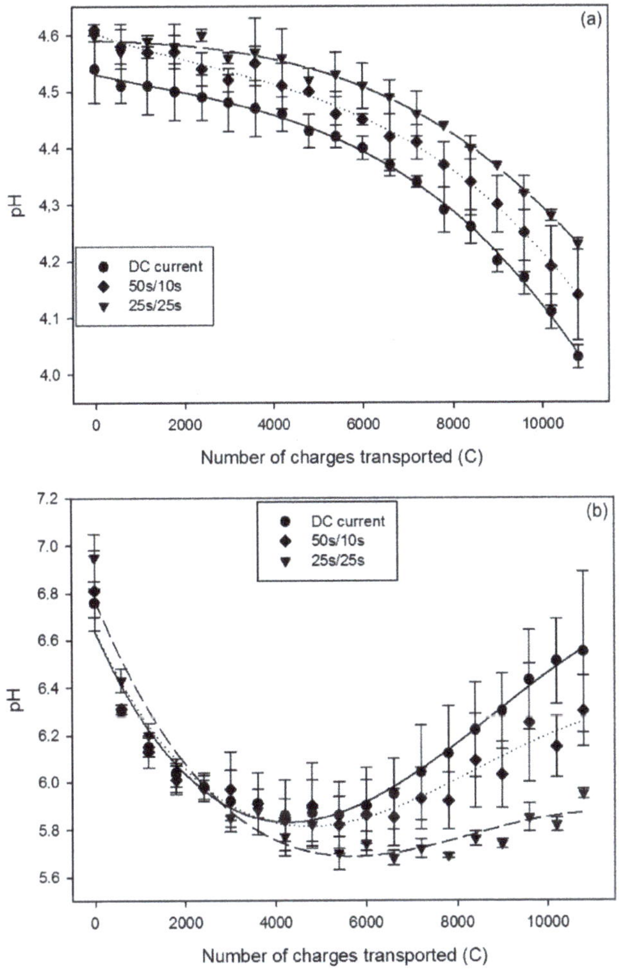

Figure 2. pH evolution for DC (dots), PEF 50 s/10 s (rhombuses), and PEF 25 s/25 s (inverted triangles) current conditions in (**a**) AWComp and (**b**) OAComp.

2.1.4. Conductivity

In the AWComp solution, the conductivity decreased significantly during the 3-hour effective treatment for all three conditions ($P < 0.003$). It dropped from an averaged initial value of 7.09 ± 0.35 mS/cm to final values of 2.62 ± 0.06 mS/cm for the DC current condition, 2.49 ± 0.08 mS/cm for PEF 50 s/10 s, and 2.30 ± 0.03 mS/cm for PEF 25 s/25 s (Figure 3a). The demineralization rate of the DC current and PEF 50 s/10 s current conditions were very similar: 64.0% ± 2.5% and 64.2% ± 3.0% respectively ($P = 0.09$), while the PEF 25 s/25 s condition showed a statistically higher demineralization rate of 67.1% ± 0.7% ($P < 0.02$ with both DC and PEF 50 s/10 s current conditions). Regarding OAComp, the conductivity also increased in a similar way regardless of the current condition applied, from an initial averaged value of 7.81 ± 0.09 to 13.12 ± 0.15, 13.19 ± 0.23, and 13.56 ± 0.28 mS/cm for the DC, PEF 50 s/10 s, and PEF 25 s/25 s current conditions respectively ($p < 0.001$) (Figure 3b). Here, again, there was no significant difference between the DC and PEF 50 s/10 s current conditions ($P = 0.92$) with mineralization rates of 40.5% ± 0.3% and 40.6% ± 0.9%

respectively, but the PEF 25 s/25 s current condition showed a better rate of 42.7% ± 0.1% ($P < 0.02$ and $P < 0.03$ with DC and PEF 50 s/10 s current conditions respectively).

The AWComp demineralization and OAComp mineralization were directly correlated to the ions' migration from AWComp to OAComp through the membranes. The better demineralization/mineralization rates for PEF 25 s/25 s current condition confirmed the previous results on lactate migration and the hypothesis of migration enhancement thanks to an appropriate PEF condition. However, as reported by Lin Teng Shee et al. (2008) [24] regarding solution demineralization using bipolar membranes, the H^+ and OH^- produced during the process contribute more to the conductivity than other ions. Water splitting phenomenon occurring during the experiments in our study, producing such ions, might thus have impacted the conductivity measurements and not reflect the real migration of mineral ions by overestimating the final mineral concentration.

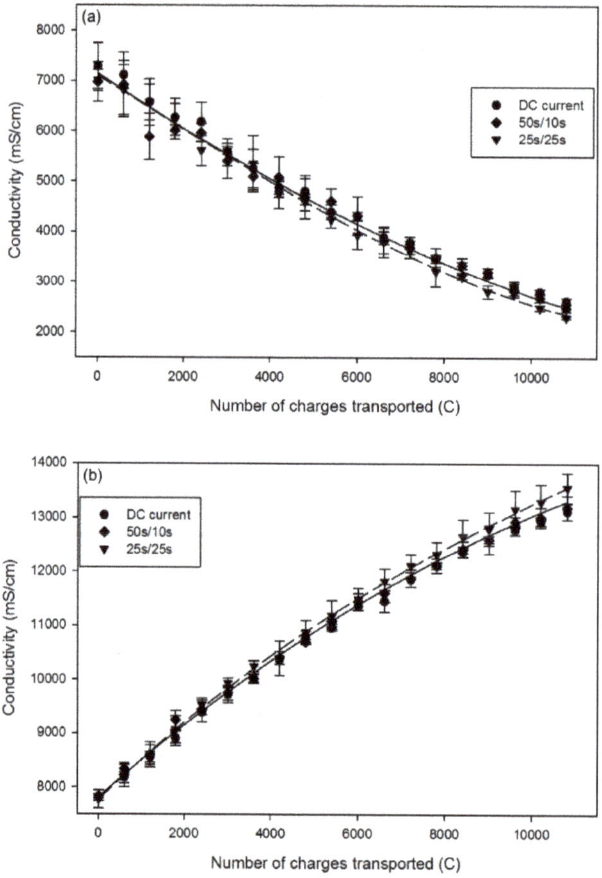

Figure 3. Conductivity evolution for DC (dots), PEF 50 s/10 s (rhombuses) and PEF 25 s/25 s (inverted triangles) current conditions in (**a**) AWComp and (**b**) OAComp.

2.2. Global System Resistance and Relative Energy Consumption

The global system resistance was significantly affected by the application of PEF. It increased from an initial value of 10.0 ± 0.4 Ω to values of 42.0 ± 6.7 Ω (increase of 4.2-folds), 36.0 ± 6.5 Ω (3.6-times increase), and 28.6 ± 0.2 Ω (2.8-folds increase) for the application of DC current, PEF 50 s/10 s, and PEF 25 s/25 s, respectively (Figure 4). Due to high values of standard deviation for DC and PEF 50 s/10 s current conditions, no significant difference was observed between them (P = 0.31). However, the PEF 25 s/25 s current condition showed a significantly lower resistance value in comparison with the DC current condition (P = 0.024). Moreover, depending on the applied current condition, the increase in global resistance appeared at a different number of charges during the process, with PEF conditions delaying the appearance of the resistance increase: around 6000 C for the DC current condition, 6600 C for the PEF 50 s/10 s condition, and 7200 C for the PEF 25 s/25 s condition. Regarding relative energy consumption, only power supply consumption was taken into account as the pumps' energy consumption represents a negligible part of the global consumption, even if the process time is doubled. The DC current condition consumed 9.33 ± 1.38 Wh/g of lactic acid recovered, while the PEF 50 s/10 s and PEF 25 s/25 s consumed 7.88 ± 0.64 and 6.21 ± 0.30 Wh/g of lactic acid recovered, respectively: this corresponds to a 33% less relative energy consumption for the 25 s/25 s condition. Here, again, only the PEF 25 s/25 s was significantly different from the DC current condition (P = 0.03).

These different values in number of charges transported, where the global resistance increased during the process, can be related to the inflection points previously observed for the pH, and would correspond to the formation of a significant scaling/fouling on the membranes [11], resulting in further water dissociation. In the present study, the application of PEF would have slowed down the appearance of scaling/fouling and, consequently, of water splitting, as demonstrated by Cifuentes-Araya (2013) [25] and Andreeva et al. (2018) [16] in studies on the same model salt solutions enriched with scale-forming ions (Ca^{2+}, Mg^{2+}, ...). Such a delay in scaling formation allowed, consequently, a decrease in the final global system resistance of 32% for PEF 25 s/25 s condition in comparison with DC current but, also, a significant saving in relative energy consumption of around 33%.

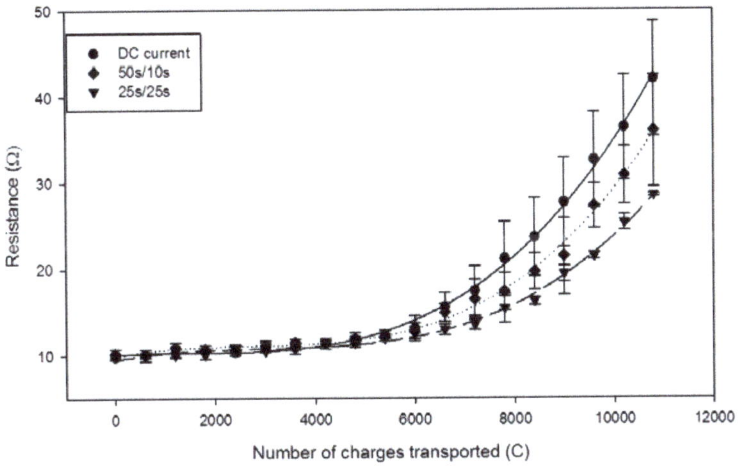

Figure 4. Global system resistance evolution for DC (dots), PEF 50 s/10 s (rhombuses), and PEF 25 s/25 s (inverted triangles) current conditions.

2.3. Membrane Analysis

2.3.1. Membrane Photographs

To evaluate whether there is formation of scaling/fouling on membranes, after each ED run and after rinsing the cell during 5 min with water, the stack was dismantled, and photographs of each membrane were taken (Figure 5). For all three current conditions, the AWComp side of the AEMs were all free of any visible fouling or scaling. However, on the OAComp side of the AEMs, a white plaster-like scaling was visible for the three conditions, with decreasing quantity from DC to PEF

50 s/10 s and then to PEF 25 s/25 s current conditions. Regarding the CEMs, no scaling or fouling was observed on both sides of the membranes, as previously reported [11].

As in the previous study by Dufton et al. [11], the scaling on the AEMs affected the pH variations, and the global system resistance increased during the process. The mitigation of this scaling by application of PEF was visible through the apparent reduction of scaling for the PEF current conditions: the scaling was more important after DC current application, slightly decreased by 50 s/10 s PEF application, and drastically decreased after the treatment with PEF 25 s/25 s.

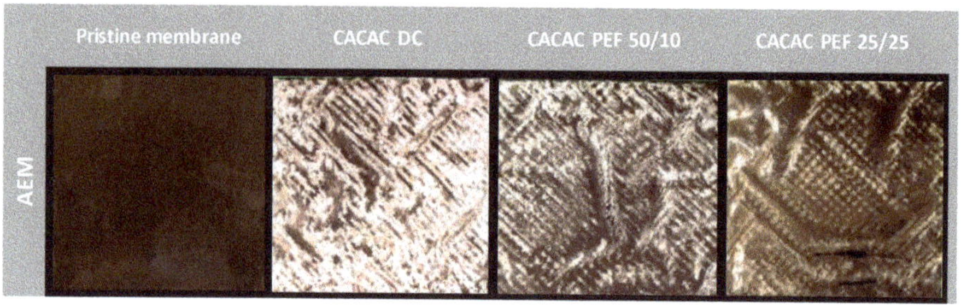

Figure 5. Entire anion-exchange membranes' (AEMs') anode sides (10 cm^2) photographs before and after ED for all three current conditions. AEM comprised of A1 and A2 with no visible differences.

2.3.2. Membrane Thickness and Conductivity

For all conditions, there was no significant difference between treatments for the AEM conductivity. However, the CEMs were similarly affected for all conditions in comparison to the pristine membranes: a decrease in conductivity of around 10% was observed for C1 ($P < 0.02$), while a 30% decrease was visible on the C2 and C3 membranes ($P < 0.006$). Regarding membrane thickness, as shown in Figure 6, the CEMs remained similar before and after ED treatment for all three current conditions ($P = 0.119$). The measurements on the AEMs had high variability caused by the irregular scaling on their surfaces, and no statistical difference was observed between the different conditions, but a clear tendency can be seen on the thickness means. The PEF 25 s/25 s AEM's thickness after ED treatment is the one closest to the pristine membrane value.

Membrane conductivity followed the same pattern as in the previous study with DC current [11], regardless of the current condition. The decrease in conductivity for the CEMs would be due to the presence in the membrane of counterions having lower electrophoretic mobilities compared to initial Na^+ counterions. Indeed, calcium and magnesium ions present in the acid whey and migrating through the CEM have lower electrophoretic mobilities (Ca^{2+}: 1.07×10^9 $cm^2/V{\cdot}s$, Mg^{2+}: 0.91×10^9 $cm^2/V{\cdot}s$) in comparison to sodium ions present in the pristine CEM (Na: 4.39×10^9 $cm^2/V{\cdot}s$) [26]. Despite the visible deposit on the AEMs, no impact was observed on conductivity, which means no significant damage was done to the membrane integrity. Regarding membrane thickness, the values for the CEMs showed that, regardless of the current condition, the ED treatment had no effect on the membranes. On the other hand, the thickness measurements for the AEMs illustrate, distinctly, the favorable effect of the application of pulsed electric field and especially of an optimized pulse/pause combination on deposit reduction.

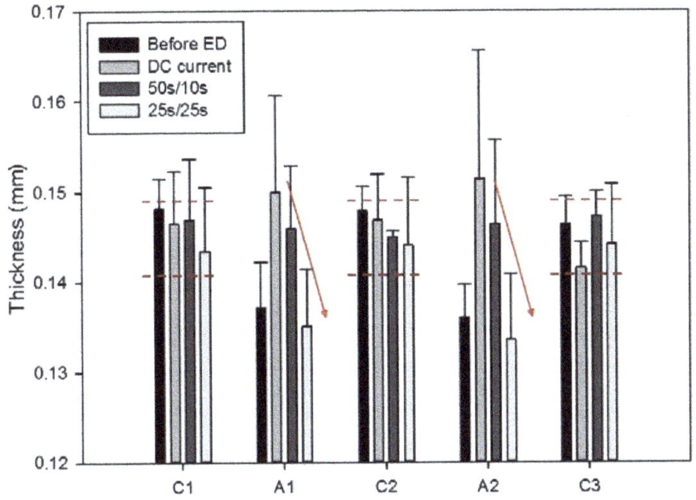

Figure 6. Membrane thickness measurements for all membranes of the ED configuration, before and after treatment, for the three different current conditions. The arrows show the thickness decreasing tendency depending on the applied current condition.

2.4. Scaling Characterization

2.4.1. Mineral Content

The mineral composition of IEMs provided information concerning the deposit nature. All membranes were similar in content for potassium (0.05 ± 0.01 and 0.01 ± 0.00 g/100 g of CEM and AEM, respectively), sodium (4.72 ± 0.23 and 0.04 ± 0.01 g/100 g of CEM and AEM, respectively) and magnesium (0.02 ± 0.00 and 0.00 ± 0.00 g/100 g of CEM and AEM, respectively) for the three different current conditions. Nevertheless, the CEMs' calcium amount after ED treatment increased around 5-fold in comparison with a pristine membrane (Figure 7a) ($P < 0.001$), and the application of PEF had no significant impact on this concentration ($P > 0.16$). However, the trend suggests that calcium concentration in the CEM was increased by PEF application. Regarding the AEMs, as shown in Figure 7b, calcium content increased significantly about 6-fold for the DC and PEF 50 s/10 s current conditions ($P = 0.013$) while remaining close to the initial value for the PEF 25 s/25 s current condition ($P = 0.99$). There were also differences in phosphorus content in the AEMs before and after processing treatment: it increased about 26 times for the PEF 25 s/25 s current condition while increasing 40-fold for the DC and PEF 50 s/10 s current conditions, in comparison with the pristine membrane ($P < 0.01$).

Once again, as already mentioned, the presence of calcium in the CEMs would be due to the residual ions inside the membrane's nanochannels due to the late and slow migration of calcium during the process [26]. Regarding the AEMs, as reported in the previous study [11], the scaling observed on their surface seems to be mainly composed of calcium phosphate. Indeed, its precipitation reaction is triggered by alkaline conditions and implied, therefore, the occurrence of water splitting to locally obtain alkaline conditions on the AEM's surface [27,28]. However, as shown in Figure 7b, the application of PEF with a pulse/pause combination of 25 s/25 s allows for significant reduction of the scaling amount. This effect was reported by other authors on saline solutions [15–17] but this is the first time that the application of an adjusted PEF is demonstrated to be efficient on scaling reduction during acid whey ED.

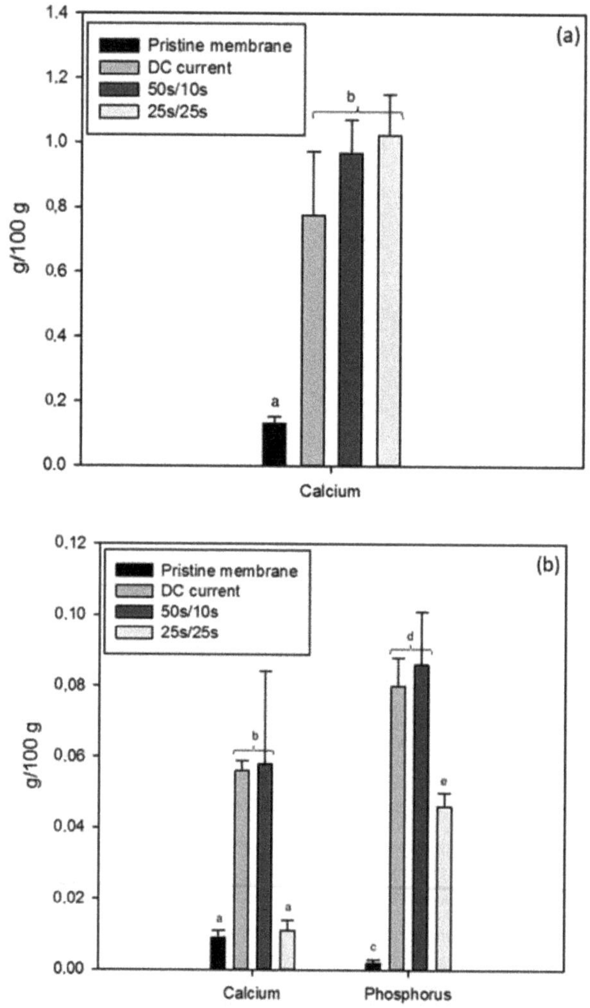

Figure 7. Calcium content of cation-exchange membranes (CEMs) (**a**) and calcium and phosphorus content of AEMs (**b**) for all three current conditions, in comparison with a pristine membrane. Different letters on histograms for each element means that there is a statistical difference between their values ($P < 0.05$).

2.4.2. Scanning Electron Microscopy (SEM), Energy Dispersive Spectroscopy (EDS), and X-ray Diffraction (XRD) Analyses

The CEMs' observation by SEM showed an absence of deposit for all current conditions and membrane's sides, with pictures similar to the pristine membrane. XRD analysis did not reveal any form of crystalline precipitate (Supplementary Data A) but very low amounts of calcium, similar for all conditions of current applied, which were visible in the elemental analysis by EDS. Regarding the AEM photographs obtained by SEM, the membranes' AWComp side appearances were similar to the pristine membrane for all current conditions, while the membranes' OAComp side from the DC current condition showed massive deposits (Figure 8a). This scaling can also be observed, to a lesser extent, on the AEMs' OAComp side for the PEF 50 s/10 s condition, but is absent from the membranes from the PEF 25 s/25 s condition. The analysis of these membranes by XRD did not reveal any crystalline deposit, but the EDS showed high concentrations of calcium and phosphorus on the OAComp′ side for the DC current condition membranes and were relatively lower for the PEF 50 s/10 s current condition (Figure 8b).

Just as reported in the previous study [11], the CEMs were free from any form of fouling or scaling, and the traces of calcium observed were most likely due to the presence of free ions migrating into the membrane nanochannels. As for the AEMs, the major scaling visible on the DC current condition membranes' OAComp side is similar to the one described in a previous study [11] and is known as an amorphous form of calcium phosphate (ACP: $Ca_xH_y[PO_4]_z$ nH_2O, n = 3–4.5; 15%–20% H_2O) [27,29]. Regarding the PEF 50 s/10 s current condition, the AEMs' OAComp side was also subject to similar scaling but at a relatively lower level (around 3 times), while the PEF 25 s/25 s current condition AEMs were completely free of apparent scaling. It was necessary to moderate these SEM observations since the scaling for PEF 25 s/25 s current condition was extremely uneven (Figure 5) and already in lesser quantity than other conditions. Furthermore, between SEM observation and stack dismantling, the membranes went through several manipulations (thickness and conductivity measurements, drying, SEM coating) where the weakly-bonded scaling potentially detached from the membrane, explaining the visible difference between SEM and macroscopic photographs taken just after ED treatment. However, the observations correlate with the above-mentioned results and the drastic decrease in membrane scaling, which can be attributed to the application of an appropriate combination of PEF.

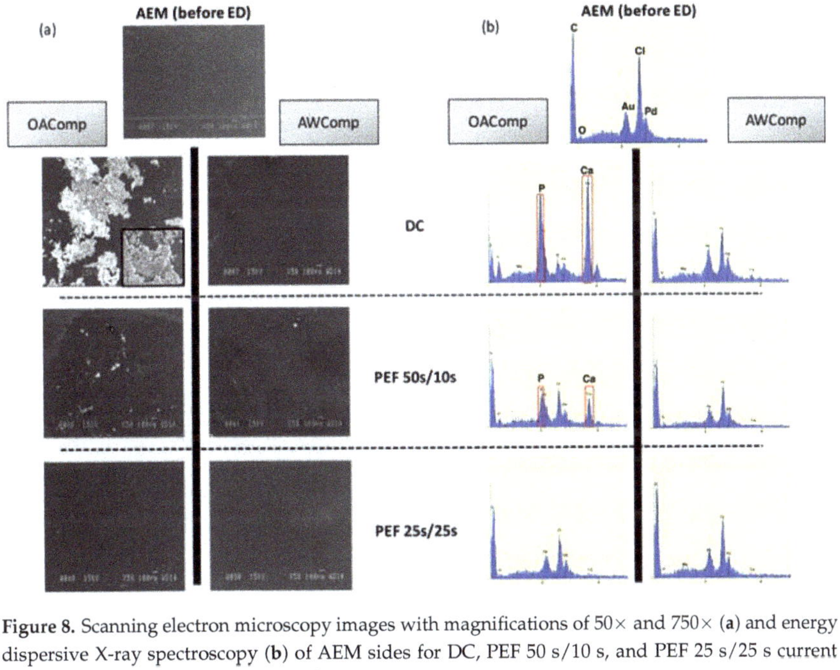

Figure 8. Scanning electron microscopy images with magnifications of 50× and 750× (**a**) and energy dispersive X-ray spectroscopy (**b**) of AEM sides for DC, PEF 50 s/10 s, and PEF 25 s/25 s current conditions compared to pristine membranes.

3. Materials and Methods

3.1. Whey

The raw acid whey samples were obtained from Parmalat-Canada (Victoriaville, QC, Canada) processing plant by refrigerated transport at 4 °C, and then stored at −30 °C. For each run, 2 L of whey were thawed by free convection at 4 °C prior to experiments. Table 2 describes the whey composition in comparison to the one reported in literature and from our previous study.

Table 2. Raw acid whey composition and physicochemical characteristics.

Composition	Unit	Acid Whey	Acid Whey from First Study [11]	Values Reported in the Literature [3,9,10,30]
Total solids	g/L	57.2 ± 1.5	59.8 ± 4.2	50.0–70.0
Total protein	g/L	7.5 ± 1.1	6.5 ± 0.7	4.2–10.0
Lactose	g/L	34.9 ± 1.0	41.2 ± 0.9	38–49
Minerals	g/L	6.9 ± 0.1	5.1 ± 1.1	4.7–7.0
P	g/L	0.76 ± 0.02	0.55 ± 0.01	0.44–0.90
Ca	g/L	1.08 ± 0.02	0.86 ± 0.02	0.43–1.60
K	g/L	1.65 ± 0.03	1.26 ± 0.05	1.28–1.82
Mg	g/L	0.10 ± 0.00	0.09 ± 0.00	0.09–0.19
Na	g/L	0.53 ± 0.02	0.39 ± 0.03	0.40–0.61
Lactate	g/L	7.12 ± 0.11	7.00 ± 0.14	5.18–8.00
Ratio Lactate/Lactose	No unit	0.20	0.17	0.12–0.15
pH	No unit	4.6	4.4	4.0–4.6
Conductivity	mS/cm^2	7.09 ± 0.35	7.05 ± 0.24	8.27 ± 0.42

3.2. *Electrodialytic Configuration*

Electrodialysis experiments were performed using an MP type cell (Multi-Purpose cell from ElectroCell AB, Täby, Sweden) with an effective surface area of 100 cm^2. The configuration used (Figure 9) was selected due to its common use in the dairy industry and in recent studies on acid whey deacidification [10,11]. Two deacidification units were set up as to follow the previous studies, and the anode was a dimensionally stable electrode (DSA-O_2) while the cathode was a stainless steel electrode. The potential difference was generated by a power supply (Model HPD 30-10, Xantrex, Burnaby, BC, Canada), the solutions circulated using centrifugal pumps (Baldor Electric Co., Fort Smith, AR, USA), and the flow rates controlled by flowmeters (Aalborg Instruments and Controls, Inc., Orangeburg, SC, USA).

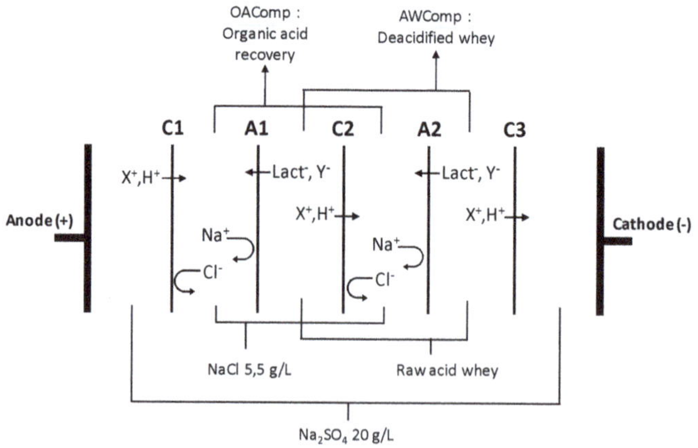

Figure 9. Electrodialysis (ED) configuration (CACAC, letters corresponding to the membrane's stacking) used for acid whey deacidification. C refers to cation-exchange membrane and A to anion-exchange membrane. X^+ and Y^- respectively refer to positively and negatively charged ionic species present in the whey (modified from Dufton et al. 2018 [11]).

The configuration consisted of stacking five commercial food grade cation- and anion-exchange membranes (respectively CEM and AEM). Diluate (acid whey compartment: AWComp) and concentrate (organic acid recovery compartment: OAComp) solutions were circulated between the membranes defining three closed loops. The solutions used were a 20 g/L Na_2SO_4 electrolyte solution (volume of 2 L, flow rate of 4 L/min), a 5.5 g/L NaCl aqueous concentrate solution (2 L, 4 L/min), and acid whey (2 L, 4 L/min). To ensure a continuous recirculation, external tanks containing the solutions were connected to each closed loop. The closest cation-exchange membrane (CEM) to the cathode (C3, see Figure 9) was added in order to avoid any anion migration in the electrolyte solution compartment, specifically lactate.

3.3. Protocol

In order to apply a similar driving force as in previous studies, a constant current density of 100 A/m^2 was applied. This current density was selected as 80% of the limiting current density after its determination according to Cowan & Brown method [31]. In addition to the experiments conducted in DC current as control, two different pulse/pause combinations were tested: 25 s/25 s (ratio pulse duration/pause duration r = 1) and 50 s/10 s (r = 5) using a Pulsewave^{TM} 760 Switcher (Bio-Rad Laboratories, Richmond, CA, USA). These pulse/pause combinations were chosen with relatively long periods and high ratios, in order to allow the solution flow to rinse the membrane's surfaces and remove the already-formed scaling, since it has been found to be plaster-like, crumbly, and easily removable by hand [11]. Such long period combinations were found to be efficient for scaling or fouling mitigation in several studies [15,16]. All experiments were conducted with different durations, according to the pulse/pause combination, or not, to correspond to an effective treatment of three hours, or to a number of charges transported of 10800 C: 3, 3.5, and 6 h for DC, 50 s/10 s and 25 s/25 s current conditions respectively. The solution tanks were kept at room temperature around 20 °C, and three replicates were performed for each current condition. At the end of each run, before dismantling the cell, the whole system (tanks, tubing, and ED cell) was rinsed for 5 min with water to remove all superficial or non adsorbed scaling.

The raw acid whey composition was analyzed in terms of minerals, protein, lactic acid, and lactose contents before and after ED. During the ED process, the electrical conductivity and pH values were recorded every 10 minutes for each solution as well as the applied voltage. Samples were taken in both OAComp and AWComp at 0, 1800, 3600, 7200, and 10800 C (corresponding to 0, 30, 60, 120 and 180 min in DC current) for organic acid concentration determination by high-performance liquid chromatography (HPLC). Membrane thickness and electrical conductivity were measured before and after each run to ascertain membrane scaling, and the membranes were dried and kept for mineral analyses and microscopy observation.

3.4. Analyses

All analyses were performed on at least three technical samples.

Total solids and ash contents. According to the AOAC methods 990.20 and 945.46, raw acid whey samples were weighed before drying for one hour on a heating plate (Corning PC-420 Hot Plate Stirrer, NY, USA). The dried samples were then weighted again for total solid content determination, and further ashed in a furnace at 550 °C overnight until they turned white. The samples were weighed after cooling, and the ash content was determined as follows (m refers to the measured weights):

$$ash,\ \% = \frac{100 \times (m_{crucible+ashes} - m_{crucible})}{m_{crucible+sample} - m_{crucible}}. \quad (1)$$

Mineral composition. Calcium, potassium, magnesium, sodium, and phosphorus concentrations were determined by optical emission spectrometry with inductively coupled plasma as atomization and excitation source (ICP-OES Agilent 5110 SVDV Agilent Technologies, Victoria, Australia), using the following wavelengths (in nm): 393.366, 396.847, 422.673 (Ca); 766.491 (K); 279.553, 280.270, 285.213 (Mg); 588.995, 589.592 (Na); 177.434, 178.222, 213.618, 214.914 (P). The analyses for all ions were carried out in axial and/or radial view, directly on acid whey samples diluted 20 times. Samples of 10 mL were diluted 1:5 in distilled water and used for ion determination.

Organic acid contents. Organic acid concentrations were determined by high-performance liquid chromatography (HPLC) using a chromatograph from Waters (Waters Corp., Milford, MA, USA), equipped with a Hitachi (Foster City, CA, USA) differential refractometer detector L-7490. An ICSep ICE-ION-300 column (Transgenomic, Omaha, NE, USA) was used with 8.5 mM of H_2SO_4 (180 μL H_2SO_4/L) as the mobile phase and at a flow rate of 0.4 mL/min. The column temperature was kept constant at 40 °C. Samples were centrifuged for five minutes at 5000 rpm (Allegra™ 25R Centrifuge, Beckman Coulter, Brea, CA, USA) and filtered (0.22 μm nylon; CHROMSPEC Syringe Filter, Brockville, ON, Canada) before injection (15 μL). A mixture of lactose anhydrous (PHR1025), citric acid (251275), DL-lactic acid (L1750) and acetic acid (338826) (from Sigma-Aldrich, St. Louis, MO, USA) was used as an external standard to perform the quantification in mg/L.

Total protein content. Total protein content was determined by the measurement of the total nitrogen concentration. This was conducted according to the Dumas combustion method using a TRUSPEC LECO FP-528 (LECO, St. Joseph, MI, USA) calibrated with EDTA. The raw acid whey was analyzed in triplicate before and after ED. A conversion factor of 6.38, proposed by Hammarsten and Sebelien in 1892, was used to determine the protein content.

pH. The pH of acid whey (AWComp) and organic acid recovery (OAComp) solutions were measured using a pH-meter model SP20 (VWR Symphony, Thermo Orion West Chester, PA, USA).

Conductivity. A YSI conductivity meter (Model 3100, Yellow Springs Instrument, Yellow Springs, OH, USA) equipped with an immersion probe (Model 3252, cell constant K = 1 /cm) was used for measuring values in acid whey (AWComp) and organic acid recovery (OAComp) solutions.

Global system resistance. The global system resistance (R, in Ω) was calculated according to Ohm's law (R = U/I). The voltage (U, in V) and current intensity (I, in A) values were directly obtained from the power supply.

Relative energy consumption (REC).

$$REC = \frac{\int_{t=0}^{t=end} \frac{U \times I}{3600} dt}{m_{lact.acid}}, \quad (2)$$

where REC is the relative energy consumption (in Wh/g of lactic acid recovered), U the voltage applied (in V), I the applied current (in A), and $m_{lact.\ acid}$ the total mass of lactic acid recovered at the end of treatment in the OAComp (in g). The time taken into account here for the PEF conditions will be the effective time during the pulse periods.

Membrane electrical conductivity. Membrane conductance (G_m) was measured using a conductivity clip (Laboratoire des Matériaux Échangeurs d'Ions, Créteil, France) with a 1 cm distance between the electrodes and the conductivity meter (Model 35, Yellow Springs Instrument Co., Yellow Springs, OH, USA). Membrane electrical resistance was calculated using the same method as Lteif et al. (1999) [32] and Lebrun et al. (2003) [33]:

$$R_m = \frac{1}{G_m} = \frac{1}{G_{m+s}} - \frac{1}{G_s} = R_{m+s} - R_s, \quad (3)$$

where R_m is the transverse electric resistance of the membrane (Ω), G_m is the membrane conductance (S), G_{m+s} is the conductance of the membrane and the solution measured together (S), G_s is the solution conductance (S), R_{m+s} is the resistance of both membrane and solution measured together (Ω), and R_s is the solution resistance (Ω). The membrane electrical conductivity was then calculated using the following relation [32]:

$$K = \frac{L}{R_m \times A} \quad (4)$$

where K is the membrane conductivity (mS/cm), L is the membrane thickness (cm), R_m the transversal resistance of the membrane (Ω), and A the electrode surface (1 cm^2).

Membrane mineral composition. The same elemental concentrations as for the liquid solutions were measured for the membranes by ICP-OES, as described previously. For each configuration, the analysis was conducted in triplicate on 18.75 cm^2 pieces of AEM (CACAC the bold character highlights the membrane analyzed) and CEM (CACAC). Pristine AEMs and CEMs were also analyzed as controls. Membranes pieces of 18.75 cm^2 were cut, weighed, and dried at 60 °C overnight in an oven (VWR Gravity Convection Oven, Radnor, PA, USA). The dried samples were then ashed in a furnace at 550 °C overnight until they turned white. The samples were weighed after cooling and the ash content was determined according to Equation (1). The ashes were resolubilized in 1 mL 25% nitric acid and diluted in 50 mL total volume with demineralized water (PURELAB®Ultra, ELGA, High Wycombe, UK). The solutions were then filtered with membranes of 0.45 μm pore size before the ICP-OES analysis.

Scanning electron microscopy (SEM) and Energy Dispersive X-ray spectroscopy (EDS) analysis. The dried samples (same protocol as for membrane mineral composition before carbonization) were coated with a thin layer of gold to improve the image quality (Technics Hummer II Sputter Coater, Anatech Ltd., Hayward, CA, USA). Images were then registered using a field emission gun scanning electron microscope with a magnification of 50× (JMS840A SEM, JEOL, Peabody, MA, USA). The microscope was equipped with a spectrometer using the energy dispersive X-ray spectroscopy (EDS, Bruker Analysis, Billerica, MA, USA) at a 15 kV accelerating voltage and a 15 mm working distance.

X-ray diffraction (XRD). The analysis was performed using a D5000 Siemens diffractometer (Montreal, QC, Canada). The radiation source (CuKα) was a copper lamp with a wavelength of 0.154 nm. The K_α radiation of copper was generated at 30 mA and 40 kV. The scan rate of 0.02° 2θ was applied to record patterns for 2θ ranging between 15° and 65°. Results were analyzed using JADE software version 2.1 with JCPDS database from the ICDD (International Centre for Diffraction Data) version 2001.

Statistical analyses. Analyses of variance (ANOVA) were performed on data and Tukey tests ($\alpha = 0.05$ as probability level) were used to compare treatments (SigmaPlot software, version 12.0 for Windows, MilliporeSigma, Burlington, MA, USA).

4. Conclusions

There are very few reported applications of PEF on complex food matrices during ED treatments, and this was the first time that such a current mode was used on acid whey to both demineralize and deacidify it. In the present study, two pulse/pause combinations were tested: 50 s/10 s and 25 s/25 s, and were compared with DC current. Regarding migration enhancement, the PEF 25 s/25 s current condition showed a lactate migration or deacidification 16% higher than the DC condition. Moreover, supplementary analysis on whey solution showed the migration enhancement also applied to mineral ions such as calcium, magnesium, and potassium, with removal rates of 53%, 32%, and 85%, respectively, 4% to 25% higher than the rates obtained with DC current. In addition to better deacidification and demineralization efficiencies, the application of PEF conditions appeared to have a significant effect on scaling mitigation. Indeed, membrane observations and characterization allowed to correlate the drastic reduction of calcium phosphate scaling intensity with the use of pulse/pause combination of 25 s/25 s. This could also be confirmed by the 32% major reduction in final global system resistance observed in comparison with the DC condition (Supplementary Data B), as well as the 33% decrease in relative energy consumption in term of Wh per gram of lactic acid recovered. These positive effects of the 25 s/25 s PEF condition were mainly due to the reduction of concentration polarization on the membranes' boundary layers. By decreasing such a phenomenon, (1) water splitting occurrence responsible for the local pH variations leading to scaling formation [16,21,25] was mitigated, and (2) ionic availability for migration through the membranes was improved, explaining the enhanced deacidification/demineralization rates [18–20]. However, it appeared crucial to pay particular attention to the pulse/pause combination applied since, in the present study, the two different PEF current conditions led to very different results regarding demineralization, deacidification, and scaling. Finally, this was the first time that PEF was demonstrated to have a simultaneous positive impact on these main concerns of ED application to complex matrices.

Supplementary Materials: Supplementary materials can be found at http://www.mdpi.com/1422-0067/20/4/797/s1.

Author Contributions: Conceptualization, L.B.; Methodology, G.D. and L.B.; Software, G.D.; Validation, G.D. and L.B.; Formal Analysis, G.D.; Investigation, G.D. and L.B.; Resources, S.G., S.M. and L.B.; Data Curation, G.D., S.M. and L.B.; Writing—Original Draft Preparation, G.D.; Writing—Review & Editing, G.D., S.M., S.G. and L.B.; Visualization, G.D.; Supervision, L.B.; Project Administration, L.B.; Funding Acquisition, L.B.

Funding: This research was funded by The Natural Sciences and Engineering Research Council of Canada (NSERC) Industrial Research Chair on ElectroMembrane processes aiming the ecoefficiency improvement of biofood production lines [IRCPJ 492889-15 to Laurent Bazinet] and the NSERC Discovery Grants Program [RGPIN-2018-04128 to Laurent Bazinet] without any internal conflict of interest.

Acknowledgments: The Natural Sciences and Engineering Research Council of Canada (NSERC) financial support is acknowledged. The authors thank Jacinthe Thibodeau and Diane Gagnon, research professionals at Université Laval, for their everyday help on the project. Also, the authors thank André Ferland and Jean Frenette, research technicians at Université Laval for their assistance in SEM and X-ray diffraction analyses as well as Alain Brousseau and Véronique Richard, research professionals at Université Laval for their respective involvement in ICP-OES and HPLC analyses.

Conflicts of Interest: The authors declare no conflict of interest.

References

1. Prazeres, A.R.; Carvalho, F.; Rivas, J. Cheese whey management: A review. *J. Environ. Manag.* **2012**, *110*, 48–68. [CrossRef] [PubMed]
2. Ganju, S.; Gogate, P.R. A review on approaches for efficient recovery of whey proteins from dairy industry effluents. *J. Food Eng.* **2017**, *215*, 84–96. [CrossRef]
3. Saffari, M.; Langrish, T. Effect of lactic acid in-process crystallization of lactose/protein powders during spray drying. *J. Food Eng.* **2014**, *137*, 88–94. [CrossRef]
4. Chandrapala, J.; Vasiljevic, T. Properties of spray dried lactose powders influenced by presence of lactic acid and calcium. *J. Food Eng.* **2017**, *198*, 63–71. [CrossRef]
5. Houldsworth, D.W. Whey technology Demineralization of whey by means of ion exchange and electrodialysis. *Int. J. Dairy Technol.* **1980**, *33*, 45–51. [CrossRef]
6. Hoppe, G.K.; Higgins, J.J. Demineralization. In *Whey and Lactose Processing*; Zadow, J.G., Ed.; Springer: Dordrecht, The Netherlands, 1992; pp. 91–131. [CrossRef]
7. Greiter, M.; Novalin, S.; Wendland, M.; Kulbe, K.D.; Fischer, J. Desalination of whey by electrodialysis and ion exchange resins: Analysis of both processes with regard to sustainability by calculating their cumulative energy demand. *J. Membr. Sci.* **2002**, *210*, 91–102. [CrossRef]
8. Chandrapala, J.; Chen, G.Q.; Kezia, K.; Bowman, E.G.; Vasiljevic, T.; Kentish, S.E. Removal of lactate from acid whey using nanofiltration. *J. Food Eng.* **2016**, *177*, 59–64. [CrossRef]
9. Bédas, M.; Tanguy, G.; Dolivet, A.; Méjean, S.; Gaucheron, F.; Garric, G.; Senard, G.; Jeantet, R.; Schuck, P. Nanofiltration of lactic acid whey prior to spray drying: Scaling up to a semi-industrial scale. *LWT Food Sci. Technol.* **2017**, *79*, 355–360. [CrossRef]
10. Chen, G.Q.; Eschbach, F.I.I.; Weeks, M.; Gras, S.L.; Kentish, S.E. Removal of lactic acid from acid whey using electrodialysis. *Sep. Purif. Technol.* **2016**, *158*, 230–237. [CrossRef]
11. Dufton, G.; Mikhaylin, S.; Gaaloul, S.; Bazinet, L. How electrodialysis configuration influences the acid whey deacidification and membrane scaling. *J. Dairy Sci.* **2018**, *101*, 7833–7850. [CrossRef]
12. Balster, J.; Krupenko, O.; Pünt, I.; Stamatialis, D.F.; Wessling, M. Preparation and characterisation of monovalent ion selective cation exchange membranes based on sulphonated poly(ether ether ketone). *J. Membr. Sci.* **2005**, *263*, 137–145. [CrossRef]
13. Casademont, C.; Pourcelly, G.; Bazinet, L. Effect of magnesium/calcium ratios in solutions treated by electrodialysis: Morphological characterization and identification of anion-exchange membrane fouling. *J. Colloid Interface Sci.* **2008**, *322*, 215–223. [CrossRef] [PubMed]

14. Mikhaylin, S.; Bazinet, L. Fouling on ion-exchange membranes: Classification, characterization and strategies of prevention and control. *Adv. Colloid Interface Sci.* **2016**, *229*, 34–56. [CrossRef] [PubMed]
15. Cifuentes-Araya, N.; Pourcelly, G.; Bazinet, L. Impact of pulsed electric field on electrodialysis process performance and membrane fouling during consecutive demineralization of a model salt solution containing a high magnesium/calcium ratio. *J. Colloid Interface Sci.* **2011**, *361*, 79–89. [CrossRef] [PubMed]
16. Andreeva, M.A.; Gil, V.V.; Pismenskaya, N.D.; Dammak, L.; Kononenko, N.A.; Larchet, C.; Grande, D.; Nikonenko, V.V. Mitigation of membrane scaling in electrodialysis by electroconvection enhancement, pH adjustment and pulsed electric field application. *J. Membr. Sci.* **2018**, *549*, 129–140. [CrossRef]
17. Mikhaylin, S.; Nikonenko, V.; Pourcelly, G.; Bazinet, L. Intensification of demineralization process and decrease in scaling by application of pulsed electric field with short pulse/pause conditions. *J. Membr. Sci.* **2014**, *468*, 389–399. [CrossRef]
18. Mishchuk, N.A.; Koopal, L.K.; Gonzalez-Caballero, F. Intensification of electrodialysis by applying a non-stationary electric field. *Colloids Surf. A Physicochem. Eng. Asp.* **2001**, *176*, 195–212. [CrossRef]
19. Malek, P.; Ortiz, J.M.; Richards, B.S.; Schäfer, A.I. Electrodialytic removal of NaCl from water: Impacts of using pulsed electric potential on ion transport and water dissociation phenomena. *J. Membr. Sci.* **2013**, *435*, 99–109. [CrossRef]
20. Sistat, P.; Huguet, P.; Ruiz, B.; Pourcelly, G.; Mareev, S.A.; Nikonenko, V.V. Effect of pulsed electric field on electrodialysis of a NaCl solution in sub-limiting current regime. *Electrochim. Acta* **2015**, *164*, 267–280. [CrossRef]
21. Pelletier, S.; Serre, É.; Mikhaylin, S.; Bazinet, L. Optimization of cranberry juice deacidification by electrodialysis with bipolar membrane: Impact of pulsed electric field conditions. *Sep. Purif. Technol.* **2017**, *186*, 106–116. [CrossRef]
22. Zabolotskii, V.I.; Bugakov, V.V.; Sharafan, M.V.; Chermit, R.K. Transfer of electrolyte ions and water dissociation in anion-exchange membranes under intense current conditions. *Russ. J. Electrochem.* **2012**, *48*, 650–659. [CrossRef]
23. Krol, J.J.; Wessling, M.; Strathmann, H. Concentration polarization with monopolar ion exchange membranes: Current-voltage curves and water dissociation. *J. Membr. Sci.* **1999**, *162*, 145–154. [CrossRef]
24. Lin Teng Shee, F.; Arul, J.; Brunet, S.; Bazinet, L. Performing a three-step process for conversion of chitosan to its oligomers using a unique bipolar membrane electrodialysis system. *J. Agric. Food Chem.* **2008**, *56*, 10019–10026. [CrossRef] [PubMed]
25. Cifuentes-Araya, N.; Pourcelly, G.; Bazinet, L. Water splitting proton-barriers for mineral membrane fouling control and their optimization by accurate pulsed modes of electrodialysis. *J. Membr. Sci.* **2013**, *447*, 433–441. [CrossRef]
26. Firdaous, L.; Malériat, J.P.; Schlumpf, J.P.; Quéméneur, F. Transfer of monovalent and divalent cations in salt solutions by electrodialysis. *Sep. Sci. Technol.* **2007**, *42*, 931–948. [CrossRef]
27. Destainville, A.; Champion, E.; Laborde, E. Synthesis, characterization and thermal behavior of apatitic tricalcium phosphate. *Mater. Chem. Phys.* **2003**, *80*, 269–277. [CrossRef]
28. Dorozhkin, S.V. Amorphous calcium (ortho)phosphates. *Acta Biomater.* **2010**, *6*, 4457–4475. [CrossRef]
29. Dorozhkin, S.V. Biphasic, Triphasic, and Multiphasic Calcium Orthophosphates. In *Advanced Ceramic Materials*; John Wiley & Sons: Hoboken, NJ, USA, 2016; pp. 33–95. [CrossRef]
30. Panesar, P.S.; Kennedy, J.F.; Gandhi, D.N.; Bunko, K. Bioutilisation of whey for lactic acid production. *Food Chem.* **2007**, *105*, 1–14. [CrossRef]
31. Cowan, D.A.; Brown, J.H. Limiting Current in Electrodialysis Cells. *Ind. Eng. Chem.* **1959**, *51*, 1445–1448. [CrossRef]
32. Lteif, R.; Dammak, L.; Larchet, C.; Auclair, B. Conductivitéélectrique membranaire: Étude de l'effet de la concentration, de la nature de l'électrolyte et de la structure membranaire. *Eur. Polym. J.* **1999**, *35*, 1187–1195. [CrossRef]
33. Lebrun, L.; da Silva, E.; Pourcelly, G.; Métayer, M. Elaboration and characterisation of ion-exchange films used in the fabrication of bipolar membranes. *J. Membr. Sci.* **2003**, *227*, 95–111. [CrossRef]

International Journal of
Molecular Sciences

MDPI

Communication

An Investigation on the Application of Pulsed Electrodialysis Reversal in Whey Desalination

Arthur Merkeland Amir M. Ashrafi *

MemBrain s. r. o. (Membrane Innovation Center), Pod Vinicí 87, 471 27 Stráž pod Ralskem, Czech Republic; Arthur.merkel@membrain.cz

* Correspondence: Amirmansoor.ashrafi@gmail.com; Tel.: +420-776-130-043

Received: 9 March 2019; Accepted: 15 April 2019; Published: 18 April 2019

Abstract: Electrodialysis (ED) is frequently used in the desalination of whey. However, the fouling onto the membrane surface decreases the electrodialysis efficiency. Pulsed Electrodialysis Reversal (PER), in which short pulses of reverse polarity are applied, is expected to decrease the fouling onto membrane surface during ED. Three (PER) regimes were applied in the desalination of acid whey (pH ≤ 5) to study their effects on the membrane fouling and the ED efficiency. The PER regimes were compared to the conventional ED as the control. For each regime, two consecutive runs were performed without any cleaning step in-between to intensify the fouling. After the second run, the membranes were subjected to the Scanning electron microscope (SEM) imaging and contact angle measurement to investigate the fouling on the membrane surface in different regimes. The ED parameters in the case of conventional ED were almost the same in the first and the second runs. However, the parameters related to the ED efficiency including ED capacity, ash transfer, and ED time, were deteriorated when the PER regimes were applied. The contact angle values indicated that the fouling on the diluate side of anion exchange membranes was more intensified in conventional ED compared to the PER regimes. The SEM images also showed that the fouling on the diluate side of both cation and anion exchange membranes under PER regimes was reduced in respect to the conventional ED. However, the back transfer to the diluate compartment when the reverse pulse was applied is dominant and lowers the ED efficiency slightly when the PER is applied.

Keywords: whey; electrodialysis; pulsed electric field; pulsed electrodialysis reversal; fouling

1. Introduction

Whey is a by-product of cheese and curd production. It is separated from casein during the manufacture of cheese or casein. Due to its high content of proteins, minerals, vitamins, and lactose, it is a potential source of nutrients. However, in its normal form, whey is not considered as foodstuff due to its high salt content. Whey is categorized into sweet whey (pH is around 6), that is produced from rennet-coagulated casein or cheese, and acid whey (pH ≤ 5) that is produced from mineral or lactic acid-coagulated casein. Considering its content of proteins and vitamins in the natural functional form, whey is a valuable product which can be used as an additive in baby food, cheese products, and candies [1–4]. Therefore, a method to desalinate the whey and utilize the demineralized whey is in high demand. It is worth noting that the decomposition of the proteins and vitamins must be avoided during its demineralization process. Membrane processes including pressure-driven and electrically-driven membranes are two main solutions for the desalination of whey. Considering that the ED is based on the electrical voltage difference as the driving force, it is a more efficient method for demineralization, particularly in the case of charged ionic species with a small size [5]. However, the application of ED is accompanied by inherent limitations, including concentration polarization and fouling on the membranes [6].

Fouling on membranes is a serious problem in which ion exchange membranes are fouled by ionic materials of medium molecular weight such as ionic surface active agents having the charge opposite to the fixed charges of the membrane. Scaling is another type of fouling that occurs when salts of limited solubility precipitate from the concentrate stream as scale [6]. It must be mentioned that pH change caused by the water splitting in solution—membrane interface results in scaling of the ions with low solubility on the membrane surface [7].

The pore size of the ion exchange membrane is approximated to be 10A; therefore, ions of medium molecular weight permeate with difficulty through the membrane. Consequently, the electrical resistance of the membrane increases during electrodialysis due to clogging of the membrane pores with the medium molecular weight ions [8]. To remove the fouling from the membrane, the cleaning process or even the membrane replacement is required which may cost about 40%–50% in electro-membrane processes [6].

The conventional method to partially avoid the fouling during ED is the reversal of the concentrate and diluate streams. The modification of the membranes used in ED is another strategy to avoid fouling [9–11]. Furthermore, using the cleaning agents can also be applied to remove the film attached to the membrane surface during ED [12]. Due to the complexity of the described methods, it is desired to find an alternative method which is easy to perform.

The use of pulsed electric field (PEF) was shown to be an alternative for fouling prevention. The PEF procedure consists of application of consecutive pulse and pause lapses of a certain duration (Ton/Toff).

The use of PEF, particularly when the pause period is extended results in the electrophoretic movement of the substances that form the screening film on the membrane surface. Furthermore, the water splitting in the solution—membrane interface caused by concentration polarization reduces due to restoration during the pause period. Consequently, the scaling of ions with low solubility also decreases. However, Sistat et al. explained that the efficiency in PEF relies on the frequency of applied potential where the efficiency in PEF increases with increasing of the frequency of the potential [13,14]. Desalination of whey has been of great importance in the food industry and therefore many studies have been carried out in this field [15–18]. The effect of the PEF on the electrodialysis of acid whey to remove the lactate was also investigated [16]. Dufton et al. applied the PEF for the desalination of acid whey and confirmed its antifouling effect. However, the time of the desalination was increased in PEF by several times to reduce the fouling on to the membrane surface [15]. In pulsed electrodialysis reversal (PER) short pulses of reverse polarity are applied instead of a long pause period in PEF. Thus, it is expected that the short period of reverse pulses the ions and the film on the membrane re-dissolves in the solution. In addition, because of the short period of the reverse pulse the ED process will not be much longer compared to the conventional ED. The change of polarity occurs without reversal of diluate and concentrate streams (in contrary to electrodialysis reversal) [6,13,19–23]. This work aimed to study the effect of PER on the membrane fouling in electrodialysis of acid whey.

2. Results and Discussion

2.1. Electrodialysis

Three potential regimes were applied in ED and compared in terms of the efficiency (the regimes are defined as, regime *I*: conventional ED; 50 V applied on the membrane stack, regime *II*: 50 V applied for 180 s and then −50 V for 3 s, regime *III*: 50 V applied for 30 s and then −50 V for 5 s). The change in the diluate conductivity during ED with different regimes is shown in Figure 1. As seen, in conventional ED the time required to reach the desired conductivity in diluate is the same in the first and the second runs. In contrast, in the case of PER more time is required to achieve a given conductivity in the second run compared to that of the first runs. The ED parameters for different regimes are also represented in Table 1. As can be observed, the ED parameters in the case of conventional ED are almost the same in the first and the second runs. However, the parameters related to the ED efficiency including

ED capacity, ash transfer, and ED time, deteriorated when the PER regimes were applied. Since the ED continued to obtain a certain conductivity in diluate, the degree of the demineralization value is almost the same in all the applied regimes. The electrodialysis capacity of PER regimes is reduced compared to that of conventional ED, indicating that more time is required to achieve a given degree of the demineralization in PER regimes. The obtained values of the ash transfer rate and the energy consumption also show that to achieve a given diluate conductivity, in PER more energy must be consumed due to lower ash transfer. Recalling only the potential regime was different in the ED processes, as the deteriorated efficiency of the PER regimes could be due to either the fouling on the membrane surface or the back transfer to the dilute when the reverse pulse was applied. The highest difference between the first and the second run was observed in the case of regime III in which the ratio of $T_{working}/T_{reverse}$ was the least and the reverse pulse duration was the highest. Evidently, by increasing the duration of the reverse pulse, the back transfer of ions to the diluate increases. The change in current on the membrane stack and the pH change in diluate and concentrate during electrodialysis are provided in Supplementary Materials (Figures S1 and S2 respectively).

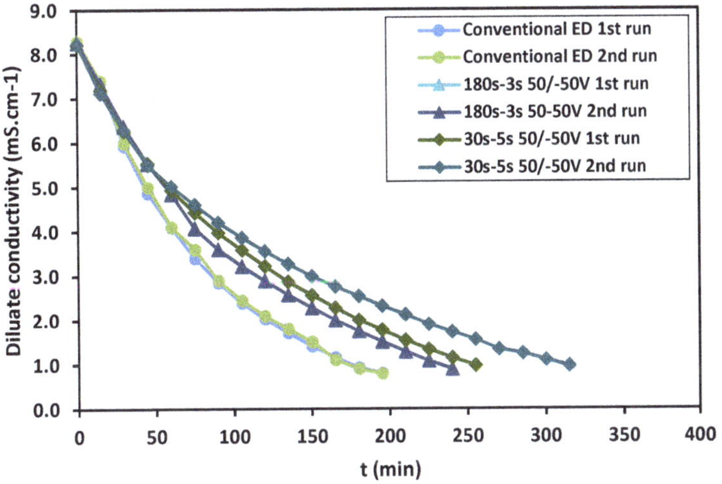

Figure 1. The change in diluate conductivity in different regimes.

Table 1. Parameters of acid whey electrodialysis with different regimes of applied potential (two consecutive runs without CIP).

Test	T (min)	κ_F ($mS{\cdot}cm^{-1}$)	$\kappa_{D, final}$ ($mS{\cdot}cm^{-1}$)	DD (%)	J ($g{\cdot}m^{-2}{\cdot}h^{-1}$)	C_F ($kg{\cdot}h^{-1}$)	E (Wh/kg_F)
Conventional ED 1	195	8.30	0.89	89.3	50	4.0	8.8
Conventional ED 2	195	8.31	0.88	89.4	50	4.0	8.9
180–3 50/−50	240	8.26	0.89	89.2	46	3.8	9.1
180–3 50/−50	285	8.16	0.75	90.8	39	3.2	9.6
30–5 50/−50	255	8.21	0.97	88.2	46	3.5	12.0
30–5 50/−50	315	8.23	0.97	88.3	36	2.9	12.9

2.2. Fouling Analysis

As shown in the SEM images (Figure 2), obviously when the pulsed regimes were applied the fouling decreased on the membrane surface. In particular, the film attached to the diluate sides of both AEM and CEM can be observed. The observed film contains particles which can be the organic molecules as well as the scaling layer. In our previous work we analyzed film attached to the membrane after the whey demineralization and it was found out that the film contains mainly Ca^{2+} and Mg^{2+}

and Al^{3+} (the ions with less solubility natural pH). With the electrodialysis process proceeding, the ion concentration near the diluate side of the membrane becomes zero, causing the water splitting and generation of OH^- and H_3O^+ ions. Consequently, the pH changes which brings about the scaling of minerals (multivalent ions) and fouling of organic molecules on the membranes including amino acids, vitamins, and polypeptides existing in the whey. The organic fouling also might be caused because of sorption of whey components including the residue of whey protein after nanofiltration, amino acids, and polypeptides [19]. In our previous work the scaling on the ion exchange membranes was analyzed and it was found that the scaling is mainly composed of sulfate and phosphate of Ca^{2+} and Mg^{2+} cations [24]. Thus, it is expected that during the reversal of the applied potential, the precipitated ions on the membrane surface partially detach from the membrane surface and dissolve in the feed.

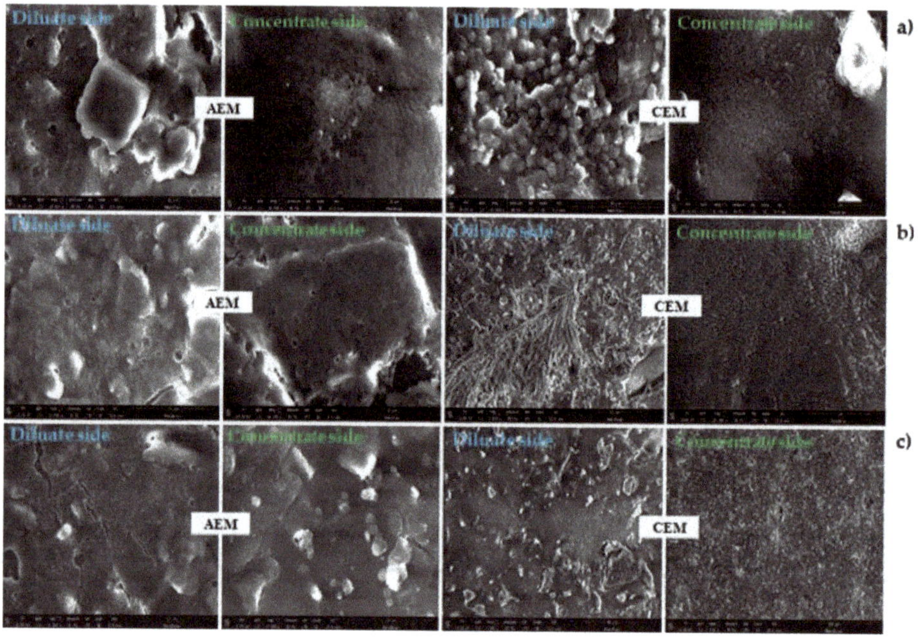

Figure 2. SEM images of the membranes after different regimes, (**a**) regime *I*, (**b**) regime *II*, (**c**) regime *III*, (AEM = anion exchange membrane, CEM = cation exchange membrane). 16000 × magnification

Therefore, the restoration of the ion concentration at the membrane interfaces is expected to occur in PER electrodialysis during the reverse pulse, resulting in a decrease in scaling and fouling. The values of the membranes contact angle after the ED are represented in Figure 3. The contact angles measurement allows measuring of the surface hydrophobicity. The surface hydrophobicity of the membranes is affected by the fouling or scaling. An increase in the hydrophobicity of the surface results in the increase of contact angles. Thus, fouling on membranes increases the hydrophobicity and consequently the contact angles of the membranes [22,25,26]. Considering the fact that most of the foulants are negatively charged, the fouling is a problem in the case of anion exchange membranes compared to the cation exchange membranes.

As seen, among the anion exchange membranes the highest values of contact angle on both diluate and concentrate sides were achieved when the conventional ED was utilized. The most significant differences can be seen between the contact angles on diluate side of anion exchange membranes in regime *I* compared to those of regime II and regime III, which indicates the prominent accumulation of the foulants in regime *I* onto diluate side of anion exchange membranes. The results indicate that the PER might result in a reduction of the fouling on the surface of anion

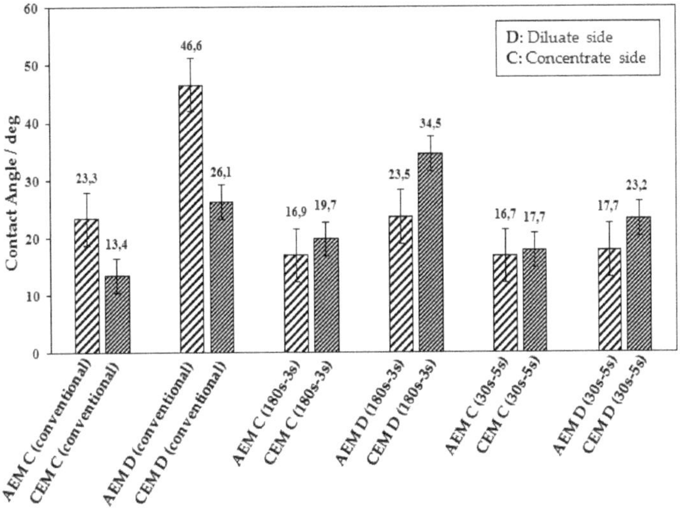

Figure 3. The values of the measured contact angles on the membranes surface after each ED of whey with different regimes.

Overall, the obtained results show that the PER could decrease the fouling on the membrane surface. Consequently, the ED operation becomes more convenient and the membrane maintenance becomes more cost-effective. However, the back transfer to the diluate lowers the ED efficiency.

3. Materials and Methods

3.1. Whey

The nanofiltrated acidic whey (NFW) was obtained in curd producing and provided by the Madeta milk factory (Jindřichův Hradec, Czech Republic) which specializes in the production of milk-based desserts as well as yogurts, fermented milk products, curd and yogurt deserts. (see Table 2).

Table 2. Feed (nanofiltrated acid whey) composition and physicochemical characteristics.

Composition	Unit	Feed Stream
Conductivity	$mS \cdot cm^{-1}$	8.22
pH	No unit	4.40
Total solids	%	18.6
Ash	%	1.37
Ash	%ODB	7.4
Acidity	°SH	60.0
Density	g/cm^3	1.0794
Lactose	$g \cdot kg^{-1}$	143.1
Total proteins	$g \cdot kg^{-1}$	14.9
True proteins	$g \cdot kg^{-1}$	5.3
NPN	$g \cdot kg^{-1}$	1.5
α-LA	$g \cdot L^{-1}$	1.46
β-LG A	$g \cdot L^{-1}$	2.40
β-LG B	$g \cdot L^{-1}$	0.62
CMP	$g \cdot L^{-1}$	2.63

Table 2. *Cont.*

Composition	Unit	Feed Stream
Lactates	mg·kg^{-1}	15706.87
Citrates	mg·kg^{-1}	7014
Na^+	mg·kg^{-1}	362.79
K^+	mg·kg^{-1}	1404.65
Mg^{2+}	mg·kg^{-1}	320.93
Ca^{2+}	mg·kg^{-1}	3106.97
S	mg·kg^{-1}	221.39
P total	mg·kg^{-1}	1925.58
Cl^-	mg·kg^{-1}	812.09
NO_3	mg·kg^{-1}	4.65

3.2. Reagents

The chemicals used in the experiments were of analytical grade and purchased from Sigma Aldrich (Germany). The demineralized water (κ≤ 10 μS·cm^{-1}) is produced in MemBrain Ltd., (Stráž pod Ralskem, Czech republic) by reverse osmosis.

3.3. Membranes

The food grade membranes were used in ED. The monopolar membranes used in ED processes for demineralization of whey were CEM-PES and AEM-PES cation and anion exchange membrane, respectively. These are heterogeneous membranes based on polyethylene as polymer and sulfonated groups as cation exchanger and quaternary ammonium groups as anion exchanger groups. Furthermore, both types of the membranes were reinforced with two polyesters (PES) fabrics. The reinforcement was performed by repressing at 150 °C and 5.06 × 10^{-6} Pa. The membranes were produced in MemBrain s.r.o., (Stráž pod Ralskem). The membranes properties including the resistivity and the permselectivity were studied and reported in previous works and presented in Table 3 [24].

Table 3. Properties of the membranes used in ED.

Membrane	d[1] Dry (mm)	d Swallowed (mm)	ρ^2 (Ω·cm)	P[3] (%)
AEM-PES	0.45	0.75	120	>90
CEM-PES	0.45	0.70	120	>95

[1] d: the thickness of the membrane; [2] ρ: specific resistivity; [3] P: the apparent permselectivity.

The electromotive force emf method [8], was used to measure the apparent permselectivity of the membrane. To briefly explain, a two-compartment cell was used whose chambers were filled with 0.5 and 0.1 M KCl, respectively. The membrane was placed in a hole between the compartments. Two Ag/AgCl (1 M/KCl) reference electrodes were inserted into the solutions close to the membrane. After 1 h of stirring the solution with magnetic stirrers the potential between two electrodes was measured and the apparent permselectivity was calculated as a ratio of the measured potential to the theoretical potential which corresponds to a 100% permeselective membrane Equation (1):

$$P\% = \frac{U_{measured}}{U_{theoritical}} \times 100 \quad (1)$$

where (P) is the apparent permselectivity, ($U_{measured}$) is the measured potential across the membrane and ($U_{theoritical}$) is the theoretical potential which is calculated for a membrane with 100% permselectivity [27,28].

For measuring the resistance of the membrane the same type of cell was used despite that both compartments were filled with 0.5 M NaCl. Two Pt wire electrodes were inserted into the solution while two Ag/AgCl (1 M/KCl) reference electrodes were placed next to the membrane on each side.

The dc current of 10 mA amplitude was applied to the Pt electrodes and the resulted potential drop between the reference electrodes was measured. The same measurement was carried out without the membrane. The resistance of membrane was calculated using the ohm law, Equation (2):

$$\rho_{sm} - \rho_s = \rho_m \ (\Omega.cm) \quad (2)$$

where (ρ_{sm}) is the specific resistivity of the membrane and solution layers trapped between the membranes and the references electrodes, (ρ_s) is the specific resistivity of the solution and (ρ_m) indicates the specific resistivity of the membrane [27,28].

3.4. *Electrodialysis*

The ED processes is shown in Figure 4. The ED was performed with modified electrodialysis unit P1 EDR-Y/50-0.8 (manufactured by MemBrain s.r.o.). The pH and the conductivity of the solutions were measured by SenTix® 940 glass electrode and TetraCon 925 conductivity cell, respectively. The probes were connected to the WTW multi 3420. It must be mentioned that the conductivity cell also possesses the temperature sensor. The stack contained 50 pairs of membranes AEM-PES and CEM-PES assembled in C-A-C (cation exchange membrane–anion exchange membrane–cation exchange membrane) configuration. The active area of each membrane was 400 cm^2. The unit was additionally equipped with a device which introduces the potential pulse and pause. The minimum length of a pulse which could be applied was 1 s. The pulse consisted of a working period and a cleaning (reverse) period. Diluate was desalinated during working period, whereas fouling was expected to be removed during cleaning period.

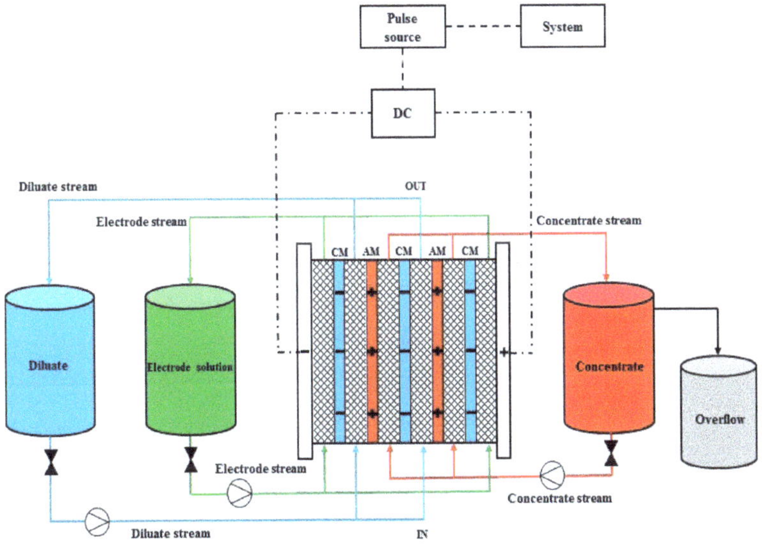

Figure 4. The configuration used for electrodialysis of whey.

Fouling could be removed due to diffusion and electric migration in electric field of reverse polarity in PER. The potential in working period was 50V (1.0V/pair). Three different regimes were applied differing in the length of working and cleaning periods and the applied potential during cleaning period. Total voltage (voltage on the whole unit) and the voltage on the stack without electrode compartments were monitored. The voltage on polarizing electrodes was adjusted so that the voltage on the stack was (50 ± 1) V. In ED of whey, diluate container was filled with the 30.0 kg of nanofiltrated whey (NFW) while 7.0 kg of tap water was poured into the concentrate chamber. The flow rate and the linear velocity through the membranes of solutions are given in Table 4. The electrodes solution was 10 g·L^{-1} $NaNO_3$ 7.0 kg. The ED was performed in batch mode. The ED regimes which were used for desalination of whey are shown in Table 5. To compare the effect of the different regimes on the membrane fouling two consecutive runs of each regime were performed without any cleaning step between. It must be mentioned that each regime was continued until the conductivity in diluate reached 1.0 mS·cm^{-1}.

Table 4. Process conditions.

Parameter	Unit	Diluate	Concentrate	Electrolyte
Utilized solution	-	Acidic whey	Tap water	Sodium nitrate
Concentration	%	20.0	-	1.0
Initial mass	kg	30.0	7.0	7.0
Solution flow rates	L/h	700	700	500
Thickness of spacers	mm	0.8	0.8	1.0
pH	-	4.4	5.5	3.0
Ending status	mS·cm^{-1}	1.1	15.0	-
Temperature	°C	15 ± 2	15 ± 2	15 ± 2

Table 5. The ED regimes which were applied for the desalination of whey.

Electrodialysis	Working Voltage	Reverse Voltage	Working Period	Reverse Period
Regime I (conventional ED)	50	Not used	Not used	Not used
Regime II	50	−50	180 s	3 s
Regime III	50	−50	30 s	5 s

3.5. Fouling Analysis

The membranes samples were submitted for the SEM, immediately after the second run of each ED regime. Images were taken on an uncoated sample with a scanning electron microscope (SEM) (Quanta FEG 450, FEI, Hillsboro, OR, USA). The potential of 5 KV was applied and the working distance was 15 mm. The hydrophobicity of the membrane was studied by measuring the contact angle using (Theta QC, Attension, Espoo, Finland). For measuring the contact angle, a drop of distilled water was placed on a surface and the contact angles between the drop and the membrane surface were measured. The contact angles ranged from 0° to 180°.

3.6. Calculations

The Degree of demineralization in ED was obtained as Equation (3) [18]:

$$Degree\ of\ demineralization\ \% = \left(1 - \frac{\kappa_{final\ of\ diluate}\left(S\ m^{-1}\right)}{\kappa_{initial\ of\ diluate}\left(S\ m^{-1}\right)}\right) \times 100 \quad (3)$$

where ($\kappa_{initial}$) and (κ_{final}) are the initial and final conductivity of the diluate.

Ash content %ODB (on dry basis) was calculated as Equation (4) where the ash content and the total solids are unit less parameters:

$$\text{Ash content \%ODB} = \frac{Ash\ content\ (\%)}{Total\ solids\ (\%)} \times 100 \quad (4)$$

The electrodialysis capacity is defined as Equation (5):

$$C_F = \frac{m_F}{N.A.t} \quad (5)$$

where (m_F) is the mass of the feed, (*A*) is the active surface of the membranes; (*N*) is the number of membrane pairs, and (*t*) is the total time of electrodialysis process.

Average ash transfer rate was determined using Equation (6):

$$J\left(kgm^{-2}h^{-1}\right) = \frac{(m_F \times W_F) - \left(m_{D,\ final} \times W_{D,final}\right)}{N \times S \times t} \quad (6)$$

where (m_F) and $(m_{D,final})$ are initial and final mass of diluate, (w_F) and $(w_{D,final})$ are initial and final ash concentration (g/kg), (N) number of membrane pairs, (S) effective membrane area (m^2) and t time (h).

Energy consumption was calculated Equation (7):

$$E = \frac{\int_{t0}^{t1} U_{avg} I dt}{m_F} \sim \frac{\sum_{t0}^{t1} U_{avg} I \Delta t}{m_F} \left(\mathrm{Wh/\ kg_F}\right) \quad (7)$$

where (U_{avg}) is average voltage on the stack (V), $\sum_{t0}^{t1} I\Delta t$ amount of transported charge (Ah) and (m_F) initial weight of diluate (kg).

4. Conclusions

Comparing the conventional ED and PER in two consecutive batch experiments without cleaning in place (CIP) between them, the electrodialysis parameters are almost the same in the first and second runs of conventional ED while in PER the parameters of the second run are evidently worse than the parameters of the first run. Since CIP was not applied, deterioration of the ED parameters such as the electrodialysis capacity and the energy consumption in PER might be attributed to the fouling on the surface of the membranes and/or to the back transfer of mass during the reversal period. Considering that the SEM analysis and the contact angle values indicate that fouling on cation exchange membranes and on concentrate side of anion exchange membranes were comparable under all regimes and fouling on diluate side of anion exchange membranes was even reduced under PED regimes, it can be concluded that the back transfer to the diluate compartment when the reverse pulse was applied is dominant. However, due to the PER the fouling (scaling) was reduced in PER regimes without significant prolongation of the ED process. As shown [15] in PEF, the efficiency of the ED was improved and the fouling/scaling was decreased. However, to achieve this the time of the ED was prolonged to around 6 × that of conventional ED. In the present work, even though the parameters of the ED efficiency were slightly decreased in PER compared to the conventional ED, the duration of the ED process was only slightly increased. The decrease in ED efficiency in PER was mainly because of the back transport of minerals in reverse pulse which occurs due to the applying a high magnitude of voltage (−50 V). The back transport of minerals in a solution containing multivalent ions was also pointed out by Tufa *et al.* [29]. Despite this, in long term application the effect of the PER can be highlighted further due to a reduction of fouling and scaling on the membranes. Therefore, research must be continued to find an optimum regime in which the back transfer does not play an important role and the fouling also decreases when the optimum pulse/pause is used.

Supplementary Materials: Supplementary materials can be found at http://www.mdpi.com/1422-0067/20/8/1918/s1.

Author Contributions: The experiments were designed and carried out by both Authors (A.M.; A.M.A.). The obtained results were discussed and the manuscript was written by both authors.

Funding: This work was supported by the program NPU I Ministry of Education Youth and Sports of the Czech Republic [project No. LO1418]; Progressive development of Membrane Innovation Centre using the infrastructure of the Membrane Innovation Centre.

Conflicts of Interest: The authors declare no conflict of interest.

Abbreviations

U/V	Cell voltage
m/kg	Mass
A/m^2	Membrane geometric surface area
N	Number of membrane pairs

I/*A*	Current
P	Apparent permselectivity
m_F/ kg	Mass of the feed in electrodialysis
N	Number of moles
t/*s*	Time
F/96500 C·mol^{-1}	Faraday constant
κ/mS·m^{-1}	Specific conductivity
j_i/mole. s^{-1}·m^{-2}	Mass flux of i through the membrane
C_F/kg·m^{-2}·h^{-1}	Electrodialysis capacity
g/cm^3	Density
E Wh·kg$_F$$^{-1}$	Electrical energy used in electrodialysis per mass of the feed
α-LA	α-lactalbumin (whey protein)
β-LG A	β-lactoglobulin A (whey protein)
β-LG B	β-lactoglobulin B (whey protein)
AEM	Anion exchange membrane
CEM	Cation exchange membrane
CMP	Casein macropeptide
CIP	Cleaning in place
°C	Temperature
DD	Degree of demineralization
ED	Electrodialysis
F	Feed (raw material)
NFW	Nanofiltered acidic whey
ODB	On dry basis in percent (Ash)
P	Product
PED	Pulsed electrodialysis
PER	Pulsed electrodialysis reversal
°SH	Acidity, Soxlet Henkel degrees (0.25N NaOH)

References

1. El-Sayed, M.M.; Chase, H.A. Trends in whey protein fractionation. *Biotechnol. Lett.* **2011**, *33*, 1501–1511. [CrossRef]
2. Mikhaylin, S.; Bazinet, L. Fouling on ion-exchange membranes: Classification, characterization and strategies of prevention and control. *Adv. Colloid Interface Sci.* **2016**, *229*, 34–56. [CrossRef]
3. Panesar, P.S.; Kennedy, J.F.; Gandhi, D.N.; Bunko, K. Bioutilisation of whey for lactic acid production. *Food Chem.* **2007**, *105*, 1–14. [CrossRef]
4. Uzdenova, A.M.; Kovalenko, A.V.; Urtenov, M.K.; Nikonenko, V.V. Effect of electroconvection during pulsed electric field electrodialysis. *Numerical experiments. Electrochem. Commun.* **2015**, *51*, 1–5. [CrossRef]
5. Lee, H.-J.; Oh, S.-J.; Moon, S.-H. Recovery of ammonium sulfate from fermentation waste by electrodialysis. *Water Res.* **2003**, *37*, 1091–1099. [CrossRef]
6. Ruiz, B.; Sistat, P.; Huguet, P.; Pourcelly, G.; Araya-Farias, M.; Bazinet, L. Application of relaxation periods during electrodialysis of a casein solution: Impact on anion-exchange membrane fouling. *J. Membr. Sci.* **2007**, *287*, 41–50. [CrossRef]
7. Asraf-Snir, M.; Gilron, J.; Oren, Y. Gypsum scaling of anion exchange membranes in electrodialysis. *J. Membr. Sci.* **2016**, *520*, 176–186. [CrossRef]
8. Strathmann, H.; Giorno, L.; Drioli, E. *Introduction to Membrane Science and Technology*; Wiley-VCH Weinheim: Weinheim, Germany, 2011; Volume 544.
9. Mulyati, S.; Takagi, R.; Fujii, A.; Ohmukai, Y.; Maruyama, T.; Matsuyama, H. Improvement of the antifouling potential of an anion exchange membrane by surface modification with a polyelectrolyte for an electrodialysis process. *J. Membr. Sci.* **2012**, *417–418*, 137–143. [CrossRef]
10. Mulyati, S.; Takagi, R.; Fujii, A.; Ohmukai, Y.; Matsuyama, H. Simultaneous improvement of the monovalent anion selectivity and antifouling properties of an anion exchange membrane in an electrodialysis process, using polyelectrolyte multilayer deposition. *J. Membr. Sci.* **2013**, *431*, 113–120. [CrossRef]

11. Vaselbehagh, M.; Karkhanechi, H.; Mulyati, S.; Takagi, R.; Matsuyama, H. Improved antifouling of anion-exchange membrane by polydopamine coating in electrodialysis process. *Desalination* **2014**, *332*, 126–133. [CrossRef]
12. Garcia-Vasquez, W.; Ghalloussi, R.; Dammak, L.; Larchet, C.; Nikonenko, V.; Grande, D. Structure and properties of heterogeneous and homogeneous ion-exchange membranes subjected to ageing in sodium hypochlorite. *J. Membr. Sci.* **2014**, *452*, 104–116. [CrossRef]
13. Lee, H.-J.; Moon, S.-H.; Tsai, S.-P. Effects of pulsed electric fields on membrane fouling in electrodialysis of NaCl solution containing humate. *Sep. Purif. Technol.* **2002**, *27*, 89–95. [CrossRef]
14. Sistat, P.; Huguet, P.; Ruiz, B.; Pourcelly, G.; Mareev, S.; Nikonenko, V. Effect of pulsed electric field on electrodialysis of a NaCl solution in sub-limiting current regime. *Electrochim. Acta* **2015**, *164*, 267–280. [CrossRef]
15. Dufton, G.; Mikhaylin, S.; Gaaloul, S.; Bazinet, L. Positive impact of pulsed electric field on lactic acid removal, demineralization and membrane scaling during acid whey electrodialysis. *Int. J. Mol. Sci.* **2019**, *20*, 797. [CrossRef]
16. Lemay, N.; Mikhaylin, S.; Bazinet, L. Voltage spike and electroconvective vortices generation during electrodialysis under pulsed electric field: Impact on demineralization process efficiency and energy consumption. *Innov. Food Sci. Emerg. Technol.* **2019**, *52*, 221–231. [CrossRef]
17. Dufton, G.; Mikhaylin, S.; Gaaloul, S.; Bazinet, L.J. How electrodialysis configuration influences acid whey deacidification and membrane scaling. *J. Dairy Sci.* **2018**, *101*, 7833–7850. [CrossRef] [PubMed]
18. Chen, G.Q.; Eschbach, F.I.; Weeks, M.; Gras, S.L.; Kentish, S.E. Removal of lactic acid from acid whey using electrodialysis. *Sep. Purif. Technol.* **2016**, *158*, 230–237. [CrossRef]
19. Bleha, M.; Tishchenko, G.; Šumberová, V.; Kůdela, V. Characteristic of the critical state of membranes in ED-desalination of milk whey. *Desalination* **1992**, *86*, 173–186. [CrossRef]
20. Cifuentes-Araya, N.; Pourcelly, G.; Bazinet, L. Impact of pulsed electric field on electrodialysis process performance and membrane fouling during consecutive demineralization of a model salt solution containing a high magnesium/calcium ratio. *J. Colloid Interface Sci.* **2011**, *361*, 79–89. [CrossRef] [PubMed]
21. Kumar, P.; Sharma, N.; Ranjan, R.; Kumar, S.; Bhat, Z.; Jeong, D.K. Perspective of membrane technology in dairy industry: A review. *Asian-Australas J. Anim. Sci.* **2013**, *26*, 1347–1358. [CrossRef]
22. Lee, H.-J.; Hong, M.-K.; Han, S.-D.; Cho, S.-H.; Moon, S.-H. Fouling of an anion exchange membrane in the electrodialysis desalination process in the presence of organic foulants. *Desalination* **2009**, *238*, 60–69. [CrossRef]
23. Lee, H.-J.; Moon, S.-H. Enhancement of electrodialysis performances using pulsing electric fields during extended period operation. *J. Colloid Interface Sci.* **2005**, *287*, 597–603. [CrossRef]
24. Merkel, A.; Ashrafi, A.M.; Ečer, J. Bipolar membrane electrodialysis assisted pH correction of milk whey. *J. Membr. Sci.* **2018**, *555*, 185–196. [CrossRef]
25. Guo, H.; Xiao, L.; Yu, S.; Yang, H.; Hu, J.; Liu, G.; Tang, Y. Analysis of anion exchange membrane fouling mechanism caused by anion polyacrylamide in electrodialysis. *Desalination* **2014**, *346*, 46–53. [CrossRef]
26. Liu, C.X.; Zhang, D.R.; He, Y.; Zhao, X.S.; Bai, R. Modification of membrane surface for anti-biofouling performance: Effect of anti-adhesion and anti-bacteria approaches. *J. Membr. Sci.* **2010**, *346*, 121–130. [CrossRef]
27. Strathmann, H. *Ion-Exchange Membrane Separation Processes*; Elsevier: Amsterdam, the Netherlands, 2004; Volume 9.
28. Sata, T. *Ion Exchange Membranes: Preparation, Characterization, Modification and Application*; Royal Society of Chemistry: London, UK, 2007.
29. Tufa, R.A.; Pawlowski, S.; Veerman, J.; Bouzek, K.; Fontananova, E.; di Profio, G.; Velizarov, S.; Goulão Crespo, J.; Nijmeijer, K.; Curcio, E. Progress and prospects in reverse electrodialysis for salinity gradient energy conversion and storage. *Appl. Energy* **2018**, *225*, 290–331. [CrossRef]

International Journal of
Molecular Sciences

MDPI

Article

Membrane Deformation and Its Effects on Flow and Mass Transfer in the Electromembrane Processes

Giuseppe Battaglia, Luigi Gurreri *, Girolama Airò Farulla, Andrea Cipollina, Antonina Pirrotta, Giorgio Micale and Michele Ciofalo

Dipartimento di Ingegneria, Università degli Studi di Palermo, viale delle Scienze Ed. 6, 90128 Palermo, Italy; giuseppe.battaglia03@unipa.it (G.B.); girolama.airofarulla@unipa.it (G.A.F.); andrea.cipollina@unipa.it (A.C.); antonina.pirrotta@unipa.it (A.P.); giorgiod.maria.micale@unipa.it (G.M.); michele.ciofalo@unipa.it (M.C.)

* Correspondence: luigi.gurreri@unipa.it; Tel.: +39-091-2386-3788

Received: 7 March 2019; Accepted: 11 April 2019; Published: 13 April 2019

Abstract: In the membrane processes, a trans-membrane pressure (TMP) may arise due to design features or operating conditions. In most applications, stacks for electrodialysis (ED) or reverse electrodialysis (RED) operate at low TMP (<0.1 bar); however, large stacks with non-parallel flow patterns and/or asymmetric configurations can exhibit higher TMP values, causing membrane deformations and changes in fluid dynamics and transport phenomena. In this work, integrated mechanical and fluid dynamics simulations were performed to investigate the TMP effects on deformation, flow and mass transfer for a profiled membrane-fluid channel system with geometrical and mechanical features and fluid velocities representative of ED/RED conditions. First, a conservatively high value of TMP was assumed, and mechanical simulations were conducted to identify the geometry with the largest pitch to height ratio still able to bear this load without exhibiting a contact between opposite membranes. The selected geometry was then investigated under expansion and compression conditions in a TMP range encompassing most practical applications. Finally, friction and mass transfer coefficients in the deformed channel were predicted by computational fluid dynamics. Significant effects of membrane deformation were observed: friction and mass transfer coefficients increased in the compressed channel, while they decreased (though to a lesser extent) in the expanded channel.

Keywords: electrodialysis; reverse electrodialysis; ion exchange membrane; profiled membrane; CFD; pressure drop; mass transfer; structural mechanics; fluid-structure interaction

---

1. Introduction

Processes based on ion exchange membranes are increasingly being adopted in industrial applications, from water treatment [1] to food processing [2] and energy harvesting [3], as both environmentally friendly and economically attractive. In electrodialysis (ED) [4], ions are driven by an imposed electric field from a dilute electrolyte solution to a concentrate one. Conversely, reverse electrodialysis (RED) [5] harvests electrical energy from the controlled mixing of two solutions at different salt concentration. ED and RED units are built by alternately stacking anion- and cation-exchange membranes, separated by net spacers or built-in profiles creating the fluid channels where the two solutions (concentrate and diluate) flow. The two membranes and the two solutions form the repeating unit, referred to as cell pair. Spacers cover part of the membrane surface, thus reducing the actual active area, and increase the electrical resistance, as they are electrically non-conductive.

Profiled membranes have recently been presented as an innovative solution to overcome net spacers drawbacks [4,6,7]. Profiled membranes simplify the stack assembly avoiding the use of spacers, and may improve the process performance. Numerical simulations [8–10] and experimental lab scale tests [4,11–15] have confirmed their potential benefits. However, the actual performance of profiled membranes stacks depends on the specific profile geometry. Simple geometries (e.g., pillar or ridges profiles) are characterised by reduced hydraulic friction, but may exhibit lesser mixing properties than spacers [10,12–14,16]; on the other hand, improved profile shapes may provide better trade-off solutions among pressure drops, mixing and Ohmic resistance, thus improving the stack performance [8,15,16].

In membrane-based processes, a trans-membrane pressure (TMP) between the different solutions flowing through a module may be a design feature or may arise for various reasons (e.g., flow arrangement or differences in geometry, flow rate or physical properties). This may lead to local deformations of membranes and membrane-bounded channels. As a result, the channel geometry (shape and average size) may be modified with respect to the nominal one, affecting fluid dynamics and transport mechanisms (of mass, heat, ions) and, thus, the process performance.

The effects of membrane/channel deformation have been studied in the context of different processes. She et al. [17] tested pressure retarded osmosis (PRO) modules at pressures up to 16 bar. Experimental performance became worse than theoretical predictions as the hydrostatic pressure increased; this difference was attributed to a more severe membrane deformation at high pressures. Later, She et al. [18] studied in detail the influence of spacer geometry on PRO efficiency under pressure loads up to 20 bar. The spacer with the largest mesh pitch gave the poorest performance in terms both of power density and of pressure drop.

Karabelas et al. [19] investigated the influence of the compressive stresses that arise in reverse osmosis (RO) spiral wound membrane modules, provided with spacers, during the assembly stage. The stresses localized at the membrane-spacer contact regions were systematically addressed as functions of spacer compaction, channel gap, membrane indentations and pressure drop. Interestingly, mild applied pressures (1–2 bar) were sufficient to cause significant effects. Correlations for the frictional losses were obtained for various applied pressures and were implemented into a process model predicting the performance of RO units.

Huang [20] simulated flow and heat transfer in deformed channels for liquid-to-air membrane energy exchanger (LAMEE) units. Membrane deformation was not actually computed, and the deformed membrane was modelled as a spherical surface. As membrane deformation increased, the friction coefficient was found to increase in the compressed (air) channel and to decrease in the expanded (liquid) channel. Heat transfer was affected by deformation in a complex way.

The influence of channel deformation on the performance of proton exchange membrane fuel cells (PEMFC) was assessed in several studies following similar approaches. Shi and Wang [21] predicted the compression of the porous gas diffusion layer due to the clamping (assembly) force, and simulated fluid dynamics, mass transport and electrochemical phenomena in the deformed geometries. The authors considered a serpentine channel and found that the assembly compression of the units enhanced pressure drop in the fluid channels, and that the process performance was particularly affected by deformation at high current densities. Zhou et al. [22] simulated a unit with a single straight channel including the membrane. As expected, most of the deformation was found to occur in the porous gas diffusion layer due to its lower mechanical stiffness. The spatial distributions of porosity and permeability were computed and the effects of assembly pressure, gas diffusion layer thickness and membrane features were assessed.

Hereijgers et al. [23] measured membrane deflection and mass transfer coefficients in membrane microcontactors using round and diamond-shaped pillar spacers of different pitch. They found that trans-membrane pressure exhibited a minimum as the spacer pitch was made to vary, and that membrane deflection had a positive or negative impact on mass transfer depending on the diffusion coefficients in the two immiscible phases.

Time-dependent membrane deformation has recently been considered as a possible means to improve process performance. Moreno et al. [24] introduced the concept of "breathing cell" for reverse electrodialysis systems. In the breathing cell, the channels thickness changes dynamically due to the intermittent (5–15 cycles per minute) closure of an outlet valve in the concentrate channels. As a result, the Ohmic resistance of the diluate compartment (which is the predominant one) decreases.

Some effects on concentration polarization are also expected. This cyclic operation was shown to yield higher net power densities in a range of flow rates.

Some ED/RED practical applications are poorly affected by these issues (TMP ≈ 0). However, in prototype and industrial size stacks with non-parallel flow layouts (cross flow, counter flow) and/or with asymmetric channels (different geometries, fluid properties, flow rates), where the pressure distribution in the two compartments is different, appreciable values of TMP may arise. In particular, when some factors enhancing pressure drop are present, TMP values amounting to some tenths of a bar can be exhibited (higher TMP levels can cause severe risks of leakages [25–27]).

For example, in the cross-flow RED prototype units (44 × 44 cm^2) installed within the REAPower project [28], pressure drops from ~0.2 to ~0.9 bar were measured at flow velocities up to 1 cm/s [29]. Despite some of the pressure drop can be supposed to occur in the manifolds, a significant part of it is expected to occur in the channels, thus causing the onset of non-negligible TMP values. Moreover, the compartments were asymmetric, because the viscosity of the concentrated solution (brine) was almost twice that of the dilute feed, thus causing an unbalanced pressure distribution in the two solutions. Larger *TMP* values (up to ~1.5 bar) were measured by Hong et al. [27] in a cross-flow RED stack (35.5 × 35.5 cm^2) fed with inlet velocities up to ~5 cm/s, which provided a significantly lower electrical power (less than half) compared to an equivalent parallel-flow stack. Although the authors attributed this decline in performance to issues of internal leakage, an important effect of deformation can be supposed.

ED units operate with fluid velocities higher than those typical of RED (in order to increase the limiting current density) and, despite the usually higher channel thickness, exhibit large pressure drops [1]. For example, Wright et al. [30] performed ED tests in a bench-scale unit and in a commercial-scale unit with parallel flow, measuring pressure drops up to ~0.65 bar and ~1.30 bar, respectively, at fluid velocities up to ~9 cm/s. If such operating conditions were adopted in non-parallel flow arrangements, they would lead to significant levels of *TMP*.

Recent studies showed that asymmetric channels are optimal for RED applications [31,32]. However, they can be affected by *TMP*-related issues. For example, in ref [32] it was shown that for the couple of NaCl solutions 15–500 mol/m^3 fed with parallel flow in a stack 50 cm long, the optimum thickness and fluid velocity are ~400 μm and ~1.4 cm/s for the concentrate and ~217 μm and ~2.6 cm/s for the diluate. The pressure drop predicted by Computational Fluid Dynamics (CFD) correlations is 0.07 bar for the concentrate and 0.46 bar for the diluate, thus giving a maximum TMP located at the inlet equal to ~0.39 bar.

It must also be added that ion exchange membranes may have very different mechanical features. The Young modulus (E) may vary within a broad range from 10 MPa to 1 GPa [33–43] or even to higher values in some cases [44,45], but decreases with ageing due to membrane usage [34–36,44]. Moreover, the new generation membranes are manufactured with low thickness, e.g., from 80 to 250 μm [46,47]; even lower values can be found among commercial membranes and experimental membranes prepared in laboratory [48]. A theoretical study [49] has recently found optimal thicknesses of 15–20 and 50–70 μm for ED and RED applications, respectively. Therefore, it is quite common that ion exchange membranes exhibit a low stiffness, due to the combined effects of a low E and a low thickness. This feature makes the membranes susceptible to large deformations in stacks with a non-negligible TMP, depending also on the spacer features.

In particular, a fluid-membrane mechanical interaction will be triggered, which will find an equilibrium state characterized by some distribution of pressure, geometry, flow rate, hydraulic friction, mass transfer coefficient, current density, Ohmic and non-Ohmic resistances in both compartments. Compared to the nominal conditions, the values of any of the above quantities under deformed conditions may be: (i) either higher or lower in the whole channel (e.g., in asymmetric configurations); (ii) higher in some parts of the channel and lower in other ones (e.g., in non-parallel flow arrangements). In both cases, these deviations from the undeformed conditions may impair the process performance due to the lack of compensation of effects between compressed zones and expanded zones (in the same or in different channels). For instance, an increase in the thickness of the diluate (which often provides the predominant resistance), in the whole channel or in a part of it, especially where the solution is less conductive, causes an increase in the average Ohmic resistance. Imbalances may also affect hydraulic friction, increasing the overall pressure drop. An increment in non-Ohmic resistance is another well-known detrimental effect of uneven flow rate distributions [13].

All the aspects of practical interest examined in this section have provided the motivation to the present work. In particular, this paper goes inside the unexplored field of the TMP effects, taking a first step concerning mechanical response (deformation), flow and mass transfer characteristics at the local scale of a periodic unit. For this purpose, simulation tools implementing well-established and validated physical models and numerical methods were developed. Profiled membranes of the Overlapped Crossed Filaments (OCF) type were simulated. They are made by an array of semi-cylinders on both membrane sides, placed at 90° each other, as shown in Figure 1.

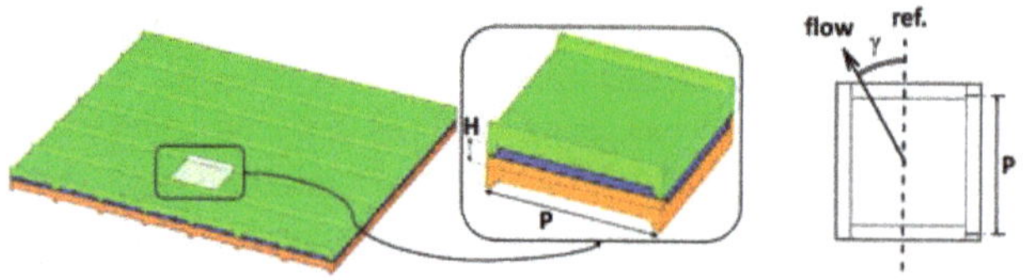

Figure 1. Profiled membranes of the Overlapped Crossed Filaments (OCF) type. The repetitive periodic unit of a cell pair is shown, enlarged, in the inset. The geometric parameters *H* (channel thickness), *P* (pitch) and γ (flow attack angle) are indicated.

2. Results and Discussion

2.1. Mechanical Results

2.1.1. Influence of Pitch to Height Ratio (*P*/*H*) and Limiting Values

Computational results for the deformation of cells with different pitch (*P*, distance between two profiles on the same membrane side) to channel height (*H*, distance between the two undeformed membranes) ratios under the conservative value of TMP = + 0.8 bar (the "+" sign refers to compression) are presented in Figure 2. Three values of *P*/*H* are considered (7, 8 and 9). The first contact between the two membranes approximately occurs for *P*/*H* = 9 and is located at the centres of the side ridges (note that an inter-membrane clearance of ~60 μm is still preserved at the centre of the periodic unit). Therefore, the value *P*/*H* = 8 was chosen as the largest admissible one.

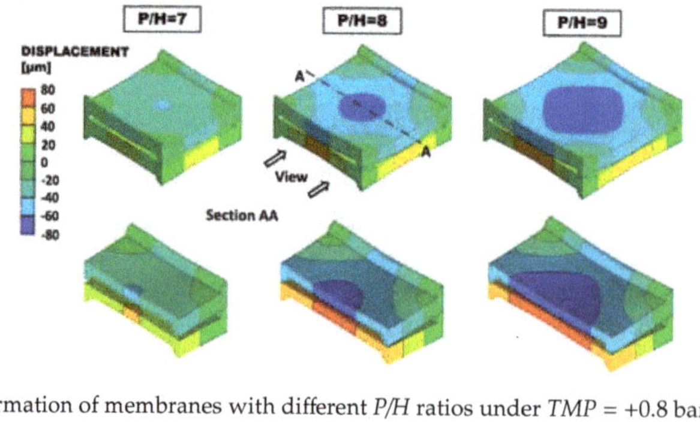

Figure 2. Deformation of membranes with different *P*/*H* ratios under *TMP* = +0.8 bar. The quantity shown is the displacement in the direction orthogonal to the undeformed membranes (*y*). Top: external view; bottom: view after sectioning by a mid-plane A-A.

2.1.2. Membrane and Channel Deformation for the Selected Geometry ($P/H = 8$)

The geometry characterized by the maximum admissible P/H ratio (8) was subjected to TMP varying in 0.1 bar steps from −0.4 bar (expansion) to +0.4 bar (compression), and the corresponding deformation was computed. Figure 3 shows the deformed configuration for $P/H = 8$ under TMP = ±0.4 bar. The insets on the right show the deformed fluid volumes. The maximum relative variation of the clearance occurs at the centres of the ridges. Here, the distance between the two opposite membranes (thickness of the fluid passage), which is $H/2 = 100$ µm in the undeformed configuration, decreases to ~53.7 µm in the compression case and increases to ~148.4 µm in the expansion case. The distance between opposite membranes at the centre of the domain, which is $H = 200$ µm in the undeformed configuration, decreases to ~130 µm under compression or increases to ~272 µm in expansion (i.e., the maximum deflection at the centre is ~ ±70 µm).

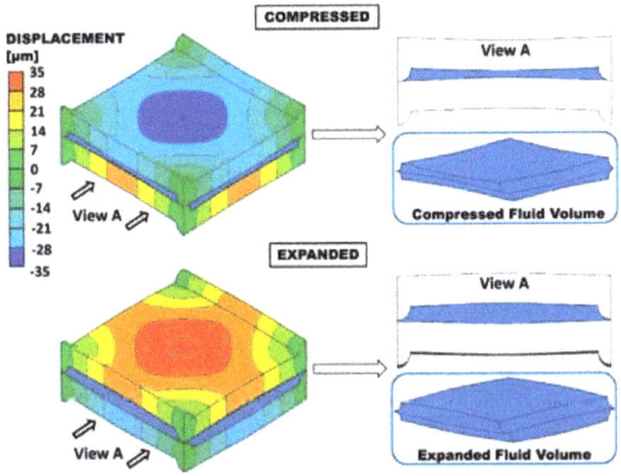

Figure 3. Deformation of membranes with $P/H = 8$ for the compressed and the expanded cases at TMP = ±0.4 bar. The quantity shown is the displacement in the direction orthogonal to the undeformed membranes (y). The corresponding deformed fluid volume is shown in the insets.

Figure 4 provides information concerning the TMP effects, in the whole range studied, on the fluid volume. The volume follows a linear trend and exhibits an almost perfect symmetry between compression and expansion; the volume changes by ±25% for TMP = ±0.4 bar.

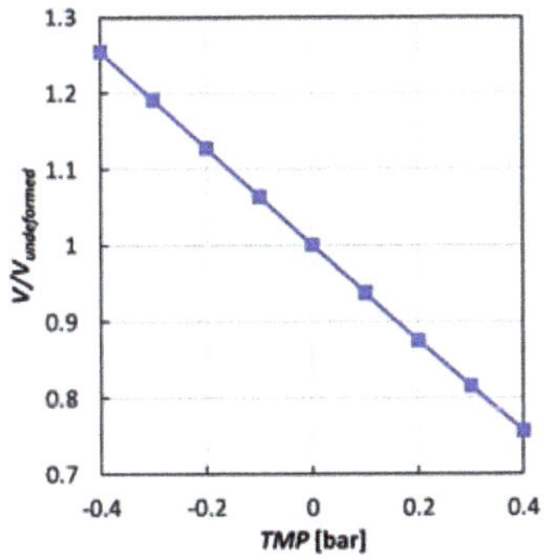

Figure 4. Fluid volume (normalized by the undeformed volume) as a function of trans-membrane pressure for $P/H = 8$.

2.2. CFD Results for P/H = 8

2.2.1. Undeformed Configuration

Figure 5 shows 3-D streamlines and maps of the polarization coefficient $\theta = c_b/c_w$ (bulk to wall concentration ratio) in the undeformed configuration characterized by $P/H = 8$ for a friction velocity Reynolds number $Re_\tau = 5.2$ (bulk Reynolds number $Re \approx 17.6$, approach velocity ~4 cm/s) and all three values of the flow attack angle (angle formed by the flow direction with the membrane ridges belonging to the upper wall) investigated ($\gamma = 0°$, 45° and 90°). The flow direction is indicated by arrows. Definitions of approach velocity and friction velocity Reynolds number are provided in Section 3.3.2, Equations (11) and (13).

The streamlines show that the flow is regular and parallel at this low value of Re. The corresponding plots for $\gamma = 0°$ and $\gamma = 90°$ are identical apart from a 90° rotation and a top-bottom reflection. For $\gamma = 45°$, streamlines were shown in two colours according to the face from which they enter the unit cell; the graph shows that there is no mixing between the two inlet streams.

The maps of $\theta = c_b/c_w$ in the bottom row show that the case $\gamma = 45°$ provides a more uniform distribution of the wall salt concentration, while the other two cases exhibit a very strong spanwise non-uniformity; the concentration is lower in the central region of the wall, where it becomes less than the bulk value despite the net overall salt flux being into the channel, and larger in the lateral regions of the channel walls, where low fluid velocities (stagnation zones) occur. Please note that the distribution of θ on the upper wall for $\gamma = 0°$, once rotated by 90°, would become the corresponding lower wall distribution for $\gamma = 90°$ and *vice versa*. Also, remember that the values of the polarization coefficient depend on the flux imposed at the boundary and on the bulk concentration considered. Therefore, for example, much lower values would be obtained for dilute solutions.

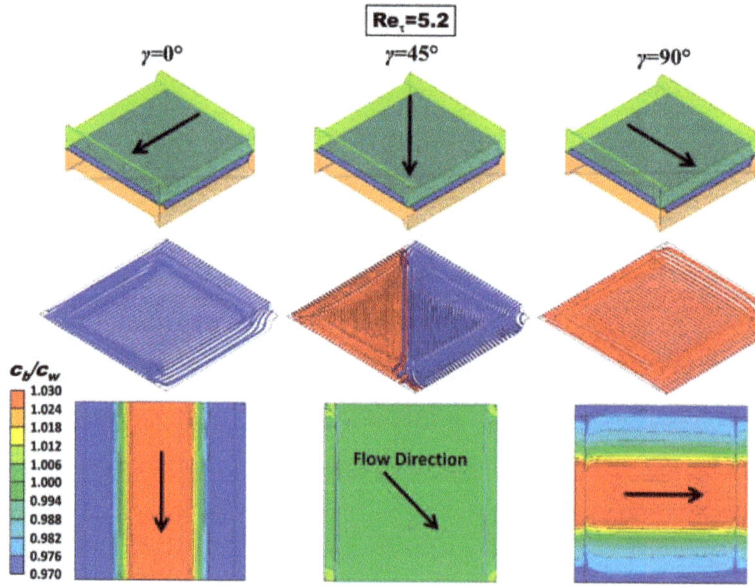

Figure 5. CFD results for the undeformed configuration with $P/H = 8$ at $Re_\tau = 5.2$ (approach velocity ~4 cm/s). Top row: sketches illustrating the flow direction; middle row: 3-D streamlines; bottom row: maps of the concentration polarization coefficient $\theta = c_b/c_w$ on the upper wall. $c_b = 500$ mol/m^3, flux corresponding to a current density of 50 A/m^2 entering the fluid domain (dilute channel of RED or concentrate channel of ED).

2.2.2. Deformed Configurations

For the sake of brevity, the influence of deformation on flow and mass transfer in OCF membranes with $P/H = 8$ is illustrated here in Figure 6 only for a friction velocity Reynolds number $Re_\tau = 5.2$ (corresponding to bulk Reynolds numbers between ~7 and ~35, approach velocity ~1.6 and ~7.8 cm/s, depending on the load conditions) and a flow attack angle $\gamma = 90°$ (flow orthogonal to the profile ridges adjacent to the upper wall of the fluid channel), as evidenced in the inset.

Three configurations are examined: compressed by a trans-membrane pressure TMP = +0.4 bar (left column), undeformed (middle column), and expanded by a trans-membrane pressure TMP = −0.4 bar (right column). The top row reports contour plots of the velocity component along the main flow direction in the central cross section of the channel, while the middle and bottom rows report contour plots of the polarization coefficient $\theta = c_b/c_w$ on both the upper and the lower wall of the fluid-filled channel, as clarified by the sketches in the rightmost part of the figure. The corresponding values of the F ratio (friction coefficient normalized by that for laminar flow in an undeformed void plane channel of indefinite width, 96/Re) and of the Sherwood number are also reported.

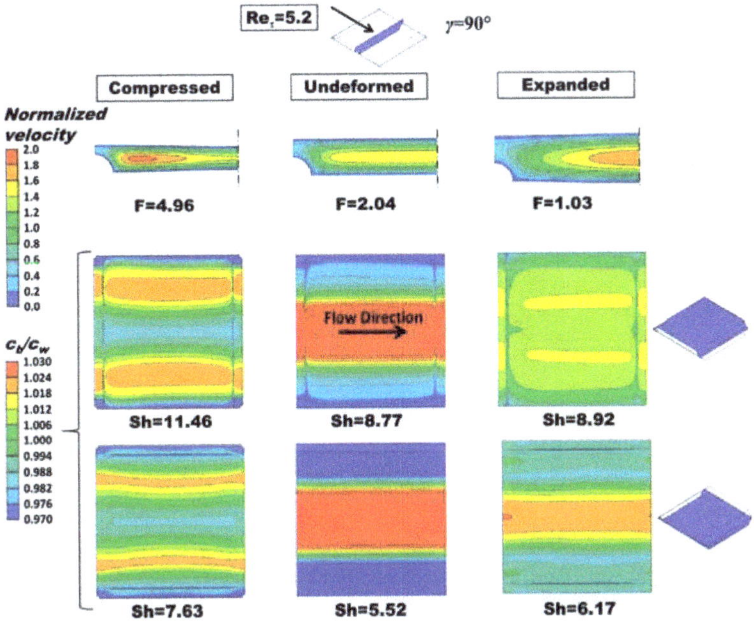

Figure 6. Influence of deformation on flow and mass transfer for $P/H = 8$, $\gamma = 90°$. Left column: compressed (TMP = +0.4 bar); middle column: undeformed; right column: expanded (TMP = −0.4 bar). Top row: distribution of the streamwise velocity component in the central cross section of the channel (for symmetry reasons, only half map is shown); middle and bottom rows: distribution of the polarization coefficient on the upper and lower walls (see sketches on the right). $c_b = 500$ mol/m^3, flux corresponding to a current density of 50 A/m^2 entering the fluid domain (dilute channel of RED or a concentrate channel of ED). F ratio and Sherwood number are also reported.

In the deformed channels, the normalized axial velocity component exhibits larger maximum values, which are located closer to the longitudinal ridges in the case of the compressed channel.

The F ratio increases from ~2.04 to ~4.96 with compression and decreases from ~2.04 to ~1.03 with expansion. In regard to the Sherwood numbers that on the upper wall (flow orthogonal to the profile ridges) increases significantly with compression (from 8.77 to 11.46, i.e., by ~30%) and increases, but negligibly, also with expansion (from 8.77 to 8.92, i.e., by ~2%). That on the lower wall (flow parallel to the profile ridges) increases significantly with compression (from 5.52 to 7.63, i.e., by ~38%) and increases less, but still appreciably, also with expansion (from 5.52 to 6.17, i.e., by ~12%).

For greater readability of the results, the dimensioned values of the approach velocity U and of the mass transfer coefficient k for the three conditions in Figure 6 (TMP = 0 or ±0.4 bar) are summarized in Table 1. In all three cases the friction velocity Reynolds number Re_τ is ~5.2, corresponding to an inlet-outlet pressure drop in a unit cell (1.6 mm in side) of ~34.36 Pa.

Table 1. Approach velocity and mass transfer coefficients for the load conditions in Figure 6.

Quantity	Compressed +0.4 Bar	Undeformed	Expanded −0.4 Bar
[cm/s]	~1.6	~4	~7.8
$\langle k \rangle$, upper wall [m/s]	~3.72 × 10^{-5}	~2.84 × 10^{-5}	~2.89 × 10^{-5}
$\langle k \rangle$, lower wall [m/s]	~2.47 × 10^{-5}	~1.78 × 10^{-5}	~2.00 × 10^{-5}

Distributions of the polarization coefficient $\theta = c_b/c_w$ are deeply affected by deformation. In the compressed configuration, both on the upper and on the lower wall the region of high θ (i.e., low concentration) observed in the undeformed case splits into two smaller regions, symmetrically located about the midline parallel to the flow direction, whereas the central region of the wall close to this midline exhibits low values of θ (i.e., high values of concentration). In the expanded configuration, the concentration distribution on the lower wall remains similar to that observed in the undeformed case, with a single large central strip where $c_w < c_b$, which is consistent with the fact that the longitudinal velocity exhibits a single central maximum as in the undeformed case (see top row). The θ distribution on the upper wall becomes flat, with two shallow θ maxima (i.e., c_w minima) symmetrically located about the longitudinal midline.

By comparing the polarization coefficient maps and the velocity maps in Figure 6, it can be observed that under the present assumption of mass flux entering the channel, higher concentration levels on the wall correspond to stagnation regions, whereas low values of concentration occur in regions of high streamwise velocity as an effect of axial advection.

2.2.3. Global Parameters

Among the performance parameters of greatest interest which can be affected by deformation, we selected the friction coefficient and the Sherwood number [16,50].

Figure 7 reports the normalized Darcy friction coefficient, i.e., the F ratio, as a function of Re for $P/H = 8$ at different values of *TMP*. Graph (a) is for flow attack angles γ of 0° or 90° (equivalent in regard to friction), while graph (b) is for γ = 45°. Please note that the results of each series of simulations performed at a given Re_τ appear as an inclined row of symbols since they correspond to different values of Re.

For any γ and applied TMP, F is flat up to Re ≈ 10, indicating that inertial effects are negligible (self-similar flow). A significant departure from the void channel behaviour is observed only for Re»10. The influence of TMP is to enhance friction under compression and to reduce it under expansion. This effect is expected because, for any given Re, in a compressed channel the cross section is reduced, local velocities increase and thus pressure drops are higher (the opposite occurs in an expanded channel). For the same absolute value of TMP, the influence of compression is slightly larger than that of expansion: TMP = +0.4 bar leads to an increase in F by a factor of ~2.5, while TMP = −0.4 bar leads only to a halving of F. The influence of the angle γ is negligible (graphs (a) and (b) are practically identical), indicating a substantial isotropy of the profiled membrane lattice in terms of hydraulic friction. This behaviour is typical in the case of low Reynolds numbers, as largely documented in the literature [10,16,50,51].

Figure 8 reports the Sherwood number on the upper channel wall, for $P/H = 8$ as a function of the Reynolds number and for different values of the trans-membrane pressure. Graphs (a), (b) and (c) are for flow attack angles γ of 0°, 45° and 90°, respectively. Please note that the cases γ = 0° and 90° are equivalent in regard to friction but not in regard to mass transfer on a specified wall. However, for symmetry reasons, the Sherwood number on the lower wall of the channel at a given γ is identical to that on the upper wall at the complementary flow attack angle $90° - \gamma$ (also the distributions of wall quantities such as concentration and mass transfer coefficient would be the same, apart from rotations and reflections). Therefore, values of Sh for the lower wall were not separately reported.

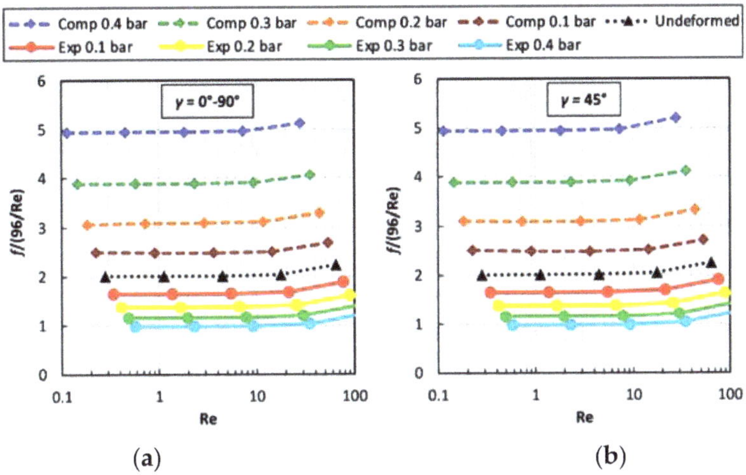

Figure 7. Normalized Darcy friction coefficient (*F* ratio) as a function of Re for *P/H* = 8, different values of the trans-membrane pressure TMP and two values of the flow attack angle γ. (**a**) $\gamma = 0°$ or $90°$; (**b**) $\gamma = 45°$.

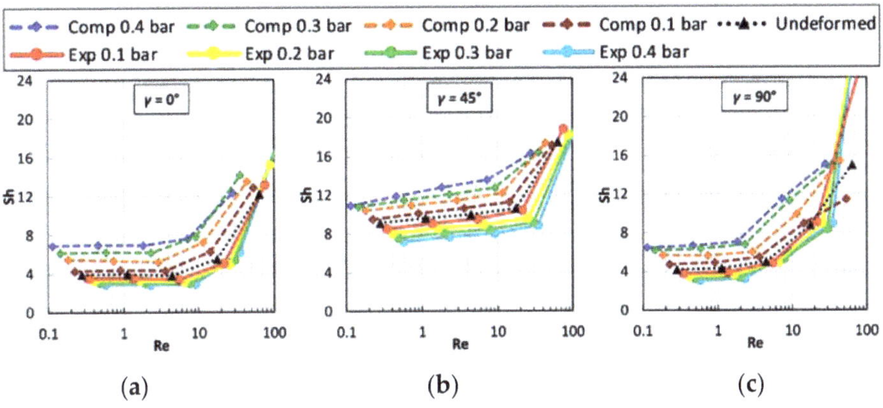

Figure 8. Sherwood number on the upper wall as a function of the Reynolds number for *P/H* = 8 and different values of the trans-membrane pressure and of the flow attack angles. (**a**) $\gamma = 0°$; (**b**) $\gamma = 45°$; (**c**) $\gamma = 90°$.

When $\gamma = 0°$, Figure 8(a), for any applied *TMP* the Sherwood number on the upper wall changes little with Re up to ~10, while for $\gamma = 90°$, Figure 8(c), the departure from this flat behaviour occurs earlier (Re ≈ 2). For $\gamma = 0°$ or $90°$, the Sherwood number at low Reynolds numbers ranges between ~3 and ~7 and thus is less than the theoretical value for a void plane channel of indefinite width (~8.24 under uniform mass flux conditions [52]). This indicates that in this Reynolds number range, the "shadow" effects of the profiles hinder mass transfer. The behaviour of Sh is different for a flow attack angle of 45°, Figure 8b, for which, even at very low Reynolds numbers, Sh increases with Re and is larger than in a void channel for all compressed configurations, while it becomes slightly lower only for the expanded ones. Under all conditions, Sh increases rapidly as Re exceeds some critical value and, at Re ≈ 30–100, it becomes much larger than in a void channel. The most peculiar behaviour is exhibited by the upper wall Sherwood number in the expanded cases and $\gamma = 90°$, which jumps to very high values (up to ~40 for TMP = −0.4 bar) as Re exceeds ~50 due to the increasing importance of flow recirculation.

The influence of trans-membrane pressure on Sh is more complex than that on *F*. On the whole, compression enhances mass transfer and expansion reduces it; the influence of channel deformation on mass transfer is less marked than on friction. Some anomalous behaviour of Sh is observed only in the cases characterized by $\gamma = 90°$ and Re > ~50, in which the highest values of Sh are obtained for the largest expansion. Under all deformation conditions, the flow orientation $\gamma = 45°$ yields the highest values of Sh. This is in contrast with the behaviour of the friction coefficient, see Figure 7, which is only minimally affected by the flow attack angle.

3. Materials and Methods

3.1. Simulation Strategy

In the present study, membrane deformation was computed by the Finite Element (FE) Ansys-Mechanical software (ANSYS, Inc., Canonsburg, PA, USA, version 18.1), while fluid dynamics and mass transfer in the deformed channels were computationally investigated by the Finite Volume (FV) Ansys-CFX software (ANSYS, Inc., Canonsburg, PA, USA, version 18.1). To reduce the computational effort, only one fluid channel at a time was simulated and the local pressure difference between adjacent channels was applied as a boundary condition. Simulations were conducted for a periodic unit as shown in Figure 1, exploiting the periodic nature of the profiled membrane geometry by the approach discussed in detail in previous papers [51] and summarized in Section 3.3.1. The analysis was carried out in three steps:

1. First, the influence of the pitch-to-height ratio (*P*/*H*) was addressed by mechanical simulations. A TMP of 0.8 bar was applied, and the geometry with the largest value of *P*/*H* still able to withstand this load without collapsing (i.e., without exhibiting a contact between opposite membranes) was identified. The figure of 0.8 bar was conservatively chosen as a value comfortably larger than the highest TMP actually expected in real RED/ED applications. The search for the largest admissible *P*/*H* was motivated by the fact that small values of *P*/*H* are associated with large pressure drops: many studies [13,16,30,53–55] have highlighted the importance of reducing pressure drop and thus mechanical power losses in the channels, especially in RED applications. It is true that the increase of *P*/*H* may also cause a reduction in mass transfer coefficients, but its effect on stack performance is usually less important.
2. The geometry thus identified was then investigated under expansion and compression conditions corresponding to TMP varying from −0.4 to +0.4 bar. As discussed in the Introduction, this range encompasses most of the conditions that are likely to occur in actual ED/RED applications. For each load condition, the deformed configuration was computed by mechanical simulations.
3. Finally, for each deformed configuration, fluid flow and mass transfer in the expanded or compressed channel were numerically simulated by CFD; in particular, friction coefficients and Sherwood numbers were computed as functions of the Reynolds number.

Steps 2–3 can be schematically represented by the flow chart in Figure 9. It is worth noting that the present modelling approach can also be applied to the simulation of membrane/spacer systems, provided suitable changes and adjustments are made in some specific aspects (e.g., concerning the boundary conditions).

Note also that in the present study, the coupling between fluid and structure is only one-way (open-loop model): a given TMP causes channel deformation, which, in its turn, causes changes in the flow field and thus in pressure losses and mass transfer coefficients. Changes in TMP induced by the flow field, which would close the fluid-structure interaction loop, are not considered at this stage: they cannot be predicted at periodic unit level but only by considering a larger scale (stack level), where they are mainly caused by inlet-to-outlet pressure gradients.

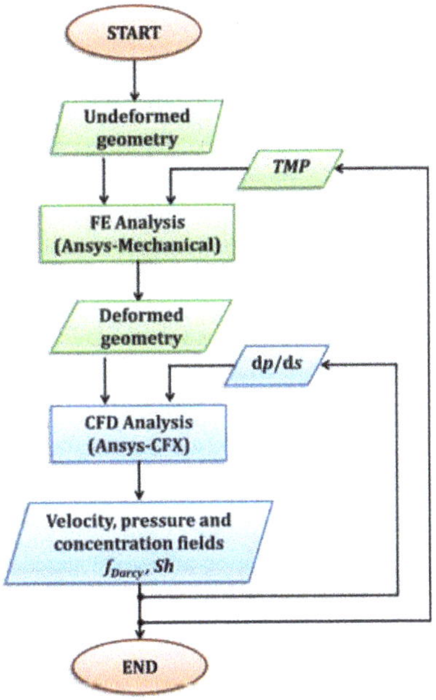

Figure 9. Flow chart of the computational process for any chosen geometry.

3.2. The Mechanical Problem

The mechanical properties of a membrane depend on manufacturing method, nature of co-polymers, cross-linking degree, ageing, etc. In the present study, cation and anion exchange membranes were assumed to exhibit the same mechanical properties and, for the sake of simplicity, were treated as linearly elastic, homogeneous and isotropic media. Representative values of 150 MPa for the Young modulus (E) and 0.4 for the Poisson ratio (ν) were chosen for the membranes among literature data broadly ranging from 10 MPa to 1 GPa for E (see the Introduction) and from 0.25 to 0.4 for ν [39,41].

The linearly elastic hypothesis is quite reasonable for ED/RED membranes within the mild load conditions considered in this study: for TMP = ±0.4 bar, the maximum computed von Mises stress is ~2 MPa, which is below the limit stress for linearly elastic behaviour generally exhibited by ion exchange membranes [39–43], including those tested in our experiments (see Figure S1 in the Supplementary Materials).

The homogeneity assumption is based on the membrane structure and on the preparation technique adopted, as described by the manufacturer (FujiFilm Europe, Tilburg, The Netherlands).

In regard to the isotropy assumption, tests conducted by the authors evidenced a maximum difference of ~15% between the values of the Young modulus along the MD (machine direction) and CD (cross direction) axes, both for cation and anion exchange membranes. In view of this modest degree of anisotropy, we felt that taking this feature into account would have unnecessarily complicated the computations and multiplied the number of representative test cases without changing the results to any significant extent.

3.2.1. Governing Equations

Equilibrium, compatibility and constitutive equations were numerically solved in order to find the deformed configuration of the body [56]. They are quite standard but, for the sake of completeness, are reported here below in Cartesian tensor notation (no summation over repeated indexes).

Equilibrium:

$$\frac{\partial \sigma_i}{\partial x_i} + \frac{\partial \tau_{ij}}{\partial x_j} + \frac{\partial \tau_{ik}}{\partial x_k} + F_i = 0 \tag{1}$$

Compatibility:

$$\frac{\partial^2 \varepsilon_i}{\partial x_j^2} + \frac{\partial^2 \varepsilon_j}{\partial x_i^2} = \frac{\partial^2 \gamma_{ij}}{\partial x_i \partial x_j}; \quad 2\frac{\partial^2 \varepsilon_k}{\partial x_i \partial x_j} = \frac{\partial}{\partial x_k}\left(\frac{\partial \gamma_{jk}}{\partial x_i} + \frac{\partial \gamma_{ik}}{\partial x_j} - \frac{\partial \gamma_{ij}}{\partial x_k}\right) \tag{2}$$

Constitutive:

$$\varepsilon_i = \frac{1}{E}\left[\sigma_i - \nu\left(\sigma_j + \sigma_k\right)\right]; \quad \gamma_{ij} = \frac{1}{G}\tau_{ij} \tag{3}$$

where σ_i are normal stresses, τ_{ij} are shear stresses, F_i are body forces, ε_i are normal strains, γ_{ij} are shear strains, and $G = E/[2 \times (1 + \nu)]$ (shear modulus). The small deformation approximation was not used.

3.2.2. Computational Domain and Boundary Conditions

The computational domain for the mechanical simulations is the periodic unit shown in the central inset of Figure 1. Please note that the planform of this unit is a square. The undeformed channel thickness H was assumed to be 0.2 mm and the undeformed membrane thickness 0.12 mm; these values are representative of advanced membrane-channel configurations currently being considered for ED and RED applications [1,46,47,57]. Profiles of adjacent membranes were assumed to be aligned on top of one another; in practice, this arrangement may not be precisely achieved since, in operation, shifts would be likely to occur.

For the sake of clarity, geometrical and mechanical quantities are summarized in Table 2. The computational domain is also shown, enlarged, in Figure 10, where the mechanical boundary conditions are evidenced:

1. Each of the four segments representing the external vertical edges of the domain (1) was clamped, i.e., zero displacement and rotation were imposed to all points belonging to it.
2. Each of the four side faces of the domain (2) was imposed zero displacement in the direction normal to itself, so that a single computational domain is representative of a periodic array of repetitive units.
3. The trans-membrane pressure TMP (relative to that of the internal fluid channel) was applied to the whole outer surface of the domain (3). Please note that TMP > 0 for compression conditions, while TMP < 0 for expansion conditions.

Table 2. Geometrical and mechanical quantities.

Quantity	Value	Units
Membrane Young modulus, E	150	MPa
Membrane Poisson ratio, ν	0.4	-
Membrane thickness	120	μm
Channel thickness, H	200	μm
Pitch-to-height ratio, P/H	7–9	-
Angle between filaments	90	deg

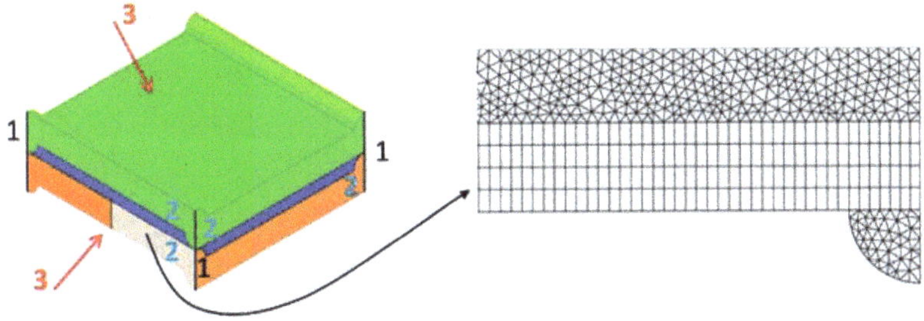

Figure 10. Computational domain. Numbers 1–3 indicate the mechanical boundary conditions (see text). A detail of the finite element mesh is shown on the right.

3.2.3. FE mesh for Mechanical Simulations

A detail of the finite element mesh is shown in Figure 10. The use of a hybrid (hexahedral-tetrahedral) grid was necessary. Grid dependence was preliminary assessed; Table 3 reports the maximum displacement at the outer surface of the domain computed for TMP = 0.8 bar and $P/H = 8$ with increasingly fine meshes.

Table 3. Grid dependence results (TMP = +0.8 bar).

FE Mechanical Mesh	No. Elements ($P/H = 8$)	Maximum Displacement at Membrane Surface [μm]
OCF-I	200×10^3	67.04
OCF-II	500×10^3	67.38
OCF-III	1 million	67.53

On the basis of these results, computational meshes of 500×10^3 elements (OCF-II) were used in all following simulations as a compromise between accuracy and computational effort.

3.2.4. Mechanical Model Validation

The FE model was first validated by comparison with existing experimental bulge test results, obtained in the context of an independent study for 10×10 cm^2 square samples of flat ion exchange membranes. Details of the experiments are reported in Section 3 of the Supplementary Material (Figure S2). Figure 11a compares the predicted and experimental maximum displacements (placed at the central point of the membrane) as functions of the trans-membrane pressure. Error bars are reported for the experimental data. A good agreement can be observed, with a maximum relative discrepancy of a few percent.

A further validation of the FE model was performed by comparing numerical predictions with an analytical solution of structural mechanics for a two-dimensional domain [58]. A square body loaded with a uniform pressure and with all the edges clamped was considered. Since the membrane deflection overcomes the "small deflection" range, a suitable analytical solution was used for the comparison. In particular, Figure 11b reports the maximum deflection as a function of the trans-membrane pressure. The broken line is the first-order approximated analytical solution reported by Iyengar and Naqvi [58], the solid line is the present FE numerical solution. In this case, the square membrane is 2 mm wide and 0.120 mm thick and Young's modulus is 150 MPa. The Poisson ratio is 0.316, as the analytical solution proposed in reference [58] was specifically obtained for this value. Figure 11b shows that our numerical simulations are in good agreement with the approximate theoretical solution, the discrepancy increasing with TMP and being only ~3% at 0.8 bar.

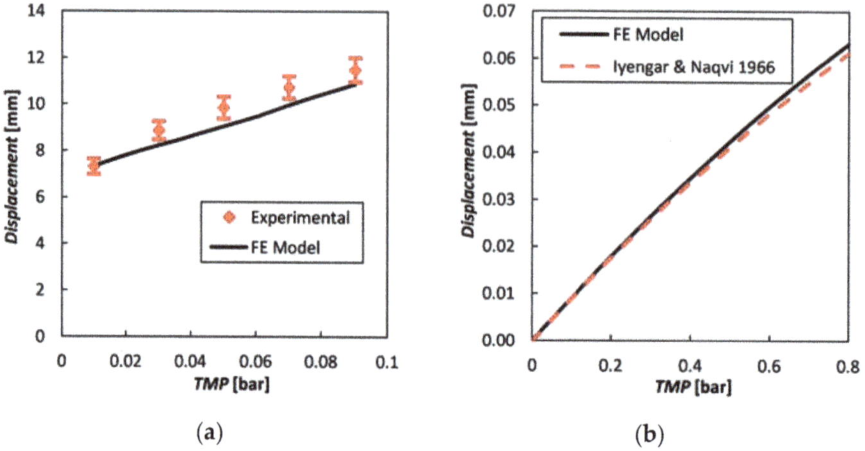

Figure 11. Maximum displacement for a square, edge-clamped membrane as a function of the trans-membrane pressure. Comparison of FE predictions (solid line) with (**a**) experimental results (symbols) of bulge tests on a 10 × 10 cm^2 sample and (**b**) the first-order analytical solution by Iyengar and Naqvi [58] for a 2 × 2 mm^2 membrane (dashed line).

3.3. The Fluid Dynamics and Mass Transfer Problem

3.3.1. Computational Approach

The fluid channel was simulated by CFD in the undeformed, compressed and expanded configurations in order to evaluate its pressure drop and mass transport performance. The ionic transport was simulated assuming the local electroneutrality condition in the whole fluid domain. Under this hypothesis, from the Nernst-Planck equations and the mass balances of the two ions of a binary electrolyte, a convective-diffusive transport equation can be derived [51,59–61]. This simplifies the calculations, requiring only the need for a choice concerning the boundary condition at the membrane-solution interface (uniform concentration, uniform flux, or mixed condition); however, the influence of the boundary conditions on the mass transfer coefficient is small [8,52]. Please note that the potential is eliminated from the transport equation, and therefore the electric field and associated phenomena (e.g., Ohmic resistance) are not calculated by this simulation approach. Moreover, electroneutrality conditions are assumed and the electric double layer at the membrane-solution interface is not simulated, so that special conditions in which an extended space charge region occurs (e.g., electroconvection under overlimiting conditions) are not taken into consideration.

The governing equations were solved by the finite volume code Ansys-CFX®. From the numerical solution of these equations, velocity, pressure and electrolyte concentration fields are obtained. Raw results are then elaborated in order to calculate friction factor and Sherwood number. This simulation method is particularly suitable for the implementation of integrated multi-scale process simulators [1], where basic data produced by CFD are merged with higher-scale simulation tools [8,16,62].

3.3.2. Governing Equations and Definitions

For the fluid dynamics/mass transfer problem the assumptions are those of steady laminar flow and constant-property fluid. The former assumption is amply justified by the low Reynolds numbers (<100 in most cases), the latter by the small concentration changes occurring in a generic unit cell (polarization coefficients c_b/c_w ranging from 0.97 to 1.03, see Figures 5 and 6). Under these assumptions, we used the most complete possible model, i.e., that made up the full, three-dimensional continuity, Navier-Stokes and scalar transport equations.

The continuity equation (with implicit summation) is simply

$$\frac{\partial u_i}{\partial x_i} = 0 \tag{4}$$

where u_i is the i-th velocity component of the fluid.

As anticipated in Section 3.1., all simulations were carried out under the hypothesis of fully developed flow and concentration field, thus simulating the Unit Cell [51]. In this approach, periodic boundary conditions are imposed to all variables between the inlet and outlet faces of the computational domain. At the same time, it is necessary to allow for the variation of pressure and bulk concentration along the main flow direction **s**, due to frictional losses and solute inflow or outflow through the channel walls, respectively. These apparently contradictory requirements are reconciled as follows.

- Consider pressure p first. In the fully developed region of a channel, p can be decomposed into a periodic component $\widetilde{p}$, whose spatial distribution repeats itself identically in each unit cell, and a large-scale component $-K_p(\mathbf{x}\cdot\mathbf{s})$ which decreases linearly along the main flow direction whose unit vector is **s** (**x** is the position vector of components x_i). By substituting $\widetilde{p} - K_p(\boldsymbol{x}\cdot\boldsymbol{s}) = \widetilde{p} - K_p x_i s_i$ for p in the i-th steady-state Navier-Stokes equation:

$$\frac{\partial \rho u_j u_i}{\partial x_j} = -\frac{\partial p}{\partial x_i} + \frac{\partial}{\partial x_j}\mu\frac{\partial u_i}{\partial x_j} \tag{5}$$

(where ρ and μ are the fluid's density and viscosity), it becomes

$$\frac{\partial \rho u_j u_i}{\partial x_j} = -\frac{\partial \widetilde{p}}{\partial x_i} + \frac{\partial}{\partial x_j}\mu\frac{\partial u_i}{\partial x_j} + K_p s_i \tag{6}$$

Equation (6) is similar to Equation (5), but (a) the "true" pressure p is replaced by its periodic component $\widetilde{p}$, and (b) a body force per unit volume (mean pressure gradient) acting along the main flow direction **s** appears at the right hand side. If required, the "true" pressure p can always be reconstructed from the simulation results as $p = \widetilde{p} - K_p(\boldsymbol{x}\cdot\boldsymbol{s})$.
- In regard to the concentration c, by definition of fully developed conditions it can be decomposed into a periodic component $\widetilde{c}$ and a large-scale component $K_c(\mathbf{x}\cdot\mathbf{s})$, where K_c can now be either positive (net inflow of electrolyte into the channel) or negative (net outflow of electrolyte from the channel). By substituting $\widetilde{c} + K_c(\boldsymbol{x}\cdot\boldsymbol{s}) = \widetilde{c} + K_c x_i s_i$ for c in the transport equation

$$\frac{\partial u_j c}{\partial x_j} = \frac{\partial}{\partial x_j} D \frac{\partial c}{\partial x_j} \tag{7}$$

(where D is the electrolyte diffusivity), after some manipulation one obtains:

$$\frac{\partial u_j \widetilde{c}}{\partial x_j} = \frac{\partial}{\partial x_j} D \frac{\partial \widetilde{c}}{\partial x_j} - K_c u_s \tag{8}$$

in which $u_s = \mathbf{u}\cdot\mathbf{s}$ is the local velocity component along the main flow direction **s**. The large-scale gradient K_c can be obtained by an elementary balance as:

$$K_c = \frac{A}{V}\frac{j}{u_s} \tag{9}$$

in which j is the molar salt flux at walls (imposed in the simulation), A is the membrane surface active area in a fluid unit cell (so that jA is the molar flow per unit time, mol/s), V is the cell volume and $\langle u_s \rangle$ is the volume average of u_s.

In the present study, the (bulk) Reynolds number was conventionally defined as:

$$\mathrm{Re} = \frac{\rho U 2H}{\mu} \tag{10}$$

i.e., it was based on the hydraulic diameter $2H$ of a void (profile-less) and undeformed channel of thickness H in the limit of indefinite width, and on the approach velocity

$$U = \frac{Q}{S} \tag{11}$$

in which Q is the volume flow rate through a cross section of the channel orthogonal to the main flow direction and S is the cross sectional area of a void (profile-less) and undeformed channel of thickness H. The above definitions of U and Re are consistent with those adopted in our previous works on undeformed spacer-filled channels or profiled membranes [10,16,50,51].

The Darcy friction coefficient f_{Darcy} was defined with reference to the above approach velocity U and hydraulic diameter $2H$, i.e., as:

$$f_{Darcy} = \left|\frac{dp}{ds}\right|\frac{4H}{\rho U^2} \tag{12}$$

In the simulations, the driving pressure gradient $K_p = |dp/ds|$ in Equation (6) was imposed, while the flow rate was obtained as part of the solution. Please note that $|dp/ds|$ can be expressed in terms of the friction velocity Reynolds number

$$\mathrm{Re}_\tau = \frac{u_\tau \rho}{\mu}\frac{H}{2} \tag{13}$$

in which u_τ is the friction velocity,

$$u_\tau = \sqrt{\frac{H}{2\rho}\left|\frac{dp}{ds}\right|} \tag{14}$$

Therefore, in the parametrical analyses illustrated above, results were obtained for a given Re_τ (friction velocity Reynolds number) rather than for a given Re (bulk Reynolds number). Please note that according to the present definitions, between Re, Re_τ and f_{Darcy} the following relation holds:

$$f_{Darcy} = 128\left(\frac{\mathrm{Re}_\tau}{\mathrm{Re}}\right)^2 \tag{15}$$

To separate the effects of profile shape and channel deformation from the effects of varying the flow rate (and thus Re), the Darcy friction coefficient was normalized by that holding for parallel laminar flow in a void plane channel of indefinite width, i.e., 96/Re. Therefore, the following quantity (F-ratio) was reported:

$$F = \frac{f_{Darcy}}{96/\mathrm{Re}} \tag{16}$$

The local concentration polarization coefficient θ was defined as:

$$\theta = \frac{c_b}{c_w} \tag{17}$$

where c_b is the molar bulk concentration and c_w is the local molar concentration at the membrane surface. Please note that defining the average polarization coefficient in such a way that it is lower than 1 [10,16,51], the local polarization coefficient in Equation (17) refers to the case of either a dilute channel of RED or a concentrate channel of ED, where the flux enters from membrane's walls.

The Sherwood number was defined as

$$Sh = \frac{jA}{(c_b - \langle c_w \rangle)A_{proj}}\frac{2H}{D} \tag{18}$$

in which A_{proj} is the projected membrane surface area and $\langle c_w \rangle$ is the area average of c_w on the same membrane. Please note that the Sherwood numbers on the two membranes facing a channel may differ depending on the flow direction.

3.3.3. Flow Attack Angle, Boundary Conditions and Simulation Settings

The flow attack angle γ is defined (Figure 1) as the angle formed by the flow direction with the membrane ridges belonging to the upper wall of the channel under consideration. In this study, the two values $\gamma = 0°$ and 45° were considered; for symmetry reasons, the case $\gamma = 90°$ exhibits the same friction coefficient as the case $\gamma = 0°$, while the surface distributions of local quantities such as concentration, mass transfer coefficient (Sherwood number) and polarization coefficient on the upper and lower walls of the channel simply exchange place and rotate by 90° with respect to the case $\gamma = 0°$.

As mentioned in discussing the unit cell approach, translational periodicity was imposed for **u**, p and c between opposite inlet-outlet boundaries. At the membrane surfaces, no slip conditions were imposed for velocity and a uniform value of 2.6×10^{-4} mol/(m^{2}s) for the molar salt flux entering the fluid, corresponding to a current density of 50 A/m^2. An NaCl aqueous solution at a bulk concentration of 500 mol/m^3 was considered (i.e., seawater, see physical properties reported in Table 4). Please note that these choices on flux and bulk concentration affect directly the polarization coefficient (Equation (17)), while, the Sherwood number depends only on geometry, Re and Sc, due to the linearity of the transport Equation (8).

Table 4. Physical properties of the 500 mol/m^3 NaCl solution at 25 °C.

Property	Value	Units
Density, ρ	1017	kg m^{-3}
Viscosity, μ	0.931×10^{-3}	N s m^{-2}
Salt diffusivity, D	1.47×10^{-9}	m^2 s^{-1}
Schmidt number, $(\mu/\rho)/D$	622	-

3.3.4. FV Mesh for CFD Simulations

Mainly hexahedral meshes were adopted in the CFD simulations. Small regions at the corners of the domain were discretized by tetrahedra (~4.4% of the total volumes), pyramids (~0.13%) and wedges (~0.03%). Grid dependence was evaluated for $P/H = 8$ in the undeformed configuration at $Re_\tau \approx 5$, corresponding to a bulk Reynolds number of ~20. Therefore, the test case selected for the grid-independence assessment lies well above the creeping flow range and close to the highest Reynolds numbers investigated. Results are shown in Table 5, where the computed values of the Darcy friction coefficient f_{Darcy} and of the Sherwood number Sh are reported as functions of the number of finite volumes.

Table 5. Grid dependence results (CFD).

FV CFD Mesh	No. Finite Volumes ($P/H = 8$, $Re_\tau = 5.2$, $\gamma = 0°$)	Darcy Friction Coefficient	Sherwood Number (Upper Wall)	Sherwood Number (Lower Wall)
OCF-A	2.252×10^6	10.985	5.685	9.122
OCF-B	3.833×10^6	11.062	5.519	8.771
OCF-C	7.502×10^6	11.117	5.491	8.596

The mesh adopted for the final simulations (OCF-B) was characterized by about 4 million volumes. The channel height H was resolved by 40 finite volumes. Details of the same mesh are presented in Figure 12.

All simulations were conducted in double precision. The "High Resolution" (higher-order upwind) interpolation scheme was adopted for the advection terms. In regard to convergence, iterations were interrupted when the residuals of all variables became less than 10^{-10}.

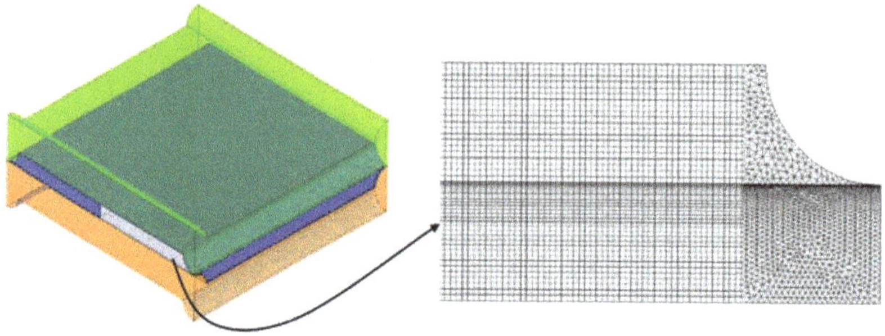

Figure 12. Details of the mesh chosen for the final simulations (undeformed domain, $P/H = 8$).

4. Conclusions

Integrated mechanical and fluid dynamics simulations were performed for profiled membranes of the OCF ("overlapped crossed filaments") type. The membranes were treated as linearly elastic, homogeneous and isotropic, and values of the Young modulus and of the Poisson ratio representative of ion exchange membranes' features ($E = 150$ MPa, $\nu = 0.4$) were adopted.

Under these assumptions, the largest value of the pitch to height ratio withstanding a trans-membrane pressure of 0.8 bar without collapsing (i.e., without exhibiting a contact between opposite membranes) was found to be $P/H = 8$.

The influence of *TMP* (which was investigated here in the range −0.4–+0.4 bar) is an increase of friction under compression conditions and a reduction of it (although to a lesser extent) under expansion conditions. This imbalance of effects may produce an increase of total pressure drop in the stack. The influence of the flow attack angle is negligible, indicating a substantial hydrodynamic isotropy of the profiled membrane lattice at low Reynolds numbers.

The influence of TMP on the Sherwood number is more complex. On the whole, compression enhances mass transfer and expansion reduces it; the influence of channel deformation on mass transfer is less marked than on friction. Some anomalous behaviour of Sh is observed in the cases characterized by $\gamma = 90°$ and Re > ~50, in which the highest Sh is obtained for the largest expansion.

This study shows that *TMP* values of practical interest for ED/RED units can produce significant effects on deformation, flow and concentration fields (and, thus, on hydraulic friction and mass transfer coefficient). In general, other important quantities, e.g., Ohmic resistance, non-Ohmic resistance and limiting current density, will also be affected.

As far as membrane manufacturers are concerned, the main suggestion arising from this study is probably that stiffer membranes should be preferred in order to reduce the undesirable effects of membrane deformation, which result in an impairment of the ED/RED equipment performance. On the other hand, the current trend is towards the reduction of membrane thickness, mainly sought in order to reduce Ohmic losses. Therefore, producing membranes with the highest possible Young modulus (provided the membrane's electrochemical behaviour is not impaired) will probably become a priority.

At a larger scale, combinations of geometric/mechanical features and operating conditions expose ED/RED stacks to the risk of severe local deformations, providing a new set-up, different from the nominal undeformed one, where all parameters are distributed as a result of a fluid-membrane mechanical interaction. Therefore, the actual process performance may be heavily affected by an imbalance of effects between compressed and expanded zones. This implies that the design and the performance prediction of devices under realistic operating conditions should take into account the membranes' mechanical features.

From this perspective, correlations describing (a) the dependence of deformation on trans-membrane pressure and (b) the dependence of friction coefficients and Sherwood numbers on deformation, derived from the results of the present modelling approach, will be implemented into higher-scale (stack-level) models in order to close the fluid-structure interaction loop and characterize the amount and effects of maldistribution phenomena. Moreover, the simulation method presented in this study can be applied to the more traditional configurations of flat membranes with net spacers by suitable adjustments.

Supplementary Materials: Supplementary materials can be found at http://www.mdpi.com/1422-0067/20/8/1840/s1. Figure S1. Engineering Stress vs Strain curve of the anion exchange membrane tested immersed in water. The linear elastic region is marked by a dashed line. The stress and strain at break are also highlighted. Figure S2. Experimental setup. (a) Overall layout; (b) Bulge test cell and laser head; (c) Detail of the Plexiglas plates making up the bulge test cell.

Author Contributions: Conceptualization, A.C., G.M. and M.C.; methodology, G.B., L.G., G.A.F. and M.C.; software, G.B., L.G., G.A.F. and M.C.; validation, L.G., A.P. and M.C.; formal analysis, G.B., L.G., G.A.F. and M.C.; investigation, G.B., L.G. and G.A.F.; resources, A.C., A.P. and G.M.; data curation, G.B. and L.G.; writing—original draft preparation, G.B., L.G. and M.C.; writing—review and editing, G.B., L.G., A.C. and M.C.; visualization, G.B., L.G. and G.A.F.; supervision, A.C., A.P., G.M. and M.C.; project administration, A.C., G.M. and M.C.; funding acquisition, A.C., A.P. and G.M.

Funding: This research was funded by the RED-Heat-to-Power (Conversion of Low Grade Heat to Power through closed loop Reverse Electro-Dialysis) and REvivED water (Low energy solutions for drinking water production by a REvival of ElectroDialysis systems) projects. These projects have received funding from the European Union's Horizon 2020 research and innovation programme under Grant Agreements no. 640667 and 685579, www.red-heat-to-power.eu, www.revivedwater.eu.

Acknowledgments: The authors are grateful to FUJIFILM Manufacturing Europe for hosting the experimental activities aimed at performing uniaxial tensile tests. A particular acknowledgment goes to Mr. Maarten Meijlink.

Conflicts of Interest: The authors declare no conflict of interest.

Abbreviations

CFD	Computational Fluid Dynamics
ED	ElectroDialysis
FE	Finite Element
FV	Finite Volume
OCF	Overlapped Crossed Filaments
RED	Reverse ElectroDialysis

References

1. Campione, A.; Gurreri, L.; Ciofalo, M.; Micale, G.; Tamburini, A.; Cipollina, A. Electrodialysis for water desalination: A critical assessment of recent developments on process fundamentals, models and applications. *Desalination* **2018**, *434*, 121–160. [CrossRef]
2. Dufton, G.; Mikhaylin, S.; Gaaloul, S.; Bazinet, L. Positive Impact of Pulsed Electric Field on Lactic Acid Removal, Demineralization and Membrane Scaling during Acid Whey Electrodialysis. *Int. J. Mol. Sci.* **2019**, *20*, 797. [CrossRef] [PubMed]
3. Tufa, R.A.; Hnát, J.; Němeček, M.; Kodým, R.; Curcio, E.; Bouzek, K. Hydrogen production from industrial wastewaters: An integrated reverse electrodialysis—Water electrolysis energy system. *J. Clean. Prod.* **2018**, *203*, 418–426. [CrossRef]
4. Strathmann, H. Electrodialysis, a mature technology with a multitude of new applications. *Desalination* **2010**, *264*, 268–288. [CrossRef]
5. Mei, Y.; Tang, C.Y. Recent developments and future perspectives of reverse electrodialysis technology: A review. *Desalination* **2018**, *425*, 156–174. [CrossRef]
6. Pawlowski, S.; Crespo, J.; Velizarov, S. Profiled Ion Exchange Membranes: A Comprehensible Review. *Int. J. Mol. Sci.* **2019**, *20*, 165. [CrossRef] [PubMed]
7. Nikonenko, V.V.; Kovalenko, A.V.; Urtenov, M.K.; Pismenskaya, N.D.; Han, J.; Sistat, P.; Pourcelly, G. Desalination at overlimiting currents: State-of-the-art and perspectives. *Desalination* **2014**, *342*, 85–106. [CrossRef]
8. Pawlowski, S.; Geraldes, V.; Crespo, J.G.; Velizarov, S. Computational fluid dynamics (CFD) assisted analysis of profiled membranes performance in reverse electrodialysis. *J. Membr. Sci.* **2016**, *502*, 179–190. [CrossRef]

9. Tadimeti, J.G.D.; Kurian, V.; Chandra, A.; Chattopadhyay, S. Corrugated membrane surfaces for effective ion transport in electrodialysis. *J. Membr. Sci.* **2016**, *499*, 418–428. [CrossRef]
10. Gurreri, L.; Ciofalo, M.; Cipollina, A.; Tamburini, A.; Van Baak, W.; Micale, G. CFD modelling of profiled-membrane channels for reverse electrodialysis. *Desalin. Water Treat.* **2015**, *55*, 3404–3423. [CrossRef]
11. Larchet, C.; Zabolotsky, V.I.; Pismenskaya, N.; Nikonenko, V.V.; Tskhay, A.; Tastanov, K.; Pourcelly, G. Comparison of different ED stack conceptions when applied for drinking water production from brackish waters. *Desalination* **2008**, *222*, 489–496. [CrossRef]
12. Vermaas, D.A.; Saakes, M.; Nijmeijer, K. Power generation using profiled membranes in reverse electrodialysis. *J. Membr. Sci.* **2011**, *385–386*, 234–242. [CrossRef]
13. Vermaas, D.A.; Saakes, M.; Nijmeijer, K. Enhanced mixing in the diffusive boundary layer for energy generation in reverse electrodialysis. *J. Membr. Sci.* **2014**, *453*, 312–319. [CrossRef]
14. Güler, E.; Elizen, R.; Saakes, M.; Nijmeijer, K. Micro-structured membranes for electricity generation by reverse electrodialysis. *J. Membr. Sci.* **2014**, *458*, 136–148. [CrossRef]
15. Pawlowski, S.; Rijnaarts, T.; Saakes, M.; Nijmeijer, K.; Crespo, J.G.; Velizarov, S. Improved fluid mixing and power density in reverse electrodialysis stacks with chevron-profiled membranes. *J. Membr. Sci.* **2017**, *531*, 111–121. [CrossRef]
16. La Cerva, M.L.; Di Liberto, M.; Gurreri, L.; Tamburini, A.; Cipollina, A.; Micale, G.; Ciofalo, M. Coupling CFD with a one-dimensional model to predict the performance of reverse electrodialysis stacks. *J. Membr. Sci.* **2017**, *541*, 595–610. [CrossRef]
17. She, Q.; Jin, X.; Tang, C.Y. Osmotic power production from salinity gradient resource by pressure retarded osmosis: Effects of operating conditions and reverse solute diffusion. *J. Membr. Sci.* **2012**, *401–402*, 262–273. [CrossRef]
18. She, Q.; Hou, D.; Liu, J.; Tan, K.H.; Tang, C.Y. Effect of feed spacer induced membrane deformation in the performance of pressure retarded osmosis (PRO): Implications for PRO process operation. *J. Membr. Sci.* **2013**, *445*, 170–182. [CrossRef]
19. Karabelas, A.J.; Koutsou, C.P.; Sioutopoulos, D.C. Comprehensive performance assessment of spacers in spiral-wound membrane modules accounting for compressibility effects. *J. Membr. Sci.* **2018**, *549*, 602–615. [CrossRef]
20. Huang, S.M. Laminar flow and heat transfer in plate membrane channels: Effects of the deformation heights. *Int. J. Therm. Sci.* **2016**, *109*, 44–53. [CrossRef]
21. Shi, Z.; Wang, X. A numerical study of flow crossover between adjacent flow channels in a proton exchange membrane fuel cell with serpentine flow field. *J. Power Sources* **2008**, *185*, 985–992. [CrossRef]
22. Zhou, Y.; Jiao, K.; Du, Q.; Yin, Y.; Li, X. Gas diffusion layer deformation and its effect on the transport characteristics and performance of proton exchange membrane fuel cell. *Int. J. Hydrog. Energy* **2013**, *38*, 12891–12903. [CrossRef]
23. Hereijgers, J.; Ottevaere, H.; Breugelmans, T.; De Malsche, W. Membrane deflection in a flat membrane microcontactor: Experimental study of spacer features. *J. Membr. Sci.* **2016**, *504*, 153–161. [CrossRef]
24. Moreno, J.; Slouwerhof, E.; Vermaas, D.A.; Saakes, M.; Nijmeijer, K. The Breathing Cell: Cyclic Intermembrane Distance Variation in Reverse Electrodialysis. *Environ. Sci. Technol.* **2016**, *50*, 11386–11393. [CrossRef] [PubMed]
25. Tanaka, Y. Pressure distribution, hydrodynamics, mass transport and solution leakage in an ion-exchange membrane electrodialyzer. *J. Membr. Sci.* **2004**, *234*, 23–39. [CrossRef]
26. Tanaka, Y. Overall mass transport and solution leakage in an ion-exchange membrane electrodialyzer. *J. Membr. Sci.* **2004**, *235*, 15–24. [CrossRef]
27. Hong, S.K.; Kim, C.S.; Hwang, K.S.; Han, J.H.; Kim, H.K.; Jeong, N.J.; Choi, K.S. Experimental and numerical studies on pressure drop in reverse electrodialysis: Effect of unit cell configuration. *J. Mech. Sci. Technol.* **2016**, *30*, 5287–5292. [CrossRef]
28. Tamburini, A.; Cipollina, A.; Tedesco, M.; Gurreri, L.; Ciofalo, M.; Micale, G. The REAPower Project: Power Production From Saline Waters and Concentrated Brines. In *Current Trends and Future Developments on (Bio-) Membranes*; Basile, A., Curcio, E., Inamuddin, I., Eds.; Elsevier: Amsterdam, The Netherlands, 2019; pp. 407–448.

29. Tedesco, M.; Cipollina, A.; Tamburini, A.; Micale, G. Towards 1 kW power production in a reverse electrodialysis pilot plant with saline waters and concentrated brines. *J. Membr. Sci.* **2017**, *522*, 226–236. [CrossRef]
30. Wright, N.C.; Shah, S.R.; Amrose, S.E.; Winter, A.G. A robust model of brackish water electrodialysis desalination with experimental comparison at different size scales. *Desalination* **2018**, *443*, 27–43. [CrossRef]
31. Long, R.; Li, B.; Liu, Z.; Liu, W. Performance analysis of reverse electrodialysis stacks: Channel geometry and flow rate optimization. *Energy* **2018**, *158*, 427–436. [CrossRef]
32. Ciofalo, M.; La Cerva, M.; Di Liberto, M.; Gurreri, L.; Cipollina, A.; Micale, G. Optimization of net power density in Reverse Electrodialysis. *Energy* **2019**. under review.
33. Bdiri, M.; Dammak, L.; Larchet, C.; Hellal, F.; Porozhnyy, M.; Nevakshenova, E.; Pismenskaya, N.; Nikonenko, V. Characterization and cleaning of anion-exchange membranes used in electrodialysis of polyphenol-containing food industry solutions; comparison with cation-exchange membranes. *Sep. Purif. Technol.* **2019**, *210*, 636–650. [CrossRef]
34. Garcia-Vasquez, W.; Dammak, L.; Larchet, C.; Nikonenko, V.; Pismenskaya, N.; Grande, D. Evolution of anion-exchange membrane properties in a full scale electrodialysis stack. *J. Membr. Sci.* **2013**, *446*, 255–265. [CrossRef]
35. Garcia-Vasquez, W.; Ghalloussi, R.; Dammak, L.; Larchet, C.; Nikonenko, V.; Grande, D. Structure and properties of heterogeneous and homogeneous ion-exchange membranes subjected to ageing in sodium hypochlorite. *J. Membr. Sci.* **2014**, *452*, 104–116. [CrossRef]
36. Garcia-Vasquez, W.; Dammak, L.; Larchet, C.; Nikonenko, V.; Grande, D. Effects of acid-base cleaning procedure on structure and properties of anion-exchange membranes used in electrodialysis. *J. Membr. Sci.* **2016**, *507*, 12–23. [CrossRef]
37. Klaysom, C.; Moon, S.H.; Ladewig, B.P.; Lu, G.Q.M.; Wang, L. Preparation of porous ion-exchange membranes (IEMs) and their characterizations. *J. Membr. Sci.* **2011**, *371*, 37–44. [CrossRef]
38. Křivčík, J.; Neděla, D.; Válek, R. Ion-exchange membrane reinforcing. *Desalin. Water Treat.* **2015**, *56*, 3214–3219. [CrossRef]
39. Kusoglu, A.; Karlsson, A.M.; Santare, M.H.; Cleghorn, S.; Johnson, W.B. Mechanical behavior of fuel cell membranes under humidity cycles and effect of swelling anisotropy on the fatigue stresses. *J. Power Sources* **2007**, *170*, 345–358. [CrossRef]
40. Safronova, E.Y.; Golubenko, D.V.; Shevlyakova, N.V.; D'yakova, M.G.; Tverskoi, V.A.; Dammak, L.; Grande, D.; Yaroslavtsev, A.B. New cation-exchange membranes based on cross-linked sulfonated polystyrene and polyethylene for power generation systems. *J. Membr. Sci.* **2016**, *515*, 196–203. [CrossRef]
41. Solasi, R.; Zou, Y.; Huang, X.; Reifsnider, K. A time and hydration dependent viscoplastic model for polyelectrolyte membranes in fuel cells. *Mech. Time-Depend. Mater.* **2008**, *12*, 15–30. [CrossRef]
42. Tang, Y.; Karlsson, A.M.; Santare, M.H.; Gilbert, M.; Cleghorn, S.; Johnson, W.B. An experimental investigation of humidity and temperature effects on the mechanical properties of perfluorosulfonic acid membrane. *Mater. Sci. Eng. A* **2006**, *425*, 297–304. [CrossRef]
43. Vandiver, M.A.; Caire, B.R.; Carver, J.R.; Waldrop, K.; Hibbs, M.R.; Varcoe, J.R.; Herring, A.M.; Liberatore, M.W. Mechanical Characterization of Anion Exchange Membranes by Extensional Rheology under Controlled Hydration. *J. Electrochem. Soc.* **2014**, *161*, H677–H683. [CrossRef]
44. Ghalloussi, R.; Garcia-Vasquez, W.; Bellakhal, N.; Larchet, C.; Dammak, L.; Huguet, P.; Grande, D. Ageing of ion-exchange membranes used in electrodialysis: Investigation of static parameters, electrolyte permeability and tensile strength. *Sep. Purif. Technol.* **2011**, *80*, 270–275. [CrossRef]
45. Narducci, R.; Chailan, J.F.; Fahs, A.; Pasquini, L.; Di Vona, M.L.; Knauth, P. Mechanical properties of anion exchange membranes by combination of tensile stress-strain tests and dynamic mechanical analysis. *J. Polym. Sci. Part B Polym. Phys.* **2016**, *54*, 1180–1187. [CrossRef]
46. Nagarale, R.K.; Gohil, G.S.; Shahi, V.K. Recent developments on ion-exchange membranes and electro-membrane processes. *Adv. Colloid Interface Sci.* **2006**, *119*, 97–130. [CrossRef]
47. Ran, J.; Wu, L.; He, Y.; Yang, Z.; Wang, Y.; Jiang, C.; Ge, L.; Bakangura, E.; Xu, T. Ion exchange membranes: New developments and applications. *J. Membr. Sci.* **2017**, *522*, 267–291. [CrossRef]
48. Hong, J.G.; Zhang, B.; Glabman, S.; Uzal, N.; Dou, X.; Zhang, H.; Wei, X.; Chen, Y. Potential ion exchange membranes and system performance in reverse electrodialysis for power generation: A review. *J. Membr. Sci.* **2015**, *486*, 71–88. [CrossRef]

49. Tedesco, M.; Hamelers, H.V.M.; Biesheuvel, P.M. Nernst-Planck transport theory for (reverse) electrodialysis: III. Optimal membrane thickness for enhanced process performance. *J. Membr. Sci.* **2018**, *565*, 480–487. [CrossRef]
50. Gurreri, L.; Tamburini, A.; Cipollina, A.; Micale, G.; Ciofalo, M. Pressure drop at low Reynolds numbers in woven-spacer-filled channels for membrane processes: CFD prediction and experimental validation. *Desalin. Water Treat.* **2017**, *61*, 170–182. [CrossRef]
51. Gurreri, L.; Tamburini, A.; Cipollina, A.; Micale, G.; Ciofalo, M. CFD prediction of concentration polarization phenomena in spacer-filled channels for reverse electrodialysis. *J. Membr. Sci.* **2014**, *468*, 133–148. [CrossRef]
52. Ciofalo, M.; La Cerva, M.; Di Liberto, M.; Tamburini, A. Influence of the boundary conditions on heat and mass transfer in spacer-filled channels. *J. Phys. Conf. Ser.* **2017**, *923*, 012053. [CrossRef]
53. Chiapello, J.-M.; Bernard, M. Improved spacer design and cost reduction in an electrodialysis system. *J. Membr. Sci.* **1993**, *80*, 251–256. [CrossRef]
54. Vermaas, D.A.; Saakes, M.; Nijmeijer, K. Doubled power density from salinity gradients at reduced intermembrane distance. *Environ. Sci. Technol.* **2011**, *45*, 7089–7095. [CrossRef]
55. von Gottberg, A. New High-Performance Spacers in Electro- Dialysis Reversal (EDR) Systems. In Proceedings of the 1998 AWWA Anual Conference, Dallas, TX, USA, 21–25 June 1998.
56. Timoshenko, S.; Goodier, J.N. *Theory of Elasticity*; McGraw-Hill: New York, NY, USA, 1951.
57. Tufa, R.A.; Pawlowski, S.; Veerman, J.; Bouzek, K.; Fontananova, E.; di Profio, G.; Velizarov, S.; Goulão Crespo, J.; Nijmeijer, K.; Curcio, E. Progress and prospects in reverse electrodialysis for salinity gradient energy conversion and storage. *Appl. Energy* **2018**, *225*, 290–331. [CrossRef]
58. Iyengar, K.T.S.R.; Naqvi, M.M. Large deflections of rectangular plates. *Int. J. Non-Linear Mech.* **1966**, *1*, 109–122. [CrossRef]
59. Sonin, A.A.; Probstein, R.F. A hydrodynamic theory of desalination by electrodialysis. *Desalination* **1968**, *5*, 293–329. [CrossRef]
60. Newman, J.; Thomas-Alyea, K. *Electrochemical Systems*, 3rd ed.; John Wiley & Sons, Inc.: Hoboken, NJ, USA, 2004; pp. 277–280.
61. Nakayama, A.; Sano, Y.; Bai, X.; Tado, K. A boundary layer analysis for determination of the limiting current density in an electrodialysis desalination. *Desalination* **2017**, *404*, 41–49. [CrossRef]
62. La Cerva, M.; Gurreri, L.; Tedesco, M.; Cipollina, A.; Ciofalo, M.; Tamburini, A.; Micale, G. Determination of limiting current density and current efficiency in electrodialysis units. *Desalination* **2018**, *445*, 138–148. [CrossRef]

International Journal of
Molecular Sciences

MDPI

Review

Profiled Ion Exchange Membranes: A Comprehensible Review

Sylwin Pawlowski, João G. Crespo and Svetlozar Velizarov *

Associated Laboratory for Green Chemistry - Clean Technologies and Processes (LAQV), REQUIMTE, Chemistry Department, FCT, Universidade NOVA de Lisboa, 2829-516 Caparica, Portugal; s.pawlowski@fct.unl.pt (S.P.); jgc@fct.unl.pt (J.G.C.)
* Correspondence: s.velizarov@fct.unl.pt; Tel.: +351-21-294-8385

Received: 6 December 2018; Accepted: 23 December 2018; Published: 4 January 2019

Abstract: Profiled membranes (also known as corrugated membranes, micro-structured membranes, patterned membranes, membranes with designed topography or notched membranes) are gaining increasing academic and industrial attention and recognition as a viable alternative to flat membranes. So far, profiled ion exchange membranes have shown to significantly improve the performance of reverse electrodialysis (RED), and particularly, electrodialysis (ED) by eliminating the spacer shadow effect and by inducing hydrodynamic changes, leading to ion transport rate enhancement. The beneficial effects of profiled ion exchange membranes are strongly dependent on the shape of their profiles (corrugations/patterns) as well as on the flow rate and salts' concentration in the feed streams. The enormous degree of freedom to create new profile geometries offers an exciting opportunity to improve even more their performance. Additionally, the advent of new manufacturing methods in the membrane field, such as 3D printing, is anticipated to allow a faster and an easier way to create profiled membranes with different and complex geometries.

Keywords: ion exchange membranes; profiled membranes; corrugated membranes; electrodialysis; reverse electrodialysis; membrane capacitive deionization; hydrodynamic; mass transfer; thermal pressing; 3D printing

1. Introduction

Electro-membrane processes range from classical membrane electrolysis and electrodialysis to emerging applications, such as reverse electrodialysis, membrane capacitive deionization, redox flow batteries, microbial, and enzymatic fuel cells, and ion exchange membrane (bio)reactors. There is an increasing worldwide interest in their use for clean water and renewable electrochemical energy harvesting and storage [1–7], which has inspired fundamental and applied research on improving the properties of their "key" components: the ion exchange membranes (IEMs).

Academia and industrial companies are usually focused on the development of tailored membrane products best suited to satisfy specific process requirements, such as low membrane electric resistance, monovalent ion perm-selectivity and/or antifouling properties [8–10]. Thus, material properties of IEMs have been mainly studied, while the design and hydrodynamic of electromembrane devices have suffered few alterations since the traditional plate-and-frame arrangement of flat membranes and spacers have been, so far, the most common choice [11,12].

Recently, profiled membranes (also known as corrugated membranes, micro-structured membranes, patterned membranes, membranes with designed topography or notched membranes) are gaining more attention, mostly due to advances in their manufacturing, as well as improvement of their performance, which depend on the shape, dimensions, orientation, and distribution of the microscopic profiles (corrugations) on their surfaces. According to our best knowledge, based on the information provided by Zabolotskii et al. [13], the idea of creating profiled ion-exchange membranes (i.e., with both membrane and spacer functionalities) emerged and was firstly implemented in the decades of 70–80 of the 20th century in the former Union of Soviet Socialist Republics (U.S.S.R) with the objective (successfully achieved) of accelerating the mass transfer processes in a electrodialyzer, by increasing the available membrane surface area. Later, in 2009, Brauns [14] suggested to use profiled ion exchange membranes also in reverse electrodialysis stacks in order to decrease their height, and consequently, the pressure drop. Veerman et al. [15] proposed a fractal profiled membranes design for reverse electrodialysis (RED) application to reduce the flow path in order to maximize salinity gradient.

In this review, we will focus especially on the developments and applications of profiled ion exchange membranes during the last 10 years, which is also the most productive period regarding their investigation. A critical discussion about the advantages and disadvantages of their use will be provided in the context of their application in electrodialysis (ED) and reverse electrodialysis (RED). We are also propose and discuss the possibility of using profiled membranes in membrane capacitive deionization (MCDI), which could open opportunities for improving this novel process. Additionally, a summary of the possible patches of preparing profiled membranes is presented, with a special focus on additive manufacturing (3D printing) and its possible symbiosis with computational fluid dynamics (CFD) studies.

2. Profiled Ion-Exchange Membranes (10 Years Ago)

In 2008, Larchet et al. [16] compared the performance of electrodialysis and electrodeionization stacks with non-conductive spacers, profiled ion exchange membranes and ion-exchange resins. Utilization of profiled membranes allowed for an effective production of drinking water from brackish waters with larger salt concentration ranges than that achieved by other stack arrangements. Utilization of profiled membranes obviates the need of using non-conductive spacers, thus increasing the membrane surface area available for mass transfer. Also, a curved shape of profiled membrane surface could facilitate initiation of electroconvection, which is favoured in dilute saline solutions and allows for a better fluid mixing near the profiled membrane surface. Utilization of profiled membranes for desalination of solutions with low conductivity showed to be more advantageous than the use of non-conductive spacers, due to the reduction of electric resistance in the compartment. Grabowski [17] reported final reduction of Reverse Osmosis (RO) permeate conductivity to approximately 2 μS/cm with profiled membranes, in comparison to approximately 5 μS/cm when flat membranes and spacers were used. Moreover, although there might exists a bipolar junction in the contact points between profiled anion and cation exchange membranes where dissociation of water could be catalysed, it has been pointed out that since the maximum local current density becomes lower when profiled membranes are used, due to a higher available membrane area, the water splitting rate decreases, which is very promising for production of high purity water [18]. Overall, the rate of mass transfer increased significantly (by a factor of four) when profiled membranes were used, while the pressure drop was similar to the one when non-profiled membranes were used.

In 2011, Vermaas et al. [19] first performed experimental tests in RED with profiled membranes (reliefs were straight ridges oriented parallelly to the flow direction, see Figure 1a). A 10% higher net power density was obtained when profiled membranes were used in comparison to the stack with non-conductive spacers. The main improvements of using those profiled membranes were the decrease by 30% of total electric resistance (mainly due to elimination of spacer shadow effect) and a four times lower pressure drop. However, the diffusion boundary layer resistance (at a Reynolds number of 5, when the highest net power density was obtained) was around 4.25 times higher when profiled membranes were used (in comparison to the stack with spacers), since the solution mixing promoted by profiled membranes was less efficient than the one offered by the spacers. Despite that, the stacks with this particular profiled membrane were significantly less sensitive to fouling than stacks with spacers [20], and easy to be cleaned [21] (Figure 1b), since the straight and parallel to the flow relief of profiled membranes created open channels for fluid flow, where the foulants struggled to get deposited.

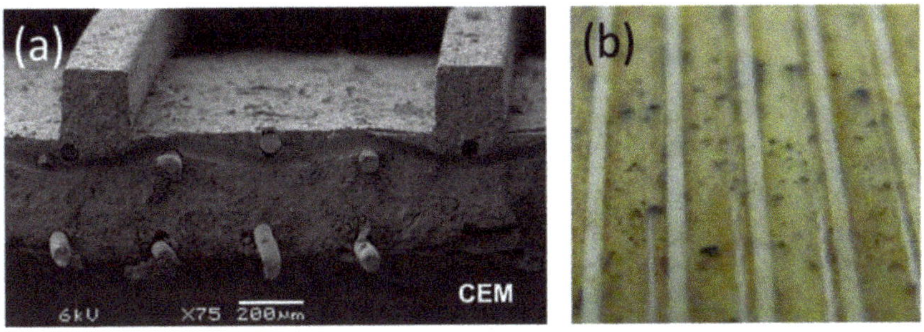

Figure 1. Profiled membranes with straight ridges parallel to the flow direction: (**a**) SEM-image of a cross-section of the profiled cation-exchange membrane (CEM). Reproduced with permission from Reference [19]. Copyright 2011 Elsevier; (**b**) Photo of 1 × 1 cm^2 section of profiled CEM after more than 60 days of reverse electrodialysis (RED) stack operation, in which air sparging was applied as an antifouling strategy. Reproduced with permission from Reference [21]. Copyright 2016 American Chemical Society.

The comparison between profiled membranes and spacers, as well as the advantages and disadvantages of each one, obviously strongly depend on the shape of the reliefs, as well as the quality of the spacers mesh. Moreover, the specific process operating conditions, a relatively high current density, and higher flow rates used in ED, and a low current density and lower flow rates applied in RED, influence the differences in hydrodynamic and mass transfer existing when spacers or profiled membranes are used. Thus, for each application, and conditions, a different reliefs shape might prove to be the optimal one, since the objectives of different electromembrane processes are distinct. In ED, the main objective is to intensify the mass transfer, while in RED, it is desired to obtain high net power density values, i.e., to maintain a low pressure drop (requiring less energy demand for pumping) at a reasonably mass transfer rate.

3. Progresses in the Performance of Profiled Ion-Exchange Membranes

Guler et al. [22] investigated experimentally the performance of profiled membranes with corrugations similar to straight-ridges, waves, and round pillars in RED stacks (Figure 2a,b). A 20% higher net power density (in comparison to a stack with spacers) was obtained when pillar profiled membranes were used, which also outperformed the other studied profiles, mainly due to the lowest pressure drop observed in channels with pillar profiled membranes, and the lower increase of the diffusion boundary layer resistance (it was still 1.5 times higher but, when straight-ridges were used at the same operating conditions, it was 2.25 higher). Similar beneficial effects of hemispherical protrusions, similar to pillar profiled membranes, were observed by Liu et al. [23] when used in a microbial RED stack. Gurreri et al. [24] simulated, by CFD tools, the performance of profiled membranes with squared and round pillars. Round pillars were anticipated to perform better than squared pillars and their simulations are in accordance with previous findings, since the pumping power consumption is expected to be lower when profiled membranes are used in comparison to the use of spacers. However, they also observed the development of stagnant zones near the pillar corrugations, which can be responsible for concentration polarization (and consequently, a higher diffusion boundary layer resistance). Moreover, it was observed that pillar-kind corrugations, independently of their exact shape (round, diamond, tear, etc.), lead to development of zones with poor mixing upstream of the corrugations, which are very prone for deposition of foulants [25] (Figure 2c–e).

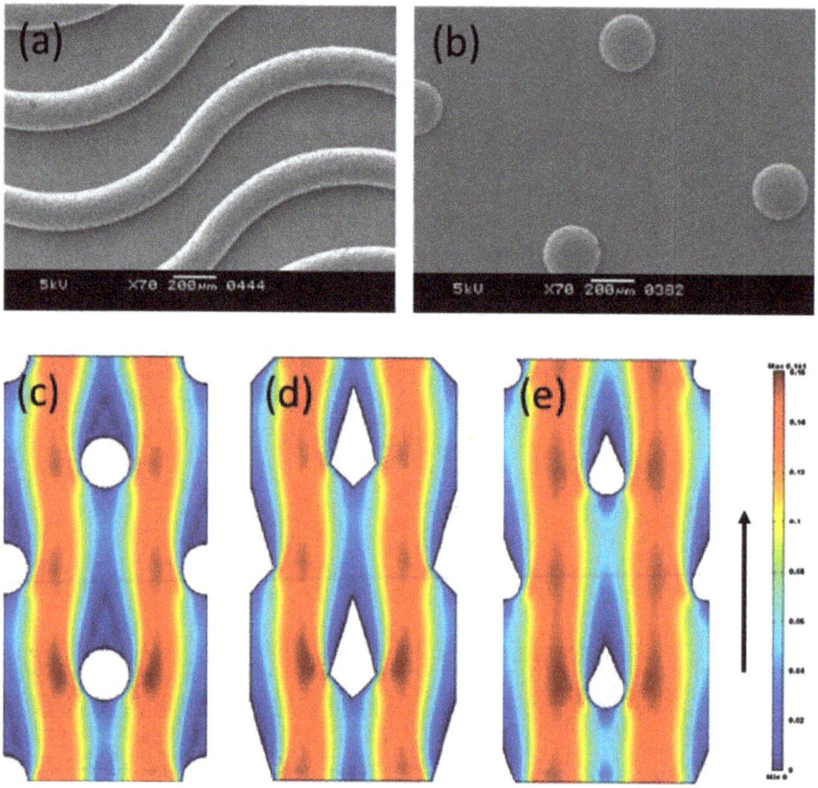

Figure 2. Surface morphology of tailor-made profiled membranes with: (**a**) waves and (**b**) pillars corrugations. Reproduced with permission from Reference [22]. Copyright 2014 Elsevier.; velocity fields, obtained by computational fluid dynamics (CFD), around pillar structures with (**c**) round, (**d**) diamond, and (**e**) tear shape, for a flow direction indicated by the upward arrow on the right. Reproduced with permission from Reference [25]. Copyright 2010 Elsevier.

Fouling is a major issue in any membrane-based process, including the electromembrane ones [26–29]. However, there are almost no reported experimental studies performed with natural waters using profiled ion-exchange membranes, except for the already previously mentioned works of Vermaas et al. [20,21]. In pressure-driven membrane-based processes, it has been seen that surface patterning by nanoimprints made from the same material as membranes [30,31], or an anti- or less-prone-to-fouling material such as titanium dioxide (TiO_2) [32], has fouling mitigation effects. Also, based on computational and experimental studies, some trends about the possible relation between fouling and flow behaviour in channels formed by profiled membranes were obtained. For example, utilization of prism corrugations equidistant by 400 μm in cross-flow filtration enhanced vortex formation and reduced the mass of deposited foulants by around seven times in comparison to flat membranes at Re = 1600 [33]. Increasing the wall shear stress was another effective strategy for minimization of foulants deposition, which can be achieved by optimizing the shape of corrugations [34] and/or increasing the linear flow velocity (Figure 3).

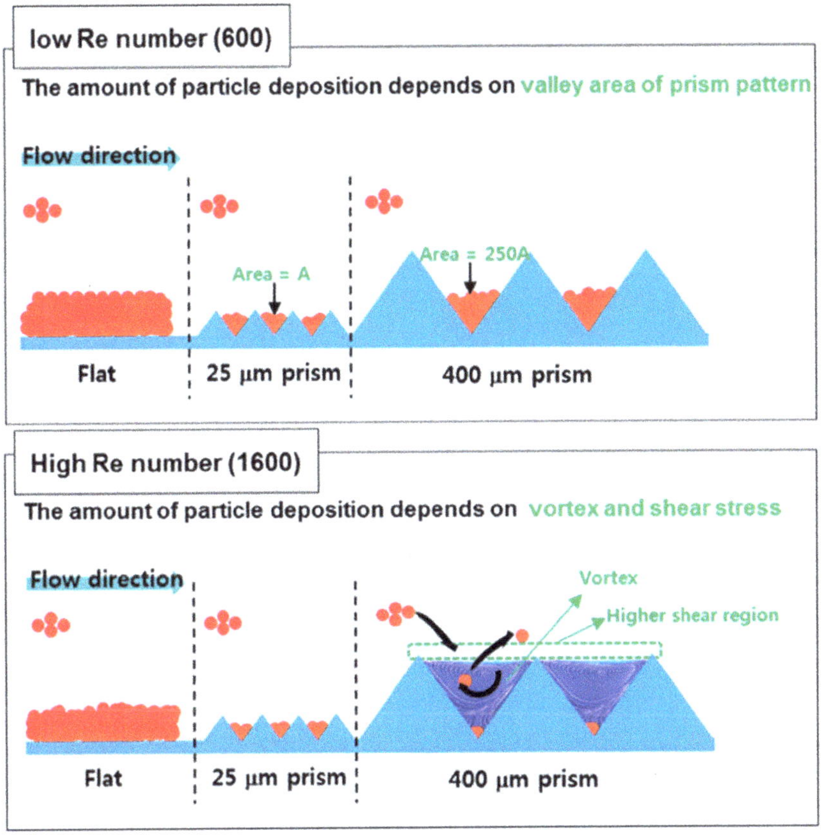

Figure 3. An illustrative scheme of correlation of membrane fouling with topography of profiled membranes in cross-flow filtration for water treatment. Re stands for Reynolds number. Particles are represented in red and the prismatic membrane surface pattern by blue triangles. The bending arrows illustrate particles entering and leaving locally created flow vortexes Reproduced with permission from Reference [33]. Copyright 2016 Elsevier.

Zhao et al. [35] performed CFD simulations of notched membranes with different distribution of squared corrugations on the membrane surface used in ED and observed that better hydrodynamic conditions result from an increase of the flow velocity perpendicular to the membrane, which leads to reduction of concentration polarization. However, they also observed that if the corrugations are very close to each other, the flow behaviour becomes very similar to that of flat membranes, thus, the best distance between corrugations and the size of their side was found to be equal to 120 μm.

Vermaas et al. [36] included sub-corrugations perpendicular to the flow in the channels formed by straight-ridge profiles in a RED stack. However, such sub-corrugations did not create any vortices since the velocity range in RED is too low. Instead, dead zones were formed in front and behind the sub-corrugations, as observed by particle tracking velocimetry (PTV) (Figure 4). However, in ED, higher flow rates and current densities (close to limiting and/or over-limiting current density) are used [37], thus utilization of such sub-corrugations might promote electro-osmotic chaotic fluid instabilities, and consequently enhance mass transfer [38–40]. Imprinting microgel patterns on membranes surfaces [41], screening them with nonconductive strips [42,43] or coating by hydrophobic materials [44] can also promote formation of such electro-convective vortices.

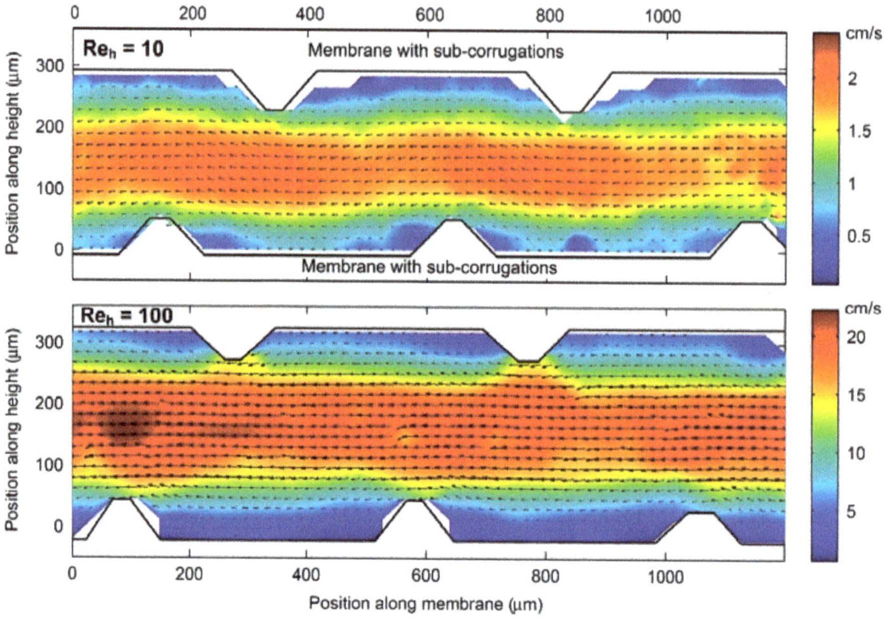

Figure 4. Velocity fields, obtained by particle tracking velocimetry (PTV), of fluid flow between sub-corrugated membranes for Reynold numbers (based on half the channel height), Re_h of 10 and 100 in a RED application. Reproduced with permission from Reference [36]. Copyright 2014 Elsevier.

Pawlowski et al. [45], based on CFD studies, proposed the use of profiled membranes with chevron (V-shaped) profiles for RED (Figure 5a). Instead of focusing on decreasing pressure drop, the approach followed was mainly focused on improving the mass transfer. To create a channel for fluid flow, two profiled membranes (bottom and top) should be aligned and stacked with the chevron tips pointed into the opposite directions. In such a way, a very specific fluid pathway is created (the fluid goes periodically up and down, and it is also distributed in the lateral directions), which promotes convective phenomena (Figure 5b). Additionally, the linear flow velocity upstream to the corrugations is high, which intensifies the mass transfer and should avoid accumulation of foulants in such critical areas (Figure 5c). Although the pressure drop in channels with chevron profiled membranes was anticipated to be higher (in comparison with pillar profiled membranes), the net power density values were anticipated to be higher (for linear flow velocities lower than 1.5 cm/s) (Figure 5d), since chevron corrugations promotes fluid mixing more effectively near the membrane interface, which enhances local mass transfer. Besides the profiles shape, it was predicted that the maximum net power density values also depend on the stack length. For a 10-cm long stack, chevron profiled membranes and linear flow velocity between 0.5 and 1.0 cm/s were expected to be the best options. These predictions were confirmed experimentally and an 8–14% increase of net power density was observed when chevron profiled membranes were used instead of pillar profiled membranes or non-conductive spacers [46] (Figure 5e). The pillar profiled membranes and non-conductive spacers had a similar performance, since state-of-the art spacers (with very fine, elastic, and flexible filaments) were used in that study.

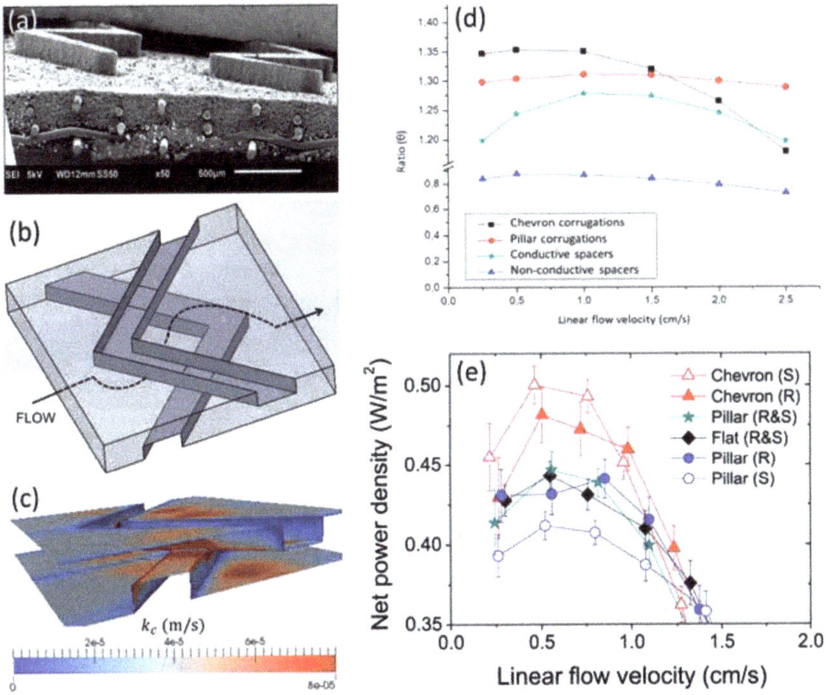

Figure 5. Chevron profiled membranes used in RED: (**a**) SEM image of the surface morphology, (**b**) segment of a channel formed by chevron profiled membranes; (**c**) map of mass transfer coefficient, obtained by CFD, at two linear flow velocity of 1.0 cm/s, (**d**) comparison, based on CFD simulations, of the ratio (θ) between net power density obtainable in RED stacks with different configurations and the one obtainable in RED stack with empty channels, (**e**) experimental net power density obtained in RED stacks with different configurations. Reproduced with permission from References [45,46]. Copyright 2016 and 2017, respectively, Elsevier.

Additionally, Pawlowski et al. [46] proposed a further simplification of chevron corrugations into "chevron-crossed" corrugations, which is expected to facilitate assembling of large stacks, although at "ideal" conditions, simulated by CFD, they would promote less convective phenomena than the original chevron profiled membranes (but still more than pillar profiled membranes). La Cerva et al. [47] simulated by CFD tools the performance of profiled membranes similar to the "chevron-crossed" corrugations (the corrugations instead of rectangular are semi-cylindric), and referred to them as "Overlapped Crossed Filaments" (OCF). The expected Sherwood number for chevron-crossed and OCF corrugations, at flow angle attack of 45°, and at the lowest simulated Re number is 10.7 [46] and 12.6 [47], respectively, thus semi-cylindrical corrugations are expected to offer a slightly better performance. Further optimizations for RED applications must also have in attention that the same profiled membranes might not always have the same beneficial effects, as, e.g., the dilute feed stream concentration influences the global stack performance [48]. For example, when compared with empty channels, inclusion of profiled membranes leads to higher net power density values if the dilute saline stream concentration is 0.01 M or lower, as this decreases significantly the electric resistance of the dilute saline solution compartment.

Melnikov et al. [49] tested profiled membranes, which surface appears to be all covered by corrugations with a shape of hemispheres (in the previously discussed cases there are clear gaps and the corrugations are spaced between each other), in an ED treatment of secondary steam condensate obtained during production of ammonium nitrate. Again, the profiled membranes allowed for an enhanced mass transfer, in comparison to that of using flat membranes made from the same material, due to the reasons already discussed in Section 2. Moreover, the total cost of ammonium nitrate treatment decreased from 0.106 €/kg to 0.061 €/kg when profiled membranes were used (in comparison with flat membranes), since the required membrane area decreased by 30% and the estimated additional cost of preparing such profiled membranes by thermal pressing is low (just 7 €/m^2) (Figure 6). Vasil'eva et al. [50] tested such profiled membranes in a diffusion dialysis process for separation and purification of amino acids. A maximum 8-fold increase of phenylalanine flux was documented, when profiled instead of flat membranes were used, since the available membrane area for mass transfer increased 2.3 times and the diffusion boundary layer thickness was reduced due to more favourable hydrodynamic conditions, resulting from the curvature of the membrane surface.

Membrane	Ralex AMH	MA-41	MA-40	MA-41$_p$
Membrane cost (Euro/m^2)	70[a]	40	40	47
Required membrane area (m^2)	916	1253	1338	892
Total investment costs (Euro/kg)	0.038	0.041	0.044	0.032
Energy cost (Euro/kg)	0.036	0.065	0.074	0.029
Total cost (Euro/kg)	0.074	0.106	0.118	0.061

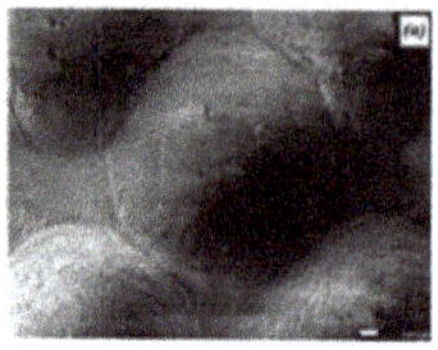

Figure 6. Economic evaluation of an ammonium nitrate treatment process using different anion-exchange membranes (AEM), in which Ralex AMH, MA-41 and MA-40 are flat, MA-41p is a profiled (subscript "p") version of MA-41, which surface denoted by *(a)* is shown on the right with a scale bar of 100 μm. [a]Prices are given for the year 2011. (Reproduced with permission from Reference [49]. Copyright 2016 Elsevier).

4. Profiled Membranes for Membrane Capacitive Deionization (MCDI)

As it can be seen, utilization of profiled membranes in ion-exchange membrane processes such as electrodialysis, electrodeionization, reverse electrodialysis, and diffusion dialysis have led to improvements in their performance in comparison to the utilization of flat membranes. We are therefore suggesting that the next electromembrane process, in which profiled ion-exchange membranes might have a bright future could be membrane capacitive deionization (MCDI). Membrane capacitive deionization cell architecture consists in modifying the classical capacitive deionization (CDI) cell design by placing an appropriate ion-exchange membrane in front of each electrode (CEM near cathode and AEM near anode). In such arrangement, the respective membranes block co-ions transport (i.e., occurrence of parasitic current), thus leading to improvement of charge efficiency and an increase of salt storage in the macropores of the corresponding electrodes [51]. However, the inclusion of membranes adds also an additional electric resistance (of the membranes themselves and of the associated diffusion boundary layers) to the system. Therefore, utilization of profiled ion-exchange membranes, besides granting a selective counter-ion transport, might also offer a larger area for mass transfer and a lower diffusion boundary layer resistance. Moreover, such beneficial effects are enhanced (as seen for the case of electrodialysis) at high feed streams flow rates, low solutions concentrations and high current densities, which are also the operating conditions relevant to MCDI.

5. Preparation of Profiled Ion Exchange Membranes

Profiled ion exchange membranes can be prepared by three main approaches: hot (thermal) pressing (Figure 7a), membrane casting (Figure 7b), and 3D printing (photopolymerization) (Figure 7c,d).

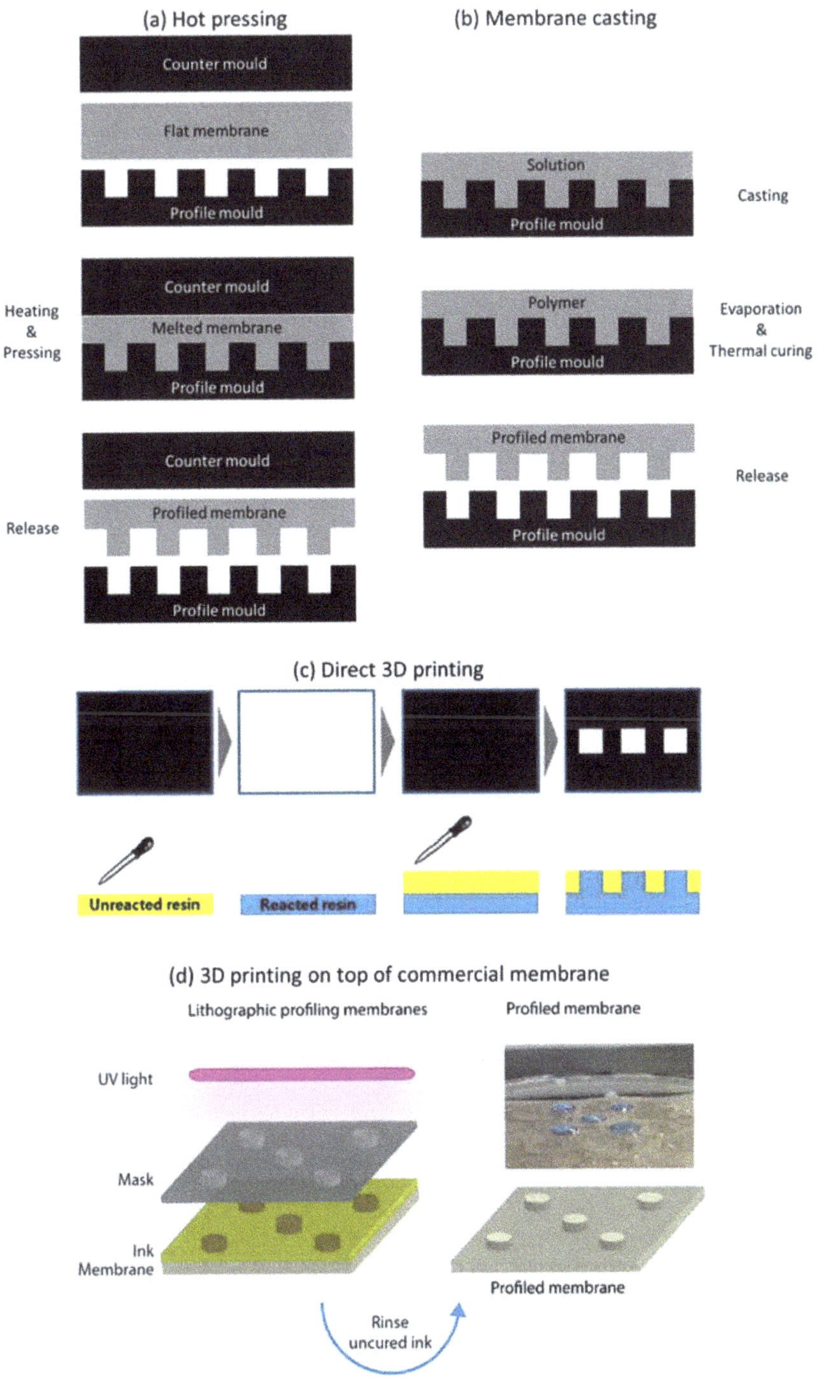

Figure 7. Different possible pathways for preparation of profiled ion exchange membranes: (**a**) hot pressing; (**b**) membrane casting (Adapted from Reference [22]); (**c**) Direct 3D printing (Reproduced with permission from Reference [52]. Copyright 2016 American Chemical Society.); (**d**) 3D printing on top of a commercial membrane (Reproduced with permission from Reference [53]. Copyright 2018 Timon Rijnaarts).

Hot pressing (Figure 7a) is a relatively easily preparation method, in which a dry flat heterogenous ion exchange membrane is first sandwiched between moulds with a desired subsequent corrugations shape and then placed in a thermal press, where the whole set-up is heated. While the membrane material melts, the moulds are pressed against each other, thus originating formation of the corrugations while the area between them is compressed [19,46]. In this way, when cooled, the new shape of the profiled membrane is formed. Releasing a profiled membrane from the mould is a delicate procedure, but it can be facilitated by using releasing agents, which are initially sprayed on the flat membrane surface [19]. Also, in order to assure an easier and complete release of the membranes from the moulds, without breaking the corrugations, their height should not be bigger than their width [46]. Alternatively, instead of using moulds, spacers can be pressed against the membranes to create channels on their surface [35]; however, the resolution of the corrugations shape and their dimensions are lower.

Since membranes dry when heated and pressed, if they are then brought into contact with an aqueous medium, they will swell. The corrugations overall shape remains the same, but their dimensions increase (differently for AEM and CEM, according to their materials swelling tendency) [46]. The concentration of saline streams also influences membranes swelling (higher volumetric changes are observed for membranes in contact with dilute solutions) [54]. Therefore, this phenomenon must be considered when the moulds' dimensions are designed. Recently, hot pressing of previously swollen flat membranes was tested but, unfortunately, the data regarding corrugations fidelity are not available [49]. Nevertheless, since the fraction of the surface area occupied by ion exchange resin particle was 25% for a profiled membrane swollen a priori, and just 12% for a profiled membrane obtained from a dry material (for a flat membrane the mentioned fraction was 15%), the mass transfer surface was twice higher [50].

Hot pressing of the membranes might lead to some alteration of their material properties [13,55], but usually they are beneficial. For example, the conductivity and effective specific resistance of Ralex AMH-PES heterogeneous anion-exchange membrane increased from 9.1 to 13.9 mS/cm [19] and decreased from 195 to 111 Ωcm [46], respectively, but no significant alterations were observed after hot pressing of a Ralex CMH-PES heterogeneous cation-exchange membrane. Most probably, during melting, the percolating pathway through the Ralex AMH-PES membrane was improved. Only heterogeneous ion exchange membranes could be profiled by hot pressing since homogeneous ion exchange membranes are often cross-linked, thus cannot be melted.

Membrane casting (Figure 7b) offers the possibility of preparing profiled membranes from a homogeneous material. The materials and operating conditions used for preparation of a profiled homogeneous ion exchange membrane are the same as for a flat membrane with the following difference: the initial membrane solution is casted onto a corrugated mould (could be the same as the one used for hot pressing) instead on the surface of a flat plate [22,23]. The thickness of the non-profiled parts of the membrane (therefore the membrane base) can be adjusted by changing the volume or concentration of the membrane-forming solution spread on the mould, while the size of the corrugations depends from the mould shape. The so far used membrane casting of profiled ion exchange membranes was always based on phase inversion (meaning solvent evaporation and solidification of membrane material in the mould) [22,23]. The disadvantage of membrane casting over hot pressing is the difficulty of releasing the profiled membranes formed from the mould, as many of them break during the process. Alternatively, a spacer structure can be replicated on a membrane by "capillary force induced surface structuring" [56], in which the top layer (of an adjustable thickness) is modified by placing a spacer on its surface, while the solvent is evaporated (therefore the membrane is created by a phase inversion method as in the previous examples). The main disadvantage of this method is the same as the one previously mentioned regarding the use of spacers in hot pressing.

Finally, 3D printing enables almost infinite possibilities for rapid prototyping of target materials, with different shapes and in different fields, by a number of available methods [57,58]. Considering fabrication of profiled ion exchange membranes, the most promising technique seems to be photopolymerization due to an easier and "greener" way of preparing ion exchange membranes by this method [59]. Seo et al. [52] prepared profiled AEMs via a photoinitiated free radical polymerization and quaternization process. A photocurable formulation (a mixture of, commercially available, oligomers, functional monomers and photoinitiators), can be directly cured into patterned films using a 3D photolithographic printer (Figure 7c). Depending of the initial composition of photocurable formulation, the final membrane material properties such as water uptake, permselectivity (which are all independent of a membrane being flat or profiled) and the ionic resistance (which is lower for profiled patterned membranes when compared with that of flat membranes of the same material volume) can be optimized [52]. Such an approach of preparing profiled ion exchange membranes is fast, solvent-free and occurs at ambient temperature. In any case, there is still a large margin for future progress in 3D printing of profiled membranes since, due to bleeding during curing, replication errors and a low-profile resolution are currently frequent [60]. Alternatively, the corrugations can be deposited on the top of a commercially-available membrane [53] (Figure 7d). 3D printing opens also an exciting new opportunity to create homogeneous profiled membranes with different shapes (Figure 8), which can be firstly designed and optimized by CFD simulations. 3D printers receive information about an object to be printed in a form of an .stl (stereolithography) file, which can be also used to define the geometry of computational domain in CFD studies [61]. Therefore, it becomes possible to change the information saved in .stl files and simulate fluid behaviour in channels formed by different corrugations, and then just print those profiled membranes, which CFD simulated performance would be promising.

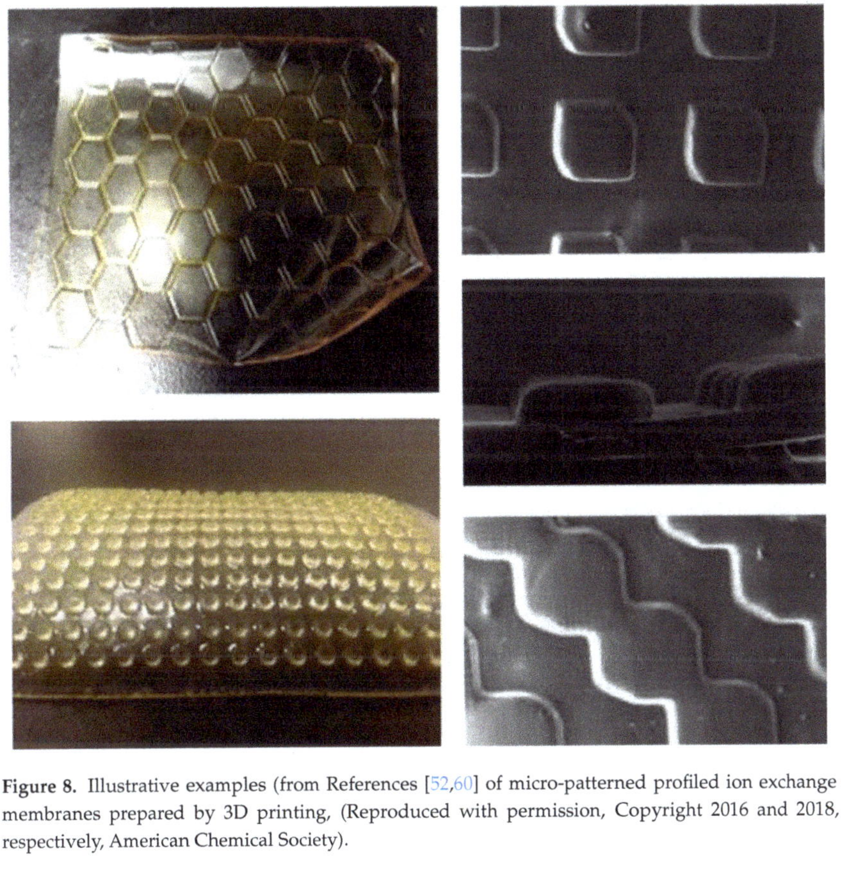

Figure 8. Illustrative examples (from References [52,60] of micro-patterned profiled ion exchange membranes prepared by 3D printing, (Reproduced with permission, Copyright 2016 and 2018, respectively, American Chemical Society).

6. Conclusions

Development of profiled ion exchange membranes have seen significant improvements over the last decade. So far, they have been already applied in research studies for electrodialysis, electrodeionization, reverse electrodialysis, and diffusion dialysis. In all these electromembrane-based processes, the utilization of profiled membranes improved their performance but, especially notorious, are their beneficial effects in electrodialysis. At high flow rates and current densities, profiled membranes facilitate formation of electroconvective vortexes, which enhance mass transfer and reduce deposition of foulants. Other mechanisms, which also promote ion transport enhancement are: lower water splitting rate; intensified hydrodynamic conditions, which lead to a decreased diffusion boundary layer thickness; and increase of active membrane area due to elimination of non-conductive spacers presence which also leads to a lower electric resistance. Thus, ultimately, profiled membranes should be designed with the objective of promoting convective phenomena. To achieve this, there is a large freedom in designing and testing (experimentally and/or by CFD simulations) novel forms, geometries, and shapes of corrugations. In a near future, manufacturing of such profiled membranes, with or without complex geometries, might see a significant progress due to the rapid advent of methods such as 3D printing. Novel applications of profiled ion exchange membranes are also envisaged in emerging processes, such as membrane capacitive deionization (MCDI).

Author Contributions: Writing—original draft preparation, S.P.; writing—review and editing J.G.C.; writing—review and editing S.V.; Funding Acquisition, S.V.

Funding: This research was funded by "Programa Operacional Regional de Lisboa, na componente FEDER" and by "Fundação para a Ciência e Tecnologia, I.P." (Project PTDC/EQU-EPQ/29579/2017).

Acknowledgments: This work was also supported by the Associated Laboratory for Sustainable Chemistry – Clean Processes and Technologies (LAQV), which is financed by Portuguese national funds from FCT/MEC (UID/QUI/50006/2013) and co-financed by the ERDF under the PT2020 Partnership Agreement (POCI-01-0145-FEDER – 007265).

Conflicts of Interest: The authors declare no conflict of interest.

Abbreviations

AEM	Anion exchange membrane
CDI	Capacitive deionization
CEM	Cation exchange membrane
CFD	Computational fluid dynamics
ED	Electrodialysis
IEM	Ion exchange membrane
MCDI	Membrane capacitive deionization
OCF	Overlapped crossed filaments
PTV	Particle tracking velocimetry
Re	Reynolds number
RED	Reverse electrodialysis
SEM	Scanning electron microscope

References

1. Strathmann, H. Electrodialysis, a mature technology with a multitude of new applications. *Desalination* **2010**, *264*, 268–288. [CrossRef]
2. Moon, S.H.; Yun, S.H. Process integration of electrodialysis for a cleaner environment. *Curr. Opin. Chem. Eng.* **2014**, *4*, 25–31. [CrossRef]
3. Paidar, M.; Fateev, V.; Bouzek, K. Membrane electrolysis—History, current status and perspective. *Electrochim. Acta* **2016**, *209*, 737–756. [CrossRef]
4. Pawlowski, S.; Crespo, J.G.; Velizarov, V. Sustainable power generation from salinity gradient energy by reverse electrodialysis. In *Electrokinetics across Disciplines and Continents*; Ribeiro, A.B., et al., Eds.; Springer International Publishing AG: Basel, Switzerland, 2016; pp. 57–80. [CrossRef]

5. Tufa, R.A.; Pawlowski, S.; Veerman, J.; Bouzek, K.; Fontananova, E.; di Profio, G.; Velizarov, S.; Goulão Crespo, J.; Nijmeijer, K.; Curcio, E. Progress and prospects in reverse electrodialysis for salinity gradient energy conversion and storage. *Appl. Energy* **2018**, *225*, 290–331. [CrossRef]
6. Hassanvand, A.; Wei, K.; Talebi, S.; Chen, G.Q.; Kentish, S.E. The role of ion exchange membranes in membrane capacitive deionisation. *Membranes (Basel).* **2017**, *7*, 54. [CrossRef] [PubMed]
7. Jaroszek, H.; Dydo, P. Ion-exchange membranes in chemical synthesis-a review. *Open Chem.* **2016**, *14*, 1–19. [CrossRef]
8. Strathmann, H.; Grabowski, A.; Eigenberger, G. Ion-exchange membranes in the chemical process industry. *Ind. Eng. Chem. Res.* **2013**, *52*, 10364–10379. [CrossRef]
9. Ran, J.; Wu, L.; He, Y.; Yang, Z.; Wang, Y.; Jiang, C.; Ge, L.; Bakangura, E.; Xu, T. Ion exchange membranes: New developments and applications. *J. Memb. Sci.* **2017**, *522*, 267–291. [CrossRef]
10. Luo, T.; Abdu, S.; Wessling, M. Selectivity of ion exchange membranes: A review. *J. Memb. Sci.* **2018**, *555*, 429–454. [CrossRef]
11. Pawlowski, S.; Crespo, J.G.; Velizarov, S. Pressure drop in reverse electrodialysis: Experimental and modeling studies for stacks with variable number of cell pairs. *J. Memb. Sci.* **2014**, *462*, 96–111. [CrossRef]
12. Pawlowski, S.; Sistat, P.; Crespo, J.G.; Velizarov, S. Mass transfer in reverse electrodialysis: Flow entrance effects and diffusion boundary layer thickness. *J. Memb. Sci.* **2014**, *471*, 72–83. [CrossRef]
13. Zabolotskii, V.I.; Loza, S.A.; Sharafan, M.V. Physicochemical properties of profiled heterogeneous ion-exchange membranes. *Russ. J. Electrochem.* **2005**, *41*, 1053–1060. [CrossRef]
14. Brauns, E. Salinity gradient power by reverse electrodialysis: Effect of model parameters on electrical power output. *Desalination* **2009**, *237*, 378–391. [CrossRef]
15. Veerman, J.; Saakes, M.; Metz, S.J.; Harmsen, G.J. Reverse electrodialysis: A validated process model for design and optimization. *Chem. Eng. J.* **2011**, *166*, 256–268. [CrossRef]
16. Larchet, C.; Zabolotsky, V.I.; Pismenskaya, N.; Nikonenko, V.V.; Tskhay, A.; Tastanov, K.; Pourcelly, G. Comparison of different ED stack conceptions when applied for drinking water production from brackish waters. *Desalination* **2008**, *222*, 489–496. [CrossRef]
17. Grabowski, A. *Electromembrane Desalination Processes for Production of Low Conductivity Water*; Logos Verlag: Berlin, Germany, 2010.
18. Nikonenko, V.V.; Pismenskaya, N.D.; Belova, E.I.; Sistat, P.; Huguet, P.; Pourcelly, G.; Larchet, C. Intensive current transfer in membrane systems: Modelling, mechanisms and application in electrodialysis. *Adv. Colloid Interface Sci.* **2010**, *160*, 101–123. [CrossRef] [PubMed]
19. Vermaas, D.A.; Saakes, M.; Nijmeijer, K. Power generation using profiled membranes in reverse electrodialysis. *J. Memb. Sci.* **2011**, *385–386*, 234–242. [CrossRef]
20. Vermaas, D.A.; Kunteng, D.; Saakes, M.; Nijmeijer, K. Fouling in reverse electrodialysis under natural conditions. *Water Res.* **2013**, *47*, 1289–1298. [CrossRef]
21. Vermaas, D.A.; Kunteng, D.; Veerman, J.; Saakes, M.; Nijmeijer, K. Periodic feed water reversal and air sparging as anti fouling strategies in reverse electrodialysis: Supporting Information. *Environ. Sci. Technol.* **2014**, *48*, 3065–3073. [CrossRef]
22. Güler, E.; Elizen, R.; Saakes, M.; Nijmeijer, K. Micro-structured membranes for electricity generation by reverse electrodialysis. *J. Memb. Sci.* **2014**, *458*, 136–148. [CrossRef]
23. Liu, J.; Geise, G.M.; Luo, X.; Hou, H.; Zhang, F.; Feng, Y.; Hickner, M.A.; Logan, B.E. Patterned ion exchange membranes for improved power production in microbial reverse-electrodialysis cells. *J. Power Sources* **2014**, *271*, 437–443. [CrossRef]
24. Gurreri, L.; Ciofalo, M.; Cipollina, A.; Tamburini, A.; Van Baak, W.; Micale, G. CFD modelling of profiled-membrane channels for reverse electrodialysis. *Desalin. Water Treat.* **2015**, *55*, 3404–3423. [CrossRef]
25. Ngene, I.S.; Lammertink, R.G.H.; Wessling, M.; Van der Meer, W.G.J. Particle deposition and biofilm formation on microstructured membranes. *J. Memb. Sci.* **2010**, *364*, 43–51. [CrossRef]
26. Pawlowski, S.; Galinha, C.F.; Crespo, J.G.; Velizarov, S. Prediction of reverse electrodialysis performance by inclusion of 2D fluorescence spectroscopy data into multivariate statistical models. *Sep. Purif. Technol.* **2015**, *150*, 159–169. [CrossRef]
27. Pawlowski, S.; Galinha, C.F.; Crespo, J.G.; Velizarov, S. 2D fluorescence spectroscopy for monitoring ion-exchange membrane based technologies—Reverse electrodialysis (RED). *Water Res.* **2016**, *88*, 184–198. [CrossRef]

28. Mikhaylin, S.; Bazinet, L. Fouling on ion-exchange membranes: Classification, characterization and strategies of prevention and control. Adv. *Colloid Interface Sci.* **2016**, *229*, 34–56. [CrossRef]
29. Luque Di Salvo, J.; Cosenza, A.; Tamburini, A.; Micale, G.; Cipollina, A. Long-run operation of a reverse electrodialysis system fed with wastewaters. *J. Environ. Manage.* **2018**, *217*, 871–887. [CrossRef]
30. Xie, M.; Luo, W.; Gray, S.R. Surface pattern by nanoimprint for membrane fouling mitigation: Design, performance and mechanisms. *Water Res.* **2017**, *124*, 238–243. [CrossRef] [PubMed]
31. Ding, Y.; Maruf, S.; Aghajani, M.; Greenberg, A.R. Surface patterning of polymeric membranes and its effect on antifouling characteristics. *Sep. Sci. Technol.* **2017**, *52*, 240–257. [CrossRef]
32. Choi, W.; Chan, E.P.; Park, J.; Ahn, W.; Jung, H.W.; Hong, S.; Lee, J.S.; Han, J.; Park, S.; Ko, D.; et al. Nanoscale pillar-enhanced tribological surfaces as antifouling membranes. *ACS Appl. Mater. Interfaces* **2016**, *8*, 31433–31441. [CrossRef] [PubMed]
33. Won, Y.J.; Jung, S.Y.; Jang, J.H.; Lee, J.W.; Chae, H.R.; Choi, D.C.; Hyun Ahn, K.; Lee, C.H.; Park, P.K. Correlation of membrane fouling with topography of patterned membranes for water treatment. *J. Memb. Sci.* **2016**, *498*, 14–19. [CrossRef]
34. Choi, D.C.; Jung, S.Y.; Won, Y.J.; Jang, J.H.; Lee, J.W.; Chae, H.R.; Lim, J.; Ahn, K.H.; Lee, S.; Kim, J.H.; et al. Effect of Pattern shape on the initial deposition of particles in the aqueous phase on patterned membranes during crossflow filtration. *Environ. Sci. Technol. Lett.* **2017**, *4*, 66–70. [CrossRef]
35. Zhao, Y.; Wang, H.; Jiang, C.; Wu, L.; Xu, T. Electrodialysis with notched ion exchange membranes: Experimental investigations and computational fluid dynamics simulations. *Sep. Purif. Technol.* **2014**, *130*, 102–111. [CrossRef]
36. Vermaas, D.A.; Saakes, M.; Nijmeijer, K. Enhanced mixing in the diffusive boundary layer for energy generation in reverse electrodialysis. *J. Memb. Sci.* **2014**, *453*, 312–319. [CrossRef]
37. Nikonenko, V.V.; Kovalenko, A.V.; Urtenov, M.K.; Pismenskaya, N.D.; Han, J.; Sistat, P.; Pourcelly, G. Desalination at overlimiting currents: State-of-the-art and perspectives. *Desalination* **2014**, *342*, 85–106. [CrossRef]
38. Wessling, M.; Morcillo, L.G.; Abdu, S. Nanometer-thick lateral polyelectrolyte micropatterns induce macrosopic electro-osmotic chaotic fluid instabilities. *Sci. Rep.* **2014**, *4*, 1–5. [CrossRef] [PubMed]
39. Nikonenko, V.V.; Mareev, S.A.; Pis'menskaya, N.D.; Uzdenova, A.M.; Kovalenko, A.V.; Urtenov, M.K.; Pourcelly, G. Effect of electroconvection and its use in intensifying the mass transfer in electrodialysis (Review). *Russ. J. Electrochem.* **2017**, *53*, 1122–1144. [CrossRef]
40. De Valença, J.; Jõgi, M.; Wagterveld, R.M.; Karatay, E.; Wood, J.A.; Lammertink, R.G.H. Confined Electroconvective Vortices at Structured Ion Exchange Membranes. *Langmuir* **2018**, *34*, 2455–2463. [CrossRef]
41. Roghmans, F.; Evdochenko, E.; Stockmeier, F.; Schneider, S.; Smailji, A.; Tiwari, R.; Mikosch, A.; Karatay, E.; Kühne, A.; Walther, A.; et al. 2D Patterned ion-exchange membranes induce electroconvection. *Adv. Mater. Interfaces* **2018**, *1801309*, 1–11. [CrossRef]
42. Nebavskaya, K.A.; Butylskii, D.Y.; Moroz, I.A. Enhancement of mass transfer through a homogeneous anion-exchange membrane in limiting and overlimiting current regimes by screening part of its surface with nonconductive strips. *Petroleum Chem.* **2018**, *58*, 780–789. [CrossRef]
43. Davidson, S.M.; Wessling, M.; Mani, A. On the dynamical regimes of pattern-accelerated electroconvection. *Sci. Rep.* **2016**, *6*, 1–10. [CrossRef]
44. Belashova, E.D.; Melnik, N.A.; Pismenskaya, N.D.; Shevtsova, K.A.; Nebavsky, A. V; Lebedev, K.A.; Nikonenko, V.V. Overlimiting mass transfer through cation-exchange membranes modified by Nafion film and carbon nanotubes. *Electrochim. Acta* **2012**, *59*, 412–423. [CrossRef]
45. Pawlowski, S.; Geraldes, V.; Crespo, J.G.; Velizarov, S. Computational fluid dynamics (CFD) assisted analysis of profiled membranes performance in reverse electrodialysis. *J. Memb. Sci.* **2016**, *502*, 179–190. [CrossRef]
46. Pawlowski, S.; Rijnaarts, T.; Saakes, M.; Nijmeijer, K.; Crespo, J.G.; Velizarov, S. Improved fluid mixing and power density in reverse electrodialysis stacks with chevron-profiled membranes. *J. Memb. Sci.* **2017**, *531*, 111–121. [CrossRef]
47. La Cerva, M.L.; Liberto, M. Di; Gurreri, L.; Tamburini, A.; Cipollina, A.; Micale, G.; Ciofalo, M. Coupling CFD with a one-dimensional model to predict the performance of reverse electrodialysis stacks. *J. Memb. Sci.* **2017**, *541*, 595–610. [CrossRef]
48. Gurreri, L.; Battaglia, G.; Tamburini, A.; Cipollina, A.; Micale, G.; Ciofalo, M. Multi-physical modelling of reverse electrodialysis. *Desalination* **2017**, *423*, 52–64. [CrossRef]

49. Melnikov, S.; Loza, S.; Sharafan, M.; Zabolotskiy, V. Electrodialysis treatment of secondary steam condensate obtained during production of ammonium nitrate. Technical and economic analysis. *Sep. Purif. Technol.* **2016**, *157*, 179–191. [CrossRef]
50. Vasil'eva, V.; Goleva, E.; Pismenskaya, N.; Kozmai, A.; Nikonenko, V. Effect of surface profiling of a cation-exchange membrane on the phenylalanine and NaCl separation performances in diffusion dialysis. *Sep. Purif. Technol.* **2019**, *210*, 48–59. [CrossRef]
51. Suss, M.E.; Porada, S.; Sun, X.; Biesheuvel, P.M.; Yoon, J.; Presser, V. Water desalination via capacitive deionization: What is it and what can we expect from it? *Energy Environ. Sci.* **2015**, *8*, 2296–2319. [CrossRef]
52. Seo, J.; Kushner, D.I.; Hickner, M.A. 3D Printing of Micropatterned Anion Exchange Membranes. *ACS Appl. Mater. Interfaces* **2016**, *8*, 16656–16663. [CrossRef]
53. Rijnaarts, T. The Role of Membranes in the Use of Natural Salinity Gradients for Reverse Electrodialysis. Ph.D. Thesis, University of Twente, Enschede, The Netherlands, 3 May 2018.
54. Svoboda, M.; Beneš, J.; Vobecká, L.; Slouka, Z. Swelling induced structural changes of a heterogeneous cation-exchange membrane analyzed by micro-computed tomography. *J. Memb. Sci.* **2017**, *525*, 195–201. [CrossRef]
55. Sheldeshov, N.V.; Zabolotskii, V.I.; Loza, S.A. Electric conductivity of profiled ion-exchange membranes. *Pet. Chem.* **2014**, *54*, 664–668. [CrossRef]
56. Balster, J.; Stamatialis, D.F.; Wessling, M. Membrane with integrated spacer. *J. Memb. Sci.* **2010**, *360*, 185–189. [CrossRef]
57. Ambrosi, A.; Pumera, M. 3D-printing technologies for electrochemical applications. *Chem. Soc. Rev.* **2016**, *45*, 2740–2755. [CrossRef]
58. Low, Z.X.; Chua, Y.T.; Ray, B.M.; Mattia, D.; Metcalfe, I.S.; Patterson, D.A. Perspective on 3D printing of separation membranes and comparison to related unconventional fabrication techniques. *J. Memb. Sci.* **2017**, *523*, 596–613. [CrossRef]
59. Yang, S.; Kim, W.-S.; Choi, J.; Choi, Y.-W.; Jeong, N.; Kim, H.; Nam, J.-Y.; Jeong, H.; Kim, Y.H. Fabrication of photocured anion-exchange membranes using water-soluble siloxane resins as cross-linking agents and their application in reverse electrodialysis. *J. Memb. Sci.* **2018**. [CrossRef]
60. Capparelli, C.; Fernandez Pulido, C.R.; Wiencek, R.A.; Hickner, M.A. Resistance and Permselectivity of 3D Printed Micropatterned Anion Exchange Membranes. *ACS Appl. Mater. Interfaces* **2018**. [CrossRef] [PubMed]
61. Pawlowski, S.; Nayak, N.; Meireles, M.; Portugal, C.A.M.; Velizarov, S.; Crespo, J.G. CFD modelling of flow patterns, tortuosity and residence time distribution in monolithic porous columns reconstructed from X-ray tomography data. *Chem. Eng. J.* **2018**, *350*, 757–766. [CrossRef]

International Journal of
Molecular Sciences

MDPI

Article

Electro-Kinetic Instability in a Laminar Boundary Layer Next to an Ion Exchange Membrane

Pierre Magnico

Aix Marseille Univ, CNRS, Centrale Marseille, M2P2 UMR 7340, 13451 Marseille, France; pierre.magnico@univ-amu.fr; Tel.: +33-04-13-55-40-69

Received: 7 March 2019; Accepted: 28 April 2019; Published: 14 May 2019

Abstract: The electro-kinetic instability in a pressure driven shear flow near an ion exchange membrane is considered. The electrochemical system, through which an electrical potential drop is applied, consists in a polarization layer in contact with the membrane and a bulk. The numerical investigation contained two aspects: analysis of the instability modes and description of the Lagrangian transport of fluid and ions. Regarding the first aspect, the modes were analyzed as a function of the potential drop. The analysis revealed how the spatial distribution of forces controls the dynamics of vortex association and dissociation. In particular, the birth of a counter-clockwise vortex between two clockwise vortices, and the initiation of clusters constituting one or two envelopes wrapping a vortex group, were examined. In regards to the second aspect, the trajectories were computed with the fourth order Runge Kutta scheme for the time integration and with the biquadratic upstream scheme for the spatial and time interpolation of the fluid velocity and the ion flux. The results for the periodic mode showed two kinds of trajectories: the trochoidal motion and the longitudinal one coupled with a periodic transverse motion. For the aperiodic modes, other mechanisms appeared, such as ejection from the mixing layer, trapping by a growing vortex or merging vortices. The analysis of the local velocity field, the vortices' shape, the spatial distribution of the forces and the ion flux components explained these trajectories.

Keywords: ion exchange membranes; electro-convective instability; overlimiting current; concentration polarisation; particle tracking

1. Introduction

Ion exchange membranes play an important role for a wide range of engineering applications [1]. These applications can be divided in two main groups: separation processes and energy production and conversion. The main processes gathered in the separation field are electrodialysis (water desalination and salt pre-concentration), diffusion dialysis (acid and base recovery from industrial waste water), bipolar membrane electrodialysis (production of acids and bases), continuous electro-deionization (production of ultrapure water), and capacitive deionization (water desalination and water softening). Energy production and storage include reverse electrodialysis, fuel cells and redox flow batteries [2,3]. The recent development of microscale devices, such as microfuel cells, lab-on-a-chip biomolecule sensing and micropumps, use electro-kinetic phenomena in the presence of ion exchange membranes [4–6]. These membranes are mainly polymeric matrices composed of backbones on which cationic or anionic groups are attached. Owing to the key role of the ion exchange membrane, the role of the functional groups, polymer architecture and chemical procedure of functionalization have been extensively studied and have led to a very wide variety of membranes [1,7]. In the case of microscale devices, a new class of membrane seems to be promising. The structure of these materials is inspired by biological processes, such as enzymatic catalysis or transport regulation of species through membrane proteins [8–11].

Electro-convective instability involving an ion exchange membrane was predicted for the first time by Rubinstein and Zaltzman [12–16]. The authors showed by means of an asymptotic expansion that the deformation of the non-equilibrium electric double layer (EDL) induces a longitudinal force (along the membrane) initiating fluid motion [13,14]. This new kind of instability, confirmed by experimental visualizations [17–23], enhances the ion flux through the membrane system. This phenomenon is all the more important because electro-convective instability occurs at the microscale as it is not bound to the Reynolds number but to the non-linear coupling between the ionic charge density and the electric field.

Three regimes are commonly observed in a current–voltage response when a potential drop ($\Delta\Phi$) is applied normal to the membrane system (see Figure 1). For a small potential drop (region (1)), the ionic conductivity is constant (Ohmic regime). The polarization layer constitutes an electroneutral diffusion layer, over which the concentration gradient is constant, and a quasi-equilibrium EDL. The concentration gradient increases with $\Delta\Phi$, which means that the width of the diffusion layer decreases. It follows that beyond a first potential threshold $\Delta\Phi^*$ (region (2)), the conductivity cancels and the current density reaches a plateau (limiting current regime) [24–26]. A new layer, called the extended space charge (ESC) region, appears between the diffusion layer and the quasi-equilibrium EDL [12]. The ESC region and the quasi-equilibrium EDL form the non-equilibrium EDL. In the ESC region, the electro-neutrality is not fulfilled. The ESC width increases with $\Delta\Phi$. Beyond a second threshold $\Delta\Phi^{**}$ (region (3)), the non-equilibrium EDL becomes unstable. A longitudinal electric charge gradient, located in this layer, induces an electric and a pressure force along the membrane surface. The enhanced ionic transport leads to an increase in current density [27–30]. This last regime is called the over-limiting current regime. The diffusion layer is moved away from the membrane and a new layer, called the mixing layer, takes place between the diffusion one and the non-equilibrium EDL. The width of the mixing layer is much higher than the vortex size owing to a plume of small ion concentration in the region of the rising fluid [21,28,31].

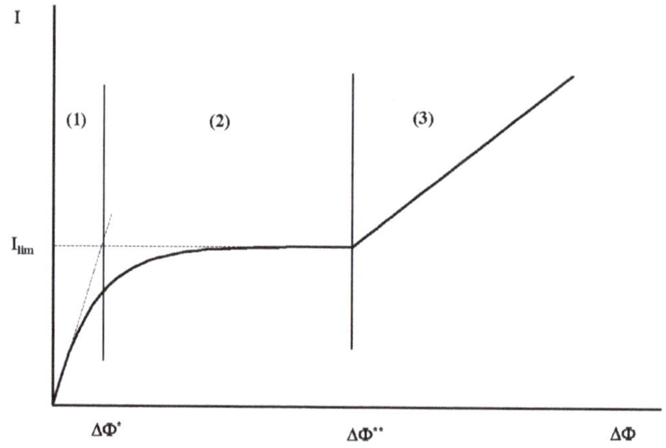

Figure 1. Schematic drawing of the current density (I) vs. potential drop ($\Delta\Phi$) over a membrane system. The regions (1), (2) and (3) localize the Ohmic regime, the limiting current regime and the overlimiting current regime respectively.

Direct numerical simulations are necessary to reach a complete understanding of the electro-kinetic instability mechanism. With this aim, the Navier–Stokes equation, coupled with the Poisson–Nernst–Planck systems, has been solved in previous works for the case of one or two membranes immersed in a stationary reservoir. In this context, instability modes, transition to chaotic motion, 3D electro-convective pattern, bifurcation, non-ideal selectivity and chaotic mixing were investigated [28,31–37]. Recently, the influence of surface heterogeneity on the electro-instability has also been investigated by direct simulations and experimental visualizations [38,39]. In particular, the authors [38] have studied how the electric field, bended by an array of charged P2VP microgels coated on Nafion membranes, enhances the ionic transfer.

Few studies have been published in the case of electro-kinetic instabilities in the presence of an imposed inflow. Kwak et al. [18] found a scaling law governing the vortex thickness. The authors also observed that the vortex advection velocity is dependent on the imposed mean velocity. In a 3D approach, experiments and simulations reveal parallel helical vortices and a merging process along the channel [40]. Nikonenko's group [28,41] described the instability mode appearing successively during a linear potential sweep. The authors observed first a periodic mode with a time period about 0.9 s, when $\Delta\Phi$ is close to the instability threshold. For a higher potential drop, they observed regularly spaced clusters of counter-rotating vortices. They noticed that vortices can emerge in these clusters. This research group also analyzed the width of different zones of the diffusion layer as a function of the current density. In particular, they found an agreement between experiments and simulations in spite of the difference of inlet concentration.

Despite the quality of the published results, the mechanisms controlling the instability modes remain only partially understood. The goal of the current work is to intend to explain some processes like vortex dissociation/association, counter-clockwise vortex birth and also time evolution of the cluster structure by considering the spatial distribution of the electric and pressure force. As in previous publications [28,33,40,41], the fluid flow is visualized by means of streamlines computed at a given time. Therefore, the time evolution of the vortex layer is a snapshot series of these streamlines named 'static' (SSL) in the article. However, considering the unsteady fluid flow, the fully Lagrangian approach provides access to the real hydrodynamics. This is visualized by trajectories called 'dynamic' (DSL) streamlines in the present work. So, in a second step, the close relation and complementarity between these 'dynamic' streamlines and the vortices computed with the SSL are clearly identified and analyzed with the spatial distribution of force and kinetic energy. This gives a new insight into the role of the vortices in the fluid motion and of the relation between the diffusion layer and the real instability layer. This approach is also extended to ion transfer in the periodic mode. Finally, the results are summarized in the conclusion.

2. Theory

2.1. Model and Governing Equations

The elementary electrodialytic cell consists of a channel between two parallel ion exchange membranes. The electrolyte solution enters one end of the channel and exits from the other. The transverse electric potential drop, imposed at the two outer sides of the membranes, induces two polarization layers at their inner sides [41]. In the current work (Figure 2), the membrane system consists of one polarization layer close to a cation exchange membrane of length L_z. The bulk is located at L_x from the membrane. At the inlet ($z = 0$), the ionic concentration is c_o, a constant transverse potential gradient $\Delta\Phi/L_x$ is imposed and a part of the Poiseuille profile is prescribed depending on the distance between the two membranes $L_m(> L_x)$. At the membrane surface ($x = 0$), the cation concentration is c^+_{interf}, the anion flux and the electric potential are zero. These two last conditions mean that the membrane is ideally selective and its electrical resistivity is zero as well. At the bulk limit ($x = L_x$), the potential and the ion concentration are imposed constant and equal to $\Delta\Phi$ and c_o, respectively. The fluid velocity is longitudinal and equal to U_{slip}. Figure 3 displays a snapshot of cationic profiles and streamlines for two values of $\Delta\Phi$ in the case of the monovalent electrolyte (Na^+, Cl^-).

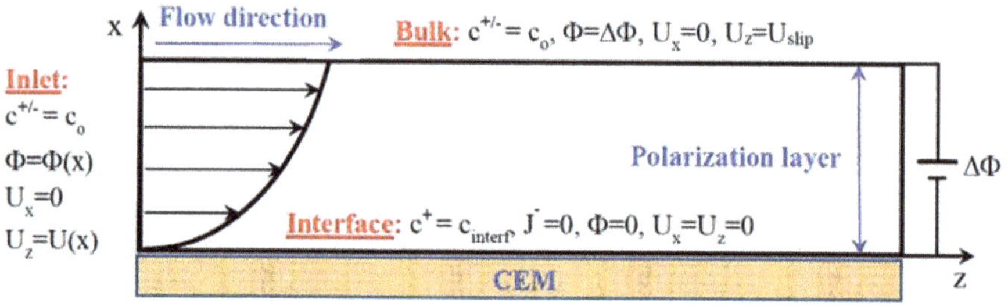

Figure 2. Schematic representation of the numerical model: an electric potential drop $\Delta\Phi$ is imposed over the width of the electrolyte layer (polarization layer) lying between a bulk, containing an electrolyte solution of concentration c_o and a cation exchange membrane at the surface of which the cation concentration c^{+}_{interf} and a non-flux of co-ions are imposed. A Poiseuille flow is imposed at the inlet.

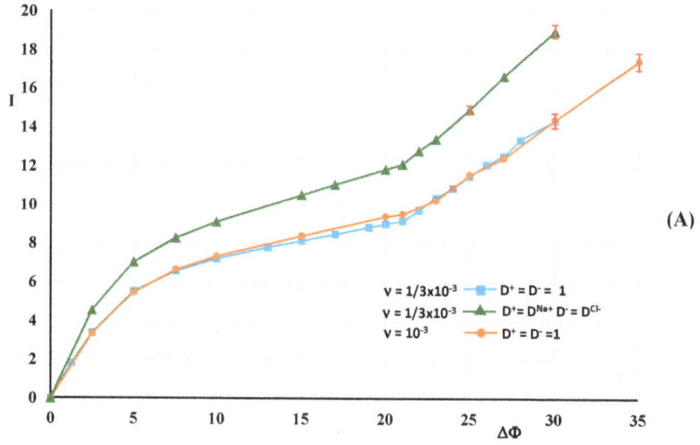

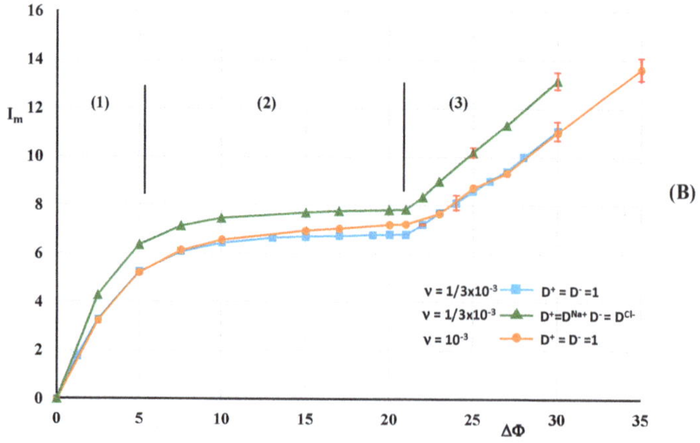

Figure 3. Current density vs. potential drop. (**A**): mean current density through the domain (I); (**B**): current density at the membrane surface (I_m). The red vertical bars represent the variance of the current density signal.

The incompressible fluid motion is described by the Stokes equation (Equation (1)) coupled with the continuity equation (Equation (2)). The ionic transport is governed by the Nernst–Planck equation (Equation (3)) and the electric potential is related to the local ionic concentration by the Poisson equation (Equation (4)):

$$0 = -\vec{\nabla}P + \mu\Delta\vec{U} - \rho_f\left(z^+c^+ + z^-c^-\right)\vec{\nabla}\Phi \tag{1}$$

$$\vec{\nabla}\cdot\vec{U} = 0 \tag{2}$$

$$\frac{d}{dt}c^\pm = \frac{\partial}{\partial t}c^\pm + \vec{U}\cdot\vec{\nabla}c^\pm = D^\pm\vec{\nabla}\cdot\left(\vec{\nabla}c^\pm + \frac{z^\pm c^\pm}{k_BT}\vec{\nabla}\Phi\right) \tag{3}$$

$$-\varepsilon_0\varepsilon_r\Delta\Phi = Q = z^+c^+ + z^-c^- \tag{4}$$

Here P is the pressure, μ is the dynamic viscosity, $\vec{U}$ is the fluid velocity, ρ_f is the fluid density, $z^\pm$ is the charge of the cation and the anion, $c^\pm$ is the concentration of the cation and of the anion, Φ is the electric potential, t is the time, $D^\pm$ is the diffusion coefficient of the cation and of the anion, Q is the electric charge density, ε_o and ε_r are the permittivity of the vacuum and the relative permittivity of the solution, respectively, and k_B and T are the Boltzmann constant and the temperature, respectively.

These equations are scaled as in Druzgalski et al. [31]. The characteristic parameters are the transverse length L_x, the diffusional time $t_{diff} = L_x^2/D_{ref}$, the diffusional velocity $U_{diff} = D_{ref}/L_x$, the thermodynamic potential $\Phi_T = k_BT/e$, the 'osmotic' pressure $P_o = \mu D_{ref}/L_x^2$ and the bulk concentration c_o. Here e is the elementary charge, D_{ref} is a reference diffusion coefficient whose value is arbitrary. For a symmetric univalent electrolyte, Equations (1)–(4), in their dimensionless form, become:

$$0 = -\vec{\nabla}P + \Delta\vec{U} - \frac{Pe}{2\nu^2}\left(c^+ - c^-\right)\vec{\nabla}\Phi \tag{5}$$

$$\vec{\nabla}\cdot\vec{U} = 0 \tag{6}$$

$$\frac{d}{dt}c^\pm = \frac{\partial}{\partial t}c^\pm + \vec{U}\cdot\vec{\nabla}c^\pm = D^\pm\vec{\nabla}\cdot\left(\vec{\nabla}c^\pm \pm c^\pm\vec{\nabla}\Phi\right) \tag{7}$$

$$-2\nu^2\Delta\Phi = Q = c^+ - c^- \tag{8}$$

Here Pe and ν are the Peclet number and the dimensionless Debye length, respectively:

$$Pe = \frac{\varepsilon_0\varepsilon_r}{\mu D_{ref}}\left(\frac{k_BT}{e}\right)^2 \tag{9}$$

$$\nu = \frac{\lambda_d}{L_x} \text{ with } \lambda_d = \left(\frac{\varepsilon_0\varepsilon_r k_BT}{2e^2c_o}\right)^{1/2} \tag{10}$$

For simple ions, the diffusion coefficient is roughly 10^{-9} m^2/s. Hence, this value will be used to define D_{ref} and the value of Pe is imposed equal to 0.5. Therefore, if $L_x = 0.1$ mm and $\nu = 10^{-3}$, then $\lambda_d = 100$ nm, $c_o = 10^{-5}$ M ($T = 300$ K) and the diffusional time is 10 s. In what follows, the dimensionless variables will be used. Therefore, the value of dimensionless transverse length L_x and of the dimensionless bulk concentration c_o are equal to 1.

The boundary conditions used at the fluid/membrane interface are the no-slip condition, and the imposed value of the cationic concentration c^+_{interf} is assumed to be equal to c_o. At the inlet, the imposed velocity profile has the following expression:

$$U_z = 6\langle U_z\rangle\frac{x}{L_m}\left(\frac{x}{L_m} - 1\right) \tag{11}$$

In this expression, $\langle U_z \rangle$ is the longitudinal velocity averaged over the distance between the two membranes and x ranges from 0 to L_x. Therefore, the slip velocity imposed at the bulk boundary is $U_{slip} = U_z(z = 0, x = L_x) = 6\langle U_z \rangle L_x / L_m (L_x / L_m - 1)$. At the outlet ($z = L_z$), the equality of the velocity, of the ion concentration and of the electric potential are imposed on the two sides of the boundary [18]. As the Stokes equation is solved with the finite volume method, the pressure is linearly extrapolated at all the domain boundaries.

At the membrane surface, the no-slip boundary condition is usually applied. However, the slip condition can be justified by the hydrophobic property of membranes. The velocity slip is modeled by the Navier equation, which is characterized by the slip length. The value of this parameter depends on the degree of hydrophobicity and heterogeneity. It is admitted that it is lower than 100 nm for a homogeneous surface and may reach a few micrometers for a grooved surface (superhydrophobicity) [42,43]. Experiments on cation exchange membranes have shown that surface modification by hydrophobization enhances the electro-convection [44,45]. The instability threshold is displaced to lower values of $\Delta\Phi$, and at a given value of $\Delta\Phi$, the current density increases. The consequence is that water splitting and salt precipitation are reduced. Numerical simulations [41,46] confirm the experimental observations. Shelistov et al. [46] have shown that moderate hydrophobicity increases the critical wavelength and the maximal growth rate, and the unsteady mode appears at lower values of $\Delta\Phi$ as the hydrophobicity increases.

In this study, $L_z = 6L_x$ and $L_m = (10/3)L_x$, $\langle U_z \rangle = 240$ (8 mm/s), and $\Delta\Phi$ varies up to 35 (0.9 V). Two values of ν were used (10^{-3} and $1/3 \times 10^{-3}$) and two cases were also considered for $D^{\pm}$ ($D^+ = D^- = 1$ and $D^+ = D^{Na+} = 1.33$, $D^- = D^{Cl-} = 2.06$). In the following, the ionic flux and the current density are defined in their dimensionless form:

$$\vec{J}^{\pm} = c^{\pm}\vec{U} - D^{\pm}\left(\vec{\nabla}c^{\pm} \pm c^{\pm}\vec{\nabla}\Phi\right) \text{ and } I = \left\langle \left(\vec{I}\right)_x \right\rangle_{\Omega} = \frac{1}{A}\int_{\Omega}\left(\left(\vec{J}^{+}\right)_x - \left(\vec{J}^{-}\right)_x\right)dA \qquad (12)$$

In this equation, A is the area of the computational domain Ω.

2.2. Numerical Method

2.2.1. Stokes and Poisson–Nernst–Planck Equations

The same discretization method and structured grid refinement as the previous article [37] were used. The number of nodes is 180 and 384 in the transverse and longitudinal direction, respectively. In the transverse direction, the common ratio of the geometric series is 0.0234. In the longitudinal direction, the refinement threshold is 2.5 Equations (5)–(8) are integrated in time with a time step δt of 10^{-4}.

2.2.2. Streamlines

In References [18,28,30–32,40,41] the hydrodynamic streamlines were computed with the velocity field at a given time. In the present work, these streamlines are named static streamlines (SSL) and are computed with Matlab. In Section 3.2, the instability structure is analyzed with the SSL. They are visualized in black in Figure 4 for example. In Section 3.3, the real trajectories of the fluid elements and ions are computed with the numerical method described below. These trajectories will be named dynamic streamlines (DSL) owing to the use of the unsteady fluid and ion transfer. The trajectories are displayed in red in Section 3.3.

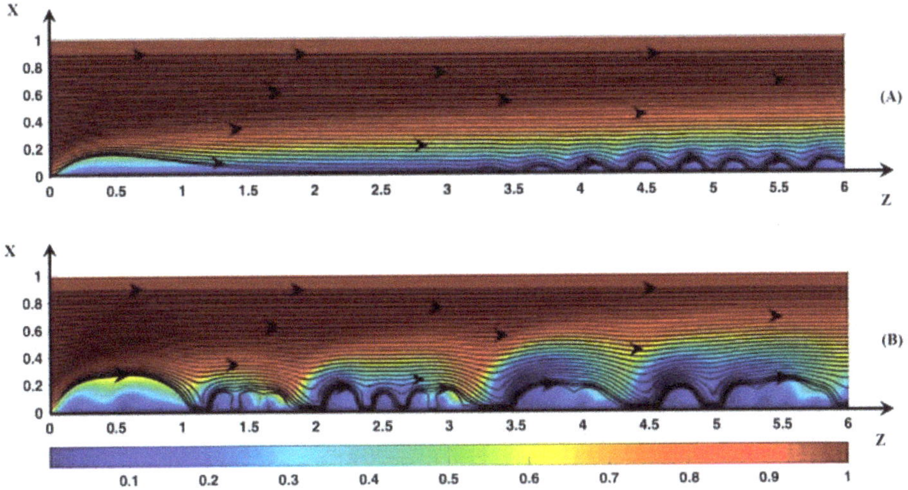

Figure 4. Color plot of the cationic concentration. (**A**): $\Delta\Phi = 22$, (**B**): $\Delta\Phi = 30$. The black lines represent the static streamlines.

Two kinds of dynamic streamlines are computed in this work: the hydrodynamic and the ionic one. As regards to the ions, their local velocity is computed by means of the ratio between the flux density and the concentration: $\vec{J}^{\pm}/c^{\pm}$. As the velocity field and the spatial distribution of concentrations and potential are unsteady, the time integration is performed by means of the fourth order Runge Kutta method.

Consider an elementary volume of fluid located at the point (z_o, x_o) at time t_o. Four Eulerian values of velocity are needed to compute the Lagrangian one:

$$
\begin{aligned}
\vec{U}_1 &= \vec{U}(z_o,\ x_o,\ t_o) \\
\vec{U}_2 &= \vec{U}\left(z_o + \left(\vec{U}_1\right)_z \times \delta t/2,\ x_o + \left(\vec{U}_1\right)_x \times \delta t/2,\ t_o + \delta t/2\right) \\
\vec{U}_3 &= \vec{U}\left(z_o + \left(\vec{U}_2\right)_z \times \delta t/2,\ x_o + \left(\vec{U}_2\right)_x \times \delta t/2,\ t_o + \delta t/2\right) \\
\vec{U}_4 &= \vec{U}\left(z_o + \left(\vec{U}_3\right)_z \times \delta t,\ x_o + \left(\vec{U}_3\right)_x \times \delta t,\ t_o + \delta t\right)
\end{aligned}
\quad (13)
$$

The Eulerian fluid velocity $\vec{U}_1$ at a point (z_o, x_o) at time t_o is computed by a spatial interpolation with an upstream biquadratic polynomial. $\vec{U}_2$ is computed in two steps. First a quadratic time interpolation of the nine velocities, used for the spatial interpolation, must be carried out because $\vec{U}_2$ is determined at the intermediate time $t_o + \delta t/2$. Then the spatial upstream biquadratic interpolation is performed with the velocities interpolated in time. The time interpolation is performed with the velocity fields at time $t_o - \delta t$, t_o, and $t_o + \delta t$. The velocities $\vec{U}_3$ and $\vec{U}_4$ are computed in the same way as $\vec{U}_2$ and $\vec{U}_1$, respectively. The computation of the ionic Lagrangian velocity is carried out in the same manner with the three flux contributions ($\vec{J}^{\pm}_{Co} = c^{\pm}\vec{U}, \vec{J}^{\pm}_{Fi} = -D^{\pm}\vec{\nabla}c^{\pm}, \vec{J}^{\pm}_{Em} = \mp D^{\pm}c^{\pm}\vec{\nabla}\Phi$) and with the ion concentration; here Co, Fi and E_m mean convective, Fickian and electro-migration contribution, respectively. The interpolated flux is divided by the interpolated concentration to determine the four $\vec{U}^{\pm}$.

3. Numerical Results

3.1. Comparison with Published Results

To validate the numerical method, the results were compared to those published in References [18,28,41]. To my knowledge, no analytical solution is available in the literature. Several analyses of the polarization layer structure in an over-limiting regime have been carried out by assuming 1D ion transfer at steady state (see Zaltzman et al. [14] and references therein). With these assumptions, the electric field is a solution of the 2nd kind Painlevé equation which does not admit a simple analytical solution.

Figure 3 shows the mean current density through the boundary layer (I) and through the membrane surface (I_m) as a function of the potential drop. Three series of computations were carried out. Each point of the curves corresponds to the value of the current density averaged in time at a prescribed potential drop. In the Ohmic and limiting regimes, the potential drop increases by a step of 2.5 until it reaches a value of 21. In the over-limiting regime, a first series is carried out with $D^{\pm} = 1$, $\nu = 1/3 \times 10^{-3}$. Beyond the value of 22, the initial condition is the unsteady solution at the previous value of $\Delta\Phi$. Concerning the other series, the initial condition at a given potential drop is the unsteady solution of the first series with the same value of $\Delta\Phi$ in order to reduce the transitory period. The series 1 and 2 ($D^+ = D^{Na+}$, $D^- = D^{Cl-}$, $\nu = 1/3 \times 10^{-3}$) will be used for the comparison with published numerical results and theoretical ones. The 3rd series is computed with $D^{\pm} = 1$ and $\nu = 10^{-3}$. The time average of the current density is computed after a stabilization period of the current signal. By stabilization it is not meant that steady state is recovered. The transitory period can take several hundred time steps, especially after a jump of the potential. The current density is averaged over one to two thousand time steps. In Sections 3.2 and 3.3, the instability structure and the dynamic streamlines are analyzed after the transitory period.

These series show that the over-limiting threshold values lie between $\Delta\Phi = 21$ and 22 (0.57 V). This value, slightly sensitive to the diffusion coefficient and the length ratio ν, is higher than the value found in the case of periodic conditions [31,37]. In Nikonenko et al. [28], an expression of the local limiting current density, in the case of a laminar flow between two plates, was proposed. Averaged over the domain length, the limiting current density I_{lim} has the following expression (Equation (17.56) in Newmann et al. [47]):

$$I_{\text{lim}} = \frac{2}{L_x}\left[1.47\left(\frac{L_x^2\langle U_z\rangle}{L_z D}\right)^{1/3}\right] \tag{14}$$

where $D = 2D^+D^-/(D^+ + D^-)$. The expression (14) is equivalent to Equation (3b) in Nikonenko et al. [28] because the cation transfer numbers T_1 and t_1 are equal to 1 and 0.5, respectively. In Equation (14) L_x is used instead of L_m because the potential drop is imposed over L_x. In the same idea, $\langle U_z\rangle$ is the inlet velocity averaged over the computational domain width. This relation gives a value of 9 and 12.4 for the series 1 and 2, respectively. Therefore, it seems from Figure 3 that I is the relevant quantity in the present model.

Let us compare the cation profiles obtained in the present work and those published in References [28,41]. The authors investigated the ion transfer between an anion exchange membrane and a cation one. In Urtenov et al. [41], the two membranes were non-ideal, i.e., co-ions could cross the membranes. The potential drop is imposed at the outer membrane surfaces. In this publication, Figure 5 shows the static streamlines and the map of the cation concentration with the parameter values $L_x = L_m = 1$ mm, $L_z = 2$ mm, $\nu = 10^{-4}$, $\langle U_z\rangle = 8 \times 10^{-4}$ ms^{-1}, $D^{+/-} = D^{Na+/Cl-}$. Therefore, if a part of the domain is modeled, i.e., $L_x = 0.3$ mm, one obtains $\nu = 1/3 \times 10^{-3}$, which corresponds to series 2 if $D^{+/-} = D^{Na+/Cl-}$. However, the value of the potential threshold is close to 1 V instead of 0.57 V. Two reasons may explain this difference in value: (1) The potential gradient is located mainly in high resistance regions, i.e., where the ion concentration is low. So, the transverse gradient takes place in the two polarization layers [28,41]. Assuming the problem is symmetric, the potential drop through the cationic polarization layer would be equal to about half of the overall potential drop. (2) The membranes have a non-zero electrical resistivity and the potential drop is imposed outside the electro-dialytic cell. These two conditions lead to an increase in the potential threshold. However, it also seems that the potential threshold depends on L_m [28,41]. At the threshold, the authors [41] found that the outlet cation concentration was equal to 0.6 at $x = x_{c^+=0.6} = L_m/10$ (100 µm) from the cation exchange membrane, which corresponds to $L_x/3$ in the present work at $\Delta\Phi = 22$ (see Figure 4A). Their values $\Delta\Phi = 1.8$ V, $c^+ = 0.6$ at $x_{c^+=0.6} = L_m/7$ (140 µm) correspond to $\Delta\Phi = 30 = 1.4 \times 21.5$, $c^+ = 0.6$ at $x_{c^+=0.6} = 0.5L_x$ (150 µm) found here. The counterion concentration color map shown in Figure A2 (Nikonenko et al. [28]) come from simulations carried out with the same type of membrane cell ($L_x = L_m = 0.5$ mm, $L_z = 2$ mm, $\langle U_z \rangle = 8 \times 10^{-4}$ ms^{-1}, $D^{+/-} = D^{Na^+/Cl^-}$) and with several inlet ion concentrations. Two cases may be used for comparison: $v = 2 \times 10^{-4}$ (Figure A2a) and $v = 6 \times 10^{-5}$ (Figure A2b). In these Figures $\Delta\Phi = 1.43\Delta\Phi^{**}$ with $\Delta\Phi^{**} = 0.7$V. In the present work, $L_m = 0.5$ mm and $\Delta\Phi = 1.43\Delta\Phi^{**}$ correspond to $L_x = 0.15$ mm and $\Delta\Phi = 31$, respectively. In Figure A2a,b, $c^+/c_o = 0.6$ at $x_{c^+=0.6} = 0.6L_x$ (90 µm), which is close to $x_{c^+=0.6} = 0.5L_x$ (75 µm) obtained with $v = 1/3 \times 10^{-3}$ (Figure 4B). The color bar range in terms of dimensionless concentration is assumed to be identical to the range in Figure 5 (Urtenov et al. [40]). The values of $x_{c^+=0.6}$ are given in Table 1.

Table 1. Width cation boundary layer defined by $c^+ = 0.6$ at the outlet of the domain. 1st row: published values (Nikonenko et al. [28] and Urtenov et al. [41]). 2nd row: this work.

$x_{c^+=0.6}$	100 µm [41]	140 µm [41]	90 µm [28]
$x_{c^+=0.6}$	100 µm	150 µm	75 µm

Kwak et al. [18] found a linear correlation between the vortex size δ_v and the potential drop $\Delta\Phi_v$ over δ_v: $\delta_v/L_m = a\left(\Delta\Phi_v^2/\langle U_z \rangle\right)^{1/3} + b$ where $\langle U_z \rangle$ is the mean inlet velocity (see Table 2). With their numerical model, Urtenov et al. [41] verified this linear behavior. In the present work, as $\Delta\Phi$ increases from 23 to 30 (Figure 4), at the outlet $\Delta\Phi_v$ and δ_v/L_m increase from 19.4 to 25.7 and from 0.04 to 0.07, respectively. If the velocity is averaged over L_x, the values of a and b are very close to those found by Urtenov et al. [41] (Table 2). Similar values are found if the velocity is averaged over L_m (a = 0.0215, b = 0.107).

Table 2. Parameter values of the correlation $\delta_v/L_m = a\left(\Delta\Phi_v^2/\langle U_z \rangle\right)^{1/3} + b$. 1st and 2nd columns: published values. 3rd columns: this work.

	Kwak et al. [18]	**Urtenov et al. [41]**	**This Work**
a	0.0101	0.0196	0.0194
b	−0.1714	−0.1084	−0.0107

3.2. Hydrodynamic Instability

The Nikonenko research group has mainly investigated the structure of the polarization layer. However, the instability modes are briefly studied, and the cluster dynamics and structure are not addressed. On the contrary, in this section, the key role of the force spot upstream of the vortex front in the vortex shape and in the birth of counter-clockwise vortices and clusters is shown as an example. In Section 3.3, the role of SSL bending, induced by the layer $\left(\vec{F}^{Tot}\right)_z < 0$ at the vortex rear, in the fluid flux through the outer edge of the unsteady vortex layer will be also pointed out. The investigation is performed with $D^{\pm} = 1$ and $v = 1/310^{-3}$. Another mode of instability found with $D^{\pm} = 1$ and $v = 10^{-3}$ is briefly discussed in the Appendix A.

3.2.1. Periodic Mode Flow ($\Delta\Phi = 22$)

Before going further, the main results described in the previous article [37] about the spatial distribution of the electric and pressure forces and of the fluid kinetic energy are recalled briefly. In forced flow, the periodic mode close to the instability threshold is very similar to the marginal instability because the rolls are stable during their travel to the outlet. In periodic systems, the steady instability is characterized by two counter-rotating vortices. The non-equilibrium EDL has a minimum width in the downward flow region and presents a hump at the upward one. The hump extends between the two vortex centers. Q decreases mainly in the hump because its transverse mean value remains constant. It is recalled that the non-equilibrium EDL gathers the ESC region and the quasi-equilibrium EDL. Two force layers along the membrane induce a longitudinal flow. In the first layer placed in the quasi-equilibrium EDL, the pressure force is dominant, whereas the electric one is dominant in the second layer located around the local maximum of electric charge density ($Q_{\max}$) in the ESC region. Two spots of kinetic energy are located in the second layer and close to the ESC hump. A third spot is located between the two vortices in the upward region due to the convergence of the fluid streamlines. The pressure and potential disequilibrium on both sides of the vortices is explained first by the transverse profile of charge density, which controls the variation of the potential gradient, and also by the equilibrium between the electric force and the pressure one in the transverse direction leading to the quadratic dependence of the pressure on the transverse potential gradient.

Figure 5 displays the vortices (SSL) and the color map of the two components of the total force ($\vec{F}^{Tot}$), which is the sum of the electric ($\vec{F}^{El}$) and pressure force ($\vec{F}^{Pr}$). All vortices rotate clockwise. The region of upward fluid motion defines the front of the vortex. Inversely, the region of downward fluid motion defines the rear of the vortex. The space between two successive vortices, i.e., the inter-vortex region, replaces the counter-clockwise vortex in the periodic model. The circumstances in which a counter-clockwise vortex occurs this space will be discussed. The minimum width of the non-equilibrium EDL is located at the rear of the vortex and the hump takes place around the vortex front (upward flow). The two spots of force (Figure 5A) are equivalent to the spots shown in Reference [37] but with a much higher intensity. In these spots, the pressure force is dominant.

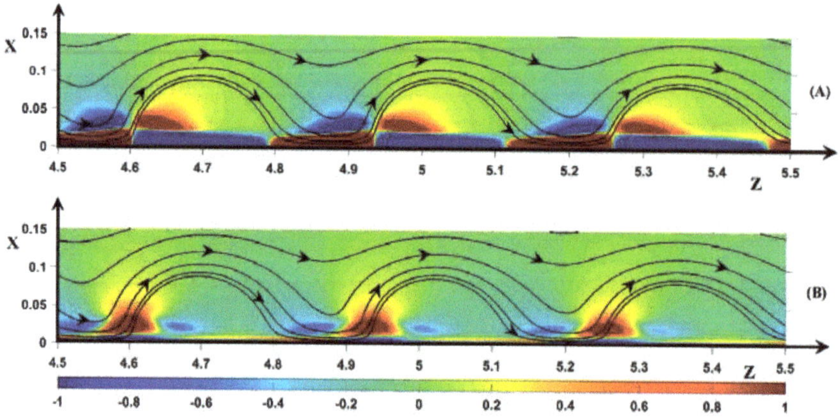

Figure 5. Color plot of: (**A**) $(\vec{F}^{Tot})_z$; (**B**) $(\vec{F}^{Tot})_x$. Black lines: static streamlines starting from the inlet. $\Delta\Phi = 22$, $v = 1/3 \times 10^{-3}$, $D^{\pm} = 1$.

A layer and a spot of the same force sign, located at two different transverse positions, may be linked as shown in Figure 5A. This means that a continuous path of the same force sign may go through the front or the rear of the vortex. This controls the inclination of the static streamlines. For this reason, the streamlines are oriented downstream of the front. When there is no link, the front is vertical. At the rear, in the vortex located at $z \approx 5.3$ for example, the spot $(\vec{F}^{Tot})_z > 0$ is linked to the layer of positive sign, whereas, in the upstream vortices rear ($z \approx 5$ and $z \approx 4.7$), the pair spot/layer $(\vec{F}^{Tot})_z < 0$ are linked. This explains the small difference in orientation of the streamlines at the rear of the two vortices and the small deviation from the exact periodicity. Below, the effect of these spot/layer links on the form of the vortices and their influence on the vortex dynamics will be more clearly shown and described.

Figure 5B shows the transverse component of the total force $(\vec{F}^{Tot})_x$. $(\vec{F}^{El})_x$ is always negative as a consequence of the positive sign of Q. The transverse pressure force is opposed to the electric one in the front region, whereas it is negative in the rear region. Three spots are located around the vortex front. At the vortex front, the pressure force dominates ($(\vec{F}^{Tot})_x > 0$). The spot is located inside the ESC region. On both sides of the front, the electric force is dominant ($(\vec{F}^{Tot})_x < 0$) suppressing the upward fluid motion. The bottom of these two spots are located at the outer edge of the ESC region [37].

In Figure 6, the fluid kinetic energy (E_c) also presents three spots of high magnitude. However, the kinetic energy at the front is much smaller than when there is no forced flow [37], and it induces a smaller roll size. This also induces a periodic trajectory of the anions (see Section 3.3.3). A disequilibrium of E_c is also visible on both sides of the front. As when there is no forced flow, the two E_c spots are located at the transverse position of $Q_{\max}$: the fluid is accelerated in the longitudinal direction by the electric force and reaches a maximum kinetic energy when $Q_{\max}$ and the ESC width begin to decrease and increase, respectively.

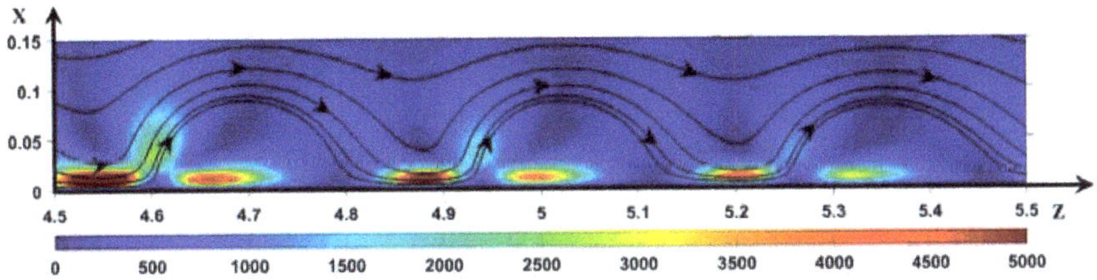

Figure 6. Color plot of the hydrodynamic kinetic energy E_c. Black lines: static streamlines starting from the inlet. $\Delta\Phi = 22$, $\nu = 1/3 \times 10^{-3}$, $D^{\pm} = 1$.

Contrary to [28,41], another periodic mode is also observed at $\Delta\Phi = 23$. The period of the vortex structure is composed of a vortex pair formed from a large and a small vortex (Figure 7). The vortex pair size is seen to increase with time. Two consecutive vortices begin to form a pair when the upstream vortex (V1) reaches the position $z = 3.5$. The pair formation consists of three steps. During the first one, the downward vortex (V2) comes closer to the downstream pair so that its size decreases. At the two ends of V2, the streamlines are vertical close to the membrane because the negative layer and the negative spots $(\vec{F}^{Tot})_z$ are linked at these ends (for an example, see a similar situation at $z = 4.7$ in Figure 8D). In the second step, when V2 reaches its minimum size, V1 begins to move closer to V2. The negative layer and the negative spot $(\vec{F}^{Tot})_z$ get closer at the rear of V1, whose size decreases during its displacement. In the third step, the negative spot $(\vec{F}^{Tot})_z$ between V1 and V2 flattens and the two layers $(\vec{F}^{Tot})_z < 0$ at the rear of V1 join leading to vertical streamlines near the interface. At the same time, at the rear of V2, the positive spot and the positive layer $(\vec{F}^{Tot})_z$ also join. The length of V2 increases and its height decreases accordingly. Figure 7 shows the third and final step.

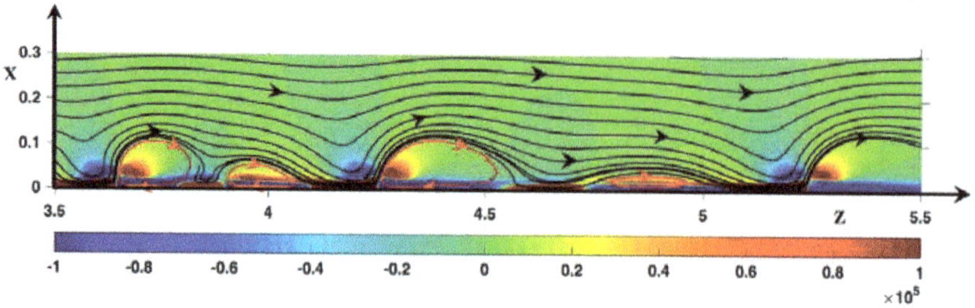

Figure 7. Color plot of $(\vec{F}^{Tot})_z$. Black lines: static streamlines. $\Delta\Phi = 23$, $\nu = 1/3 \times 10^{-3}$, $D^{\pm} = 1$.

3.2.2. Non-Periodic Mode Flow ($\Delta\Phi = 27$)

As the potential drop increases, the periodicity disappears gradually. At $\Delta\Phi = 27$ and higher, an unsteady instability mode is clearly established and the rolls associate to form clusters. The simulations with $D^{+/-} = D^{Na+/Cl-}$, $\Delta\Phi = 27$, $\nu = 1/3 \times 10^{-3}$, show vortex groups independent from one another as observed in Nikonenko et al. [28] and no roll exchange occurs. Inside the groups, dissociation and association are observed. With $D^{\pm} = 1$, the spatial periodicity of the vortex group is not observed, and the clusters are a transitory structure. They become one vortex (Figure 8) or break into several rolls (Figure 9). It must be noticed that the definition of envelope and cluster is not the same as in Nikonenko et al. [28]. In the latter article, the envelope is the black SSL around a vortex group and a cluster is a counter-rotating vortex pair. In the current work, a cluster is a vortex group wrapped in an envelope (e.g., red SSL in Figure 8C). It does not contain counter-clockwise rolls but several clockwise ones.

In Figures 8 and 9, the static streamlines (SSL) are associated with the color map of $(\vec{F}^{Tot})_z$. The blue, the green and the red SSL represent clockwise rolls, counter-clockwise rolls and the cluster envelopes, respectively. The time evolution of the cluster begins at $\delta t = 1600\delta t$ (with $\delta t = 10^{-4}$). For the sake of simplicity, this time is used as the initial one and the time is expressed in terms of δt in the comments (with $t' = t/\delta t - 1600$).

First, let us consider Figure 8. In Figure 8A ($t' = 0$), a counter-clockwise roll (V1/−) lies between the two clockwise vortices V1/+ and V2/+ located at $z = 3.4$ and 3.8, respectively. The signs + and − refer to the rotation direction. In Figure 8A at the rear of V2/+, the positive layer $(\vec{F}^{Tot})_z$ is the pressure force layer. It can be recalled that the negative layer $(\vec{F}^{Tot})_z$ is composed of the pressure layer next to the membrane and the electric one just above. Owing to the negative pressure force spot between V1/+ and V2/+, the black SSL at the rear of V1/+ and at the front region of V2/+ are vertical. V1/− is surrounded by the fluid film separating the vortex from V1/+, V2/+ and the membrane. V1/− rotates counter-clockwise due to the friction with the fluid film, despite the negative sign of $(\vec{F}^{Tot})_z$ at the roll bottom. That is the pressure force acts against the fluid motion at the bottom of V1/−. The inter-vortex region can be viewed as a cavity with a slip condition at the walls. This condition is induced by the fluid film.

In Figure 8B ($t' = 20$), the volume of V1/− increases and its bottom moves to the membrane, so that it is clearly located inside the positive layer $(\vec{F}^{Tot})_z$. However, the thin fluid film continues to separate V1/− from the membrane. The vortex V1/+ front gets closer to V2/+ which does not move. The negative pressure spot $(\vec{F}^{Tot})_z$ begins to join the layer of the same sign through the front of V2/+. A new vortex appears in the tail of V2/+.

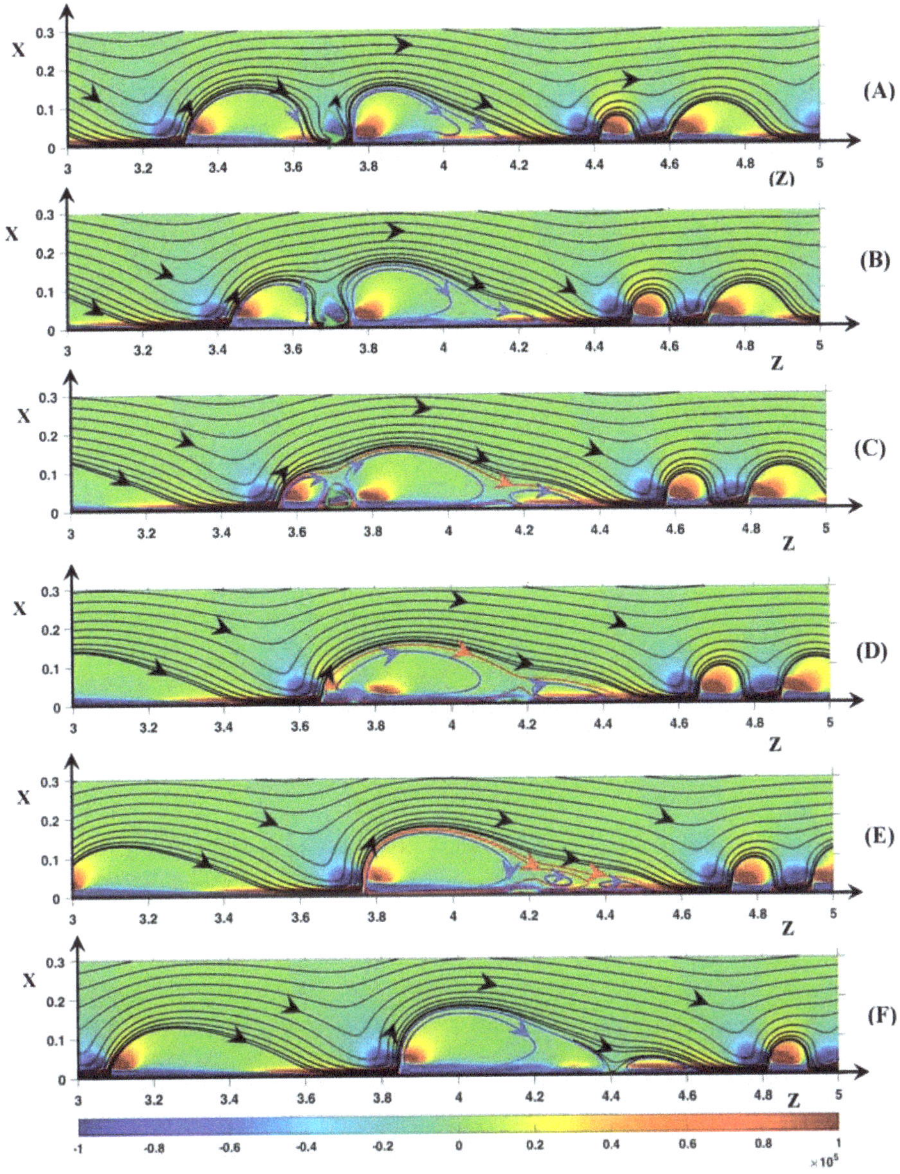

Figure 8. Temporal evolution of the cluster structure. (**A**) $t' = 0$; (**B**) $t' = 20$; (**C**) $t' = 40$; (**D**) $t' = 60$: (**E**) $t' = 80$; (**F**) $t' = 100$. Color plot: $(\vec{F}^{Tot})_z$. Black lines: static streamlines starting from the inlet. Blue streamlines: clockwise rotating rolls, green streamlines: counter clockwise rotating rolls, red streamlines: cluster envelope. $\Delta\Phi = 27$, $\nu = 1/3 \times 10^{-3}$, $D^{\pm} = 1$.

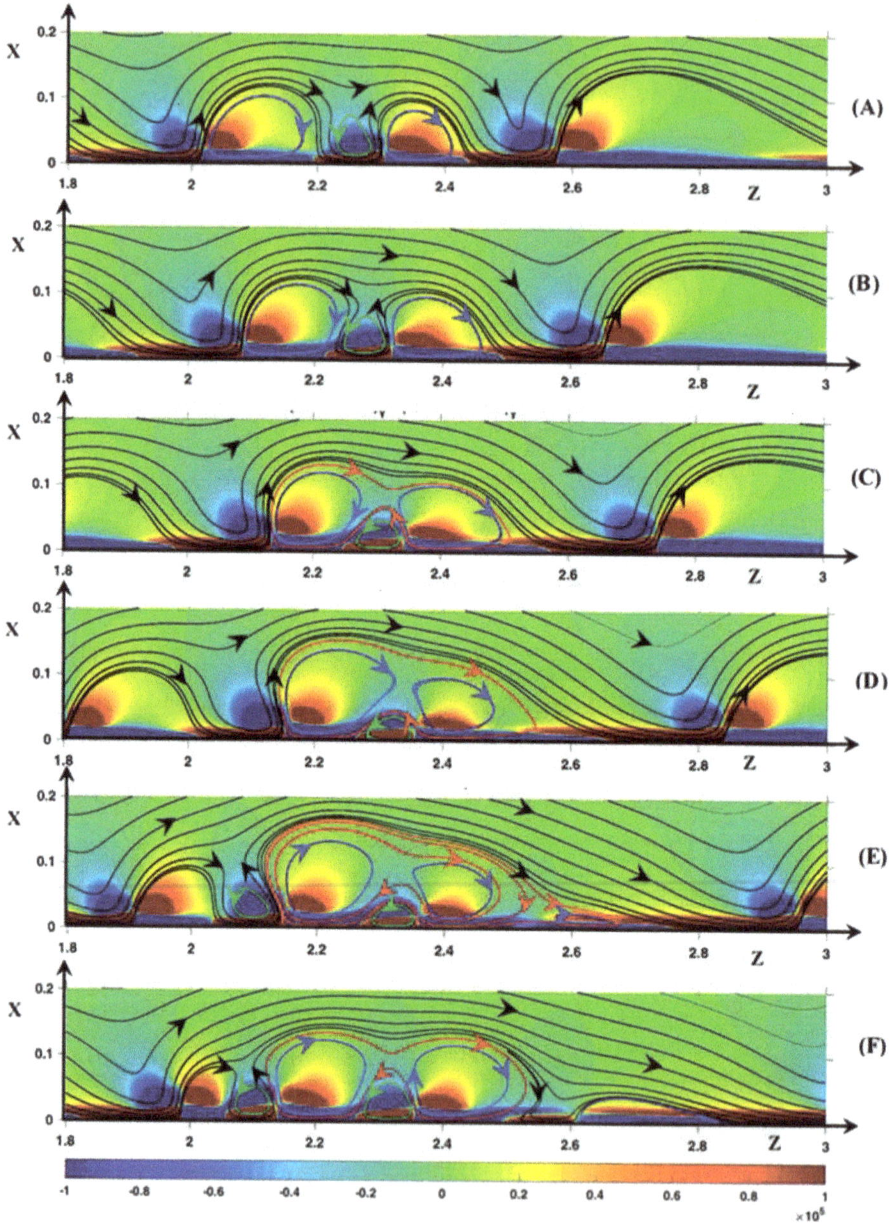

Figure 9. Temporal evolution of the cluster structure. (**A**) $t' = 0$; (**B**) $t' = 20$; (**C**) $t' = 40$; (**D**) $t' = 60$: (**E**) $t' = 80$; (**F**) $t' = 100$. Color plot: $(\vec{F}^{Tot})_z$. Black lines: static streamlines starting from the inlet. Blue streamlines: clockwise rotating rolls, green streamlines: counter-clockwise rotating rolls, red streamlines: cluster envelope. $\Delta\Phi = 27$, $\nu = 1/3 \times 10^{-3}$, $D^{\pm} = 1$.

In Figure 8C ($t' = 40$), the negative layer $(\vec{F}^{Tot})_z$ inside V1/+ connects with the negative spot $(\vec{F}^{Tot})_z$ inside V1/−. Therefore, a continuous path occurs along the membrane joining V1/+ and V2/+. This coincides with the apparition of an envelope around V1/+ and V2/+. Now, V2/+ clearly constitutes two rolls: V2a/+ (head) and V2b/+ (tail). During the merging process (Figure 8A–E), V2/+ remains at the same place, so that the distance from the downstream vortex increases. The very long non-equilibrium EDL becomes unstable and a hump of charge appears at the front of V2b/+. The SSL forming the rear of V1/+ and the front of V2/+ respectively are clearly tilted. Moving downward, V1/− gets into contact with the membrane as explained in Figure 9. However, V1/− remains isolated from V1/+ and V2/+ by the fluid film forming the cluster.

In Figure 8D ($t' = 60$), V1/− and the front of V2/+ disappear owing to the decrease in the charge hump in the non-equilibrium EDL. The spot $(\vec{F}^{Tot})_z > 0$ and the trace of the negative spot $(\vec{F}^{Tot})_z$ on both sides of the previous front of V2/+ remain. As soon as the charge hump disappears, the former spot moves upstream and joins the positive spot $(\vec{F}^{Tot})_z$ of V1/+. In Figure 8E ($t' = 80$), V2/+, i.e., the charge hump moves downstream this time. The tail is composed of two vortices rotating clockwise separated by a secondary envelope. In Figure 8F ($t' = 100$), the tail, constituting one roll, is separated completely from the head of V2/+ and the cluster disappears.

The concept of thin film can be extended to the cluster envelope. This film separates the clockwise vortices from the membrane and the surrounding electrolyte. Along the membrane surface, it spreads over a width smaller than the non-equilibrium EDL one. The presence of this film means that a kind of upstream electro-osmotic velocity is imposed between the vortex group and the membrane. However, the immobility of the cluster is not related to this velocity but to the stability of the non-equilibrium EDL.

In the second series (Figure 9), V1/− is initially located between the two rolls V1/+ ($z = 2.1$) and V2/+ ($z = 2.35$). Contrary to Figure 8A, a continuous path of negative $(\vec{F}^{Tot})_z$ extends along the membrane over the vortex group. Despite this path, no SSL envelops the vortex group because the tops of V1/+ and V2/+ are too separated, allowing V1/− to spread far from the membrane interface. V1/+ and V2/+ get closer (Figure 9B) and the SSL between the vortices are more and more tilted owing to the profile of $(\vec{F}^{Tot})_z$ (see comments in Figures 7 and 8).

At the top of V1/+ and V2/+, the imposed main flow drives the fluid located close to the vortices in the downstream direction by means of the viscous force. At the same time, in the inter-vortex region, the longitudinal pressure force pushes the fluid upstream. Therefore, at the outer edge of the inter-vortex region, the SSL changes direction and the two points of the same SSL, on both sides of the neck, are close to each other. This behavior is all the more pronounced given that the two vortices are close to each other and the magnitude of the negative pressure spot $(\vec{F}^{Tot})_z$ increases. Therefore, the SSL join at the upper region of the roll group at $t' = 40$ (Figure 9C). This induces a displacement of V1/− to the membrane and the presence of an envelope around the vortex group which becomes a cluster. At the same time, the roll (V0/+ located at $z = 1.6$) upstream of the cluster moves closer with a decreasing volume. From $t' = 40$ to 80 (Figure 9C–E), the tail of V2/+ extends as the downstream vortex goes away and a charge hump appears at $x = 2.5$ as explained in Figure 8C. Later, the cluster breaks, V0/+ fuses with V1/+, V2/+ and its tail becomes independent, and the negative layer $(\vec{F}^{Tot})_z$ in the inter-vortex region is transformed into a spot. It seems that a cluster including three clockwise vortices is not stable or may only exist during a very short time.

3.3. *Dynamic Streamlines of Fluid Flow and Ionic Transfer*

A fully Lagrangian approach to the fluid flow and ion transfer is necessary to investigate the mixing mechanisms in unsteady flow. In this section, the dynamic streamlines (DSL) display the real trajectories of the fluid and ions, and point out that the vortices visualized by the SSL are fictitious. In this way, for instance, the DSLs show that, in a periodic regime, the vortices and the region far from the membrane, i.e., the diffusion layer containing high ion concentration, are linked by a mechanism of fluid flux through the vortices. However, in an aperiodic regime, the unsteady size of the inter-vortex regions and the vortices promotes fluid exchange with the diffusion layer and enhances the efficiency of the mixing. The same parameter values as in Section 3.2 are used.

3.3.1. Fluid DSL in Periodic Regime ($\Delta\Phi = 22$)

Figure 10 shows four dynamic streamlines at five different times. The line source is located inside a vortex at $z_o = 4.7$. The vortices move with a longitudinal velocity (U_z^{roll}) of 36. As in Section 3.2, the time value will be expressed in terms of time step and one assumes that initially $t' = 0$.

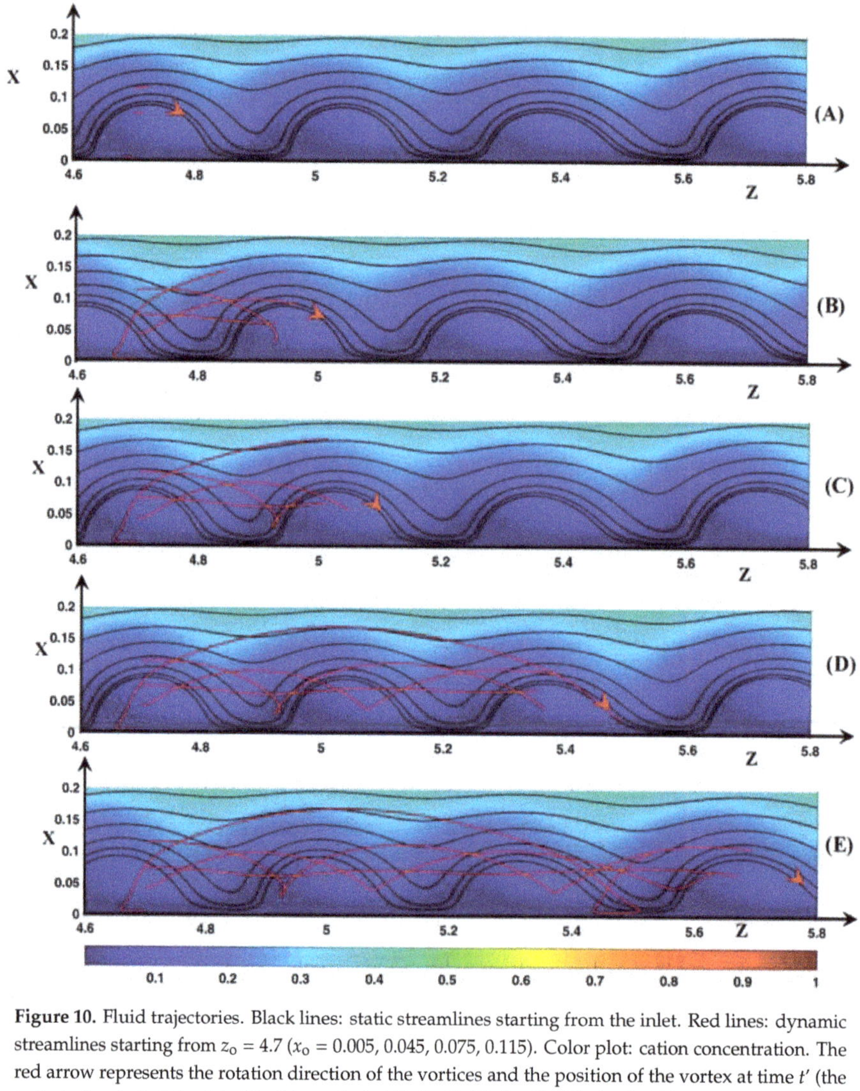

Figure 10. Fluid trajectories. Black lines: static streamlines starting from the inlet. Red lines: dynamic streamlines starting from $z_o = 4.7$ ($x_o = 0.005, 0.045, 0.075, 0.115$). Color plot: cation concentration. The red arrow represents the rotation direction of the vortices and the position of the vortex at time t' (the trajectories start at $t' = 0$ inside this vortex). (**A**) $t' = 5$; (**B**) $t' = 70$; (**C**) $t' = 95$; (**D**) $t' = 195$; (**E**) $t' = 275$. $\Delta\Phi = 22$, $\nu = 1/3 \times 10^{-3}$, $D^{\pm} = 1$.

Let us consider the starting point $x_o = 0.005$. This point is located between the membrane surface and the E_c maximum inside the E_c spot. At this point, the transverse equilibrium between the electric force and the pressure force is fulfilled. Initially, $U_z = -43$ and $U_x = -1$. At $t' = 10$, the fluid begins to move upward as it comes closer to the vortex front. Then U_x and U_z increases and decreases, respectively. Consequently, at $t' = 20$ ($x = 0.033$), U_x reaches a maximum value of 49 and $U_z = 24 < U_z^{roll}$.

Then U_z increases and becomes greater than U_z^{roll}. U_z decreases first because the fluid element moves away from the E_c spot (see Figure 6) and then increases owing to the velocity inlet boundary condition. At $t' = 95$ (Figure 10C), the fluid is just above the vortex ($x = 0.17$, $z = 5$) with $U_x \approx 0$ and $U_z = 74.5$. During the upward motion, the DSL deviates from the vortex front for two reasons: first $U_z < U_z^{roll}$ for a period of time and then the angle θ between the membrane surface and the front is lower than 90°. With regards to the 2nd reason, if the fluid element moves during a time interval Δt upward over the distance $\Delta x = U_x \Delta t$, the distance to the front becomes $\Delta z_2 = U_x \Delta t \times \tan(\pi/2 - \theta)$. Finally, the distance from the fluid element to the front is $\Delta z = \Delta z_1 + \Delta z_2$ with $\Delta z_1 = \left(U_z^{roll} - U_z\right)\Delta t$ and Δz cancels if $U_z = U_z^{roll} + U_x \times \tan(\pi/2 - \theta)$. If $U_x > 0$, U_z must be much greater than U_z^{roll} if θ is small. As Q is positive, $(\vec{F}^{El})_x$ is always negative. In the front region $(\vec{F}^{Pr})_x > 0$ (see Figure 5B). During the fluid rising, $\left|(\vec{F}^{Pr})_x\right|$ and $\left|(\vec{F}^{El})_x\right|$ decrease but the pressure contribution is higher than the electric one. Above the vortex, the force balance occurs. However, the force magnitude is very small because Q is very small. On the other side of the vortex, $(\vec{F}^{Pr})_x$ is negative. Therefore, the downward motion of the fluid is accelerated due to the two negative force contributions. The fluid enters the vortex at $t' = 190$ far from the center of rotation and close to the membrane ($x = 0.03$, $z = 5.47$—Figure 10D) (the fluid entrance can be explained by the above relation $\Delta z = \Delta z_1 + \Delta z_2$ with $U_x < 0$ and $\theta > \pi/2$). The fluid element continues to move to the membrane surface until the transverse force equilibrium is fulfilled ($U_x = 0$). Owing to the vortex displacement in the downstream direction, the E_c spot (with $U_z < 0$) moves towards the fluid element. This induces the displacement of the fluid to the front with $U_x \approx 0$, so that at $t' = 230/240$, the fluid element leaves the vortex and then moves upward.

Figure 10 shows that the fluid motion looks like a trochoid around the same vortex. The time and the spatial period are approximately 210 and 0.8, respectively. As x_o increases to 0.035, the fluid element is less and less accelerated in the ejection region at the vortex front and spends less time in it. Therefore, the maximal position in the transverse direction decreases, the loop of the trochoid appears more and more early and its size decreases. So that the period decreases to 70 and 0.39 in time and space, respectively.

If the starting point is located at ($x_o = 0.045$, $z_o = 4.7$), the streamline shape shows another kind of trochoid because the starting point is located above the center of rotation. Initially $U_x \approx -5$ and $U_z = -9$. The fluid moves upward then downward as explained above. This time, the DSL enters the vortex at $t' = 70$ at the position ($x = 0.1$, $z = 4.97$) close to the top of the vortex (Figure 10B). During the downward displacement, $|U_x|$ and U_z decrease because the fluid element moves closer to the center of rotation. U_z becomes lower than U_z^{roll} and at $t' = 115$ the center of rotation catches up with the fluid element. Then the fluid is pushed downstream with an increasing value of U_z. So, the fluid element stays above the center of rotation. As the vortex front moves closer to the position of the fluid element, the latter leaves the vortex at $t' = 160$ ($x = 0.071$, $z = 5.13$). The time and spatial periodicity are around 114 and 0.37, respectively.

As x_o increases to 0.09, the transverse amplitude of the trochoid decreases and the periodicity remains equal to about 110 and 0.37 in time and space, respectively. This time periodicity corresponds to the time needed for the vortices to move over the distance between two successive vortex fronts. Outside the vortex ($x_o > 0.09$), the trajectory is in fact a prolongation of another trajectory beginning upstream (z'_o, x'_o) with x'_o lower than the rotation center and at an earlier time given the periodic shape of the trajectory. As a last observation, Figure 10 shows that if $x_o = 0.075$, the trajectory is linear ($U_x \approx 0$). The fluid element remains close to the rotation center and moves with the same velocity as the roll.

Therefore, the figure shows that the real width of the "hydrodynamic instability" layer is much larger than the vortex one. Its outer edge is in contact with the diffusion layer. In the reference frame of the vortices, the trajectories form concentric circles. The rotation center, placed at $x \approx 0.75$, is above the SSL vortex one. The fluid exchange between the vortex layer and the diffusion layer is controlled by an inward flux at the rear and an outward one at the front. The trajectories passing close to the membrane are the most efficient for the fluid exchange. This explains the presence of the mixing layer whose width is higher than the SSL vortex one.

Now if $z_o = 4.55$, i.e., upstream of the vortex front, the DSL shape is not always a trochoid because the fluid element can remain inside the inter-vortex region as shown in Figure 11. As an example, let us consider the starting position ($x_o = 0.015$, $z_o = 4.55$). The point is located inside the E_c spot. Initially, $U_x = -6$ and $U_z = 87 > U_z^{roll} = U_z^{spot}$. Therefore, as the DSL gets closer to the front vortex, the fluid element begins to move upward, and U_z decreases to a value smaller than U_z^{roll} (see the discussion about Figure 10). Consequently, at $t' = 35$ (Figure 11B), the vertical line defined by $U_x = 0$ (black dashed line) and the DSL meet at $z = 4.67$ (the fluid element is at $x = 0.063$). This line moves with the same velocity as the vortices. Therefore, the fluid element passes on the other side of this line and moves towards the membrane with an increasing velocity. Then it enters the E_c spot, so that U_z reaches a value of 85 at $t' = 75$ (see Figure 11C). Therefore, the fluid element remains in the inter-vortex region (at $t' = 275$ $z = 5.5$, $x = 0.06$—Figure 11E). This behavior is observed until $x_o = 0.065$. If $0.075 \leq x_o \leq 0.095$, i.e., at the outer edge of the inter-vortex region, the trochoid shape is observed but the DSL does not wind around the same roll. Most of the time, the fluid element moves with $U_z << U_z^{roll}$ and with $|U_x|$ small enough for the fluid element to enter the rear of the upstream vortex close to the membrane. This is shown for example with $x_o = 0.095$ in Figure 11C ($x = 0.014$, $z = 4.76$). The loop is located at $z = 4.8$ (Figure 11D). Figure 11E shows that the same fluid element enters the even upstream vortex a second time at $t' = 255$ ($x = 0.024$, $z = 5.06$). The loop at $z = 5$ belongs to the trajectory beginning at $x_o = 0.085$. Therefore, the inter-vortex is not connected to the diffusion layer and it has a small role in the mixing process.

3.3.2. Fluid DSL in Aperiodic Regime ($\Delta\Phi = 27$)

In Figure 12 the source is located inside the vortex V0. Initially, the front of V0 and of the downstream vortex V1 are at $z = 3.7$ and $z = 4.25$, respectively. Until $t' = 25$, the volume of V0 is unchanged and its height is around 0.18. Then its volume decreases during the merging process with V1. The front of V0 moves with a longitudinal velocity of 66.7 until $t' = 85$. Furthermore, U_z^{roll} is smaller than the fluid inlet velocity at $x = 0.18$ ($U_z = 74$). The front of V1 remains at the same place in the time interval. Two starting points, ($x_o = 0.01$, $z_o = 3.9$) and ($x_o = 0.1$, $z_o = 3.9$), are chosen for the discussion because they give rise to two kinds of motion.

In the first case, the fluid element moves first upstream along the membrane to the vortex front, then moves upward in the front region at $t' = 25$ ($x = 0.01$, $z = 3.8$) and finally passes V0 at $t' = 85$, as explained in Section 3.3.1. The height difference between V0 and V1 makes the SSL tilted to the upward direction for a long time, such that at $t' = 85$ the fluid element is at a transverse position much greater than the height of V0. Reaching V1, the fluid leaves the instability region (Figure 12C). The electric force acts no more. The fluid moves mainly in the longitudinal direction with $U_z \approx 120 >> U_z^{roll}$. Therefore, the addition of the high jump at $t' = 25$ and of the increase of the height of V1 leads to the ejection of the fluid outside the vortex layer.

On the contrary, the fluid remains close to the instability layer if $x_o = 0.1$. The source is located above the center of rotation of V0. Therefore, the fluid element leaves the vortex outside the fluid ejection region as explained in Section 3.3.1 (see Figure 12A). The fluid element is just above V0 (Figure 12B) and then above V1 at $t' = 135$ because $U_z > U_z^{roll}$ (Figure 12C). It reaches its maximum transverse position $x = 0.25$ at $z = 4.77$. From $t' = 135$ to 190, as the fluid element overtakes V1, the height of V1 increases to 0.18 and its shape becomes symmetric. The static streamlines are more and more directed downward near the rear region of V1 (see Section 3.2.1 $\Delta\Phi = 23$) and the fluid element is driven towards the membrane. For the same reason as explained in the case of $\Delta\Phi = 22$, the fluid element must enter the vortex V1 and the loop of the trochoid begins at $t' = 210$ ($x = 0.03$, $z = 5.2$).

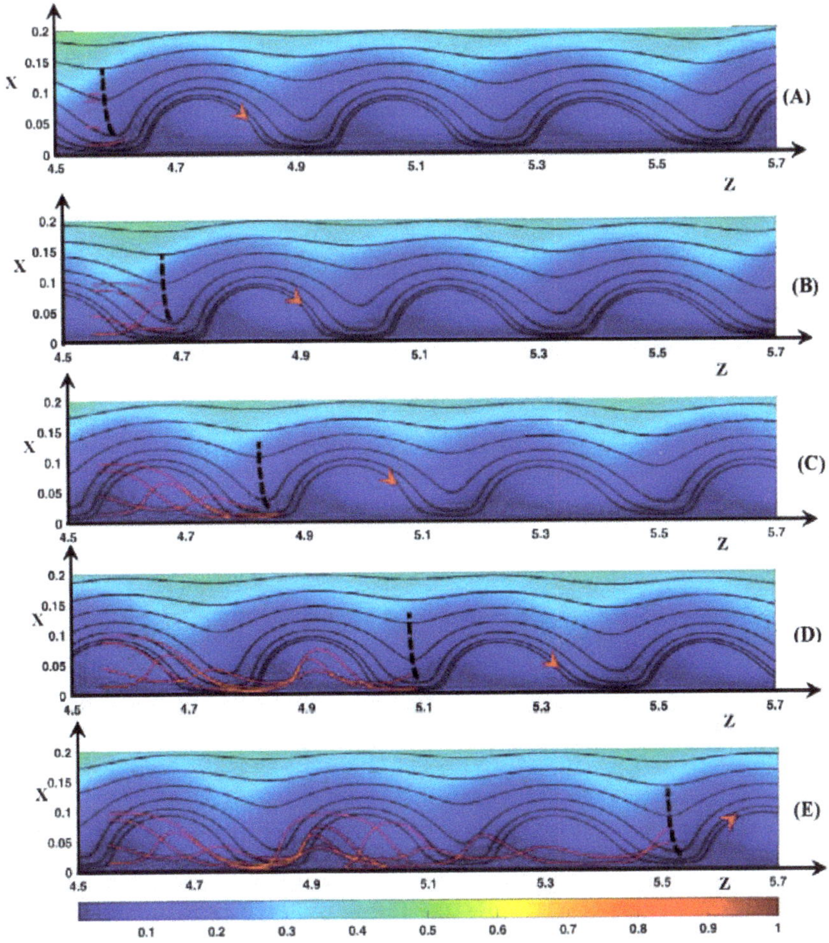

Figure 11. Fluid trajectories. Black lines: static streamlines starting from the inlet. Red lines: dynamic streamlines starting from $z_o = 4.55$ ($x_o = 0.015, 0.045, 0.085, 0.095$). Black dashed line: curve defined by $U_x = 0$. Color plot: cation concentration. The red arrow represents the rotation direction of the vortices and the position of the vortex at time t′ (this vortex is downstream the trajectories source at $t' = 0$). (**A**) $t' = 10$; (**B**) $t' = 35$; (**C**) $t' = 80$; (**D**) $t' = 155$; (**E**) $t' = 275$. $\Delta\Phi = 22$, $\nu = 1/3 \times 10^{-3}$, $D^{\pm} = 1$.

As x_o decreases from 0.1, the DSL reaches a higher transverse position during the fluid excursion, i.e., the fluid explores the diffusion layer more deeply, and it enters the vortex layer at a higher distance from the source position. This means that the high transverse amplitude of the trajectory and the large dispersion of the DSL allow a fluid exchange between different distant vortices in contrast to the periodic regime.

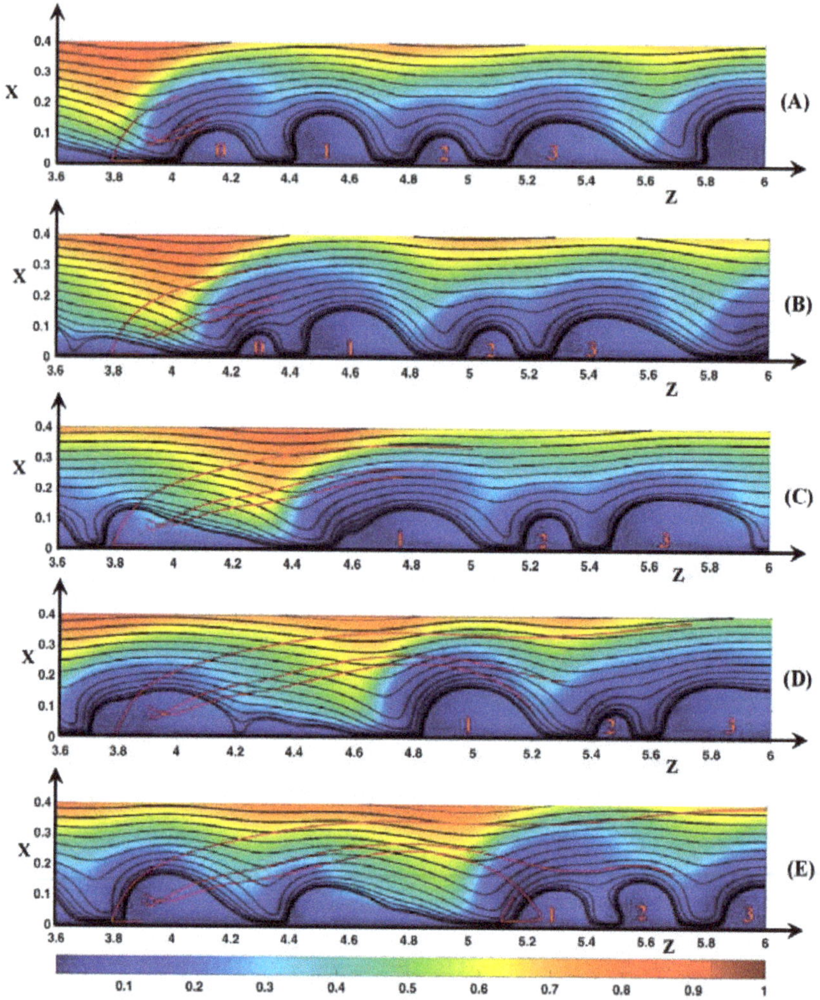

Figure 12. Fluid trajectories. Black lines: static streamlines starting from the inlet. Red lines: dynamic streamlines starting from $z_o = 3.9$ ($x_o = 0.01, 0.08, 0.09$). Color plot: cation concentration. Red numbers: vortex numbering. Vortex n°0 is the vortex in which the trajectories source is located at $t' = 0$. (**A**) $t' = 55$; (**B**) $t' = 85$; (**C**) $t' = 135$; (**D**) $t' = 190$; (**E**) $t' = 245$. $\Delta\Phi = 27$, $\nu = 1/3 \times 10^{-3}$, $D^{\pm} = 1$.

Figure 13 displays the fluid trajectories when the source is located at $z_o = 2.7$ between V0/+ and V1/+ (see Figure 13A). Between these rolls, a counter-clockwise roll (V/−) is located at $x \approx 0.04$. During the rolls' displacement, the inter-vortex region between V0/+ and V1/+ expands and V/− disappears at $t' = 55$. At $t' = 135$ (Figure 13C), V1/+ stops to move downstream ($z_{front} = 3.2$), V0/+ moves closer to V1/+ and the roll V/− appears one more time between these two vortices, which merge at $t' = 225$ (Figure 13D).

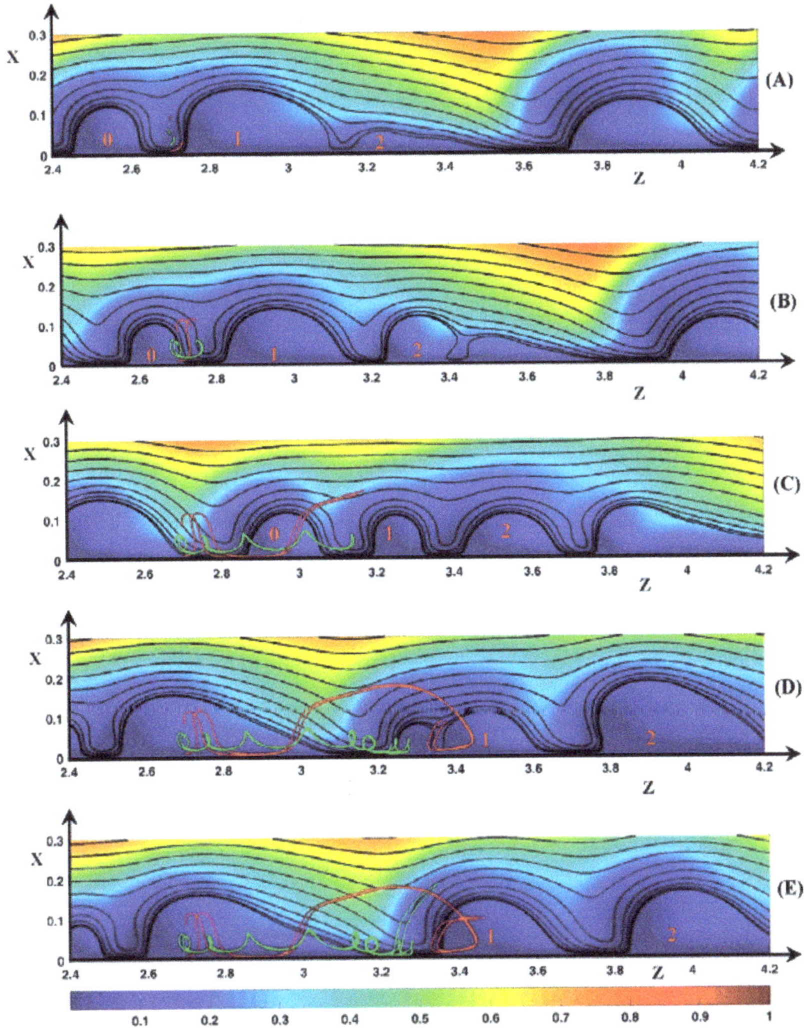

Figure 13. Fluid trajectories. Black lines: static streamlines starting from the inlet. Red and green lines: dynamic streamlines starting from $z_o = 2.7$ ($x_o = 0.01, 0.025, 0.05, 0.085$). Color plot: cation concentration. Red numbers: vortex numbering. Vortex n°0 is the vortex upstream the trajectories source location at $t' = 0$. Vortex n°1 is the vortex downstream the source. (**A**) $t' = 5$; (**B**) $t' = 45$; (**C**) $t' = 135$; (**D**) $t' = 220$; (**E**) $t' = 245$. $\Delta\Phi = 27$, $\nu = 1/3 \times 10^{-3}$, $D^{\pm} = 1$.

Let us consider the first trajectory $x_o = 0.01$ (red trajectory). This starting point is located inside the E_c spot close to the membrane, i.e., in the fluid film below V/−. If $t' < 80$ ($z < 3$), the fluid trajectory is similar to the trajectory beginning at $x_o = 0.015$, shown in Figure 11. The fluid element remains in the thin film surrounding V/− in the inter-vortex region. However, the fluid element passes above V/− during the first upward period and leaves for a very short time the fluid film at the outer edge of the inter-vortex region. This means that V/− is also isolated from the neck, a behavior which the SSL cannot not reveal (see Figures 8 and 9). At $t' = 100$ ($x = 0.12$, $z = 3$), the fluid element reaches the region where U_x is small. However, the fluid element does not cross the boundary $U_x = 0$ because the inter-vortex region expands. Therefore, the fluid element continues to move upward and leaves the inter-vortex region with an increasing longitudinal velocity, which becomes higher than U_z^{roll} (see Figure 13C). As the size of V1/+ increases when $t' > 135$, the DSL enters the V1/+ rear at $t' = 175$ as explained previously and then the fluid trajectory makes a loop inside V1/+.

For the source point $x_o = 0.025$ (green trajectory), the fluid element is initially inside V/−. The vortex is isolated from the neighborhood. So, until $t' = 55$, the fluid element remains inside V/− and rotates counter-clockwise two times. At $t' = 60$, V/− disappears; therefore, until $t' = 125$ (see Figure 13C), the fluid trajectory oscillates in the transverse direction as explained in Section 3.3.1. At $t' = 135$, V/− appears, and the fluid element makes two counter-clockwise rotations until $t' = 180$ ($x = 0.012$, $z = 3.2$). The inter-vortex region between V0/+ and V1/+ is so small that the rear of V0/+ passes the fluid element during its upward motion at $t' = 200$ ($x = 0.045$, $z = 3.25$). The fluid consequently moves upstream inside V0/+ to the front (Figure 13D), then it moves upward during the merging of V0/+ with V1/+. The volume of the new vortex increases. Therefore, as the fluid element moves with $U_z > U_z^{roll}$ at $t' = 245$, it must enter the new vortex by the rear later on and must then begin a loop motion.

The two last Figures show that several kinds of trajectories follow one another owing to the vortex dynamics. Therefore, the trajectories playing a small role in the exchange between the vortex layer and the diffusion layer have a finite lifetime. The inter-vortex regions exchange fluid with the vortices. As in the periodic mode, the passage through the vortices plays a key role in this exchange and enhances the ion exchange. The vortices trap the fluid elements by means of their changing shape. The counter-clockwise vortices do not exchange fluid through its boundary.

3.3.3. Ionic DSL in the Periodic Regime ($\Delta\Phi = 22$)

In this section, the ion transfer in the aperiodic regime is not investigated because the anions are simply ejected from the vortex region if $\Delta\Phi > 24$, and the cations follow the same kind of trajectories as in periodic regime.

Let us consider the anionic motion when the starting longitudinal position is $z_o = 4.7$, i.e., inside a vortex as shown in Figure 14. It is recalled that the ion velocity ($\vec{U}^{\pm}$) is computed by means of the ion flux density ($\vec{J}^{\pm}$) (see Section 2.2.2).

Consider first the case $x_o = 0.005$. At this distance from the membrane, the negative Fickian contribution $(\vec{J}^-_{Fi})_x$ balances the positive electric one $(\vec{J}^-_{Em})_x$. The convection $(\vec{J}^-_{Co})_z$ and the concentration gradient $(\vec{J}^-_{Fi})_z$ push the anions to the vortex front. At the beginning of the rising period (Figure 14A), the anions are pushed transversely by convection because $(\vec{J}^-_{Fi})_x = -(\vec{J}^-_{Em})_x$. Then around the position ($x = 0.1$, $z = 4.8$), the electro-migration contribution becomes dominant with $(\vec{J}^-_{Fi})_x \approx -(\vec{J}^-_{Co})_x$. Finally at $t' = 95$ Figure 14B, the ions reach the maximal transverse position with the balance $(\vec{J}^-_{Fi})_x \approx -(\vec{J}^-_{Em})_x \cdot (\vec{J}^-_{Co})_x \approx 0$) and with $U_z^- > U_z^{roll}$. In the longitudinal direction, the ions move by convection. During the downward motion (see Figure 14C), the ions enter the same vortex at $t' \approx 210$ and get closer to the vortex rotation center. The ions are pushed to the membrane by convection and diffusion and with $U_z^- < U_z^{roll}$. At $t' \approx 210$, the anions arrive at the same transverse position as the rotation center. $(\vec{J}^-_{Co})$ becomes negligible and $(\vec{J}^-_{Fi}) \approx -(\vec{J}^-_{Em})$, i.e., the anions arrive at a stagnation point. Therefore, after a while the anions cross over to the other side of the rotation center, then they begin an upward motion and leave the vortex a second time. As x_o increases from 0.035 to 0.085, the ions deviate from the ejection region as they leave the vortex. Consequently, the transverse amplitude of their trajectory decreases, and the cusps move upstream.

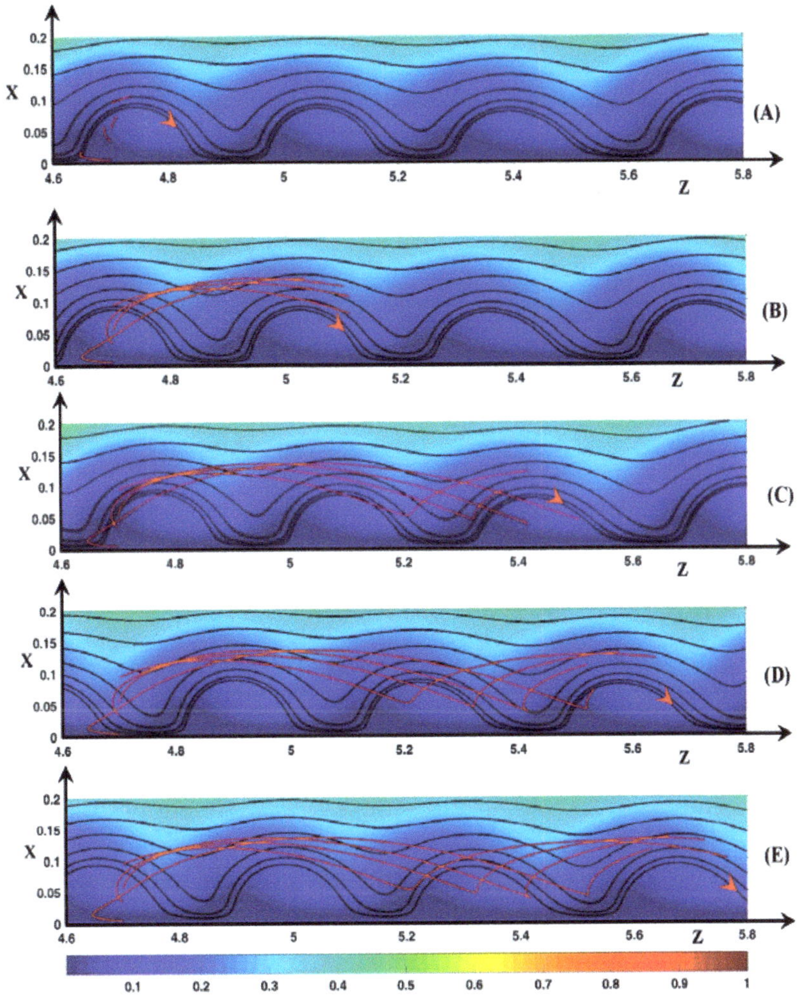

Figure 14. Fluid and anionic trajectories. Black lines: fluid static streamlines starting from the inlet. Red lines: anionic dynamic streamlines starting from $z_o = 4.7$ ($x_o = 0.005, 0.035, 0.055, 0.085$). Color plot: cation concentration. The red arrow represents the rotation direction of the vortices and the position of the vortex at time t' (the trajectories start at $t' = 0$ inside this vortex). (**A**) $t' = 10$; (**B**) $t' = 95$; (**C**) $t' = 210$; (**D**) $t' = 250$; (**E**) $t' = 275$. $\Delta\Phi = 22$, $\nu = 1/3 \times 10^{-3}$, $D^{\pm} = 1$.

Figure 15A displays the anionic DSL when the anion source is located between two vortices ($z_o = 4.55$). Two families of trajectory can be observed. The first one is identical to the trajectories shown in Figure 14 if x_o lies between 0.035 and 0.055. However, the stagnation point is replaced by an elongated and tilted loop because the anions reach a transverse position lower than the rotation center one and move to the vortex front by diffusion $(\vec{J}^{-}_{Fi})_z$ and convection $(\vec{J}^{-}_{Co})_z$. At higher values of x_o, U^{-}_z is always positive because the DSL never enter the vortex owing to higher fluid velocity and to the positive contribution of $(\vec{J}^{-}_{Fi})_z$. The sign of $(\vec{J}^{-}_{Fi})_z$ is positive because the anion position is always downstream of the location of the high ion concentration region at the vortex rear. The consequence is that the anions move faster than the vortices along the membrane if $x_o \geq 0.075$.

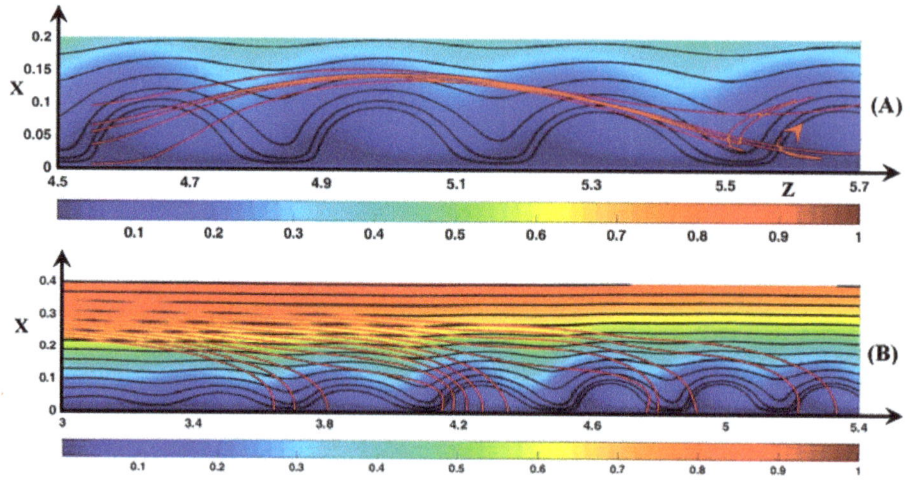

Figure 15. Fluid and ionic trajectories. Red lines: (**A**) anionic dynamic streamlines starting from z_o = 4.55 (x_o = 0.005, 0.045, 0.055, 0.065, 0.1). The red arrow represents the position of the vortex at time t' (this vortex is downstream the trajectories source at t' = 0); (**B**): cationic dynamic streamlines starting from z_o = 3, x_o = [0.23, 0.35]. Black lines: fluid static streamlines starting from the inlet. Color plot: cation concentration. t' = 275. $\Delta\Phi = 22$, $\nu = 1/3 \times 10^{-3}$

Therefore, the longitudinal Fickian contribution is of great importance to the Lagrangian anionic motion. The two last Figures also show that despite the electric force directed upward, the anions remain in the layer explored by the fluid DSL. At the outer-edge of this layer, the Fickian diffusion balances the electro-migration and keeps the anions leaving the region.

In regards to the cations (Figure 15B), the dynamic streamlines are computed with the vertical source (z_o = 3, $0.23 \leq x_o \leq 0.35$). This time, the electro-migration flux is directed downward like the Fickian one. $(\vec{J}^{+}_{Em})_x$ is the main contribution to the cation motion towards the membrane, and Figure 15B shows that the cations move monotonically. At the initial time, cation motion is mostly longitudinal ($U_z^+ > U_x^+$). As they get closer to the membrane, the DSL become vertical because $(\vec{J}^{+}_{Em})_x$ increases and becomes dominant. The convective contribution has the same order of magnitude as the electro-migration one in the region around the outer edge of the vortex layer and it leads to the increase of the cationic flux. The DSL form four groups at the vicinity of the membrane because they reach the membrane through the inter-vortex region and the rear of the rolls, i.e., regions where the transverse cationic flux is important.

4. Conclusions and Summary

In this work, the analysis of electro-convective instabilities in a laminar boundary layer along a cation exchange membrane is presented. The periodic and aperiodic modes were studied. In particular, for the latter mode, vortex association/dissociation and fusion, as well as time evolution of cluster and counter-clockwise vortex, were investigated. The influence of the instability structure and dynamics on the fluid and ion transfer was also investigated by means of a Lagrangian approach. The aperiodic mode is sensitive to the value of the parameters $D^{\pm}$ and ν. Despite this observation, the present study was performed mainly with $D^{\pm} = 1$ and $\nu = 1/3 \times 10^{-3}$.

The vortex structure and dynamics were analyzed by considering the spatial distribution of the longitudinal component of the electric and pressure force. This distribution is composed of a layer along the membrane inside the non-equilibrium EDL and of two spots on both sides of the vortex fronts and above the force layer. These two structures undergo a change of sign at the front and the rear of the vortices. The link between a spot and a layer of negative sign plays a key role in the vortex dynamics and in the birth of counter-clockwise vortices and clusters. This link bends the streamlines in the upward direction and induces a neck at the outer edge of the inter-vortex region where the fluid is pushed downward by the forced flow. The counter-clockwise vortices occur in the inter-vortex region over which the negative spot is located. This vortex is isolated from the neighboring clockwise vortices and the membrane by a thin fluid film coming from the neck. The clockwise vortices associate into a cluster when the neck closes, and a continuous path of negative longitudinal force occurs over the vortex group along the membrane. In this case, a thin film also wraps the vortex group, which is therefore separated from the membrane and the boundary layer. The counter-clockwise vortices are excluded from the clusters in contact with the membrane during the clustering process. The upstream fluid motion in the film along the membrane plays a similar role as the electro-osmotic boundary conditions.

The real motion of the fluid, as well as the ions, must be computed by considering the unsteadiness of the velocity and flux fields. In the periodic mode, the fluid trajectories passing through a vortex, computed with the SSL, form a trochoid. The SSL vortices are fictitious. In the vortex reference frame, these DSLs form rolls whose rotation center is above the SSL vortex center and whose size is much greater. The real vortex layer is therefore in contact with the diffusion layer containing high ion concentration and it controls the mixing layer expansion. This is not the case for the inter-vortex region. Surprisingly, the anions follow the same kind of trajectory with roughly the same transverse amplitude. The Fickian diffusion plays a key role in the loop formation and in the transverse amplitude. The outer edge of the region explored by the anions corresponds to the balance between the electro-migration and the Fickian diffusion fluxes in the transverse direction. Further away from the instability threshold, the electric field is high enough for the anions to be ejected from the vortex layer.

For the aperiodic mode, the real instability layer visualized by the DSL is also much larger than the SSL rolls. Its size increases with $\Delta\Phi$, imposing the upward displacement of the diffusion layer. In this case, fluid exchange is performed between distant SSL vortices. The transverse amplitude of the trajectories is high enough for the mean Lagrangian velocity to be much greater than the SSL roll one. The size increase allows the rolls to attract and trap the fluid elements. It is also observed that the inter-vortex regions contribute to the mixing process. The variation of the inter-vortex length destabilizes the fluid trajectory leading to the fluid ejection outside the SSL vortex layer and initiating a trochoid-like trajectory.

Funding: This research received no external funding.

Conflicts of Interest: The authors declare no conflict of interest.

Abbreviations

PNP	Poisson–Nernst–Planck
EDL	Electric double layer
ESC	Extended space charge
SSL	Static streamline
DSL	Dynamic streamline
Symbols	
A	Surface of the domain
$c^{\pm}$, c_o, c^{+}_{Interf}	Cation and anion concentration, bulk salt concentration, cation concentration at the membrane surface
e	Elementary charge
E_c	Kinetic energy
$D^{\pm}$, D_{ref}	Cation and anion diffusion coefficient, reference diffusion coefficient
$\vec{F}^{El}$, $\vec{F}^{Pr}$, $\vec{F}^{Tot}$	Electric force, Pressure force, total force

k_B	Boltzmann constant
I, I_m, I_{lim}	Mean current density, current density at the membrane surface, limiting current
$\vec{J}^{\pm}, \vec{J}^{\pm}_{Co}, \vec{J}^{\pm}_{Em}, \vec{J}^{\pm}_{Fi}$	Cation and anion flux, convective, electro-migration and Fickian flux contribution
L_x, L_z, L_m	Longitudinal and transverse domain length, inter-membrane distance
P, P_o	Pressure, osmotic pressure
Pe	Peclet number
$Q, Q_{\max}$	Charge density, maximum charge density in the ESC
$t, t_{diff}, \delta t$	Time, diffusional time, time step
T	Temperature
$\vec{U}, \vec{U}^{\pm}, U_{slip}, U_z^{roll}, U_z^{spot}$	Fluid velocity, cation and anion diffusional velocity, imposed velocity at the upper edge of the domain, longitudinal vortex velocity, longitudinal velocity of the kinetic energy spot
$z^{\pm}$	Cation and anion charge
Greek Symbols	
δ_V	Vortex height
$\varepsilon_o, \varepsilon_r$	Vacuum and relative permittivity
λ_d	Debye length
μ	Dynamic viscosity
ν	Debye length to the longitudinal domain length ratio
ρ_f	Fluid density
θ	Local angle between the membrane surface and the vortex front
Φ, Φ_V, Φ_T	Electric potential, potential in the vortex layer, thermodynamic potential
Ω	Domain

Appendix A. Influence of the Length Ratio on the Instability Modes

If ($D^{\pm} = 1, \nu = 10^{-3}$), the periodic instability, without roll pair formation, occurs in the range $23 \leq \Delta\Phi \leq 25$. Aperiodic mode appears at $\Delta\Phi = 27$. For this potential drop value, rolls of different sizes cross the domain. Vortex association and dissociation are also observed from time to time. However, a new kind of periodicity takes place at $\Delta\Phi \geq 30$. Figure A1 shows the time evolution of the vortex structure at $\Delta\Phi = 35$.

Three steps have been observed. Every other roll is in the same phase and the rolls between are in another one, that is two phases coexist. The Figures show clearly that, at the vortex front and rear, the SSL orientation is related to the unsteady link between a spot and a layer of longitudinal force of the same sign. Let us consider the smallest rolls at $z = 3.1$ and 4.3 in Figure A1A. As a first step, these rolls, of initial size 0.2×0.11 in the z and x direction, respectively, move to $z = 3.2$ and 4.4 and increase in volume (Figure A1B). At the beginning, the negative spot $(\vec{F}^{Tot})_z$ is transformed into a tilted layer in the downstream inter-vortex region. As the volume increases (Figure A1B), the SSL turn progressively to the upstream direction at the roll front, and a counter-clockwise vortex in the upstream inter-vortex region appears ($z = 3.05$ and 4.3). It disappears in the next step. This first step lasts $60\delta t$. In a second step (Figure A1C), the roll volume decreases and becomes elongated and symmetric ($z = 3.4$ and 4.6). The elongation occurs at the rear, the front remaining roughly at the same place. The roll dimension is about 0.6×0.16. This step lasts $30\delta t$. In the last step, the roll volume continues to decrease so that the rolls return to their initial form. This third step lasts $80\delta t$. As every second roll is in step N, the other rolls are in step N − 1 or N + 1. However, in the same roll family, a small phase shift in the shape evolution may occur from one roll to another and the growth amplitude may also differ. But the mechanism remains the same.

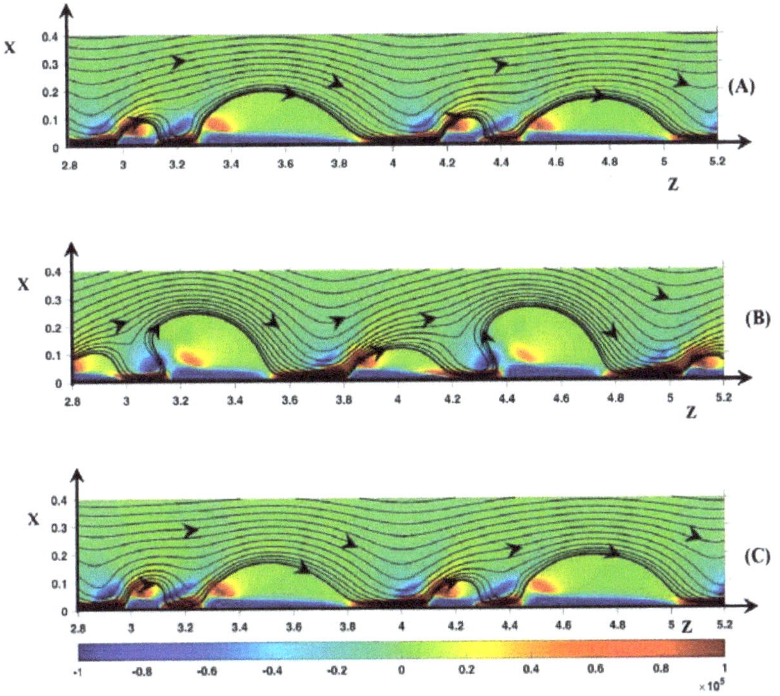

Figure A1. Temporal evolution of the vortices. (**A**) $t' = 0$; (**B**) $t' = 60$; (**C**) $t' = 90$. Color plot: $(\vec{F}^{Tot})_z$. Black lines: static streamlines starting from the inlet. $\Delta\Phi = 35$, $\nu = 10^{-3}$, $D^{\pm} = 1$.

References

1. Ran, J.; Wu, L.; He, Y.; Yang, Z.; Wang, Y.; Jiang, C.; Ge, L.; Bakangura, E.; Xu, T. Ion exchange membranes: New developments and applications. *J. Membr. Sci.* **2017**, *522*, 267–291. [CrossRef]
2. Hong, J.G.; Zhang, B.; Glabman, S.; Uzal, N.; Dou, X.; Zhang, H.; Wei, X.; Chen, Y. Potential ion exchange membranes and system performance in reverse electrodialysis for power generation: A review. *J. Membr. Sci.* **2015**, *486*, 71–88. [CrossRef]
3. Weber, A.Z.; Mench, M.M.; Meyers, J.P.; Ross, P.N.; Gostick, J.T.; Liu, Q. Redox flow batteries: A review. *J. Appl. Electrochem.* **2011**, *41*, 1137. [CrossRef]
4. De Jong, J.; Lammertink, R.G.H.; Wessling, M. Membranes and microfluidics: A review. *Lab Chip* **2006**, *6*, 1125–1139. [CrossRef]
5. Kim, S.J.; Song, Y.; Han, J. Nanofluidic concentration devices for biomolecules utilizing ion concentration polarization: Theory, fabrication and applications. *Chem. Soc. Rev.* **2010**, *39*, 912–922. [CrossRef]
6. Suss, M.E.; Mani, A.; Zangle, T.A.; Santiago, J.G. Electroosmotic jump performance is affected by concentration polarizations on both electrodes and pump. *Sens. Actuators A* **2011**, *165*, 310. [CrossRef]
7. Ulbricht, M. Advanced functional polymer membranes. *Polymer* **2006**, *47*, 2217–2262. [CrossRef]
8. Moore, C.M.; Minteer, S.D.; Martin, R.S. Microchip-based ethanol/oxygen biofuel cell. *Lab Chip* **2005**, *5*, 218–225. [CrossRef]
9. Suzuki, H.; Tabata, K.V.; Noji, H.; Takeushi, S. Highly reproducible method of planar lipid bilayer reconstruction in polymethyl methacrylate microfluidic chip. *Langmuir* **2006**, *22*, 1937–1942. [CrossRef]
10. Sandison, M.E.; Morgan, H. Rapid fabrication of polymer microfluidic systems for the production of artificial lipid bilayers. *J. Micromech. Microeng.* **2005**, *15*, S139–S144. [CrossRef]

11. Montané, X.; Bogdanowicz, K.A.; Prats-Reig, J.; Colace, G.; Reina, J.A.; Giamberini, M. Advances in the design of self-supported ion-conducting membranes-New family of columnar liquid crystalline polyamines. Part 2: Ion transport characterization and comparison to hybrid membranes. *Polymer* **2016**, *105*, 234–242. [CrossRef]
12. Rubinstein, I.; Zaltzman, B. Electro-osmotically induced convection at a permselective membrane. *Phys. Rev. E* **2000**, *62*, 2238–2251. [CrossRef]
13. Rubinstein, I.; Zaltzman, B. Electro-osmotic slip of the second kind and instability in concentration polarization at electrodialysis membranes. *Math. Models Methods Appl. Sci.* **2001**, *11*, 263–300. [CrossRef]
14. Zaltzman, B.; Rubinstein, I. Electro-osmotic slip and electro-convective instability. *J. Fluid Mech.* **2007**, *579*, 173–226. [CrossRef]
15. Rubinstein, I.; Zaltzman, B.; Lerman, I. Electroconvective instability in concentration polarization and non-equilibrium electro-osmotic slip. *Phys. Rev. E* **2005**, *72*, 011505. [CrossRef] [PubMed]
16. Abu-Rjal, R.; Rubinstein, I.; Zaltzman, B. Driving factors of electro-convective instability in concentration polarization. *Phys. Rev. Fluids* **2016**, *1*, 023601. [CrossRef]
17. Kim, S.J.; Wang, Y.C.; Lee, J.H.; Jang, H.; Han, J. Concentration polarization and nonlinear electrokinetic flow near a nanofluidic channel. *Phys. Rev. Lett.* **2007**, *99*, 044501. [CrossRef]
18. Kwak, R.; Pham, V.S.; Lim, K.M.; Han, J. Shear flow of an electrically charged fluid by ion concentration polarization: Scaling laws for electroconvective vortices. *Phys. Rev. Lett.* **2013**, *110*, 11450. [CrossRef] [PubMed]
19. Rubinstein, S.M.; Manukyan, G.; Staicu, A.; Rubinstein, I.; Zaltzman, B.; Lammertink, R.G.H.; Mugele, F.; Wessling, M. Direct observation of a nonequilibrium electro-osmotic instability. *Phys. Rev. Lett.* **2008**, *101*, 236101. [CrossRef]
20. Chang, H.C.; Yossifon, G. Understanding electrokinetics at the nanoscale: A perspective. *Biomicrofluidics* **2009**, *3*, 012001. [CrossRef]
21. Kwak, R.; Guan, G.; Peng, W.K.; Han, J. Microscale electrodialysis: Concentration profiling and vortex visualization. *Desalination* **2013**, *308*, 138–146. [CrossRef]
22. Valenca, J.C.; Wagterveld, R.M.; Lammertink, R.G.H.; Tsai, P.A. Dynamics of microvortices induced by ion concentration polarization. *Phys. Rev. E* **2015**, *92*, 031003R. [CrossRef]
23. Davidson, S.M.; Wessling, M.; Mani, A. On the Dynamical Regimes of Pattern-Accelerated Electroconvection. *Sci. Rep.* **2016**, *6*, 22505. [CrossRef] [PubMed]
24. Manzanares, J.A.; Murphy, W.D.; Mafe, S.; Reiss, H. Numerical simulation of the nonequilibrium diffuse double layer in ion-exchange membranes. *J. Phys. Chem.* **1993**, *97*, 8524–8530. [CrossRef]
25. Zabolotskii, V.I.; Manzanares, J.A.; Mafe, S.; Nikonenko, V.V.; Lebedev, K.A. Steady-state ion transport through a three-layered membrane system: A mathematical model allowing for violation of the electroneutrality condition. *Russ. J. Electrochem.* **2002**, *38*, 819–827. [CrossRef]
26. Magnico, P. Influence of the ion-solvent interactions on ionic transport through ion-exchange-membranes. *J. Membr. Sci.* **2013**, *442*, 272–285. [CrossRef]
27. Nikonenko, V.V.; Kovalenko, A.V.; Urtenov, M.K.; Pismenskaya, N.D.; Han, J.; Sistat, P.; Pourcelly, G. Desalination at overlimiting currents: State-of-the-art and perspectives. *Desalination* **2014**, *342*, 85–106. [CrossRef]
28. Nikonenko, V.V.; Vasil'eva, V.I.; Akberova, E.M.; Uzdenova, A.M.; Urtenov, M.K.; Kovalenko, A.V.; Pismenskaya, N.P.; Mareev, S.A.; Pourcelly, G. Competition between diffusion and electroconvection at an ion-selective surface in intensive current regimes. *Adv. Colloid Interface Sci.* **2016**, *235*, 233–246. [CrossRef]
29. Długołecki, P.; Anet, B.; Metz, S.J.; Nijmeijer, K.; Wessling, M. Transport limitations in ion exchange membranes at low salt concentrations. *J. Membr. Sci.* **2010**, *346*, 163–171. [CrossRef]
30. Kwak, R.; Pham, V.S.; Kim, B.; Chen, L.; Han, J. Enhanced salt removal by unipolar ion conduction in ion concentration polarization desalination. *Sci. Rep.* **2016**, *6*, 25349. [CrossRef] [PubMed]
31. Druzgalski, C.L.; Andersen, M.B.; Mani, A. Direct numerical simulation of electroconvective instability and hydrodynamic chaos near an ion-selective surface. *Phys. Fluids* **2013**, *25*, 110804. [CrossRef]
32. Druzgalski, C.; Mani, A. Statistical analysis of electroconvection near an ion-selective membrane in the highly chaotic regime. *Phys. Rev. Fluids* **2016**, *1*, 073601. [CrossRef]

33. Demekhin, E.A.; Nikitin, N.V.; Shelistov, V.S. Direct numerical simulation of electrokinetic instability and transition to chaos motion. *Phys. Fluids* **2013**, *25*, 122001. [CrossRef]
34. Demekhin, E.A.; Nikitin, N.V.; Shelistov, V.S. Three-dimensional coherent structures of electrokinetic instability. *Phys. Rev. E* **2014**, *90*, 013031. [CrossRef]
35. Ganchenko, G.S.; Kalaydin, E.N.; Schiffbauer, J.; Demekhin, E.A. Modes of electrokinetic instability for imperfect membranes. *Phys. Rev. E* **2016**, *94*, 063106. [CrossRef]
36. Pham, V.S.; Li, Z.; Lim, K.M.; White, J.K.; Han, J. Direct numerical simulation of electroconvective instability and hysteretic current-voltage response of a permselective membrane. *Phys. Rev. E* **2012**, *86*, 046310. [CrossRef]
37. Magnico, P. Spatial distribution of mechanical forces and ionic flux in electro-kinetic instability near a permselective membrane. *Phys. Fluids* **2018**, *30*, 014101. [CrossRef]
38. Roghmans, F.; Evdochenko, E.; Stockmeier, F.; Schneider, S.; Smailji, A.; Tiwari, R.; Mikosch, A.; Karatay, E.; Kühne, A.; Walther, A.; et al. 2D Patterned Ion-Exchange Membranes Induce Electroconvection. *Adv. Mater. Interfaces* **2018**, *342*, 1801309. [CrossRef]
39. Valença, J.; Jogi, M.; Wagterveld, R.M.; Karatay, E.; Wood, J.A.; Lammertink, R.G.H. Confined electroconvective vortices at structured ion exchange membranes. *Langmuir* **2018**, *34*, 2455–2463. [CrossRef]
40. Pham, S.V.; Kwon, H.; Kim, B.; White, J.K.; Lim, G.; Han, J. Helical vortex formation in three-dimensional electrochemical systems with ion-selective membranes. *Phys. Rev. E* **2016**, *93*, 033114. [CrossRef]
41. Urtenov, M.K.; Uzdenova, A.M.; Kovalenko, A.V.; Nikonenko, V.V.; Pimenskaya, N.D.; Vasil'eva, V.I.; Sistat, P.; Pourcelly, G. Basic mathematic model of overlimiting transfer enhanced by electroconvection in flow-through electrodialysis membrane cells. *J. Membr. Sci.* **2013**, *447*, 190–202. [CrossRef]
42. Tsai, P.; Peters, A.M.; Pirat, C.; Wessling, M.; Lammertink, R.G.H.; Lohse, D. Quantifying effective slip length over micropatterned hydrophobic surfaces. *Phys. Fluids* **2009**, *21*, 112002. [CrossRef]
43. Maali, A.; Bhushan, B. Measurement of slip length on superhydrophobic surfaces. *Philos. Trans. R. Soc. A* **2012**, *370*, 2304–2320. [CrossRef] [PubMed]
44. Korzhova, E.; Pismenskaya, N.; Lopatin, D.; Baranov, O.; Dammak, L.; Nikonenko, V.V. Effect of surface hydrophobization on chronopotentiometric behavior of an AMX anion-exchange membrane at overlimiting currents. *J. Membr. Sci.* **2016**, *500*, 161–170. [CrossRef]
45. Andreeva, M.A.; Gil, V.V.; Pismenskaya, N.D.; Nikonenko, V.V.; Dammak, L.; Larchet, C.; Grande, D.; Kononenko, N.A. Effect of homogenization and hydrophobization of a cation-exchange membrane surface on its scaling in the presence of calcium and magnesium chlorides during electrodialysis. *J. Membr. Sci.* **2017**, *540*, 183–191. [CrossRef]
46. Shelistov, V.S.; Demekhin, E.A.; Ganchenko, G.S. Electrokinetic instability near charge-selective hydrophobic surfaces. *Phys. Rev. E* **2014**, *90*, 013001. [CrossRef]
47. Newmann, J.; Thomas-Alyea, K.E. *Electrochemical Systems*, 3rd ed.; Electrochemical Society Series; Wiley-Interscience: Hoboken, NJ, USA, 2004.

International Journal of
Molecular Sciences

MDPI

Article

Behavior of Embedded Cation-Exchange Particles in a DC Electric Field

Lucie Vobecká [1], Tomáš Belloň [1] and Zdeněk Slouka [1,2,*]

[1] Department of Chemical Engineering, University of Chemistry and Technology Prague, Technická 3, Prague 16628, Czech Republic
[2] New Technologies–Research Centre, University of West Bohemia, Univerzitní 8, Pilsen 30614, Czech Republic
* Correspondence: sloukaz@vscht.cz; Tel.: +420-2-2044-3239

Received: 26 June 2019; Accepted: 21 July 2019; Published: 22 July 2019

Abstract: Electrodialysis and electrodeionization are separation processes whose performance depends on the quality and properties of ion-exchange membranes. One of the features that largely affects these properties is heterogeneity of the membranes both on the macroscopic and microscopic level. Macroscopic heterogeneity is an intrinsic property of heterogeneous ion-exchange membranes. In these membranes, the functional ion-exchange component is dispersed in a non-conductive binder. The functional component is finely ground ion-exchange resin particles. The understanding of the effect of structure on the heterogeneous membrane properties and behavior is thus of utmost importance since it does not only affect the actual performance but also the cost and therefore competitiveness of the aforementioned separation processes. Here we study the electrokinetic behavior of cation-exchange resin particle systems with well-defined geometrical structure. This approach can be understood as a bottom up approach regarding the membrane preparation. We prepare a structured cation-exchange membrane by using its fundamental component, which is the ion exchange resin. We then perform an experimental study with four different experimental systems in which the number of used cation-exchange particles changes from 1 to 4. These systems are studied by means of basic electrochemical characterization measurements, such as measurement of current–voltage curves and direct optical observation of phenomena that occur at the interface between the ion-exchange system and the adjacent electrolyte. Our work aims at better understanding of the relation between the structure and the membrane properties and of how structure affects electrokinetic behavior of these systems.

Keywords: ion-exchange membranes; ion-exchange particles; heterogeneity; electrokinetics; current–voltage curves

---

1. Introduction

Directed transport of ions by ion-exchange membranes when in DC (direct current) electric field is pertinent to electromembrane separation processes such as electrodialysis and electrodeionization [1–3]. These membranes possess a property called ion selectivity, which stems from a combination of two important factors [4,5]. The first factor is a so-called fixed charge that can act on mobile ions in the electrolyte solutions by exerting electrostatic forces. The second factor is the very structure of the functional part of ion-exchange membranes [6]. The internal structure has to have characteristic dimensions on the nanometer scale so that the electrostatic attraction or repulsion between the fixed charges and the mobile ions can occur. The ion-exchange membranes can then be classified into cation and anion-exchange membranes, which is given by the charge of the functional group fixed in the membrane [7]. Another classification divides the membranes into two groups on the basis of their macroscopic compositions. The first group represents membranes that are viewed as homogeneous on the macroscopic level, i.e., the fixed charge is present throughout the membrane [7]. The second group is a group of so called heterogeneous membranes having macroscopic domains of different composition and internal structure and thus different functionality. Typical heterogeneous membranes consist of functional ion-exchange material (usually in the form of finely ground resin particles) and an inert binder, which houses the functional component and provides the membrane with mechanical stability [8,9]. The functional component can be viewed as homogenous. Often polymeric fibers are laid on the surface of these membranes. Each type of the membranes brings its benefits and disadvantages. While the homogenous membranes usually surpass the heterogeneous ones in the electrochemical characteristics (resistance, selectivity, etc.) [10], the heterogeneous membranes possess better mechanical and chemical stability and their production cost is lower. This fact motivates the research focusing on better understanding of the ion transport in case of heterogeneous membranes, which is reflected in their properties such as resistance or the selectivity.

The transport across ion-exchange membranes when polarized in DC electric field results in an interesting current–voltage curve [11–13]. In a broader sense, the current–voltage curve can be viewed as a dependence of the flux (flow) on the driving force. In the case of the ion-exchange membranes, the flux is represented by the current density (electric current carried by ions) and the driving force by the electric voltage. The current–voltage curve usually consists of three major regions, which are referred to as an underlimiting, limiting, and overlimiting one [14]. The underlimiting region occurs at low voltages when enough ions able to carry the electric current are present on either side of the membrane. However, their concentration decreases with increasing voltage on one side of the membrane, which is a result of a so-called concentration polarization phenomenon. This side of the membrane is denoted as the depletion side. This side controls the ionic current going through the system at higher voltages and thus the behavior of the whole system. The concentration polarization [15–17] predicts that the concentration of ions on the depletion side drops essentially to zero at certain value of polarizing voltage and that the electric current will reach a limiting value, i.e., it is not possible to achieve a higher current by increasing the voltage. The tendency of current to level off is typical for the limiting region. However, it was found out that true saturation in most cases does not occur [18] and one can observe an overlimiting current in these systems. The overlimiting current is still under scientific scrutiny and many theories have been developed to explain its appearance [19–21]. All theories, no matter what system they apply to, predict the appearance of new or fresh ions in the depletion region, which in other words means partial destruction of the depletion region itself. There are two major mechanisms accepted in the scientific community that can lead to the appearance of new ions in the depletion region and these are: (i) Convective mixing invoked by the conditions on the depletion side of the membrane [22–24] and (ii) water splitting reaction generating hydrogen and hydroxide ions [25–27]. The convective mixing can be driven either by the electric field itself and the resulting phenomenon is called electroconvection or by natural convection, which is associated with the generation of gradients in the density of the desalted electrolyte solution. These gradients are caused by variations in concentrations or temperatures. The appearance of the overlimiting current makes ion-exchange membrane based separation unique in the area of membrane separation. The performance of other membrane separation processes is limited by the concentration polarization [28] (such as reverse osmosis). Concentration polarization does not allow one to reach flux of the given component larger than the limiting one. This fact attracts scientific attention since if used properly in electrodialysis, the exploitation of convective mixing producing overlimiting current can lead to significant intensification of these processes [20,22]. However, today most of the electrodialysis units are operated under underlimiting current conditions since appearance of the overlimiting currents can be accompanied by processes with deleterious effect on the desalination. Extensive release of Joule's heat or water splitting reaction can be named as examples. Thorough understanding of the overlimiting mechanism and their effect is thus necessary.

Electroconvection as the mechanism behind the overlimiting current always results from interaction of a strong electric field with a spatial mobile charge that is localized between the electrolyte solution and the ion-exchange membrane. There are a number of different theories as to why electroconvection occurs. A nice overview of the current understanding of electroconvection was given by Nikonenko et al. [20]. Today one acknowledges that the structure of the membrane plays a significant role in the development of electroconvection. It has been found that both surface profiling of homogenous ion-exchange membranes [29,30] and presence of heterogeneities in case of heterogeneous ion-exchange membranes [26,31] have profound effect on the developing electroconvection. In general, profiling or heterogeneity cause the electric field to deviate from the perpendicular direction towards the membrane, which results in a substantial tangent electric field component with strong electric force on the extended polarized layer on the membrane surface. This tangent component then produces electrokinetic flow on the membrane whose structure depends on the geometry of the system and the experimental conditions. Until now, electroconvection has been studied for given membranes or membranes with given profiling or structuring.

In this work we attempt to produce "membranes" containing spatially arranged whole cation-exchange resin particles. We study the effect of the number of particles on (i) the measured current–voltage curve (CVC) curves, (ii) electrokinetic effects developing around the particles during polarization, and (iii) the tendency of these systems to split water. The main aim of manuscript is to describe differences one can observe in these systems when producing membranes with different number of ion-exchange particles.

2. Results

This section is organized as follows. In the first part of the section, the individual membranes are studied experimentally with respect to the exhibited current–voltage curve (CVC) and the behavior of the electrolyte on its depletion side of the membrane. Our experimental setup allows us to detect the appearance of electroconvection when the CVC is measured and to track the main motion of the electrolyte that develops around the membranes. In the second part of this section, we test some of the systems for their susceptibility to a water splitting reaction.

2.1. Studied Membranes—Current–Voltage Curves

The membranes studied in this work are homemade membranes consisting of a certain number of cation-exchange resin particles. The number of particles, we could incorporate into our membranes, was limited by our requirement of the observation of all particles under microscope. For this reason, the membranes prepared for this study contained one, two, three, and four particles. The two- and three-particle membranes were made so that the particles were positioned horizontally in a line, the four-particle membrane had two rows of particles, i.e., upper row contained two particles and the bottom row also contained two particles. We determined the ion-exchange area of these membranes by using the strategy described in Section 4.3. The measured distances used in the evaluation of these surface areas are given in figures showing the result for individual studied systems.

2.1.1. One-Particle System

The results for the one-particle membrane are shown in Figure 1. The Figure shows the measured current–voltage curve on the left and then a set of fluorescent images, which capture the situation on the depletion side of the membrane during polarization. Specifically, the fluorescent pictures show the situation at specific points on the CVC, which are of special interest. Point 1 shows the transition of the system from the underlimiting to the limiting region, point 2 the situation in the limiting region, point 3 during the transition to the overlimiting region, and point 4 in the fully developed overlimiting region. The same notation is also used in the description of the results for the other multi-particle membranes.

The current–voltage curve shows all three regions that are typical for the ion-exchange systems, i.e., underlimiting, limiting, and overlimiting regions. The limiting current density reached a value of 38.4 A/m^2. By looking at the shape of the CVC, one clearly sees that the underlimiting region had the highest slope and the limiting one had the lowest slope. The values of the slopes for the underlimiting and overlimiting regions, which also indirectly provide a picture about the conductivity of the system in the given region, were 72.7 and 60.5 $A/m^2/V$. The value of the overlimiting slope was smaller than that for the underlimiting region, which is consistent with the fact that the mechanism driving the overlimiting current cannot fully restore the conditions corresponding to the underlimiting region. To decipher what mechanism could be responsible for the overlimiting current, we synchronized the obtained images with the measurement of the CVC. One can see from Figure 1 that at the transition from limiting to overlimiting region, there were no visible characteristic features at the interface (point 1). However, right after transitioning into the limiting region (point 2) one can see the appearance of a fluorescent wave that went from the particle outward. In the third frame, which depicts the situation at the transition from the limiting to the overlimiting region, the fluorescent zone widened and contained an array of dark spots that were evenly distributed around the particles. On analyzing the obtained images, one will see that the dark spots are small vortices that form on the surface of the particles. These vortices grow with polarizing current and merge into a few larger ones when the membrane is fully in the overlimiting region (see Figure 1 frame 4). The obtained images clearly show that electroconvection played an important role in the appearance of the overlimiting current. These observations are consistent with our previous results [14]. The role of possible water will be discussed later. This one-particle membrane system serves as a basic system to which the others will be compared.

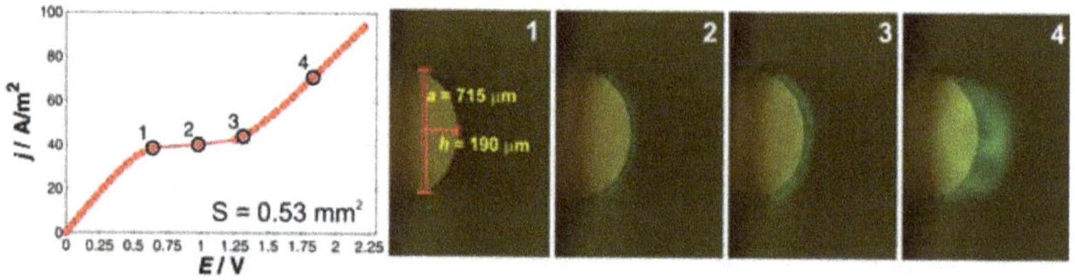

Figure 1. Current–voltage curve and a set of characteristic images for the one-particle membrane. The black circles on the current–voltage curve mark the current densities at which the depicted fluorescent images denoted as 1, 2, 3, and 4 were obtained.

2.1.2. Two-Particle System

The results for the two-particle system are shown in Figure 2. The current–voltage curve again shows three distinctive regions characteristic for the ion-exchange systems with no qualitative changes when compared to the one-particle system. The underlimiting region, which possesses the slope of 73 $A/m^2/V$, transitions into the limiting one at the current density of 31.4 A/m^2. While the slope of the underlimiting region was very similar to that of the one-particle membrane, the limiting current density was lower. This might be given by the close arrangement of the two particles. The particles share the space of depletion, thus a lower current is needed to reach the limiting conditions. The transition from the limiting to the overlimiting region was more gradual than in the previous. The slope of the overlimiting region was 43.6 $A/m^2/V$, which is approximately 30 percent lower than in the case of the one-particle system. One explanation of this observation is that the ion-exchange area is larger and thus the local electric field is not that strong, which in turn means that all overlimiting processes are driven with an electric field of lower strength.

The accompanied fluorescent pictures show how the system behaves at the transition from the underlimiting to the limiting region (point 1), in the limiting region (point 2), at the transition from the limiting to the overlimiting region (point 3), and in the overlimiting region (point 4). The description of the observed behavior given for one-particle membrane is also qualitatively valid for the two-particle system. The first fluorescence appears when the system was in the limiting regions (point 2). This fluorescence emanated from around the whole ion-exchange surface and its intensity was evenly distributed. An array of dark spots appeared within the fluorescent region and grew in size with the polarizing current (point 3). One characteristic dark spot was in between the particles. These dark spots remained there until the end of polarization. In the overlimiting region, smaller vortices merged into a smaller number of the larger ones (point 4).

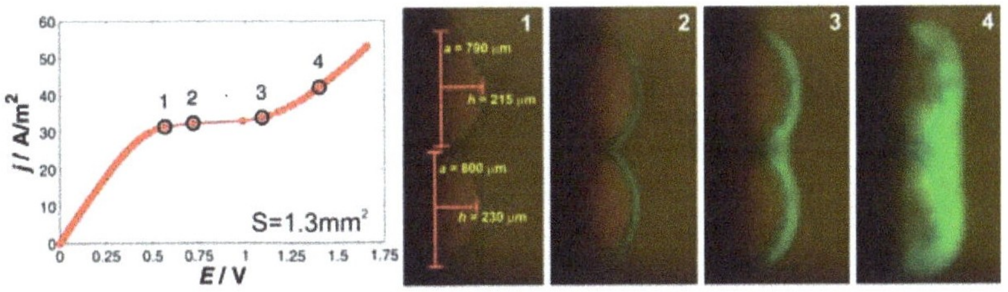

Figure 2. Current–voltage curve and a set of characteristic images for the two-particle membrane. The black circles on the current–voltage curve mark the current densities at which the depicted fluorescent images denoted as 1, 2, 3, and 4 were obtained.

2.1.3. Three-Particle System

The results for the three-particle membranes are depicted in Figure 3. The current–voltage curve shows three typical regions with no significant differences in its quality when compared to the previous cases. The slopes of the underlimiting and overlimiting regions were 41 and 33.2 A/m^2/V, respectively, the limiting current density reached value of 34.7 A/m^2. When compared to the previous cases, both the underlimiting and overlimiting regions showed smaller conductivity. The explanation for the overlimiting region might be given in a weaker electric field than that in the case of a larger area it will not attain such strength. The increased resistance of the underlimiting might have its origin in increased resistance on the other side of the membrane or between the membrane and the reference electrodes.

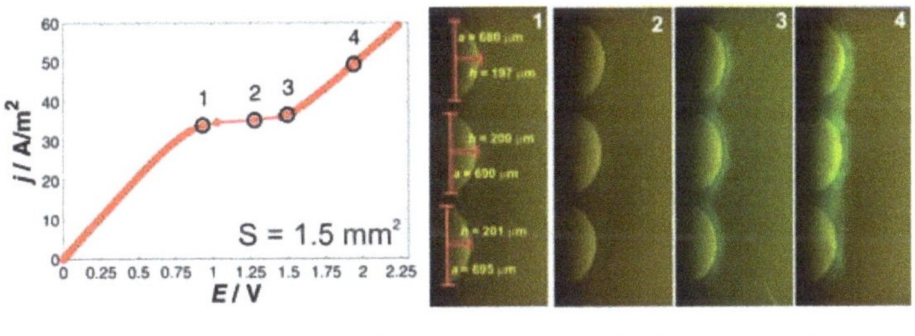

Figure 3. Current–voltage curve and a set of characteristic images for the three-particle membrane. The black circles on the current–voltage curve mark the current densities at which the depicted fluorescent images denoted as 1, 2, 3, and 4 were obtained.

The fluorescent images again showed development of the fluorescent zone with darker spots, which grew with polarizing current. Two characteristic dark spots were again seen between the particles. One qualitative difference when compared to the previous membranes was that the first fluorescence was detected at the transition from the limiting to the overlimiting current (see point 2 and 3). The appearance was thus delayed with respect to previous cases, which points at the fact that another resistance was present in the system. This resistance caused the electric field to readjust accordingly. The fluorescent images, however, clearly showed that electroconvection played an important role and was definitely one of the mechanisms driving the overlimiting current.

2.1.4. Four-Particle System

The four-particle membrane differed from the previous membranes in that the particles are arranged in two rows (top and bottom) with each row having two particles. The fluorescent images in

Figure 4 thus depict two sets of images, one for particles in the top row and the other in the bottom row. The images were obtained by selective focus on the given row of particles.

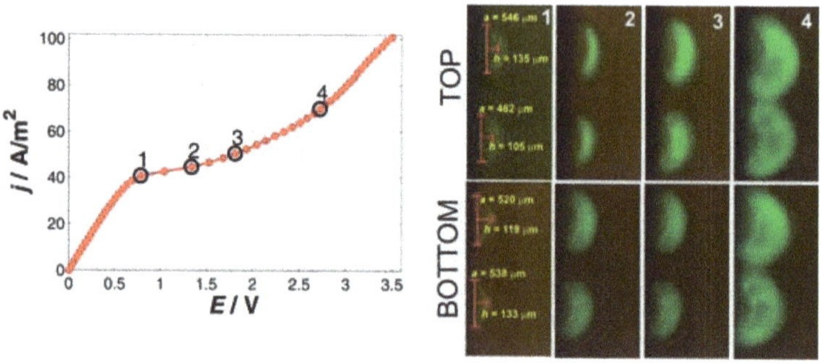

Figure 4. Current–voltage curve and a set of characteristic images for the four-particle membrane. The black circles on the current-voltage curve mark the current densities at which the depicted fluorescent images denoted as 1, 2, 3, and 4 were obtained.

The current–voltage curve for the four-particle membrane bears a shape typical for the ion-exchange systems. The underlimiting region was characterized by the slope of 64.4 $A/m^2/V$, which is very close to the slope observed for the one and two-particle systems. The transition to the limiting regions occurred at the current density of 40 A/m^2, which is a value very close to that measured for a single particle. By inspecting the images of the four-particle membrane (especially frames for point 2 in Figure 4), one will see that the system essentially consisted of four particles that are isolated, i.e., they act as individual ion-exchange domains. This is nicely documented by the fluorescent signal (point 2 in Figure 4) that is disconnected unlike the previous cases in which the fluorescent signal was continuous along the whole particle membrane. This system can thus be considered as four non-interacting single particles connected in parallel. While the transition from the underlimiting to the limiting region was very well defined, the transition from the limiting to the overlimiting current was very slow. The slope of the overlimiting region above point 4 on the CVC was 44.6 $A/m^2/V$, which is comparable to previous systems, however, this value was obtained at a much larger driving voltage around 3 V. The slope between the point 3 and 4 was much smaller. The explanation of this gradual transition from the limiting to the overlimiting region can be found in the fact that the transition from the limiting to the overlimiting region of individual particles in this system was not synchronized and occurred at different currents and voltages. This hypothetical explanation had its support in the fluorescent images (point 2 and 3) in which the upper particles seem to be behind the bottom particles in the development of the fluorescent wave with dark regions corresponding to the array of vortices. The qualitative behavior regarding the formation of the vortices and their growth was very similar to the previous systems. One specific feature, however, was the formation of very bright small dots at the particles surface (point 4). These spots look like stationary stagnant points between the individual vortices formed on the particles in which fluorescein can accumulate.

2.2. pH Changes at Studied Systems

In all the previous experiments, we were able to identify electroconvection that develops on the studied systems during the transitions from the limiting to the overlimiting regions of the CVCs. The other mechanism often mentioned with the appearance of the overlimiting current is the water splitting reaction. Although the mechanism of the reaction might be complex, its occurrence was manifested in the changes in pH values in the solutions adjacent to the ion-exchange system at which the aforementioned reaction took place. This is of course true in the case when no pH buffers were present in the solutions.

To reveal any possible contributions of water splitting reaction to the overlimiting current, we performed experiments in which various voltages were applied on the system for 5 min. After that we mixed the solutions in all chambers and measured their pH. These experiments were carried out for one- and two-particle systems and a single anion-exchange particle as a control. The results of these experiments are plotted in Figure 5a–c. Each of the graphs shows the dependence of the pH of the four solutions from different chambers on the applied voltage. As can be seen in Figure 5a,b for the one- and two-particle membranes, the pH in the chambers adjacent to the membrane did not depend on the voltage (blue curve for the depletion side, black curve for the concentration side) and was equal to the pH of the solution measured at the underlimiting voltage of 1 V (no changes expected). The same was true for the anodic compartment (green curve). The pH in the cathodic compartment (cathode) increased with increasing voltage (red curve in Figure 5a,b). This pH increase was given by a water splitting reaction that occurred on this electrode. However, the pH changes in this compartment did not affect the pH in the neighboring chamber in which the pH was constant (depletion side). The schematic in Figure 5d shows a hypothetical situation in a cation-exchange system in which the water splitting reaction would occur. The generation of H^+ and OH^- ions would result in a pH decrease on the concentration side of the membrane and pH increase on the depletion side of the membrane. Moreover, this pH change would be voltage dependent. Neither was observed for the studied particles, which led to the conclusion that water splitting reaction did not occur in these systems, or its extent was very low.

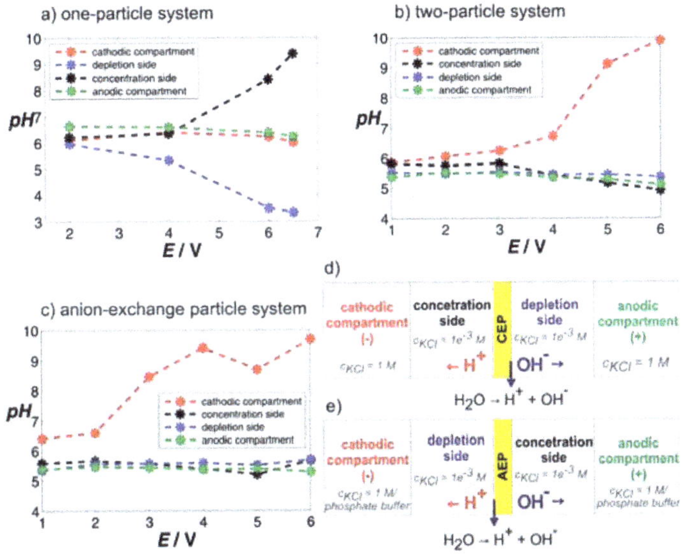

Figure 5. pH changes in the solutions taken from individual chambers of the cell after running chronoamperometric measurement for 5 min. The graphs (**a**) and (**b**) are for one- and two-particle membranes studied here, graph (**c**) for an anion-exchange resin particle serving as a control system. The schematics (**d**) and (**e**) depict the situation which would occur in a cation- and anion-exchange system with the proceeding water splitting reaction, respectively.

The results for the anion-exchange particle are plotted in Figure 5c. In this particular case, the anion-exchange particle turned out to be very efficient in splitting water. While the pH of the solution on the depletion side decreased, the pH of the solution on the concentration side increased and the pH changes were voltage dependent. The higher voltage the larger the pH change. These pH changes were in good agreement with the schematic in Figure 5e, which showed a situation in an anion-exchange system with the proceeding water splitting reaction. The pH of the solutions in the electrode compartments did not change for any of the applied voltages (we added another electrode chamber filled with phosphate buffer to our cell to prevent observed changes in the cathodic compartments). The results obtained on the anion-exchange particle, which are known to split water, were used as a positive control regarding our experimental setup for the determination of pH changes in our cell.

3. Discussion

The presented study aims at gaining a better understanding of the behavior of ion-exchange systems in which geometrical complexity is increased by addition of ion-exchange domains. The study investigates four different systems that contain 1, 2, 3, or 4 cation-exchange resin particles. While the particles were arranged in a horizontal line in one-, two-, and three-particle membranes, the particles in the four-particle membrane were arranged in a square. Since these systems are difficult to fabricate, we resorted to studying the individual systems from the point of quality, i.e., what behavior these systems exhibit and suggest what might be the cause for the observed differences.

All studied systems exhibited behavior typical for ion-exchange systems, i.e., their current–voltage characteristics bear the shape typical for these systems. The accompanying recordings of the processes developing on the membranes revealed that electroconvection develops in all systems when the systems transition from the limiting to the overlimiting region. This electroconvection presents itself as an array of vortices that grows with increasing polarizing current. From the qualitative perspective, we did not observe significant differences in the produced electroconvection for systems with a different number of particles. This evidences electroconvection as one of the major mechanisms responsible for the overlimiting current. Measurement of the pH changes around the one- and two-particle membrane showed that water splitting did not occur in the studied cation-exchange systems unlike the anion-exchange system. The anion-exchange resin particle was used as a positive control. Overall the study showed that no matter how many cation-exchange particles we had in the system, they would always show the current–voltage curve typical for ion-exchange systems and the electroconvection would play a major role in the appearance of the overlimiting current.

From the quantitative perspective, we found differences in the behavior of the four studied systems. The origin of the differences is most probably associated with the way the particles forming the membrane are positioned one to another and how much of their surface is available for the ion-exchange. Figure 6 shows the measured CVC curves for all studied systems and the Table 1 shows some of the characteristics evaluated from these CVC curves along with the surfaces of individual systems. This table documents the following observations. The limiting current density was larger for systems in which the surfaces of individual particles making up the membrane were smaller. The smaller the surface of the particles was the more it acts as a point source with a radial impact on the surrounding electrolyte. In other words the volume of the surrounding electrolyte required to be depleted to reach the limiting region was larger for smaller isolated surfaces (such as the four-particle membrane, Figure 4, point 1,2) than that for larger interacting surfaces (two-particle membrane, Figure 2, point 1,2). Larger, non-isolated particles might "share" the depletion zone and thus lower current is needed to reach depletion. The overall surface did not seem to play an essential role in evaluating the limiting current density.

These observations are consistent with experimental data of e.g., Butylskii [32] or Green et al. [33]. In [32], the authors studied systems with surface electric inhomogeneities at which they found the appearance of two transition times during the measurement of chronopotentiometric curves. The first transition time is associated with the depletion of ions at conductive domains. The value of the first transition time is strongly affected by not only normal diffusion but also tangential diffusion (diffusion from non-conductive domains to the conductive domains). The second transition time is associated with depletion of ions over the whole membrane. In this case, electroconvection developing at the interface between conductive and nonconductive regions at around the first transition time plays a significant role in ion transport to the conductive domains [34,35]. In [33] the authors studied the effect of conductive heterogeneity on systems of parallel ion-selective nanochannels. Their experimental data clearly showed that an increase in the system heterogeneity (decrease in the number of parallel nanochannels or increase in their spacing) leads to the increase in the underlimiting conductance and the limiting current.

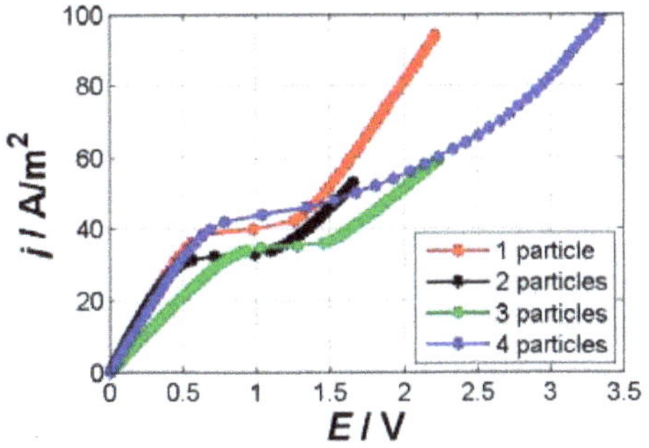

Figure 6. Current–voltage curves of the studied systems. The red curve represents the one-particle systems, the black one the two-particle system, the green one the three-particle system, and the blue one the four-particle system. We evaluated the slope of the underlimiting and overlimiting regions and the limiting current density, which are given in Table 1.

Table 1. Characteristics of the individual studied systems and their ion-exchange area.

Membrane	Particle Surfaces	Overall Surface/mm²	Limiting Current Density/A/m²	Slope of Underlimiting Region/A/m²/V	Slope of Overlimiting Region/A/m²/V
One-particle membrane	S_1 = 0.52 mm²	0.52	38.4	72.7	60.5
Two-particle membrane	S_1 = 0.63 mm² S_2 = 0.67 mm²	1.3	31.4	73	43.6
Three-particle membrane	S_1 = 0.49 mm² S_2 = 0.50 mm² S_3 = 0.51 mm²	1.5	34.7	41	33.2
Four-particle membrane	S_1 = 0.29 mm² S_2 = 0.20 mm² S_3 = 0.26 mm² S_3 = 0.28 mm²	1.03	40	64.4	44.6

The number of particles did not significantly affect the slope of the underlimiting region. Only the system with three particles had a significantly lower value of this parameter, which is most probably given by the additional resistance in the system introduced during fabrication. The slope of the over-limiting region did seem to be larger for systems with a smaller overall area, however there might be many other effects playing an important role in this case. Since electroconvection was established as the major mechanism for the overlimiting current, one should account for the actual geometry of the systems, the voltage and current at which the overlimiting region occurs, etc. The description of these effects requires further experimental work.

4. Materials and Methods

4.1. Materials

The materials used in this work include: Cation-exchange resin particle Dowex® 21K from Sigma-Aldrich (Prague, Czech Republic), anion-exchange particle Suquing 201 × 4 Cl. The chemicals used in this work include: KCl from Penta a.s. (Prague, Czech Republic), and fluorescein (acid free) from Fluka (ordering number: 46955).

4.2. Electrochemical Cell

The details about the fabrication of our custom-made electrochemical cell and preparation of the ion-exchange particle based membranes can be found in our previous papers [8,36]. Here, we only outlined the fabrication procedure. The electrochemical cell was designed as a four-chamber cell in which two outer chambers serve for imposing a source signal on the studied system and two inner cells house reference electrodes for the measurement of the electric potential differences occurring on the measured system. The two inner cells were separated with the studied membrane (see detail in Figure 7b). At the same time, the studied membrane was fixed in the system in such a way that it allowed one to observe the interface of the studied particles and the surrounding electrolyte e.g., by using fluorescence microscopy. The cell itself was made of polycarbonate foil and was glued together with the use of UV curable glue Acrifix 182. The cell prepared for measurement is depicted in Figure 7a.

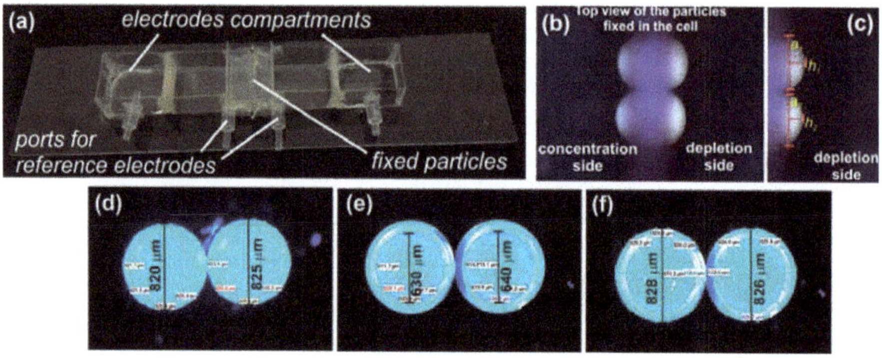

Figure 7. (**a**) Picture of the electrochemical cell used in the experiments; (**b**) picture of a two-particle system as seen under a microscope after it was fixed in the electrochemical cell and the particles were fully swollen; (**c**) detail of the swollen two-particle system on its depletion side. The depicted distances show how the surface areas used in calculating the current density were evaluated. The pictures (**d**), (**e**), and (**f**) show the top view of the two cation-exchange particles used in the preparation of the two-particle membrane system. Figure (**f**) depicts the two resin particles in the dry state before their embedding into the acrylic resin, figures (**e**) and (**f**) show dry resin particles after their embedding into the acrylic resin with corresponding measured distances.

The particle membranes were prepared by the method developed previously [36]. First, resin particles of roughly the same size were selected from a stock (see Figure 7d). The particles were checked for any cracks. A given number of particles were placed in a mold for their encapsulation in a UV curable resin. The particles were placed close to each other (they were touching) and the upper part of the mold was put in place so that the particles did not move during the encapsulation. UV curable resin Acrifix 182 was slowly injected into the mold. Special care was taken not to trap any bubbles and to firmly encapsulate all the particles. Due to the bottom and top cover of the mold the opposing poles of the particles remained free of any glue. The resin was cured under UV exposure source. The particle membrane was recovered from the mold and checked for its quality under a microscope. The top view of the embedded particles in the dry state can be seen in Figure 7e,f. Such a membrane was fixed in the electrochemical cell with the UV curable glue in such a way that (i) the chambers adjacent to the particle membrane were completely separated (no leaks) and (ii) the particles were easily observable under a microscope (see Figure 7b).

4.3. Ion-Exchange Surface Area

The active ion-exchange surface area of our membranes is determined from the microscopic images of the resin particles. An example of the surface area determination is in Figure 7c. The particles making up the membrane were geometrically characterized when fully swollen in the used solution. We measured the height and the base of the particle caps that were exposed to the surrounding electrolyte. By determining these geometrical parameters, one could calculate the surface of each particle cap as $A_i = \pi(a_i^2/4 + h_i^2)$. The overall surface of every membrane was then given as the sum of surfaces of all caps.

4.4. Experimental Set-Up

All the electrochemical measurements were performed on galvanostat/potentiostat Gamry 600 in a four-electrode set-up. Two gold wires were used as source electrodes (current circuit) and these were placed in the outer chambers of the cell. Two reference silver/silver chloride electrodes were used as micro-reference electrodes. Their distance from the membrane under study was 6 mm on either side of the membrane. The electrochemical cell was placed under a microscope Olympus BX51WI with an attached color camera Pixelink PL-D775CU. The electrochemical experiments were controlled by software Gamry Framework, the camera by PixeLink Capture OEM.

4.5. Measurements

The current–voltage curve was measured by setting a given current load and measuring the voltage response of the system. Specifically the current load started at 0 μA and was stepped up at a rate of 1 μA/s until a given final load was reached. The final load was set in such a way that the length of the overlimiting region was roughly of the same length as the length of the underlimiting region. These values strongly depended on the ion-exchange surface of the given membranes. These experiments were carried out with KCl solution of concentration of 0.01 M with addition of fluorescein at its final concentration of 10^{-5} M. When running the polarization curve, we also imaged the interface between the membrane and the electrolyte solution on its depletion side. The images were taken so that we obtained a characteristics image for each applied current.

The experiments aimed at determining the pH changes associated with a possible water splitting reaction were performed by running chronoamperometric measurements under a given voltage. The duration of these experiments was 5 min. After five minutes, the electrolytes in all four chambers were thoroughly mixed and their pH was measured. In these experiments we used 0.001 M KCl in the chambers adjacent to the membrane and 1M KCl in the source compartments. In the case of the measurement on the anion-exchange particle system we also added another compartment with a phosphate buffer to limit the possible effect of the electrode reaction on pH changes in the system.

5. Conclusions

Performed experiments with cation-exchange particle membranes of an increasing number of ion-exchange domain showed that (i) there was no significant effect of the increasing number of particles on the current–voltage curves, (ii) the main mechanism of the overlimiting current for the studied cation-exchange particle systems was electroconvection, which manifests itself very similarly in all cases, and (iii) water splitting did not contribute to the overlimiting current in the case of the cation-exchange particle systems unlike the anion-exchange particle, which showed pronounced susceptibility to water splitting. Quantitative differences in the behavior of the studied membranes were observed, e.g., different values of the limiting current density. These differences were mainly caused by the way the particles were incorporated into the membrane (mutual position, their available surface) rather than the number of the individual particles.

Author Contributions: Conceptualization, L.V. and Z.S.; methodology, L.V. and Z.S.; formal analysis, L.V. and Z.S.; investigation, L.V.; data curation, L.V. and Z.S.; writing—original draft preparation, Z.S.; writing—review and editing, Z.S.; visualization, L.V. and T.B; supervision, Z.S; project administration, Z.S.; funding acquisition, Z.S.

Funding: This research was funded by the Czech science foundation, grant number 18-13491S and specific university research funds MSMT no. 21-SVV/2019. ZS acknowledges the support of the CENTEM project, reg. no. CZ.1.05/2.1.00/03.0088, co-funded by the ERDF as part of the Ministry of Education, Youth and Sports OP RDI program and, in the follow-up sustainability stage, supported through CENTEM PLUS (LO1402) by financial means from the Ministry of Education, Youth and Sports under the National Sustainability Program I.

Conflicts of Interest: The authors declare no conflict of interest.

Abbreviations

CVC	Current–voltage curve
DC	Direct current

References

1. Strathmann, H. Electrodialysis, a mature technology with a multitude of new applications. *Desalination* **2010**, *264*, 268–288. [CrossRef]
2. Alvarado, L.; Chen, A.C. Electrodeionization: Principles, strategies and applications. *Electrochim. Acta* **2014**, *132*, 583–597. [CrossRef]
3. Strathmann, H.; Grabowski, A.; Eigenberger, G. Ion-exchange membranes in the chemical process industry. *Ind Eng Chem. Res.* **2013**, *52*, 10364–10379. [CrossRef]
4. Slouka, Z.; Senapati, S.; Chang, H.C. Microfluidic systems with ion-selective membranes. *Annu. Rev. Anal. Chem.* **2014**, *7*, 317–335. [CrossRef] [PubMed]
5. Luo, T.; Abdu, S.; Wessling, M. Selectivity of ion exchange membranes: A review. *J. Membr. Sci.* **2018**, *555*, 429–454. [CrossRef]
6. Yaroslavtsev, A.B.; Nikonenko, V.V.; Zabolotsky, V.I. Ion transfer in membrane and ion exchange materials. *Usp Khim* **2003**, *72*, 438–470. [CrossRef]
7. Nagarale, R.K.; Gohil, G.S.; Shahi, V.K. Recent developments on ion-exchange membranes and electro-membrane processes. *Adv Colloid. Interfac.* **2006**, *119*, 97–130. [CrossRef] [PubMed]
8. Vobecka, L.; Svoboda, M.; Benes, J.; Bellon, T.; Slouka, Z. Heterogeneity of heterogeneous ion-exchange membranes investigated by chronopotentiometry and x-ray computed microtomography. *J. Membr. Sci.* **2018**, *559*, 127–137. [CrossRef]
9. Svoboda, M.; Benes, J.; Vobecka, L.; Slouka, Z. Swelling induced structural changes of a heterogeneous cation-exchange membrane analyzed by micro-computed tomography. *J. Membr. Sci.* **2017**, *525*, 195–201. [CrossRef]
10. Marti-Calatayud, M.C.; Buzzi, D.C.; Garcia-Gabaldon, M.; Bernardes, A.M.; Tenorio, J.A.S.; Perez-Herranz, V. Ion transport through homogeneous and heterogeneous ion-exchange membranes in single salt and multicomponent electrolyte solutions. *J. Membr. Sci.* **2014**, *466*, 45–57. [CrossRef]
11. Krol, J.J.; Wessling, M.; Strathmann, H. Concentration polarization with monopolar ion exchange membranes: Current-voltage curves and water dissociation. *J. Membr. Sci.* **1999**, *162*, 145–154. [CrossRef]
12. Svoboda, M.; Slouka, Z.; Schrott, W.; Snita, D. Cation exchange membrane integrated into a microfluidic device. *Microelectron Eng.* **2009**, *86*, 1371–1374. [CrossRef]
13. Maletzki, F.; Rosler, H.W.; Staude, E. Ion transfer across electrodialysis membranes in the overlimiting current range - stationary voltage current characteristics and current noise power spectra under different conditions of free-convection. *J. Membr. Sci.* **1992**, *71*, 105–115. [CrossRef]
14. Belloň, T.; Polezhaev, P.; Vobecká, L.; Svoboda, M.; Slouka, Z. Experimental observation of phenomena developing on ion-exchange systems during current-voltage curve measurement. *J. Membr. Sci.* **2019**, *572*, 607–618. [CrossRef]
15. Tanaka, Y. Concentration polarization in ion-exchange membrane electrodialysis. *J. Membr. Sci.* **1991**, *57*, 217–235. [CrossRef]
16. Tanaka, Y. Concentration polarization in ion-exchange membrane electrodialysis - the events arising in a flowing solution in a desalting cell. *J. Membr. Sci.* **2003**, *216*, 149–164. [CrossRef]
17. Tanaka, Y. Concentration polarization in ion-exchange membrane electrodialysis - the events arising in an unforced flowing solution in a desalting cell. *J. Membr. Sci.* **2004**, *244*, 1–16. [CrossRef]
18. Rosler, H.W.; Maletzki, F.; Staude, E. Ion transfer across electrodialysis membranes in the overlimiting current range - chronopotentiometric studies. *J. Membr. Sci.* **1992**, *72*, 171–179. [CrossRef]
19. Dydek, E.V.; Zaltzman, B.; Rubinstein, I.; Deng, D.S.; Mani, A.; Bazant, M.Z. Overlimiting current in a microchannel. *Phys. Rev. Lett.* **2011**, *107*, 118301. [CrossRef]

20. Nikonenko, V.V.; Kovalenko, A.V.; Urtenov, M.K.; Pismenskaya, N.D.; Han, J.; Sistat, P.; Pourcelly, G. Desalination at overlimiting currents: State-of-the-art and perspectives. *Desalination* **2014**, *342*, 85–106. [CrossRef]
21. Balster, J.; Yildirim, M.H.; Stamatialis, D.F.; Ibanez, R.; Lammertink, R.G.H.; Jordan, V.; Wessling, M. Morphology and microtopology of cation-exchange polymers and the origin of the overlimiting current. *J. Phys. Chem. B* **2007**, *111*, 2152–2165. [CrossRef] [PubMed]
22. Nikonenko, V.V.; Mareev, S.A.; Pis'menskaya, N.D.; Uzdenova, A.M.; Kovalenko, A.V.; Urtenov, M.K.; Pourcelly, G. Effect of electroconvection and its use in intensifying the mass transfer in electrodialysis (review). *Russ. J. Electrochem.* **2017**, *53*, 1122–1144. [CrossRef]
23. Zabolotskii, V.I.; Nikonenko, V.V.; Urtenov, M.K.; Lebedev, K.A.; Bugakov, V.V. Electroconvection in systems with heterogeneous ion-exchange membranes. *Russ. J. Electrochem.* **2012**, *48*, 692–703. [CrossRef]
24. Belova, E.I.; Lopatkova, G.Y.; Pismenskaya, N.D.; Nikonenko, V.V.; Larchet, C.; Pourcelly, G. Effect of anion-exchange membrane surface properties on mechanisms of overlimiting mass transfer. *J. Phys. Chem. B* **2006**, *110*, 13458–13469. [CrossRef] [PubMed]
25. Simons, R. Water splitting in ion-exchange membranes. *Electrochim. Acta* **1985**, *30*, 275–282. [CrossRef]
26. Belova, E.; Lopatkova, G.; Pismenskaya, N.; Nikonenko, V.; Larchet, C. Role of water splitting in development in ion-exchange membrane of electroconvection systems. *Desalination* **2006**, *199*, 59–61. [CrossRef]
27. Slouka, Z.; Senapati, S.; Yan, Y.; Chang, H.C. Charge inversion, water splitting, and vortex suppression due to DNA sorption on ion-selective membranes and their ion-current signatures. *Langmuir* **2013**, *29*, 8275–8283. [CrossRef]
28. Sablani, S.S.; Goosen, M.F.A.; Al-Belushi, R.; Wilf, M. Concentration polarization in ultrafiltration and reverse osmosis: A critical review. *Desalination* **2001**, *141*, 269–289. [CrossRef]
29. Roghmans, F.; Evdochenko, E.; Stockmeier, F.; Schneider, S.; Smailji, A.; Tiwari, R.; Mikosch, A.; Karatay, E.; Kuhne, A.; Walther, A.; et al. 2D patterned ion-exchange membranes induce electroconvection. *Adv. Mater. Interfaces* **2019**, *6*, 1801309. [CrossRef]
30. Davidson, S.M.; Wessling, M.; Mani, A. On the dynamical regimes of pattern-accelerated electroconvection. *Sci. Rep.* **2016**, *6*, 22505. [CrossRef]
31. Vasil'eva, V.I.; Akberova, E.M.; Zabolotskii, V.I. Electroconvection in systems with heterogeneous ion-exchange membranes after thermal modification. *Russ. J. Electrochem.* **2017**, *53*, 398–410. [CrossRef]
32. Butylskii, D.Y.; Mareev, S.A.; Pismenskaya, N.D.; Apel, P.Y.; Polezhaeva, O.A.; Nikonenko, V.V. Phenomenon of two transition times in chronopotentiometry of electrically inhomogeneous ion exchange membranes. *Electrochim. Acta* **2018**, *273*, 289–299. [CrossRef]
33. Green, Y.; Park, S.; Yossifon, G. Bridging the gap between an isolated nanochannel and a communicating multipore heterogeneous membrane. *Phys. Rev. E* **2015**, *91*, 011002. [CrossRef] [PubMed]
34. Mareev, S.A.; Nebavskiy, A.V.; Nichka, V.S.; Urtenov, M.K.; Nikonenko, V.V. The nature of two transition times on chronopotentiograms of heterogeneous ion exchange membranes: 2d modelling. *J. Membr. Sci.* **2019**, *575*, 179–190. [CrossRef]
35. Nikonenko, V.; Nebavsky, A.; Mareev, S.; Kovalenko, A.; Urtenov, M.; Pourcelly, G. Modelling of ion transport in electromembrane systems: Impacts of membrane bulk and surface heterogeneity. *Appl. Sci.* **2019**, *9*, 25. [CrossRef]
36. Belloň, T.; Polezhaev, P.; Vobecká, T.; Slouka, Z. Fouling of a heterogeneous anion-exchange membrane and single anion-exchange resin particle by ssdna manifests differently. *J. Membr. Sci.* **2019**, *572*, 619–631. [CrossRef]

International Journal of
Molecular Sciences

MDPI

Correction

Correction: Abbasi, A. et al. Poly(2,6-dimethyl-1,4-phenylene oxide)-Based Hydroxide Exchange Separator Membranes for Zinc-Air Battery. *Int. J. Mol. Sci.* 2019, *20*, 3678

Ali Abbasi [1,2], Soraya Hosseini [1,2], Anongnat Somwangthanaroj [1], Ahmad Azmin Mohamad [3] and Soorathep Kheawhom [1,2,*]

1 Department of Chemical Engineering, Faculty of Engineering, Chulalongkorn University, Bangkok 10330, Thailand; abbasi.1000@gmail.com (A.A.); soraya20h@gmail.com (S.H.); anongnat.s@chula.ac.th (A.S.)
2 Computational Process Engineering Research Laboratory, Chulalongkorn University, Bangkok 10330, Thailand
3 School of Materials and Mineral Resources Engineering, Universiti of Sains Malaysia, Nibong Tebal 14300, Pulau Pinang, Malaysia; aam@usm.my
* Correspondence: soorathep.k@chula.ac.th; Tel.: +66-81-490-5280

Received: 13 December 2019; Accepted: 6 January 2020; Published: 7 January 2020

The authors would like to make the following corrections to their paper published in the International Journal of Molecular Science [1]. The ionic conductivities shown in Table 1 were wrong because of the inconsistent unit of the thickness of membranes used in the calculation. In the corrected version, we updated the ionic conductivity and added the thickness, area, and bulk resistance of each membrane. The following changes are noted. The changes do not affect the conclusions of the article.

1. Change in Table 1

Table 1 should be replaced with the following:

Table 1. Basic properties of PPO-TMA, PPO-MPy, and PPO-MIM separator membranes.

Sample	DI Water			KOH, 7M			Thickness (μm)	Area (cm^2)	R_b (Ω) [2]	Ionic Conductivity (mS/cm)-σ	Zincate Diffusion Coefficient (× 10^{-8} cm^2/min)-D
	Uptake ΔW (wt%)	Area Change ΔA (%)	Volume Change ΔV (%)	Uptake ΔW (wt%)	Area Change ΔA (%)	Volume Change ΔV (%)					
PPO-TMA	89	65	119	31	11	39	50	1.766	0.1630	17.37	1.13
PPO-MPy	78	41	76	30	11	39	40	1.766	0.1394	16.25	0.28
PPO-MIM [1]	13	14	30	3	0	0	30	1.766	5.8014	0.29	N/A [1]

[1] Due to very low electrolyte uptake of PPO-MIM and its very low ionic conductivity, this membrane was not included in the rest of the study. [2] R_b values were obtained by deducting the value obtained for the cell without using any separator (4.4606 Ω) from the resistance values measured for each sample.

2. Change of Figure 1

Figure 1 should be replaced with the following (the units of resistance in the insert image was changed to [Ω.cm^2]):

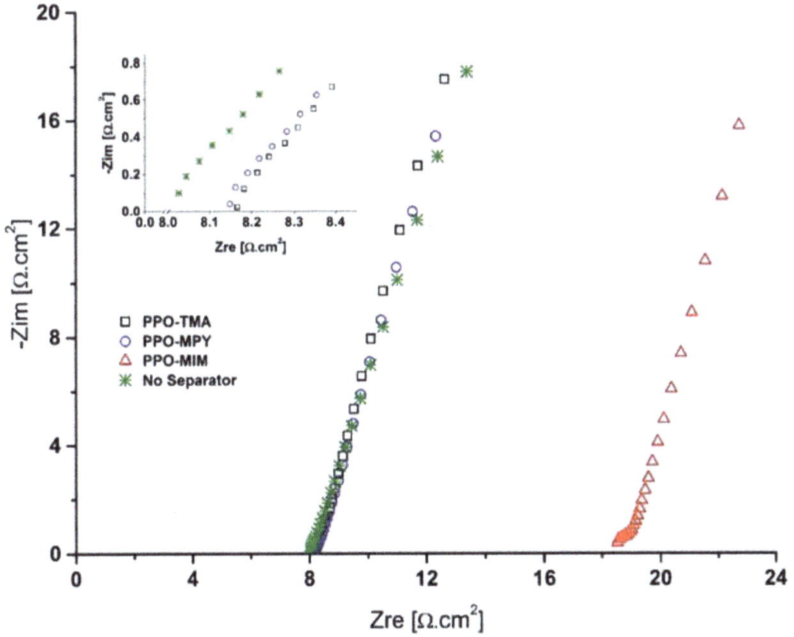

Figure 1. Nyquist plot of EIS for determining ionic conductivity of PPO-based separator membranes. The values of Zre and Zim were obtained by multiplying R_b values by the samples' area (1.766 cm^2).

3. Changes in Text

Lines 11–13 of the Abstract should be replaced with the following text:

Ionic conductivity of PPO–TMA, PPO–MPY, and PPO–MIM was determined using electrochemical impedance spectroscopy to be 17.37, 16.25, and 0.29 mS/cm, respectively.

Lines 1–11 on page 7 should be replaced with the following text:

Also, very low electrolyte uptake of PPO–MIM was reflected in the ionic conductivity measurements, showing very low conductivity of 0.29 mS/cm determined using a Nyquist plot of electrochemical impedance spectroscopy (EIS) (Figure 1). For PPO–TMA and PPO–MPy, the ionic conductivity was calculated to be 17.37 and 16.25 mS/cm. Due to deficient electrolyte uptake and low ionic conductivity of PPO–MIM, it was not included in the rest of the study.

Slightly higher ionic conductivities have been reported for the same separator membranes, which could be attributed to the higher measurement temperature and lower KOH solution concentration. In this study, the measurements were carried out in KOH, 7 M solution to mimic the real cell operation condition. As can be seen in Table 1, the separator membranes absorb much less electrolyte than they do in water, resulting in lower measured ionic conductivity.

Lines 3–5 in the Conclusion should be replaced with the following text:

They offered a good ionic conductivity of ~17 mS/cm along with very low zincate diffusion coefficient of 1.13 × 10^{-8} and 0.28 × 10^{-8} cm^2/min for PPO–TMA and PPO–MPY, respectively.

We apologize for any inconvenience caused to the readers by this error.

Reference

1. Abbasi, A.; Hosseini, S.; Somwangthanaroj, A.; Mohamad, A.A.; Kheawhom, S. Poly(2,6-Dimethyl-1,4-Phenylene Oxide)-Based Hydroxide Exchange Separator Membranes for Zinc–Air Battery. *Int. J. Mol. Sci.* **2019**, *20*, 3678. [CrossRef] [PubMed]

MDPI
St. Alban-Anlage 66
4052 Basel
Switzerland
Tel. +41 61 683 77 34
Fax +41 61 302 89 18
www.mdpi.com

International Journal of Molecular Sciences Editorial Office
E-mail: ijms@mdpi.com
www.mdpi.com/journal/ijms

www.ingramcontent.com/pod-product-compliance
Lightning Source LLC
LaVergne TN
LVHW070237100526
838202LV00015B/2140
* 9 7 8 3 0 3 6 5 1 3 5 9 1 *